Unpublished Manuscripts

© Lars Hörmander

Lars Hörmander

Unpublished Manuscripts

from 1951 to 2007

 Springer

Lars Hörmander
Department of Mathematics
Lund University
Lund
Sweden

ISBN 978-3-030-09916-9   ISBN 978-3-319-69850-2   (eBook)
https://doi.org/10.1007/978-3-319-69850-2

Mathematics Subject Classification (2010): 35A, 35G, 35H, 35J, 35M, 35P, 35S, 47G, 58G, 35L

This Springer imprint is published by the registered company Springer International Publishing AG part of
Springer Nature
The registered company address is: Gewerbestrasse 11, 6330 Cham, Switzerland

Foreword

About a year after retirement, my father Lars Hörmander received a phone call from the University Library in Lund inviting him to hand over to the library, at a time of his own choosing, correspondence and other unpublished material documenting his activities. At this time also, he was asked to write a contribution to the planned volume *Fields Medal lists' Lectures* (reprinted here). Finally, he was diagnosed with an acoustic neuroma, a benign brain tumor which might – or might not, depending on its growth – prove troublesome in years to come. These events jointly prompted him to review his mathematical life. He had saved everything related to his work, every manuscript, every note, every letter sent or received, and now the time had come to go through it all. He worked through 23 binders of correspondence as well as his entire mathematical production. He compiled a complete bibliography of all published materials (included at the end of this book), even adding references to reviews, but also carefully went through all unpublished manuscripts, checking them with his very critical eye and deciding on what should be kept for posterity and what should be thrown out. Every paper he felt should be kept was carefully retyped in AMS-TeX, unless already in that format, sometimes with added comments, and in at least two cases (manuscripts 10 and 11) also translated into English. One paper (manuscript 4) was even written from scratch as he was reminded of a paper by Petrowsky that he had struggled with during his graduate studies and now wanted to revisit. He did all this with the University Library in mind – not intending a publication – but shortly before his death he suggested I seek advice from his colleagues and do as I please.

The order of the manuscripts follows his evolution as a mathematician, starting with his licentiate degree from 1951 and – in the last update – ending with the last paper he felt was worthy of inclusion, from 2007. He did continue his mathematical work after that year, in fact he kept working on a range of problems until he became seriously ill in 2011, but did not get far enough with any of them to want to update this collection.

One paper may stand out as an unexpected item here: *Guide to the mathematical models at the Department of Mathematics in Lund* (manuscript 22). I am glad he included it, however, as this was a project he really took delight in. He always loved creating things with his hands, he loved carpentry all his life, and as a child was fascinated by Meccano. I believe his first interest in mathematics, as a young school boy, came through geometry. So when he retired he took on this project with great enthusiasm, working his way through all the models at the department. He even spent months creating a large plaster sculpture of Steiner's Roman surface (which, I am told, is a level surface for a solution of a differential equation discussed in his paper [100], listed in the Complete Bibliography). As a side project, for my then young son Sander he built all the Platonic and Archimedean solids in cardboard; they also constructed several of them together with a magnetic building toy. I can only hope that some of you readers will be inspired and come to Lund to see the models at the Department for yourselves.

You are now holding in your hands my father's *Unpublished Manuscripts*. As the title of a book, this is of course an oxymoron, but that's what my father called them, so that's what I will continue to call them. Not even a comma has been altered, this is what he gave me. Enjoy.

Lund in October 2017

Sofia Broström, daughter of Lars Hörmander

Sculpture of Steiner's Roman surface. Photographed by Lars Hörmander outside his home in Veberöd, Sweden. © Lars Hörmander

Contents

Origin of the Manuscripts[1]

1. *Applications of Helly's theorem to estimates of Tchebycheff type.* Thesis for the licentiate degree 1951.

2. *An isoperimetric inequality in homogeneous Finsler spaces.* Essentially written in Zürich 1952.

3. *On best approximation in L^∞.* Probably written in 1952 for a lecture at the Swedish Mathematical Society.

4. *Petrowsky's L^2 estimates for hyperbolic systems.* Written in 1997 when reminded of Petrowsky's paper [P1] which I never managed to understand as a graduate student in the early 50's.

5. *Proof of the existence of fundamental solutions and of some inequalities.* This is Section 3 in a manuscript from 1955 submitted in my application for professorship in Stockholm.

6. *Inequalities between normal and tangential derivatives of harmonic functions.* Incomplete manuscript from 1955–1956, with references reconstructed.

7. *The Dirichlet problem in a maximal open set.* This is a development made in 1997 of a sketch originally intended for a special project for the masters degree.

8. *The diffusion approximation in neutron transport theory.* Translation of FOA report October 1958.

9. *Some Tauberian theorems of Lewitan.* Lecture given in 1958 at a meeting in Gothenburg of the Swedish Mathematical Society, followed by a related probably earlier manuscript (Some variants of an inequality due to H. Bohr).

10. *Electromagnetic wave propagation over ground with small inhomogeneities in the electrical constants.* Translation of manuscript written at FOA in September 1961.

11. *Classes of infinitely differentiable functions.* Translation of the notes of a seminar in Stockholm 1962 written jointly with my then student Jan Boman.

12. *Approximation on totally real manifolds.* Manuscript from the 1980's simplifying my joint paper with John Wermer.

---

[1]The indications in brackets refer to the reference list in the corresponding manuscript.

13. *The work of Carleson and Hoffman on bounded analytic functions in the disc.* Lecture in the Princeton Current Literature Seminar during 1966–67.

14. *Some problems concerning linear partial differential equations.* Manuscript from the spring of 1968 in preparation for a return to graduate teaching, with comments added in 1997.

15. *Remarks on convexity with respect to operators of real principal type.* Manuscript written at IHES in 1972.

16. *Fourier multipliers with small norm.* Manuscript written in 1973 in connection with H. S. Shapiro's application for a professorship at Stockholm University.

17. *The cosine theorem on a surface and the notion of curvature.* Manuscript of lecture in March 1979 at the University of Stockholm and at the teacher instruction day of the Swedish Mathematical Society.

18. *Corrections to my paper On Sobolev spaces associated with some Lie algebras.* Corrections to [1] written in December 1986 and distributed with the preprints after that.

19. *Gårding's inequality during three decades.* Lecture at the Gårding symposium May 31, 1985 with minor corrections during retyping.

20. *Symplectic geometry and differential equations.* Lectures in Augsburg, February 25 and 26, 1991.

21. *General Mehler formulas and the Weyl calculus.* Lecture at the Wiener Centennial at MIT October 10, 1994.

22. *Guide to the mathematical models at the Department of Mathematics in Lund.* Finished for a lecture at the Lund Mathematical Society May 14, 1998.

23. *The proof of the Nirenberg-Treves conjecture according to N. Dencker and N. Lerner.* This manuscript was finished in August 2005. At the end it contains a list of major open questions in the area and also a correction of a theorem in [H2].

24. *Lower bounds for subelliptic operators.* This manuscript, essentially finished in 2006, only studies the simple case of subelliptic operators satisfying condition (P). Later work on the general case has not been completed.

25. *Approximation of solutions of constant coefficient boundary problems and of entire functions.* This manuscript contains an elaboration of ideas already hinted at in [2]. However, the main interest is probably the incomplete discussion of approximation of entire functions in the last section.

APPLICATIONS OF HELLY'S THEOREM
TO ESTIMATES OF TCHEBYCHEFF TYPE

LARS HÖRMANDER

The present paper consists of four parts

(i) a generalization of a theorem by Helly on convex bodies,
(ii) two general theorems concerning generalized polynomials,
(iii) applications of the first theorem in (ii) to the theory of approximations,
(iv) applications of the second theorem in (ii) to the theory of polynomials that differ as little as possible from zero.

A generalization of a theorem by Helly on convex bodies.

Before the first great war Helly found the following theorem[1]

A set of convex bodies in an n-dimensional Euclidean space have a common point if and only if any combination of $n+1$ of them have a common point.

By "convex body" we mean a pointset that is

1) convex
2) closed
3) bounded

The object of our generalization is to show that the third condition can be replaced by a weaker one and we shall in fact give the weakest possible.

REMARK. If we have only a finite number of point sets, the third condition is quite unnecessary. For, let $K_1, \ldots, K_N$ be convex, closed pointsets such that $K_{i_1} \cap \cdots \cap K_{i_{n+1}} \neq \emptyset$ for every combination $i_1, \ldots, i_{n+1}$ from the N numbers $1, \ldots, N$. As there is only a finite number of such combinations we can find a sphere S so large that also $S \cap K_{i_1} \cap \cdots \cap K_{i_{n+1}} \neq \emptyset$ for all combinations or, which is equivalent $(S \cap K_{i_1}) \cap \cdots \cap (S \cap K_{i_{n+1}}) \neq \emptyset$. The pointsets $(S \cap K_1), \ldots, (S \cap K_N)$ satisfy 1) – 3) and hence it follows from Helly's theorem that $\cap_{i=1}^N (S \cap K_i) \neq \emptyset$ and so we have $\cap_{i=1}^N K_i \neq \emptyset$.

Before giving the condition replacing 3) we shall study pointsets that satisfy 1) and 2) more closely.

LEMMA 1. *If a pointset K satisfies 1) and 2) but not 3) there exists a vector $\xi \neq 0$ such that, if $x \in K$ then every point $x + t\xi$ with $t \geq 0 \in K$. — We shall call such a vector a direction of infinity of K.*

PROOF. Take a point $x_0 \in K$ and a sequence of points $x_1, \ldots, x_m, \ldots$ all belonging to K so that $|x_0 - x_m| \to \infty$ as $m \to \infty$. From the sequence $\xi_n = \frac{x_m - x_0}{|x_m - x_0|}$ we can, as $|\xi_n| = 1$,

---

[1] For a rather complete bibliography on the theorem see Bonnesen-Fenchel: Theorie der konvexen Körper page 3.

© Springer International Publishing AG, part of Springer Nature 2018
L. Hörmander, *Unpublished Manuscripts*,
https://doi.org/10.1007/978-3-319-69850-2_1

choose a convergent partial sequence $\xi_{n'} \to \xi$ as $n' \to \infty$. Because all points $\xi_0 + t\xi_{n'}$ with $0 \leq t \leq |x_{n'} - x_0|$ belong to K it follows that every point $x_0 + t\xi$ with $t \geq 0$ belong to K for K is closed and we have $x_0 + t\xi = \lim_{n' \to \infty} (x_0 + t\xi_{n'})$. If we now take another point $x \in K$ and repeat the same procedure with x instead of x_0 and choose the points x_m as $x_0 + m\xi$ it immediately follows that every point $x + t\xi$ with $t \geq 0$ belongs to K.

LEMMA 2. *The set of all directions of infinity, if we add the origin, is a convex cone.*

PROOF. From the definition it is evident that if ξ is a direction of infinity so is $\lambda\xi$ one if $\lambda > 0$. If ξ_1 and ξ_2 are directions of infinity and $x \in K$ then $x + t\xi_1 \in K$ and $x + t\xi_2 \in K$ for $t \geq 0$ and hence from the convexity $x + t(\xi_1 + \xi_2)/2 \in K$ for $t \geq 0$ so that $(\xi_1 + \xi_2)/2$ is a direction of infinity.

From the condition that we want to replace 3) we shall require the property, that if two pointsets are equal except for a translation they shall either both satisfy or both not satisfy the condition. The condition must then contain the following

If ξ is a direction of infinity then $-\xi$ is also one.

PROOF. Suppose that we have a point set K satisfying 1) and 2) and such that ξ but not $-\xi$ is a direction of infinity. We can then find two points z and y such that $z \in K$ and $y \notin K$ and $z = y + \xi$. Let a plane separating y from K be $(x, u) = a$ (it is easy to prove that there exists one; e.g. see Bonnesen-Fenchel: Theorie der konvexen Körper). We have then if K lies on the positive side of the plane

$$(z, u) > a \quad (y, u) < a \quad \text{and hence as} \quad (z, u) = (y, u) + (\xi, u) \quad (\xi, u) > 0.$$

We now put $K_m = $ the pointset K translated through the vector $m\xi$. As ξ is a direction of infinity it immediately follows that $n + 1$ arbitrary of these pointsets have a point in common. However, $\cap_{i=1}^{\infty} K_i = \emptyset$. For if $x \in K_i$ for all i then $x - i\xi \in K$ and hence $(x - i\xi, u) > a$ for all i whereas $(x - i\xi, u) = (x, u) - i(\xi, u) \underset{i \to \infty}{\to} -\infty$ as we have proved that $(\xi, u) > 0$. Hence the required condition must exclude this case and the proposition is proved.

The following three conditions are equivalent:

H_1: *If ξ is a direction of infinity of K then $-\xi$ is one.*

H_2: *The set of all directions of infinity forms a linear manifold if we add the origin.*

H_3: *There exists a linear manifold L and a convex body k in the orthogonal manifold $L^{\perp}$ so that $K = k \oplus L = \{x_1 + x_2; x_1 \in k, x_2 \in L\}$.*

PROOF. H_2 follows immediately from H_1 and lemma 2. Suppose now that K satisfies H_2. Put $L = $ the linear manifold of all directions of infinity of K, and put $k = K \cap L^{\perp}$. Then I assert that $K = k \oplus L$. For it follows from the definition of direction of infinity that $K \supset k \oplus L$. Let on the other hand $x \in K$. We can write $x = x_1 + x_2$ where $x_1 \in L^{\perp}$ and $x_2 \in L$. Then $x_1 \in K \cap L^{\perp}$ so that $x_1 \in k$. But then $x = x_1 + x_2$ belongs to $k \oplus L$ and hence $K \subset k \oplus L$. k is a convex body in $L^{\perp}$. For in the contrary case it would have a direction of infinity ξ (lemma 1). Then ξ is a direction of infinity of K too and situated in $L^{\perp}$ contrary to hypothesis. Hence H_3 follows from H_2. It is trivial that H_1 follows from H_3.

We shall denote all these conditions by H.

THEOREM. *A set of convex, closed pointsets in the n-dimensional Euclidean space all satisfying H have a point in common if and only if any combination of $n+1$ of them have a common point.*

PROOF. Let the pointsets be K_α, $\alpha \in A$. According to H_3 we can write them $K_\alpha = k_\alpha \oplus L_\alpha$ where k_α are convex bodies and L_α linear manifolds. Put $L = \cap_{\alpha \in A} L_\alpha$. First suppose that L only contains the origin, $L = \{0\}$. From dimensional considerations it then follows that we can find $p \le n$ from the L_α, $L^{(1)}, \ldots, L^{(p)}$ so that already $L^{(1)} \cap \cdots \cap L^{(p)} = \{0\}$. Let the corresponding bodies be $K^{(1)}, \ldots, K^{(p)}$. $K^{(1)} \cap \cdots \cap K^{(p)}$ can have no direction of infinity for if ξ were so we should have $\xi \in L^{(1)}, \ldots, \xi \in L^{(p)}$ and so $\xi \in L$ which gives a contradiction. Hence $K = K^{(1)} \cap \cdots \cap K^{(p)}$ is a convex, bounded closed pointset. Now form $K'_\alpha = K \cap K_\alpha$. If we take $n+1$ of them, K'_{α_i}, $i = 1, \ldots, n+1$, we get

$$K'_{\alpha_1} \cap \cdots \cap K'_{\alpha_{n+1}} = K^{(1)} \cap \cdots \cap K^{(p)} \cap K_{\alpha_1} \cap \cdots \cap K_{\alpha_{n+1}} \ne \emptyset$$

according to our remark above on Helly's theorem. From Helly's theorem it now follows that $\cap_{\alpha \in A} K'_\alpha \ne \emptyset$ and hence $\cap_{\alpha \in A} K_\alpha \ne \emptyset$. — Now suppose that $L \ne \{0\}$. Put $L_\alpha = L \oplus L'_\alpha$ and consequently $K_\alpha = k_\alpha \oplus L'_\alpha \oplus L$. We have $\cap_{\alpha \in A} L'_\alpha = \{0\}$. As it immediately follows that $n+1$ arbitrary of the pointsets $k_\alpha \oplus L'_\alpha$ have a point in common it follows from the particular case treated above that all $k_\alpha \oplus L'_\alpha$ have a point in common. Hence all K_α have a common point.

EXAMPLES. Denote the dimension of L by p, $0 \le p \le n$. For $n = 2$ we get $p = 0, 1, 2$ and hence respectively a convex twodimensional body, an infinite strip and the whole plane; for $n = 3$ we get $p = 0, 1, 2, 3$ and hence respectively a convex threedimensional body, an infinite cylinder with a convex twodimensional body as base, a domain consisting of the points between and on two parallel planes, which I shall call a disk, and finally the whole space; and so on in higher dimensions. We shall especially note the sets consisting of the points between and on two parallel $(n-1)$-dimensional planes and we shall call these sets *disks*.

REMARK. We can formulate our condition H in one more way: Every convex closed set is the intersection of the halfspaces containing it whereas a convex closed set satisfying H is even the intersection of all disks that contain it.

Two theorems concerning generalized polynomials.

Consider an arbitrary set M and n realvalued functions $\varphi_1(x), \ldots, \varphi_n(x)$ defined for $x \in M$. For the sake of simplicity we suppose that these functions are linearly independent over M which in fact means very little specialization. By $\varphi(x)$ we denote linear combinations of these functions with real coefficients:

$$(1) \qquad \varphi(x) = \sum_{i=1}^{n} c_i \varphi_i(x).$$

In addition we choose two functions $A_1(x)$ and $A_2(x)$ defined for $x \in M$ such that $A_2(x) \le A_1(x)$ and consider the set of functions Γ defined by

$$(2) \qquad \Gamma = \{\varphi(x) \mid \varphi(x) \text{ of type (1) with } A_2(x) \le \varphi(x) \le A_1(x) \text{ for } x \in M\}.$$

We shall also study fixed linear combinations of the coefficients, ω, where

$$(3) \qquad \omega(\varphi) = \sum_{i=1}^{n} \omega_i c_i.$$

In particular we are interested in the range of $\omega(\varphi)$ as $\varphi \in \Gamma$. We shall denote this by W.

It is clear that we can represent $\varphi(x)$ in a one to one manner with the point $(c_1, \ldots, c_n)$ in a n-dimensional euclidean space. In the sequel we shall denote by $\varphi(x)$ and φ as well the function as the corresponding point. The condition in the point z in order that $\varphi(x)$ shall belong to Γ is that

$$(4) \qquad A_2(z) \le c_1\varphi_1(z) + \cdots + c_n\varphi_n(z) \le A_1(z).$$

Geometrically this means that the point φ shall belong to a disk $K(z)$. Hence we immediately conclude a geometric representation of the class Γ

$$(5) \qquad \varphi(x) \in \Gamma \iff \varphi \in \bigcap_{z \in M} K(z) = K.$$

As intersection of closed convex sets, K must be convex and closed. K is in fact a convex body. For in the contrary case it would have a direction of infinity $(c_1, \ldots, c_n)$; this direction is then a direction of infinity for every $K(z)$ ($z \in M$) so that $c_1\varphi_1(z) + \cdots + c_n\varphi_n(z) = 0$ for every $z \in M$. As all c_i are not zero we have reached a contradiction.

It is now obvious that W is either empty or a finite closed interval. Hence $\omega(\varphi)$ attains its upper and lower bounds as $\varphi \in \Gamma$.

We shall need some other notations. With C_m we denote a combination of m points $\in M$. Analogous to Γ we define

$$(2)' \qquad \Gamma_{C_m} = \{\varphi(x) \mid \varphi(x) \text{ of type (1) with } A_2(z) \le \varphi(z) \le A_1(z) \text{ for } z \in C_m\}.$$

It follows immediately from the above that it will be geometrically represented by the intersection $\cap_{z \in C_m} K(z)$, or

$$(5)' \qquad \varphi(x) \in \Gamma_{C_m} \iff \bigcap_{z \in C_m} K(z) \ni \varphi.$$

Hence the range W_{C_m} of $\omega(\varphi)$ as $\varphi \in \Gamma_{C_m}$ is either empty or a finite closed interval or the whole real axis.

Of course Γ may become empty. However we have

THEOREM 1. *In order that Γ should be empty it is necessary and sufficient that some of the classes $\Gamma_{C_{n+1}}$ is empty.*

PROOF. According to (5) Γ is empty if and only if $K = \emptyset$. From our generalized Helly's theorem it follows that $K = \emptyset$ if and only if there exists $x_1, \ldots, x_{n+1} \in M$ so that $\cap_{i=1}^{n+1} K(x_i) = \emptyset$ or, equivalently, $\Gamma_{C_{n+1}}$ is empty with $C_{n+1} = \{x_1, \ldots, x_{n+1}\}$.

THEOREM 2. $W = \cap W_{C_n}$ *where the intersection is to be taken over all C_n.*

PROOF. As $\Gamma \subset \Gamma_{C_n}$ we have $W \subset W_{C_n}$ for all C_n and hence $W \subset \cap W_{C_n}$. Now take a $w \in \cap W_{C_n}$. Then for every C_n there is a $\varphi(x) \in \Gamma_{C_n}$ with $\omega(\varphi) = w$. Denote by π the $(n-1)$-dimensional plane where $\omega(\varphi) = w$ (this equation cannot be inconsistent according to the assumptions). We then have

$$(6) \qquad \pi \bigcap \left(\bigcap_{x \in C_n} K(x) \right) \ne \emptyset \quad \text{for all } C_n.$$

Put $k(x) = \pi \cap K(x)$. In the $(n-1)$-dimensional space π the set $k(x)$ is a disk or the whole of π as is immediately verified ($k(x)$ cannot become empty on account of (6)). We can now write (6) in a new form

$$(6)' \qquad \bigcap_{x \in C_n} k(x) \neq \emptyset \quad \text{for all } C_n.$$

Hence by the generalized Helly's theorem $\pi \cap (\cap_{x \in M} K(x)) = \cap_{x \in M} k(x) \neq \emptyset$. Then by (5) there is a $\varphi(x) \in \Gamma$ with $w(\varphi) = w$ so that $w \in W$. Hence $W \supset \cap W_{C_n}$ and the theorem follows.

Tschebyscheff's approximation theorem.

We shall here get an application of theorem 1.

We define the distance between two bounded functions f and g on M as

$$(7) \qquad \sup_{x \in M} |f(x) - g(x)| = |f - g|.$$

We now suppose that the functions $\varphi_1(x), \ldots, \varphi_n(x)$ are bounded as $x \in M$ and take another function $f(x)$ on M which is bounded there. As a measure of the "degree of approximability" of f with functions of type (1) we introduce the distance from f to the space of these functions, hence

$$(8) \qquad \varrho = \inf |f - \varphi|$$

where we take the infimum over all functions $\varphi(x)$ of type (1). It is easy to derive another expression of ϱ. For put in (2)

$$(9) \qquad A_1(x) = f(x) + d \quad A_2(x) = f(x) - d$$

with a parameter $d \geq 0$. Then Γ, K and $K(x)$ will depend on d, we denote them by Γ_d, K_d, $K_d(x)$. We then have

$$(10) \qquad \varrho = \inf_{K_d \neq \emptyset} d = \sup_{K_d = \emptyset} d.$$

For if $K_d \neq 0$ there exists a $\varphi \in K_d$, i.e., $|f(x) - \varphi(x)| \leq d$ $(x \in M)$, and hence by the definition $d \geq \varrho$ and accordingly $\inf_{K_d \neq \emptyset} d \geq \varrho$. If $K_d = \emptyset$ there exists no φ with $|f - \varphi| \leq d$ and then $d \leq \varrho$ and $\sup_{K_d = \emptyset} d \leq \varrho$. As the last equality in (10) is trivial we have now proved (10).

THEOREM 3. *There exists a best approximating function of type (1).*

PROOF. Take a sequence $d_1 > d_2 > \cdots \to \varrho$. $K_{d_1}, K_{d_2}, \ldots$ will all be non empty closed bounded and monotonically decreasing. Hence by Cantor's theorem they have a point φ in common. Then $|f - \varphi| \leq d_m$ for every m and hence $|f - \varphi| \leq \varrho$. From (8) we have the reversed inequality so that $|f - \varphi| = \varrho$ and φ is a best approximating polynomial.

We can substitute for the norm defined by (7) another which only depends on the values of our functions in C_m, viz.

$$(7)' \qquad |f - g|_{C_m} = \max_{x \in C_m} |f(x) - g(x)|.$$

We can then repeat the above almost entirely. For example we define

$$(8)' \qquad \varrho_{C_m} = \min |f - \varphi|_{C_m}$$

where the minimum is taken over all functions φ of type (1). However, one thing requires some consideration: the functions $\varphi_1(x), \ldots, \varphi_n(x)$ may become dependent over C_m. Then the proof of theorem 3 breaks down. We then take a maximum number of linearly independent functions among $\varphi_1(x), \ldots, \varphi_n(x)$. It is obvious that the linear combinations of these give the same approximation of f in the norm (7)' as the space of all $\varphi(x)$ of type (1). It now follows from theorem 3 that there exists a best approximating function among the linear combinations mentioned and as these are all of type (1) there exists a best approximating function of this type.

We can now form our Tschebyscheffian theorem.

THEOREM 4. $\varrho = \sup \varrho_{C_{n+1}}$ *where* sup *is to be taken over all* C_{n+1}.

PROOF. Put $\sup \varrho_{C_{n+1}} = \varrho'$. As the norm (7)' is not greater than the norm (7) we have $\varrho_{C_{n+1}} \leq \varrho$ and hence $\varrho' \leq \varrho$. On the other hand we have that $(\Gamma_{\varrho'})_{C_{n+1}}$ is not empty for any C_{n+1} and hence from theorem 1 $\Gamma_{\varrho'}$ is not empty, so that $\varrho' \geq \varrho$. Hence $\varrho' = \varrho$.

In order to sharpen theorem 4 we must introduce some topological assumptions. We suppose that M is a compact topological space over which the functions $\varphi_1(x), \ldots, \varphi_n(x)$ and $f(x)$ are continuous. Then we have

THEOREM 4'. $\varrho = \max \varrho_{C_{n+1}}$.

PROOF. According to theorem 4 we choose a sequence $C_{n+1}^{(r)}$ where $C_{n+1}^{(r)} = \{x_1^{(r)}, \ldots, x_{n+1}^{(r)}\}$ and $\varrho_{C_{n+1}^{(r)}} \to \varrho$. As M is compact we can find a partial sequence r' so that $\lim x_k^{(r')}$ exists as $r' \to \infty$ $(k = 1, \ldots, n+1)$. We call this limit x_k. $x_1, \ldots, x_{n+1}$ form if we do not distinguish those which may be equal a combination C_m with $m \leq n+1$. Then we have $\varrho_{C_m} = \varrho$. For otherwise we would have $\varrho_{C_m} = \varrho' < \varrho$. Hence there exists a $\varphi(x)$ so that

$$|f(x_k) - \varphi(x_k)| \leq \varrho' \quad k = 1, \ldots, n+1.$$

Then it follows for sufficiently big values of r' that

$$|f(x_k^{(r')}) - \varphi(x_k^{(r')})| \leq \frac{\varrho + \varrho'}{2} \quad k = 1, \ldots, n+1.$$

Hence by definition $\varrho_{C_{n+1}}^{(r')} \leq (\varrho + \varrho')/2$ for big values of r' so that $\varrho = \lim \varrho_{C_{n+1}^{r'}} \leq (\varrho + \varrho')/2$ and we have reached a contradiction. If $m < n+1$ we take $C_{n+1} \supset C_m$. Then $\varrho \geq \varrho_{C_{n+1}} \geq \varrho_{C_m} = \varrho$ and hence $\varrho_{C_{n+1}} = \varrho$.

We shall now introduce a very restrictive condition, namely that $\varphi_1(x), \ldots, \varphi_n(x)$ form a Tschebyscheff-system[2].

DEFINITION. $\varphi_1(x), \ldots, \varphi_n(x)$ form a Tschebyscheff-system if a function of type (1) with some coefficient $c_i \neq 0$ has at most $n - 1$ zeros $\in M$.

This condition has a simple geometric meaning. Take n different points $x_1, \ldots, x_n \in M$. The vectors

$$(\varphi_1(x_1), \ldots, \varphi_n(x_1)), \ldots, (\varphi_1(x_n), \ldots, \varphi_n(x_n))$$

---

[2]The reason for this condition is clearly explained in Haar: Die Minkowskische Geometrie und die Annäherung an stetige Funktionen, Math. Ann. 78.

are then linearly independent i.e. they span the whole n-dimensional space. For in other case there would exist $c_1, \ldots, c_n$ not all zero so that $c_1 \varphi_1(x) + \cdots + c_n \varphi_n(x) = 0$ for $x = x_1, \ldots, x_n$ against the assumption.

Now consider the case where M consists of only $n + 1$ points $x_1, \ldots, x_{n+1}$. Put $v_i = (\varphi_1(x_i), \ldots, \varphi_n(x_i))$, $i = 1, \ldots, n+1$. These vectors must be linearly dependent, that is, $\sum_{i=1}^{n+1} \mu_i v_i = 0$ with all $\mu_i \neq 0$. If we put $c = (c_1, \ldots, c_n)$ we get identically $\sum_{i=1}^{n+1} (f(x_i) - (c, v_i)) \mu_i = \sum_{i=1}^{n+1} \mu_i f(x_i)$. Hence

$$\left| \sum_{i=1}^{n+1} \mu_i f(x_i) \right| \leq \sum_{i=1}^{n+1} |\mu_i| \cdot \max_j |f(x_j) - (c, v_j)|,$$

$$\varrho = \min_c \max_i |f(x_i) - (c, v_i)| \geq \frac{\left| \sum_{i=1}^{n+1} \mu_i f(x_i) \right|}{\sum_{i=1}^{n+1} |\mu_i|}.$$

Equality can occur if and only if $\operatorname{sgn}(\mu_i)(f(x_i) - (c, v_i)) = \text{constant}$ independently of i. But this system is solvable in one and only one way so that there exists only one best approximating polynomial and the approximation is given by

(11)
$$\varrho = \frac{\left| \sum_{i=1}^{n+1} \mu_i f(x_i) \right|}{\sum_{i=1}^{n+1} |\mu_i|}.$$

We shall apply this to the general case in

THEOREM 5. *If in addition to the conditions in theorem 4′ $\varphi_1(x), \ldots, \varphi_n(x)$ form a Tschebyscheff-system the best approximating polynomial φ is uniquely determined and $|f(x) - \varphi(x)|$ attains its maximum in at least $n + 1$ points.*

PROOF. According to theorem 4′ there exists a C_{n+1} so that $\varrho = \varrho_{C_{n+1}}$. Every best approximating polynomial for the whole of M is then best approximating in this C_{n+1}. By the special case treated above there is only one best approximating polynomial in C_{n+1} and hence the uniqueness is proved. It also follows that $|f(x) - \varphi(x)|$ attains its maximum in every point in C_{n+1}.

If in particular M is a pointset on the real axis we can take $\varphi_1(x) = 1$, $\varphi_2(x) = x$, $\ldots$, $\varphi_n(x) = x^{n-1}$. A function of type (1) is then an ordinary polynomial of degree $< n$. In this case we shall calculate μ_i explicitly. Let $P(x)$ be a polynomial of degree $< n$. By the Lagrange interpolation formula we have

$$P(x_{n+1}) = \sum_{i=1}^{n} P(x_i) \prod_{\substack{i \neq k \\ k \neq n+1}} \frac{x_{n+1} - x_k}{x_i - x_k} = -\sum_{i=1}^{n} P(x_i) \frac{\prod_{k \neq n+1}(x_{n+1} - x_k)}{\prod_{k \neq i}(x_i - x_k)}$$

or more symmetrically

$$\sum_{i=1}^{n+1} \frac{P(x_i)}{\prod_{k \neq i}(x_i - x_k)} = 0.$$

Hence $\mu_i = 1 / \prod_{k \neq i}(x_i - x_k)$ which with (11) gives a formula by de la Vallée Poussin (see his Leçons sur l'approximations des fonctions d'une variable réelle p. 80). However we shall only need that if $x_1 > x_2 > \cdots > x_{n+1}$ we have $\operatorname{sgn} \mu_i = (-1)^{i-1}$. Hence we conclude the "classical" Tschebyscheff approximation theorem:

THEOREM 6. *Let M be a compact subset of the real axis and $f(x)$ a continuous function over M. Then there is one and only one best approximating polynomial of degree $< n$ and this polynomial $P(x)$ is characterized by the fact that $P(x) - f(x)$ attains its maximum in at least $n + 1$ consecutive points with alternating signs.*

PROOF. Everything but the characterization is contained in theorem 5. Take the C_{n+1} used in the proof of this theorem. As $\operatorname{sgn} \mu_i (f(x_i) - P(x_i))$ is constant if $C_{n+1} = \{x_1, \ldots, x_{n+1}\}$, $P(x) - f(x)$ attains its absolute maximum in $n + 1$ consecutive points with alternating signs. On the other hand let $\max |P(x) - f(x)| = \varrho'$ and $P(x_i) - f(x_i) = \pm \varrho$, $i = 1, \ldots, n + 1$ with alternating signs $(x_1 > x_2 > \cdots > x_{n+1})$. Then as $\operatorname{sgn} \mu_i (f(x_i) - P(x_i))$ is independent of i $P(x)$ is best approximating function in C_{n+1} and $\varrho \geq \varrho'$. By definition $\varrho \leq \varrho'$ so that $\varrho = \varrho'$ and the theorem is proved.

Theorem 6 has essentially been proved by Tschebyscheff (Sur les questions de minima qui se rattachent à la représentation approximative des fonctions, Mém. de l'Acad. Imp. des Sciences de St Petersbourg, Sixième série VII, 1859 p. 199–291). From Theorem 6 de la Vallée Poussin has derived theorem 4 in a special case (Leçons sur l'approximations des fonctions p. 86 et p. 99). By that method we can prove theorem 4 if M is a subset of the real axis and $\varphi_1(x), \ldots, \varphi_n(x)$ form a Tschebyscheff-system. In our general form the theorem is however new.

Theorem 5 is essentially proved by Haar: loc. cit. but with a quite different method.

On some extremal properties.

We now pass to the applications of theorem 2.

Suppose that Γ is not empty. Then we can put $W = (\underline{W}, \overline{W})$ and $W_{C_m} = (\underline{W}_{C_m}, \overline{W}_{C_m})$. We can now write theorem 2 $\overline{W} = \inf \overline{W}_{C_n}$; $\underline{W} = \sup \underline{W}_{C_n}$. With suitable restrictive conditions this can be sharpened to

THEOREM 2'. *If M is a compact topological space and $A_1(x) > A_2(x)$ as $x \in M$ (i.e. we exclude the equality) and finally $\varphi_1(x), \ldots, \varphi_n(x)$ form a Tschebyscheff-system we have $\overline{W} = \min \overline{W}_{C_n}$; $\underline{W} = \max \underline{W}_{C_n}$.*

PROOF. (Analogous with that of theorem 4'.) Choose a sequence $C_n^{(r)}$ where $C_n^{(r)} = \{x_1^{(r)}, \ldots, x_n^{(r)}\}$ and $\overline{W}_{C_n^{(r)}} \to \overline{W}$. This is possible on account of theorem 2. As M is compact we can find a partial sequence r' so that $\lim x_k^{(r')}$ exists as $r' \to \infty$ $(k = 1, \ldots, n)$. We call this limit x_k. $x_1, \ldots, x_n$ form if we do not distinguish those which may be equal a combination C_m with $m \leq n$. Then we have $\overline{W}_{C_m} = \overline{W}$. For let φ be an interior point of $K(x_1) \cap \cdots \cap K(x_n)$ (such exist as these disks have independent normals and a positive thickness; then every point is the limit of interior points.) We then have

$$A_2(x_k) < \varphi(x_k) < A_1(x_k) \quad k = 1, \ldots, n$$

Then it follows from the continuity that for sufficiently big values of r'

$$A_2(x_k^{(r')}) \leq \varphi(x_k^{(r')}) \leq A_1(x_k^{(r')}) \quad k = 1, \ldots, n$$

so that $\omega(\varphi) \leq \overline{W}_{C_n^{(r')}}$ and hence $\omega(\varphi) \leq \overline{W}$. This relation follows for every $\varphi \in K(x_1) \cap \cdots \cap K(x_n)$ as it is a limit of a sequence of interior points. Hence $\overline{W}_{C_m} \leq \overline{W}$ and as the reversed inequality is evident the equality follows. If $m < n$ we take $C_n \supset C_m$. Then $\overline{W} \leq \overline{W}_{C_n} \leq \overline{W}_{C_m} = \overline{W}$ so that $\overline{W}_{C_n} = \overline{W}$. — Analogously we can prove the result about $\underline{W}$.

In his work "Über Polynome die möglichst wenig von Null abweichen", Math. Ann. 77, Wladimir Markoff gives a theorem (Satz 1, p. 216) which is equivalent with theorem 2' for the case of ordinary polynomials, $M = (-1, +1)$ and $A_1(x) = -A_2(x) = +1$ for $|x| \leq 1$. With the method employed by Markoff it is indeed possible to give a more direct proof of theorem 2' which however seems to be new. We observe that the passage from theorem 2 to theorem 2' meant extraordinarily strong specializations. Theorem 2 therefore seems to be the most valuable one and as this theorem cannot be proved with Markoff's method we have chosen the proofs given here.

In order to simplify the theory we shall confine ourselves to the case where $M = $ a compact subset of the real axis with at least n points and $\varphi_1(x) = 1$, $\varphi_2(x) = x$, ..., $\varphi_n(x) = x^{n-1}$ for $x \in M$. (Most of the theory will still be valid if $M = $ compact subset of the real axis and $\varphi_1(x), \ldots, \varphi_n(x)$ form a Tschebyscheffsystem on the whole of the real axis.) Functions of type (1) will now become ordinary polynomials of degree $< n$. $\varphi_1(x), \ldots, \varphi_n(x)$ are linearly independent over M as a polynomial of degree $< n$ with n zeros is identically zero. Therefore $\varphi_1(x), \ldots, \varphi_n(x)$ form a Tschebyscheff-system. We shall throughout this part suppose that $A_1(x) > A_2(x)$ $x \in M$. — In the sequel we write $P(x)$ instead of $\varphi(x)$.

The first question to be answered is when Γ can become empty. Take $n + 1$ points $x_1, \ldots, x_{n+1} \in M$ so that $x_1 > x_2 > \cdots > x_{n+1}$. As has been shown above we have for every $P(x)$ of degree $< n$

$$\sum_{i=1}^{n+1} \frac{P(x_i)}{\prod_{k \neq i}(x_i - x_k)} = 0.$$

As $\operatorname{sgn} \prod_{k \neq i}(x_i - x_k) = (-1)^{i-1}$ it follows if $P(x) \in \Gamma_{C_{n+1}}$ where $C_{n+1} = \{x_1, \ldots, x_{n+1}\}$ that

(12)
$$\frac{A_1(x_1)}{\prod_{k \neq 1}(x_1 - x_k)} + \frac{A_2(x_2)}{\prod_{k \neq 2}(x_2 - x_k)} + \frac{A_1(x_3)}{\prod_{k \neq 3}(x_3 - x_k)} + \cdots \geq 0$$
$$\frac{A_2(x_1)}{\prod_{k \neq 1}(x_1 - x_k)} + \frac{A_1(x_2)}{\prod_{k \neq 2}(x_2 - x_k)} + \frac{A_2(x_3)}{\prod_{k \neq 3}(x_3 - x_k)} + \cdots \leq 0$$

Conversely if (12) is satisfied we can find λ and $\mu \geq 0$ so that $\lambda + \mu = 1$ and

$$\frac{\lambda A_1(x_1) + \mu A_2(x_1)}{\prod_{k \neq 1}(x_1 - x_k)} + \frac{\mu A_1(x_2) + \lambda A_2(x_2)}{\prod_{k \neq 2}(x_2 - x_k)} + \frac{\lambda A_1(x_3) + \mu A_2(x_3)}{\prod_{k \neq 3}(x_3 - x_k)} + \cdots = 0$$

Hence there exists a polynomial $P(x)$ of degree $< n$ so that

$$P(x_{2i}) = \mu A_1(x_{2i}) + \lambda A_2(x_{2i}); \quad P(x_{2i+1}) = \lambda A_1(x_{2i+1}) + \mu A_2(x_{2i+1})$$

and $P(x)$ accordingly $\in \Gamma_{C_{n+1}}$. Now it immediately follows from theorem 1 that

THEOREM 7. *A necessary and sufficient condition in order that* Γ *should not be empty is that if* $x_1 > \cdots > x_{n+1}$ *are arbitrary points* $\in M$ *then* (12) *is satisfied. (This theorem is still valid if* M *is not compact,* $A_1(x)$ *and* $A_2(x)$ *not continuous and not necessarily* $A_1(x) > A_2(x)$.)

In what follows we suppose that Γ is not empty.

Take a non-negative integer κ and a real number z. $P^{(\kappa)}(z)$ is then of the form (3); we put $\omega(P) = P^{(\kappa)}(z)$. Hence $\max P^{(\kappa)}(z)$ and $\min P^{(\kappa)}(z)$ as $P \in \Gamma$ exist. It also follows that we can apply theorem 2, but before doing so we shall introduce some notations and study the case when M contains exactly n points more closely.

(13)   *Notations:* $\begin{cases} \overline{W}_\kappa(z) = \max_{P \in \Gamma} P^{(\kappa)}(z) & 0 \leq \kappa \leq n-1 \\ \underline{W}_\kappa(z) = \min_{P \in \Gamma} P^{(\kappa)}(z) & 0 \leq \kappa \leq n-1 \end{cases}$

where z is an arbitrary real number.

The special case when M has only n points.

The great advantages of this case are that we can use interpolation formulas.

Let $M = \{x_1, \ldots, x_n\}$. A polynomial $P(x)$ of degree $< n$ belongs to Γ if $A_2(x_i) \leq P(x_i) \leq A_1(x_i)$ $i = 1, \ldots, n$. We now introduce the interpolation polynomials of Lagrange

(14) $$P_i(x) = \prod_{k \neq i} \frac{x - x_k}{x_i - x_k}$$

A polynomial $P(x)$ of degree $< n$ can then be written $P(x) = \sum_1^n P(x_i)P_i(x)$. Geometrically this means that we have changed basis in our finite dimensional function space. Put $P(x_i) = \frac{1}{2}(A_1(x_i) + A_2(x_i)) + \eta_i$. We can then write the inequalities defining Γ

(15) $$|\eta_i| \leq \tfrac{1}{2}(A_1(x_i) - A_2(x_i)).$$

We can now get explicit formulas for $\overline{W}_\kappa(z)$ and $\underline{W}_\kappa(z)$. For we have

$$P^{(\kappa)}(z) = \sum_1^n P(x_i)P_i^{(\kappa)}(z) = \sum_1^n \frac{A_1(x_i) + A_2(x_i)}{2} P_i^{(\kappa)}(z) + \sum_1^n \eta_i P_i^{(\kappa)}(z)$$

Hence by (15) we conclude

$$\overline{W}_\kappa(z) = \sum_1^n \frac{A_1(x_i) + A_2(x_i)}{2} P_i^{(\kappa)}(z) + \sum_1^n \frac{A_1(x_i) - A_2(x_i)}{2} |P_i^{(\kappa)}(z)| \quad \text{or}$$

(16) $$\overline{W}_\kappa(z) = \sum_1^n \left\{ A_1(x_i) \frac{1 + \operatorname{sgn} P_i^{(\kappa)}(z)}{2} + A_2(x_i) \frac{1 - \operatorname{sgn} P_i^{(\kappa)}(z)}{2} \right\} P_i^{(\kappa)}(z)$$

(17) $$\underline{W}_\kappa(z) = \sum_1^n \left\{ A_1(x_i) \frac{1 - \operatorname{sgn} P_i^{(\kappa)}(z)}{2} + A_2(x_i) \frac{1 + \operatorname{sgn} P_i^{(\kappa)}(z)}{2} \right\} P_i^{(\kappa)}(z)$$

where $\operatorname{sgn} u = +1$ for $u > 0$, $\operatorname{sgn} u = -1$ for $u < 0$ and $\operatorname{sgn} 0 = 0$. Obviously we get the maximum for the polynomial

(18) $$P(x) = \sum_1^n \left\{ A_1(x_i) \frac{1 + \operatorname{sgn} P_i^{(\kappa)}(z)}{2} + A_2(x_i) \frac{1 - \operatorname{sgn} P_i^{(\kappa)}(z)}{2} \right\} P_i(x)$$

and only for this if all $P_i^{(\kappa)}(z) \neq 0$. If we agree to replace $\operatorname{sgn} 0$ by an arbitrary number λ with $-1 \leq \lambda \leq 1$ in formula (18) this will always be valid, but $P(x)$ is not uniquely determined if some $P_i^{(\kappa)}(z) = 0$. — We shall call the polynomial determined the maximum-polynomial for κ and z. Analogously we define the minimum-polynomial for κ and z, that is

(19) $$P(x) = \sum_1^n \left\{ A_1(x_i) \frac{1 - \operatorname{sgn} P_i^{(\kappa)}(z)}{2} + A_2(x_i) \frac{1 + \operatorname{sgn} P_i^{(\kappa)}(z)}{2} \right\} P_i(x)$$

We shall almost exclusively regard maximum-polynomials; the results for minimum-polynomials follow immediately.

(18) can also be formulated as follows: If $\operatorname{sgn} P_i^{(\kappa)}(z) = 1$ the maximum-polynomial passes through $(x_i, A_1(x_i))$, if $\operatorname{sgn} P_i^{(\kappa)}(z) = -1$ through $(x_i, A_2(x_i))$ and if $\operatorname{sgn} P_i^{(\kappa)}(z) = 0$ through an arbitrary point (x_i, y_i) with $A_2(x_i) \leq y_i \leq A_1(x_i)$.

a) Consider first the case where $\kappa = 0$. Let $x_1 > x_2 > \cdots > x_n$. For $x_i > z > x_{i+1}$ we have

$$\operatorname{sgn} P_1(z) = (-1)^{i-1}, \ldots, \operatorname{sgn} P_i(z) = +1,$$

$$\operatorname{sgn} P_{i+1}(z) = 1, \operatorname{sgn} P_{i+2}(z) = -1, \ldots, \operatorname{sgn} P_n(z) = (-1)^{n-i-1}$$

For $z > x_1$ we have $\operatorname{sgn} P_j(z) = (-1)^{j-1}$. For $z < x_n$ we have $\operatorname{sgn} P_j(z) = (-1)^{j-n}$.

Hence the maximum-polynomial is constant in each of these intervals and is uniquely determined. In the points x_i it is however strongly indeterminated.

(figure missing in original manuscript)

The figure shows an example with $n = 4$. The polynomial curves are dotted where they are not maximum-polynomials.

b) $0 < \kappa < n - 1$. We first have to prove two algebraic lemmas.

LEMMA 1. *Let* $G(x) = \prod_1^N (x - a_k)$ $H(x) = \prod_1^N (x - b_k)$ *where* $a_1 > \cdots > a_N$ *and* $b_1 > \cdots > b_N$ *and* $a_1 \geq b_1 \geq \cdots \geq a_N \geq b_N$ *and some* $a_i \neq b_i$. *Then the roots of* $G^{(\kappa)}(x)$ *and* $H^{(\kappa)}(x)$ *separate each other in the sharpest meaning of this word, and the roots of* $G^{(\kappa)}(x)$ *are largest.*

PROOF. $G(x)/H(x) = \sum_k c_k/(x - b_k) + C$ where $c_i = (b_i - a_i) \prod_{k \neq i} (b_i - a_k)/(b_i - b_k) \leq 0$. Every c_i cannot become zero for then we should have $G(x) = CH(x)$ and $G(x)$ and $H(x)$ would have exactly the same roots. By differentiation we get

$$\frac{G'(x)H(x) - G(x)H'(x)}{H(x)^2} = \sum_1^N -\frac{c_k}{(x - b_k)^2} > 0.$$

Consider in particular the points where $G'(x) = 0$. $H'(x)$ has there the same sign as $-G(x)$ and hence alternatingly $+$ and $-$, so that $H'(x)$ will have at least one root between two roots of $G'(x) = 0$. This is together $n - 1$ of the roots. The remaining must then lie either to the left of the smallest or to the right of the largest root of $G'(x) = 0$. As $\operatorname{sgn} H'(x) = \operatorname{sgn} -G(x) > 0$ in the largest root of $G'(x) = 0$ (that corresponds to a minimum of $G(x)$) we have no root to the right of it. Hence the lemma is proved.

LEMMA 2. *Let* $G(x) = \prod_1^N (x - a_k)$ $H(x) = \prod_1^{N-1} (x - b_k)$ *where* $a_1 > \cdots > a_N$, $b_1 > \cdots > b_{N-1}$ *and* $a_1 \geq b_1 \cdots \geq b_{N-1} \geq a_N$. *Then the roots of* $G^{(\kappa)}(x) = 0$ *and* $H^{(\kappa)}(x)$ *separate each other strongly.*

PROOF. This is a limit case of lemma 1 as $b_N \to -\infty$. However it is simplest to prove it as we proved lemma 1.

These lemmas are essentially known and proved by Markoff: loc. cit. They can be easily derived from a theorem by Kakeya.

Now consider two interpolation-polynomials $P_i(x)$ and $P_k(x)$ with $i > k$. Their roots are respectively $x_1, \ldots, x_{i-1}, x_{i+1}, \ldots, x_n$ and $x_1, \ldots, x_{k-1}, x_{k+1}, \ldots, x_n$. Hence they satisfy lemma 1 and the roots of $P_i^{(\kappa)}(x) = 0$ and $P_k^{(\kappa)}(x) = 0$ separate each other and the roots of the former are "largest".

Denote the roots of $P_i^{(\kappa)}(x) = 0$ with $x_{i,1}, \ldots, x_{i,n-1-\kappa}$ so that $x_{i,1} > x_{i,2} > \cdots > x_{i,n-1-\kappa}$. By the above we get

$$x_{n,1} > x_{n-1,1} > \cdots > x_{1,1} > x_{n,2} > \cdots > x_{2,n-1-\kappa} > x_{1,n-1-\kappa}$$

As all zeros of $P_i^{(\kappa)}(x) = 0$ are simple it is clear that $P_i^{(\kappa)}(x)$ changes signs in every zero. Hence

For $z > x_{n,1}$ is $\operatorname{sgn} P_j(z) = (-1)^{j-1}$.

For $x_{n-1,1} < z < x_{n,1}$ is $\operatorname{sgn} P_n(z) = -(-1)^{n-1}$; $\operatorname{sgn} P_j(z) = (-1)^{j-1}$ for $j < n$.

. . .

For $x_{n-k,1} < z < x_{n-k+1,1}$ is $\operatorname{sgn} P_j(z) = +(-1)$ for $j < n - k + 1$; $\operatorname{sgn} P_j(z) = -(-1)$ for $j > n - k$.

. . .

For $x_{n,2} < z < x_{1,1}$ is $\operatorname{sgn} P_j(z) = -(-1)^{j-1}$ for all j.

. . .

(The same sequence will repeat itself except for a change of signs of all $P_j(z)$.)

. . .

For $z < x_{1,n-1-\kappa}$ is $\operatorname{sgn} P_j(z) = (-1)^{n-1}(-1)^{j-1}$ for all j.

Hence the maximum-polynomials are constant and uniquely determined in all these intervals. In the point $x_{i,k}$ $P_i^{(\kappa)}(z) = 0$ and all the other $P_j^{(\kappa)}(z)$ of alternating signs. Thus the maximum-polynomial has one degree of freedom.

c) $\kappa = n - 1$. All $P_j(z)$ are now constants and we get $\operatorname{sgn} P_j(z) = (-1)^{j-1}$. There is one and only one maximum-polynomial and this is independent of z.

If we exclude the maximum-polynomials in a point x_i for $\kappa = 0$ all the maximum-polynomials obtained are of one of the following types.

   (1) All $P_j(z) \neq 0$ and of alternating signs.
   (2) $P_i(z)$ and $P_{i+1}(z)$ are of the same sign and all $P_j(z) \neq 0$ and of alternating signs for $j \leq i$ and $j \geq i + 1$.
   (3) $P_i(z) = 0$, $P_j(z) \neq 0$ for $j \neq i$ and of alternating signs for these values of j.

If $0 < \kappa < n - 1$ we really get all these variations of signs. We hence conclude the remarkable fact that the maximum-polynomials are independent of κ ($0 < \kappa < n-1$) (but of course they are maximum-polynomials in different points depending on κ.) Another fact worth noticing is that the same combination of signs will occur many times every other time with a change of signs for all $P_j^{(\kappa)}(z)$. Hence an extremal polynomial is maximum- or minimum-polynomial in

$n - \kappa$ intervals if it is of type 1).

$n - 1 - \kappa$ intervals if it is of type 2).

$n - 1 - \kappa$ points if it is of type 3). (We have here supposed $\kappa > 0$.)

After this detailed analysis we shall go to the general case where we shall repeatedly use theorem 2'. We shall therefore point out the essential feature of this theorem. — It is trivial that if the maximum-polynomial for z and κ from Γ_{C_n} belongs to Γ it will be maximum-polynomial from Γ. But conversely theorem 2' states that all maximum-polynomials are of this trivial kind.

The generalized Tschebyscheff-polynomials.

Take a number z larger than any number in M (there is surely one as M is compact). Consider the problem of finding the maximum of $P(z)$ as $P \in \Gamma$. According to theorem $2'$ there is a combination C_n of n points $x_1, \ldots, x_n \in M$ so that the maximum of $P(z)$ as $P \in \Gamma$ = the maximum of $P(z)$ as $P \in \Gamma_{C_n}$. We can then formulate the following theorem

THEOREM 8. *If z is a point to the right of M there is one and only one polynomial from Γ that gives the maximum of $P(z)$. This polynomial is characterized by the fact that it belongs to Γ and in n points $x_1 > x_2 > \cdots > x_n$ is equal respectively to $A_1(x_1)$, $A_2(x_2)$, $A_1(x_3)$, $\ldots$. The same polynomial gives the maximum for all such z.*

PROOF. As we have seen that the maximum-polynomial $P(x)$ is maximum-polynomial for Γ_{C_n} the uniqueness of the maximum-polynomial and the necessity of the condition follows by the special case treated above. Conversely if $P(x) \in \Gamma$ and $P(x_1) = A_1(x_1)$, $P(x_2) = A_2(x_2)$, $\ldots$ $P(x)$ must be maximum-polynomial from Γ_{C_n} where $C_n = \{x_1, \ldots, x_n\}$ and hence $P(x)$ is maximum-polynomial from Γ. From the characterization of the maximum-polynomial it immediately follows that we get the same polynomial for all z to the right of M.

Of course we get an analogous theorem for the minimum of $P(z)$.

THEOREM 9. *If z is a point to the right of M there is one and only one polynomial from Γ that gives the minimum of $P(z)$. This polynomial is characterized by the fact that it belongs to Γ and in n points $x_1 > x_2 > \cdots > x_n$ is equal to respectively $A_2(x_1)$, $A_1(x_2)$, $A_2(x_3)$, $\ldots$. The same polynomial gives the minimum for all such z.*

Those polynomials which are defined in theorems 8 and 9 will be called respectively the upper and the lower Tschebyscheff-polynomial associated with M, $A_1(x)$ and $A_2(x)$ of degree $< n$. They have a number of extremal properties which will be listed below:

A) They give the maximum and minimum of $P(z)$ if z is a point to the right or to the left of M.

B) Let $0 < \kappa < n - 1$. The upper Tschebyscheff-polynomial gives in $n - \kappa$ intervals (two of which are infinite in one direction) maximum or minimum of $P^{(\kappa)}(z)$ alternatingly, beginning with giving the maximum of $P^{(\kappa)}(z)$ for all z to the right of M. The lower Tschebyscheff-polynomial has analogous properites.

C) The upper Tschebyscheff-polynomial and only this gives the maximum of the highest coefficient in $P(x)$. The lower Tschebyscheff-polynomial and only this gives the minimum of the highest coefficient in $P(x)$.

PROOF. A) – C) follow immediately from the trivial remark preceding the subsection. The nontrivial result is that there exists a Tschebyscheff-polynomial as has been proved in theorems 8–9.

Examples of Tschebyscheff-polynomials.

1. Let $M = (-1, +1)$, $A_1(x) = -A_2(x) = 1$. The two polynomials

$$P(x) = \pm \cos((n - 1) \arccos x)$$

satisfy the characterizations of theorems 8–9 and are therefore the generalized Tschebyscheff polynomials for this case (that they are polynomials is well known). But these polynomials are what is generally called Tschebyscheff polynomials which justifies our notations. A) – C) immediately gives a lot of known extremal properties of those polynomials.

(figure missing in original manuscript)

2. Let M consist of the two intervals $(-b, -a)$ and (a, b) where $0 < a < b$, $A_1(x) = -A_2(x) = 1$. Suppose that n is odd, $n = 2k + 1$.

$$P(x) = \pm \cos \left(k \arccos \frac{2x^2 - a^2 - b^2}{b^2 - a^2} \right)$$

are then the generalized Tschebyscheff-polynomials.

In this manner we may generalize almost the whole of W. Markoff's results; we can study the variations of the extremal polynomials with z and so on. However, we shall abstain from that as our results above give the most valuable results.

AN ISOPERIMETRIC INEQUALITY IN
HOMOGENEOUS FINSLER SPACES

The aim of this note is to give an isoperimetric inequality for the case that the area is measured in the ordinary sense but the element of arc is "homogeneous and non-isotropic", that is measured with some non-euclidean (and not even convex) metric. — The idea of the proof is due to E. Schmidt[3]. However, as this author works quite insymmetrically our formalism has to be quite different. — We finally give some applications of the isoperimetric theorem. The inequalities which we shall derive are also due to E. Schmidt[4]. The proof is in principle identical with his, but the geometric interpretation and the presentation seem to be new.

In the sequel we denote by $x, \xi, \ldots$ vectors in a two-dimensional vector space. We suppose that this is normed with a norm $\|x\|$, which we suppose to be homogeneous, i.e. $\|ax\| = |a|\|x\|$. (We do not assume that it is convex.) Consider the area spanned by x and another vector ξ,

$$|x, \xi| = \begin{vmatrix} x_1 & x_2 \\ \xi_1 & \xi_2 \end{vmatrix}.$$

Analogously to the definition of supporting function (or dual norm) we introduce

$$(1) \qquad \langle\!\langle x \rangle\!\rangle = \max_{\|\xi\| \leq 1} \big| |x, \xi| \big|.$$

It is obvious that $\langle\!\langle x \rangle\!\rangle$ is homogeneous and as it is the maximum of convex functions it is also convex. Hence $\langle\!\langle x \rangle\!\rangle$ is a Minkowski metric.

REMARK. If we call the dual norm $\|x\|^*$ it is obvious that $\langle\!\langle x \rangle\!\rangle = \|x'\|^*$ where $x' = (-x_2, x_1)$ that is the vector x turned through an angle of $\frac{\pi}{2}$ in the positive direction. We shall however never use this remark.

As $\langle\!\langle x \rangle\!\rangle$ is convex it is differentiable except on a countable number of rays through the origin. Let it be differentiable at x. Take ξ so that $|x, \xi| = \langle\!\langle x \rangle\!\rangle$ and $\|\xi\| = 1$. We have then

$$\langle\!\langle x + ty \rangle\!\rangle - \langle\!\langle x \rangle\!\rangle \geq |x + ty, \xi| - |x, \xi| = t|y, \xi| \quad \text{and hence}$$
$$\frac{\langle\!\langle x + ty \rangle\!\rangle - \langle\!\langle x \rangle\!\rangle}{t} \gtreqless |y, \xi| \quad \text{as } t \gtreqless 0.$$

---

[3] Über das isoperimetrische Problem im Raum von n Dimensionen, Math. Z. **44**(1939).

[4] Über die Ungleichung welche die Integrale über eine Potenz einer Funktion und über eine andere Potenz ihrer Ableitung verbindet, Math. Ann. **117**(1940–1941).

© Springer International Publishing AG, part of Springer Nature 2018
L. Hörmander, *Unpublished Manuscripts*,
https://doi.org/10.1007/978-3-319-69850-2_2

But as we have supposed that the limit exists when $t \to 0$ through positive and negative values we must have

$$\lim_{t \to 0} \frac{\langle\langle x + ty \rangle\rangle - \langle\langle x \rangle\rangle}{t} = |y, \xi|$$

or in different notation

(2)
$$d\langle\langle x \rangle\rangle = |dx, \xi|.$$

We now formulate the isoperimetric problem. Let C be a curve with the parameter representation $x(t)$, described in the positive direction as t goes from 0 to 1. We suppose that $x(t)$ is absolutely continuous. The area of the curve is defined by

(3)
$$A = \tfrac{1}{2} \int |x, dx|$$

and we define the arclength by

(4)
$$L = \int \|dx\|.$$

The problem is to determine the maximum of A for fixed L.

It is very natural to guess that A should be maximum if always $|x, dx| = \langle\langle x \rangle\rangle \|dx\|$. But by the definition of ξ in formula (2) this implies that dx lies in the direction of ξ and so by (2) $d\langle\langle x \rangle\rangle = 0$. (N. B. if $\langle\langle x \rangle\rangle$ is differentiable in the point x. In this heuristic argument we can however suppose that.) Hence we may expect the curves $\langle\langle x \rangle\rangle = $ constant to give extremum.

Consider the curve $\langle\langle x \rangle\rangle = \varrho$ with some constant ϱ. Its area is $\bar{A} = \tfrac{1}{2} \int |x, dx| = \tfrac{1}{2} \int \langle\langle x \rangle\rangle \|dx\| = \tfrac{\varrho}{2} \int \|dx\| = \tfrac{\varrho}{2} \bar{L}$. For except in a denumerable number of points $\langle\langle x \rangle\rangle$ is differentiable and so $|dx, \xi| = 0$ for displacements dx along the curve. Hence dx is in the direction of ξ and $|x, dx| = \langle\langle x \rangle\rangle \|dx\|$. (Compare the unsharp argument above in the reversed direction.)

Now we associate with the curve C a curve $\bar{C} : \bar{x}(t)$ $(0 \le t \le 1)$ with $\langle\langle \bar{x} \rangle\rangle = \varrho = $ constant. $\bar{x}(t)$ need not describe this curve monotonously, but we suppose that it totally makes one circulation round the curve so that we may write

$$\bar{A} = \tfrac{1}{2} \int |\bar{x}, d\bar{x}|.$$

We shall later fix ϱ and the parametrization in a convenient manner. Now form

$$A + \bar{A} = \tfrac{1}{2} \int |x, dx| + \tfrac{1}{2} \int |\bar{x}, d\bar{x}| = \tfrac{1}{2} \int (|x, d\bar{x}| + |\bar{x}, dx|) + \tfrac{1}{2} \int |x - \bar{x}, d(x - \bar{x})|.$$

We have by partial integration $\int |x, d\bar{x}| = - \int |dx, \bar{x}| = \int |\bar{x}, dx|$ and $\int |\bar{x}, dx| \le \int \langle\langle \bar{x} \rangle\rangle \|dx\|$ $= \varrho \int \|dx\| = \varrho L$. Thus we can write

$$A + \bar{A} \le \varrho L + \tfrac{1}{2} \int |x - \bar{x}, d(x - \bar{x})|.$$

By the inequality for geometric and arithmetic means we have $A + \bar{A} \ge 2\sqrt{A\bar{A}}$ and if we eliminate ϱ by means of $\varrho = 2\bar{A}/\bar{L}$ we get

$$\sqrt{\frac{A}{\bar{A}}} \le \frac{L}{\bar{L}} + \frac{1}{4\bar{A}} \int |x - \bar{x}, d(x - \bar{x})|.$$

To obtain the theorem we have to find ϱ and the parametrization so that the conditions above are satisfied and the integral vanishes. We thus want to get $d(x - \bar{x})$ in the direction of $x - \bar{x}$ that is in a constant direction. — Choose any vector v and draw the smallest strip containing C with the direction v. (This is the essential idea of E. Schmidt.) On both of the two bounding lines L_1 and L_2 there must be at least one point of C. Choose on each L_i a point of C corresponding to the parameter value t_i. — Obviously we can find $\bar{x}_0$ and ϱ so that the curve $\bar{x}_0 + \bar{x}$ with $\langle\!\langle \bar{x} \rangle\!\rangle = \varrho$ has L_1 and L_2 as tangents. As t increases from t_1 to t_2 (or a value of t congruent with t_2) we project $x(t)$ on the arc of $\bar{C}$ which goes in the positive direction from the line L_1 to L_2, and analogously for the interval $(t_2, t_1 + 1)$. The projection $\bar{x}(t)$ then describes $\bar{C}$ one time in the positive direction. Also $x - \bar{x}$ has constant direction so that the parametrization satisfies all our conditions. Hence we have proved

$$(5) \qquad\qquad \sqrt{\frac{A}{\bar{A}}} \leq \frac{L}{\bar{L}}$$

that is the isoperimetric inequality. (As a convex curve has straight lines only in a denumerable number of directions we can really choose v so that L_1 and L_2 only have one point on the curve.) — We now pass to conditions of equality. A necessary condition is that $|\bar{x}, dx| = \langle\!\langle \bar{x} \rangle\!\rangle \|dx\|$. Suppose that $\bar{x}$ is differentiable in the point $\bar{x}$ considered. Then dx must be in the direction of $\bar{\xi}$ that is in the direction of $d\bar{x}$, $dx - \lambda d\bar{x} = 0$. But as $dx - d\bar{x}$ has the direction of v this means that $dx = d\bar{x}$ unless $d\bar{x}$ is in the direction of v, and hence dx in the direction of v. But as v is arbitrary we conclude $dx = d\bar{x}$ even in this case. The only thing to consider is what happens in the points where $\bar{x}$ is not differentiable. Those points are denumerable on $\bar{C}$. If in no neighborhood of the parameter value t_0 $\bar{x}(t)$ lies constantly in a point $\bar{X}$ where there is no derivative we conclude that also in that point $dx = d\bar{x}$ to the right and to the left. So we finally have the possibility that $\bar{x}$ could lie fixed for an interval of t-values so that C has a linear arc. But of course that also proves to be impossible if we choose another direction v. Hence we have established the

THEOREM. *With the notations of the text we have $\sqrt{A/\bar{A}} \leq L/\bar{L}$ with equality only if the curve C is homothetic with the curve $\langle\!\langle x \rangle\!\rangle = 1$.*

For the applications below we make the following remark. Let $\bar{A}$ in particular be the area of the curve $\langle\!\langle x \rangle\!\rangle = 1$. Then $\bar{L} = 2\bar{A}$ and the theorem can be stated $A \leq L^2/4\bar{A}$. This saves much calculation.

Applications. Consider the metric $\|x\| = (|x_1|^a + |x_2|^a)^{\frac{1}{a}}$. We get

$$\langle\!\langle x \rangle\!\rangle = \max_{\|\xi\| \leq 1} |x_1\xi_2 - x_2\xi_1| = \max_{\|\xi\| \leq 1} |x_1\xi_1 + x_2\xi_2| = (|x_1|^{a'} + |x_2|^{a'})^{\frac{1}{a'}}$$

where a' is the exponent conjugate to a: $\frac{1}{a} + \frac{1}{a'} = 1$. We calculate the area of $\langle\!\langle x \rangle\!\rangle = 1$. We have

$$\bar{A} = 4\int_0^1 (1 - x_1^{a'})^{\frac{1}{a'}}\,dx_1 = 4\int_0^1 (1-t)^{\frac{1}{a'}}t^{\frac{1}{a'}-1}\,dt/a' = 4B(\tfrac{1}{a'}+1, \tfrac{1}{a'})/a'$$

$$= \frac{4}{a'}\frac{\Gamma(1+\frac{1}{a'})\Gamma(\frac{1}{a'})}{\Gamma(1+\frac{2}{a'})} = 4\frac{\Gamma(1+\frac{1}{a'})^2}{\Gamma(1+\frac{2}{a'})}.$$

Writing out the isoperimetric theorem we have $2\sqrt{A\bar{A}} \leq L$ with

$$A = \left| \int_0^1 x_1(t)x_2'(t)\,dt \right|,$$

$$L = \int (|dx_1|^a + |dx_2|^a)^{\frac{1}{a}} = \int_0^1 (|x_1'|^a + |x_2'|^a)^{\frac{1}{a}}\,dt \leq \left(\int_0^1 (|x_1'|^a + |x_2'|^a)\,dt \right)^{\frac{1}{a}}$$

for the mean values increase with the exponents, as the length of the integration interval is 1. Applying the theorem to $\lambda^{\frac{2}{a}}x_1$ and $\mu^{\frac{2}{a}}x_2$ we have

$$\lambda\mu\left(2\sqrt{\bar{A}A}\right)^a \le \lambda^2 \int_0^1 |x_1'|^a\,dt + \mu^2 \int_0^1 |x_2'|^a\,dt \quad (\lambda,\mu \ge 0)$$

and hence $\quad \frac{1}{4}(4A\bar{A})^a \le \left(\int_0^1 |x_1'|^a\,dt\right)\left(\int_0^1 |x_2'|^a\,dt\right)$

or $\quad \dfrac{\left|\int_0^1 x_1(t)x_2'(t)\,dt\right|}{\|x_1'\|_a\|x_2'\|_a} \le \dfrac{\Gamma(1+\frac{2}{a'})}{4^{1+\frac{1}{a'}}\Gamma(1+\frac{1}{a'})^2}.$

This is as we shall see below almost identical with one of the inequalities by Erhard Schmidt quoted above. To get a more general theorem we must use some other norms, which I do not think have been employed before so I shall give a rather detailed discussion of them.

Obviously we need to get a simple analytic expression for the curve $\langle\!\langle x\rangle\!\rangle = 1$, whereas the norm $\|x\|$ does not appear so very much in the calculations. We therefore choose $\langle\!\langle x\rangle\!\rangle$ first and then find a $\|x\|$ corresponding to it. ($\|x\|$ is not necessarily uniquely determined, but if we search for a convex $\|x\|$ it is unique.) As the curve $\langle\!\langle x\rangle\!\rangle = 1$ we choose the curve $|x_1|^a + |x_2|^b = 1 \quad (a,b \ge 1)$. This is convex as is immediately verified. We now put

$$\|x\| = \max_{|\xi_1|^a+|\xi_2|^b=1}\left\|\begin{matrix} x_1 & x_2 \\ \xi_1 & \xi_2 \end{matrix}\right\|.$$

It then follows in a straightforward manner that $\langle\!\langle x\rangle\!\rangle = \max_{\|\xi\|=1}\left\|\begin{matrix} x_1 & x_2 \\ \xi_1 & \xi_2 \end{matrix}\right\|$ so that we get back the original metric.

Let now $f(t) \in L^{b'}(0,1)$ and $g(t) \in L^{a'}(0,1)$. We want to estimate the integral $\int_0^1 \|(f(t),g(t))\|\,dt$. For every value of t we can find $\xi_1(t)$ and $\xi_2(t)$ so that $|\xi_1|^a + |\xi_2|^b = 1$ and $\|(f(t),g(t))\| = |f(t)||\xi_2(t)| + |g(t)||\xi_1(t)|$. Hence by Hölder's inequality

$$\int_0^1 \|(f(t),g(t))\|\,dt \le \|f\|_{b'}\|\xi_2\|_b + \|g\|_{a'}\|\xi_1\|_a.$$

But $\|\xi_2\|_b^b + \|\xi_1\|_a^a = 1$ so we get by the definition of $\|\cdot\|$ that

(6) $$\int_0^1 \|(f(t),g(t))\|\,dt \le \|(\|f\|_{b'},\|g\|_{a'})\|.$$

We now return to the isoperimetric inequality. We first calculate

$$\bar{A} = 4\int_0^1 (1-\xi_1^a)^{\frac{1}{b}}\,d\xi_1 = 4\int_0^1 (1-t)^{\frac{1}{b}}t^{\frac{1}{a}-1}\,dt/a = \frac{4}{a}B(\frac{1}{b}+1,\frac{1}{a}) = 4\frac{\Gamma(1+\frac{1}{a})\Gamma(1+\frac{1}{b})}{\Gamma(1+\frac{1}{a}+\frac{1}{b})}.$$

The isoperimetric inequality can be written

$$2\sqrt{\bar{A}}\left|\int x_1 x_2'\,dt\right| \le \int \|(x_1',x_2')\|\,dt \le \|(\|x_1'\|_{b'},\|x_2'\|_{a'})\|.$$

Now we can replace x_1 by $x_1 t_1$ and x_2 by $x_2 t_2$ and get $(t_i > 0)$

$$2\sqrt{\bar{A} \left| \int x_1 x_2' \, dt \right|} \leq \min_t \frac{\|(t_1 \|x_1'\|_{b'}, t_2 \|x_2'\|_{a'})\|}{\sqrt{t_1 t_2}} = \sqrt{\|x_1'\|_{b'} \|x_2'\|_{a'}} \min_\xi \frac{\|(\xi_1, \xi_2)\|}{\sqrt{\xi_1 \xi_2}}.$$

Obviously $\min_\xi \|(\xi_1, \xi_2)\| / \sqrt{\xi_1 \xi_2} = 1 / \max_{\|\xi\|=1} \sqrt{\xi_1 \xi_2}$. To determine this we compare it with the corresponding quantity for the norm $\langle\!\langle x \rangle\!\rangle$. We have (and this is a quite general theorem)

$$\max_{\|\xi\|=1} \sqrt{\xi_1 \xi_2} \cdot \max \sqrt{\eta_1 \eta_2} = \tfrac{1}{2}.$$

For as $\|\xi\| = 1$, $\|\eta\| = 1$ $\quad 1 \geq \left| \begin{matrix} \xi_1 & \xi_2 \\ \pm\eta_1 & \pm\eta_2 \end{matrix} \right|$ and so

$$1 \geq |\xi_1 \eta_2| + |\xi_2 \eta_1| \geq 2\sqrt{|\xi_1 \xi_2 \eta_1 \eta_2|} = 2\sqrt{|\xi_1 \xi_2|}\sqrt{|\eta_1 \eta_2|}.$$

It is clear that we get equality if we choose ξ and η so that their directions are symmetrical in one of the axes. Now

$$\max_{|\eta_1|^a + |\eta_2|^b = 1} |\eta_1 \eta_2| = \max_{t_1 + t_2 = 1} t_1^{\frac{1}{a}} t_2^{\frac{1}{b}} = \max \frac{t_1^{\frac{1}{a}} t_2^{\frac{1}{b}}}{(t_1 + t_2)^{\frac{1}{a} + \frac{1}{b}}} = \max \frac{\left(\frac{b}{a+b} u_1\right)^{\frac{1}{a}} \left(\frac{a}{a+b} u_2\right)^{\frac{1}{b}}}{\left(\frac{b}{a+b} u_1 + \frac{a}{a+b} u_2\right)^{\frac{1}{a} + \frac{1}{b}}}$$

$$= \frac{\left(\frac{1}{a}\right)^{\frac{1}{a}} \left(\frac{1}{b}\right)^{\frac{1}{b}}}{\left(\frac{1}{a} + \frac{1}{b}\right)^{\frac{1}{a} + \frac{1}{b}}} \max \left[\frac{u_1^{\frac{b}{a+b}} u_2^{\frac{a}{a+b}}}{\frac{b}{a+b} u_1 + \frac{a}{a+b} u_2} \right]^{\frac{a+b}{ab}} = \frac{\left(\frac{1}{a}\right)^{\frac{1}{a}} \left(\frac{1}{b}\right)^{\frac{1}{b}}}{\left(\frac{1}{a} + \frac{1}{b}\right)^{\frac{1}{a} + \frac{1}{b}}}$$

by the inequality for geometric and arithmetic means. Hence finally

$$\frac{\left| \int x_1 x_2' \, dt \right|}{\|x_1'\|_{b'} \|x_2'\|_{a'}} \leq \frac{\left(\frac{1}{a}\right)^{\frac{1}{a}} \left(\frac{1}{b}\right)^{\frac{1}{b}}}{\left(\frac{1}{a} + \frac{1}{b}\right)^{\frac{1}{a} + \frac{1}{b}}} \frac{\Gamma(1 + \frac{1}{a} + \frac{1}{b})}{4\Gamma(1 + \frac{1}{a})\Gamma(1 + \frac{1}{b})}.$$

(Added in 1997: By the Hahn-Banach theorem the last estimate is equivalent to

$$\min_c \|x_1 - c\|_a \leq \frac{\left(\frac{1}{a}\right)^{\frac{1}{a}} \left(\frac{1}{b}\right)^{\frac{1}{b}}}{\left(\frac{1}{a} + \frac{1}{b}\right)^{\frac{1}{a} + \frac{1}{b}}} \frac{\Gamma(1 + \frac{1}{a} + \frac{1}{b})}{4\Gamma(1 + \frac{1}{a})\Gamma(1 + \frac{1}{b})} \|x_1'\|_{b'}.$$

Erhard Schmidt proved more: one can choose $c = \frac{1}{2}(\max_t x_1(t) + \min_t x_1(t))$ in the left-hand side.)

ON BEST APPROXIMATION IN L^∞

LARS HÖRMANDER

Let X be a compact space and $d\mu$ a positive measure on X. The space of equivalence classes of μ-measurable complex valued functions with μ-integrable p^{th} power we denote by L^p; when $p = \infty$ the definition is modified in the usual way. We assume that $\int_X d\mu = 1$, which implies that

$$(1) \qquad \|f\|_p \quad \text{is an increasing function of } p,$$

if f is a μ-measurable function.

Let L_i, $i = 1, \ldots, n$, be linearly independent linear functionals on L^1, which means that we can write

$$(2) \qquad L_i(f) = \int f l_i \, d\mu, \quad f \in L^1,$$

where $l_i \in L^\infty$ and are linearly independent. The linear forms L_i are of course also continuous on every L^p in view of (1). Given complex numbers $y_1, \ldots, y_n$ we now study the interpolation problem

$$(3) \qquad L_i(f) = y_i, \quad i = 1, \ldots, n.$$

Since the functions l_i are linearly independent a solution can always be found in L^∞. We set

$$(4) \qquad E_p = E_p(y) = \inf\{\|f\|_p; f \in L^p, \text{ and } f \text{ satisfies (3)}\}.$$

In view of (1) we know that $E_p(y)$ is an increasing function of p for $1 \le p \le \infty$. When $1 < p < \infty$ it follows from the uniform convexity of L^p (see Nagy, *Spektraldarstellung linearer Transformationen des Hilbertschen Raumes*, p. 7, footnote 1) that there is a unique $f \in L^p$ for which the infimum in (4) is attained. We denote this function by f_p. Our purpose is to study the convergence of f_p when $p \to \infty$. We note that f_p is the unique function in L^p satisfying (3) for which

$$(5) \quad \|f_p + g\|_p \ge \|f_p\|_p \text{ for every } g \in L^p \text{ such that } L_i(g) = \int g l_i \, d\mu = 0, \ i = 1, \ldots, n.$$

© Springer International Publishing AG, part of Springer Nature 2018
L. Hörmander, *Unpublished Manuscripts*,
https://doi.org/10.1007/978-3-319-69850-2_3

LEMMA 1. *Let f and $g \in L^p$. Then we have $(1 < p < \infty)$*

(6) $$\|f - \lambda g\|_p \geq \|f\|_p \quad \text{for every complex number } \lambda$$

if and only if

(7) $$\int \frac{|f|^p}{f} g \, d\mu = 0.$$

Here $|f|^p/f$ is defined as 0 where $f = 0$.

PROOF. That (6) implies (7) follows by differentiating (6) with λ replaced by $t\lambda$, where t is a real parameter, and then putting $t = 0$. Conversely, assume that (7) is satisfied. If p' is the exponent conjugate to p, i.e., $p^{-1} + p'^{-1} = 1$, we have with

(8) $$\varphi = |f|^p/f$$

that $|\varphi|^{p'} = |f|^p$, so that $\varphi \in L^{p'}$ and $\|\varphi\|_{p'}^{p'} = \|f\|_p^p$. Hence we have

$$\|\varphi\|_{p'} \|f + \lambda g\|_p \geq \left| \int (f + \lambda g)\varphi \, d\mu \right| = \left| \int f\varphi \, d\mu \right| = \|f\|_p \|\varphi\|_{p'},$$

which proves (6).

We now apply the lemma to the function f_p. If we set

(9) $$\varphi_p = |f_p|^p/f_p,$$

it follows from the lemma that (5) is equivalent to

(10) $$\varphi_p = \sum_1^n \alpha_p^i l_i$$

for some constants α_p^i. Hence, for $1 < p < \infty$ there is one and only one $f_p \in L^p$ satisfying (3) such that the function φ_p defined by (9) satisfies (10) for some constants α_p^i.

We shall now estimate the constants α_p^i. To do so we note that

$$\|\varphi_p\|_{p'}^{p'} = \|f_p\|_p^p = E_p^p,$$

and since $p/p' = p - 1 = 1/(p' - 1)$ this gives

(11) $$\|\varphi_p\|_{p'} = E_p^{p-1} = E_p^{1/(p'-1)}.$$

Now we have for every function φ of the form $\varphi = \sum_1^n \alpha^i l_i$

$$\|\varphi\|_1 \leq \|\varphi\|_{p'} \leq \|\varphi\|_\infty \leq \sup_i \|l_i\|_\infty \sum_1^n |\alpha^i|.$$

On the other hand, since all norms in a finite dimensional vector space are equivalent we know that $\sum_1^n |\alpha^i|$ is equivalent to $\|\varphi\|_1$. Hence we have for some constant M

(12) $$M^{-1} \sum_1^n |\alpha^i| \leq \| \sum_1^n \alpha^i l_i \|_{p'} \leq M \sum_1^n |\alpha^i|.$$

The constant M is independent of p'. In particular, we may take $\alpha^i = \alpha^i_p$, which gives

(13)
$$M^{-1}E_p{}^{p-1} \leq \sum_{i=1}^{n} |\alpha^i_p| \leq ME_p{}^{p-1}.$$

In particular it follows from (13) that $\alpha_p = (\alpha^1_p, \ldots, \alpha^n_p)$ lies in a compact set in $\mathbb{C}^n$ when p is bounded from above. It now follows immediately that α_p is a continuous function of p when $1 < p < \infty$. In fact, if α is a limit point of α_q when $q \to p$, and $1 < p < \infty$, and we set $\varphi = \sum \alpha^i l_i$, then $f = |\varphi|^{p'}/\varphi$ satisfies (3) since f is a limit of $|\varphi_q|^{q'}/\varphi_q$ when $q \to p$, and this implies that $f = f_p$ since f_p is the unique extremal function. In particular, it follows that E_p is a continuous function of p. We list these facts as a lemma.

LEMMA 2. *The coefficients α^i_p and the extremum value E_p are continuous functions of p when $1 < p < \infty$.*

We can now consider the behavior of f_p when $p \to \infty$. To do so we first renormalize the coefficients α^i_p by setting
$$\beta^i_p = \alpha^i_p/E_p{}^{p-1}.$$

By (13) we then have
$$M^{-1} \leq \sum_{i=1}^{n} |\beta^i_p| \leq M,$$

so that for a suitable sequence $p_\nu \to \infty$ the limits of $\beta^i_{p_\nu}$ all exist $= \beta^i$. Set $\psi_1 = \sum \beta^i l_i$. If we set $E = \lim_{p \to \infty} E_p \leq E_\infty$ we obtain since

$$f_p = E_p \Big| \sum_i \beta^i_p l_i \Big|^{p'} \Big/ \sum_i \beta^i_p l_i$$

that the functions f_p are uniformly bounded from above when $p \to \infty$, that

$$\varlimsup_{p \to \infty} |f_p| \leq E$$

and that

$$\lim_{\nu \to \infty} f_{p_\nu} = E\,\overline{\mathrm{sgn}}\,\psi_1$$

in the set X_1 where $\psi_1 \neq 0$, which is not a null set. Now in the complement of X_1 we have $\psi_1 = 0$ so that there is at least one linear relation between $l_1, \ldots, l_n$ there. Assume for example that $l_1, \ldots, l_m$ is a maximal subsystem which is linearly independent in L^∞ over $\complement X_1$. Over $\complement X_1$ we can express $\sum_1^n \alpha^i_p l_i$ in terms of $l_1, \ldots, l_m$ only. Again we can change the normalization of the coefficients by dividing out their sum. Repeating this procedure at most m times, making a passage to a subsequence every time, we find:

LEMMA 3. *There exist:*
  1) *a sequence $p_\nu \to \infty$;*
  2) *sets $X_1, \ldots, X_k$, $k \leq m$ which are disjoint and μ-measurable*
  3) *functions $\psi_1, \ldots, \psi_k$ which are linear combinations of $l_1, \ldots, l_n$ such that for every j we have $\psi_j = 0$ outside $X_1 \cup \cdots \cup X_j$ but $\psi_j \neq 0$ in X_j,*
  4) *numbers $e_1 \geq e_2 \geq \cdots \geq e_k > 0$ such that $e_1 = E$;*

we have with these notations:

$$\lim_{\nu \to \infty} f_{p_\nu}(x) = e_j \, \overline{\mathrm{sgn}} \, \psi_j(x) \quad \textit{if } x \in X_j$$

$$\lim_{\nu \to \infty} f_{p_\nu}(x) = 0 \quad \textit{if } x \text{ does not belong to any } X_j.$$

The function $f(x) = \lim f_{p_\nu}(x)$ satisfies the condition (3) since we have seen that the sequence f_p is uniformly bounded so that Lebesgue's theorem can be applied. Furthermore we have, with numbers, functions and sets as defined in 2), 3) and 4), that

(14) $\qquad\qquad\qquad f(x) = e_j \, \overline{\mathrm{sgn}} \, \psi_j(x) \quad \text{if } x \in X_j$

(15) $\qquad\qquad\qquad f(x) = 0 \quad \text{if } x \text{ does not belong to any } X_j.$

A first consequence is that $E_\infty \le E$ and since we know already that $E \le E_\infty$ it follows that $E = E_\infty$, that is, E_p is continuous when $1 < p \le \infty$. Furthermore, we shall prove that f is uniquely determined by the listed properties. Hence *we conclude that f_p converges to f*, for example in the weak* topology in L^∞.

To prove the uniqueness of f we have to prove the following lemma.

LEMMA 4. *Let f be a function satisfying* (3), (14) *and* (15). *If f' is another function in L^∞ satisfying* (3) *and if $f' = f$ almost everywhere in X_i when $i < j$, we have with $Y_j = \complement(X_1 \cup \cdots \cup X_{j-1})$ that*

$$e_j = \mathrm{ess} \, \sup_{Y_j} |f| \le \mathrm{ess} \, \sup_{X_j} |f'|;$$

if equality holds here we have $f = f'$ almost everywhere in X_j.

Note that it follows that ess $\sup_{Y_j} |f| \le$ ess $\sup_{Y_j} |f'|$. If f' satisfies the same hypotheses as f the roles can be interchanged and we conclude that there is equality hence that $f = f'$ and that $X_j = X_j'$. Thus recursive use of the lemma shows that f is uniquely determined by the listed properties.

PROOF OF THE LEMMA. Since both f and f' satisfy (3) we have

$$\int (f - f') \psi_j \, dx = 0.$$

The integration here need only be extended over X_j since $f - f' = 0$ or $\psi_j = 0$ in its complement. Now we have $f \psi_j = |\psi_j| e_j$ in X_j. If $|f'| \le e_j$ almost everywhere in X_j we therefore obtain that $f' \psi_j = |\psi_j| e_j$ almost everywhere in X_j, which proves that $f' = f$ almost everywhere in X_j. This proves the lemma.

PETROWSKY'S L^2 ESTIMATES FOR HYPERBOLIC SYSTEMS

LARS HÖRMANDER

1. Introduction. As a graduate student back in 1953–54 I struggled in vain to understand the basic estimates in Petrowsky's paper [P1] on hyperbolic differential equations, but 40 years later they no longer seem so hard. We shall here discuss how this work of Petrowsky can in fact be easily understood from the point of view of pseudodifferential calculus for which it may be considered a crude precursor. For comparison we shall also present the original arguments of Petrowsky in a manner which we hope makes them more transparent.

The crucial pages 820–837 in [P1] consist of a proof of essentially the following estimate. Let

$$(1.1) \qquad \frac{\partial u(t,x)}{\partial t} = \sum_{k=1}^{n} A_k(t,x) \frac{\partial u(t,x)}{\partial x_k} + B(t,x)u(t,x) + f(t,x)$$

where u and f are functions of $(t,x) \in \mathbf{R} \times (\mathbf{R}^n/l\mathbf{Z}^n)$ with values in $\mathbf{C}^N$, and A_k, B are also periodic in x, with values in $N \times N$ matrices. Suppose that A_k and B have continuous bounded derivatives of order $\leq 3n$, that u is smooth, and that the equation (1.1) is *strictly hyperbolic*. This means that the eigenvalues of $\sum_1^n \xi_k A_k(t,x)$ are real and distinct, with differences uniformly bounded from below when $\xi \in \mathbf{R}^n$ and $|\xi| = 1$. Then, if the t,x derivatives of A_k and B are small enough, there are constants C_1, C_2, C_3 such that for $t > 0$

$$(1.2) \qquad \int_Q |u(t,x)|^2 \, dx \leq e^{C_1 t} \Big[C_2 \int_Q |u(0,x)|^2 \, dx + C_3 \iint_{0 < \tau < t, x \in Q} |f(\tau,x)|^2 \, d\tau \, dx \Big],$$

where $Q = \{x \in \mathbf{R}^n; 0 \leq x_j \leq l, j = 1, \ldots, n\}$. (See (49) in [P1].)

If the system (1.1) is symmetric hyperbolic, that is, all A_k are Hermitian symmetric, the proof of (1.2) is simple and only requires that the coefficients are Lipschitz continuous. This follows from the identity

$$\frac{\partial}{\partial t} \int_Q |u(t,x)|^2 \, dx = 2 \operatorname{Re} \int_Q (\partial u(t,x)/\partial t, u(t,x)) \, dx$$

$$= 2 \operatorname{Re} \int_Q \Big(\sum_{k=1}^{n} A_k(t,x) \partial u(t,x)/\partial x_k + B(t,x)u(t,x) + f(t,x), u(t,x) \Big) \, dx,$$

where integration by parts gives

$$\int_Q \sum_{k=1}^{n} (A_k(t,x) \partial u(t,x)/\partial x_k, u(t,x)) \, dx$$

$$= -\int_Q \sum_{k=1}^{n} (A_k(t,x)u(t,x), \partial u(t,x)/\partial x_k) \, dx - \int_Q \sum_{k=1}^{n} (\partial A_k(t,x)/\partial x_k u(t,x), u(t,x)) \, dx$$

© Springer International Publishing AG, part of Springer Nature 2018
L. Hörmander, *Unpublished Manuscripts*,
https://doi.org/10.1007/978-3-319-69850-2_4

which by the Hermitian symmetry of A_k means that

$$2\operatorname{Re}\int_Q \sum_{k=1}^n (A_k(t,x)\partial u(t,x)/\partial x_k, u(t,x))\,dx = -\int_Q \sum_{k=1}^n (\partial A_k(t,x)/\partial x_k u(t,x), u(t,x))\,dx.$$

Hence, with $\|\cdot\|$ denoting the norm in $L^2(Q)$,

$$\frac{\partial \|u(t,\cdot)\|^2}{\partial t} \le 2C\|u(t,\cdot)\|^2 + 2\|f(t,\cdot)\|\,\|u(t,\cdot)\|, \quad \text{that is,}$$

$$\frac{\partial \|u(t,\cdot)\|}{\partial t} \le C\|u(t,\cdot)\| + \|f(t,\cdot)\|,$$

which after integration gives an estimate of the form (1.2),

$$\|u(t,\cdot)\| \le e^{Ct}\|u(0,\cdot)\| + \int_0^t e^{C(t-\tau)}\|f(\tau,\cdot)\|\,d\tau, \quad t > 0.$$

In the general case the strict hyperbolicity assumed just implies that the eigenvalues of $\sum_1^n \xi_k A_k(t,x)$ are real numbers $\lambda_1(t,x,\xi) < \lambda_2(t,x,\xi) < \cdots < \lambda_N(t,x,\xi)$ when $|\xi| = 1$ and $\xi \in \mathbf{R}^n$. They are C^∞ functions of ξ and as smooth as functions of (t,x) as the coefficients A_k are, with the derivatives bounded. Let $\lambda(x,\xi)$ denote the corresponding diagonal matrix. The one dimensional eigenspaces have the same differentiability properties. If the coefficients A_k are nearly constant in x they do not vary much so there is no difficulty in choosing equally smooth normalized real eigenvectors $v_1(t,x,\xi)$, ..., $v_N(t,x,\xi)$. If $K(t,x,\xi)$ is the inverse of the map $\mathbf{C}^N \ni (z_1,\dots,z_N) \mapsto \sum_1^N z_j v_j(t,x,\xi) \in \mathbf{C}^N$ then

$$(1.3) \qquad K(t,x,\xi)\sum_1^n \xi_k A_k(t,x) = \lambda(t,x,\xi)K(t,x,\xi),$$

when $|\xi| = 1$, for both sides are equal to $K(t,x,\xi)\lambda_j v_j(t,x,\xi)$, that is, λ_j times the jth unit vector when applied to $v_j(t,x,\xi)$. We extend $K(t,x,\xi)$ and $\lambda(t,x,\xi)$ to all $\xi \in \mathbf{R}^n \setminus \{0\}$ as homogeneous functions of degree 0 and 1 respectively, which preserves the equation (1.3). For $\xi = 0$ we define $K(t,x,0) = \mathrm{Id}$.

Petrowsky's proof of (1.2), given in [P1] with corrections and amplifications in [P2], used Fourier analysis. However, instead of the Fourier coefficients

$$(1.4) \qquad \hat{u}_\xi(t) = \frac{1}{l^n}\int_Q e^{-i\langle x,\xi\rangle} u(t,x)\,dx, \quad \xi \in (2\pi/l)\mathbf{Z}^n = \mathbf{Z}_l^n,$$

for which one has Parseval's formula

$$(1.5) \qquad \|u(t,\cdot)\|^2 = l^n \sum_{\xi \in \mathbf{Z}_l^n} |\hat{u}_\xi(t)|^2$$

he considered

$$(1.6) \qquad c_\xi(t) = \frac{1}{l^n}\int_Q e^{-i\langle x,\xi\rangle} K(t,x,\xi)u(t,x)\,dx, \quad \xi \in \mathbf{Z}_l^n.$$

The advantage of this is that the equation (1.1) gives

$$\frac{\partial c_\xi(t)}{\partial t} = \frac{1}{l^n} \int_Q e^{-i\langle x,\xi\rangle} \frac{\partial K(t,x,\xi)}{\partial t} u(t,x)\, dx$$

$$+ \frac{1}{l^n} \int_Q e^{-i\langle x,\xi\rangle} K(t,x,\xi) \Big(\sum_{k=1}^n A_k(t,x) \frac{\partial u(t,x)}{\partial x_k} + B(t,x)u(t,x) + f(t,x)\Big)\, dx.$$

Integration by parts in the terms where u is differentiated gives the first order contribution

$$\frac{1}{l^n} \int_Q e^{-i\langle x,\xi\rangle} K(t,x,\xi) \sum_{k=1}^n i\xi_k A_k(t,x) u(t,x)\, dx = \frac{1}{l^n} \int_Q e^{-i\langle x,\xi\rangle} i\lambda(t,x,\xi) K(t,x,\xi) u(t,x)\, dx.$$

If $B = f = 0$ and A_k does not depend on (t,x), we get $\partial c_\xi(t)/\partial t = i\lambda(\xi) c_\xi(t)$, hence

$$\frac{\partial |c_\xi(t)|^2}{\partial t} = 2\,\mathrm{Re}(\partial c_\xi(t)/\partial t, c_\xi(t)) = 0,$$

so $\sum_{\xi\in\mathbf{Z}_l^n} |c_\xi(t)|^2$ is then independent of t. Since $c_\xi(t) = K(\xi)\hat{u}_\xi(t)$ where K and its inverse are bounded for $\xi \in \mathbf{Z}_l^n$, this sum is equivalent to $\sum_{\xi\in\mathbf{Z}_l^n} |\hat{u}_\xi(t)|^2$, so Parseval's formula (1.5) gives the desired estimate $\|u(t,\cdot)\| \leq C\|u(0,\cdot)\|$. However, in the general case we only obtain

$$(1.7) \qquad \frac{\partial |c_\xi(t)|^2}{\partial t} = 2\,\mathrm{Re}(P_\xi^{(1)}(t) + P_\xi^{(2)}(t) + P_\xi^{(3)}(t), c_\xi(t)), \quad \text{where}$$

$$(1.8) \qquad P_\xi^{(1)}(t) = \frac{1}{l^n} \int_Q e^{-i\langle x,\xi\rangle} i\lambda(t,x,\xi) K(t,x,\xi) u(t,x)\, dx,$$

$$(1.9) \qquad P_\xi^{(2)}(t) = \frac{1}{l^n} \int_Q e^{-i\langle x,\xi\rangle} \big(\partial K(t,x,\xi)/\partial t - \sum_1^n \partial K(t,x,\xi)/\partial x_k A_k(t,x)$$

$$- K(t,x,\xi)\big(\sum_1^n \partial A_k(t,x)/\partial x_k - B(t,x)\big)\big) u(t,x)\, dx,$$

$$(1.10) \qquad P_\xi^{(3)}(t) = \frac{1}{l^n} \int_Q e^{-i\langle x,\xi\rangle} K(t,x,\xi) f(t,x)\, dx.$$

Note that in $P_\xi^{(1)}$ the function $\lambda(t,x,\xi)$ is of first order with respect to ξ. As in the simple constant coefficient case discussed above it is the fact that λ is a real diagonal matrix, hence Hermitian symmetric, which is crucial for the following estimate proved in pp. 824–831 in [P1]

$$(1.11) \qquad \sum_{\xi\in\mathbf{Z}_l^n} \mathrm{Re}(P_\xi^{(1)}(t), c_\xi(t)) \leq C \sum_{\xi\in\mathbf{Z}_l^n} |\hat{u}_\xi(t)|^2.$$

On pp. 831, 832 in [P1] proofs are given for the much easier estimates

$$(1.12) \qquad \sum_{\xi\in\mathbf{Z}_l^n} |P_\xi^{(2)}(t)|^2 \leq C \sum_{\xi\in\mathbf{Z}_l^n} |\hat{u}_\xi(t)|^2,$$

$$(1.13) \qquad \sum_{\xi\in\mathbf{Z}_l^n} |c_\xi(t)|^2 \leq C \sum_{\xi\in\mathbf{Z}_l^n} |\hat{u}_\xi(t)|^2,$$

$$(1.14) \qquad \sum_{\xi\in\mathbf{Z}_l^n} |P_\xi^{(3)}(t)|^2 \leq C \sum_{\xi\in\mathbf{Z}_l^n} |\hat{f}_\xi(t)|^2.$$

They imply that (see (36) in [P1])

(1.15)
$$\frac{d}{dt} \sum_{\xi \in \mathbf{Z}_l^n} |c_\xi(t)|^2 \leq C\Big(\sum_{\xi \in \mathbf{Z}_l^n} |\hat{u}_\xi(t)|^2 + \sum_{\xi \in \mathbf{Z}_l^n} |\hat{f}_\xi(t)|^2 \Big).$$

The remaining pages 833–837 in [P1] are devoted to the proof that if the derivatives of the coefficients A_k of order $1, \ldots, 2n$ are small enough, then

(1.16)
$$\sum_{\xi \in \mathbf{Z}_l^n} |\hat{u}_\xi(t)|^2 \leq C \sum_{\zeta \in \mathbf{Z}_l^n} |c_\xi(t)|^2.$$

Combined with (1.13) this completes the proof as in the constant coefficient case.

In Section 2 we shall interpret the approach of Petrowsky in the language of pseudo-differential operators. For the sake of simplicity we assume that the coefficients in (1.1) have bounded continuous coefficients of all orders although it would be sufficient to assume far less than Petrowsky did. We shall then deduce the estimates (1.11) – (1.16) from pseudodifferential calculus. In Section 3 we shall discuss Petrowsky's original proof of the estimates (1.11) – (1.16) and compare it with the pseudodifferential proof outlined in Section 2.

2. Translation to pseudodifferential calculus. We shall now study systematically operators of the form (1.6), or rather the operators mapping u to the functions with these Fouriercoefficients,

$$(t, x) \mapsto \sum_{\xi \in \mathbf{Z}_l^n} e^{i\langle x, \xi \rangle} \frac{1}{l^n} \int_Q e^{-i\langle y, \xi \rangle} K(t, y, \xi) u(t, y) \, dy.$$

We have several such operators to consider, and the parameter t is now just a notational nuisance as is also the fact that u and K are vector and matrix valued. We therefore state our preliminary problem as follows. Given a symbol $\psi \in S_{1,0}^m(\mathbf{R}^n \times \mathbf{R}^n)$[5] such that $\psi(x + y, \xi) = \psi(x, \xi)$ when $y \in l\mathbf{Z}^n$, we define

(2.1)
$$\Psi u(x) = \sum_{\xi \in \mathbf{Z}_l^n} e^{i\langle x, \xi \rangle} \frac{1}{l^n} \int_Q e^{-i\langle y, \xi \rangle} \psi(y, \xi) \, u(y) \, dy, \quad u \in C^\infty(\mathbf{R}^n/\mathbf{Z}^n).$$

The series converges to a function in C^∞ for partial integration shows that the integral can be estimated by $C_N(1 + |\xi|)^{m-N}$ for any integer N.

THEOREM 2.1. *The operator Ψ defined by (2.1) is a pseudodifferential operator on $\mathbf{R}^n/\mathbf{Z}^n$ with right ordered symbol ψ mod $S^{-\infty}$. Thus the standard (left ordered) symbol is given by the asymptotic sum*

$$\sum_\alpha D_x^\alpha \psi^{(\alpha)}(x, \xi)/\alpha!.$$

Here we have used the term "right ordered symbol" in the sense of (18.5.2) in [H2]: If $a \in S_{1,0}^m(\mathbf{R}^n \times \mathbf{R}^n)$ then the pseudodifferential operator with right ordered symbol a is defined by

$$\tilde{a}(x, D)u(x) = (2\pi)^{-n} \iint a(y, \xi) e^{i\langle x-y, \xi \rangle} u(y) \, dy \, d\xi, \quad u \in \mathcal{S}(\mathbf{R}^n).$$

---

[5]We use throughout the notation of Chapter XVIII in [H2].

Equivalently,

$$\langle \tilde{a}(x, D)u, v \rangle = (2\pi)^{-n} \iiint a(y, \xi) e^{i\langle x-y, \xi \rangle} u(y) v(x) \, dx \, dy \, d\xi$$

$$= (2\pi)^{-n} \iint e^{i\langle y, \xi \rangle} a(y, -\xi) u(y) \hat{v}(\xi) \, dy \, d\xi = \langle u, a(y, -D)v \rangle, \quad u, v \in \mathcal{S}(\mathbf{R}^n),$$

with the pseudodifferential operator $a(y, -D)$ defined in the usual way. Note that

$$(2.2) \qquad e^{i\langle y, \theta \rangle} a(y, -D) e^{-i\langle y, \theta \rangle} v = a(y, \theta - D)v \sim a(y, \theta)$$

when y is in an open set where $v = 1$. Such asymptotics characterize pseudodifferential operators.

PROOF OF THEOREM 2.1. By the preceding remarks it is natural to form the transpose of the operator Ψ. With $v \in C_0^\infty(\mathbf{R}^n)$ we therefore multiply (2.1) by $v(x)$ and integrate. With $\Pi v = \sum_{g \in l\mathbf{Z}} v(\cdot + g)$ and $\langle \cdot, \cdot \rangle_Q$ denoting the bilinear scalar product in $L^2(Q)$ we obtain when $u \in C^\infty(\mathbf{R}^n / l\mathbf{Z}^n)$

$$\langle \Psi u, \Pi v \rangle_Q = \frac{1}{l^n} \int_Q u(y) \left(\sum_{\xi \in \mathbf{Z}_l^n} e^{-i\langle y, \xi \rangle} \psi(y, \xi) \hat{v}(-\xi) \right) dy$$

which means that

$$(2.3) \qquad {}^t\Psi \Pi v(y) = \frac{1}{l^n} \sum_{\xi \in \mathbf{Z}_l^n} e^{-i\langle y, \xi \rangle} \psi(y, \xi) \hat{v}(-\xi), \quad v \in C_0^\infty(\mathbf{R}^n).$$

As in (2.2) we now form

$$(2.4) \qquad e^{i\langle y, \theta \rangle} ({}^t\Psi \Pi)(v(y) e^{-i\langle y, \theta \rangle}) = \frac{1}{l^n} \sum_{\xi \in \mathbf{Z}_l^n} e^{i\langle y, \theta - \xi \rangle} \psi(y, \xi) \hat{v}(\theta - \xi), \quad \theta \in \mathbf{R}^n,$$

and we shall study the asymptotic behavior as $\theta \to \infty$. Taylor expansion of ψ at θ with respect to the second variable gives the terms

$$(2.5) \qquad \frac{1}{l^n} \sum_{\xi \in \mathbf{Z}_l^n} e^{i\langle y, \theta - \xi \rangle} \psi^{(\alpha)}(y, \theta)(\xi - \theta)^\alpha \hat{v}(\theta - \xi)/\alpha!.$$

Here $\xi \mapsto e^{i\langle y, \xi+\theta \rangle}(\xi + \theta)^\alpha \hat{v}(\xi + \theta)$ is the Fourier transform of $x \mapsto e^{-i\langle x, \theta \rangle} D^\alpha v(x + y)$, so it follows from Poisson's summation formula (see e.g. (7.2.1)' in [H1]) that the term (2.5) can be written

$$(2.6) \qquad (-1)^{|\alpha|} \sum_{x \in l\mathbf{Z}} \psi^{(\alpha)}(y, \theta) e^{-i\langle x, \theta \rangle} D^\alpha v(x + y)/\alpha!.$$

If Ω is an open set in $\mathbf{R}^n$ such that $v(y + x) = 0$ if $y \in \Omega$ and $0 \neq x \in l\mathbf{Z}$, then (2.6) reduces to $(-1)^{|\alpha|} \psi^{(\alpha)}(y, \theta) D^\alpha v(y)/\alpha!$ when $y \in \Omega$, and in a subset of Ω where $v = 1$ this is equal to 0 when $\alpha \neq 0$ and $\psi(y, \theta)$ when $\alpha = 0$.

If N is a large integer then

$$(2.7) \qquad \left| \psi(y, \xi) - \sum_{|\alpha| < N} \psi^{(\alpha)}(y, \theta)(\xi - \theta)^\alpha / \alpha! \right| \leq C_N |\xi - \theta|^N |\theta|^{|m| - N}, \quad \xi \in \mathbf{R}^n,$$

if $|\theta|$ is large. When $|\xi - \theta| \leq |\theta|/2$ this follows from Taylor's formula. If $|\theta| \leq 2|\xi - \theta|$ then $|\xi| \leq 3|\xi - \theta|$ and the estimate is valid for every term in the left-hand side. From (2.7) it follows that the error committed in (2.4) when $\psi(y, \xi)$ is replaced by the Taylor expansion of order N is $O(|\theta|^{|m|-N})$. In view of Proposition 18.1.4 in [H2] it follows that $^t\Psi$ is a pseudodifferential operator with symbol $\psi(y, -\eta)$, which completes the proof of the theorem for the last statement follows from Theorem 18.5.10 in [H2].

The estimates (1.11) – (1.16) are easy consequences of Theorem 2.1. If we write

$$\mathcal{K}(t)u(x) = \sum_{\xi \in \mathbf{Z}_l^n} c_\xi(t)e^{i\langle x, \xi\rangle}$$

with the notation in (1.6) then (1.13) and (1.14) are equivalent to $\|\mathcal{K}(t)\| \leq C$. This follows at once from Theorem 2.1 since pseudodifferential operators of order 0 are bounded in L^2. Also (1.12) follows in the same way since it involves an operator of the same form as $\mathcal{K}(t)$, with K replaced by the more complicated symbol in (1.9). To prove (1.16) we note that the symbol $K(t, x, \xi) - K(t, 0, \xi)$ is small if the derivatives of the coefficients A_k are small. If $\mathcal{K}_0(t)$ is defined as $\mathcal{K}(t)$ but with the symbol $K(t, x, \xi)$ replaced by $K(t, 0, \xi)$ it follows that the norm of $\mathcal{K}(t) - \mathcal{K}_0(t)$ is small. Now we saw in the discussion of the constant coefficient case that there is a bound for the inverse of $\mathcal{K}_0$, so we obtain

$$\|u\| \leq C\|\mathcal{K}_0(t)u\| \leq C\|\mathcal{K}(t)u\| + C\|\mathcal{K}_0(t) - \mathcal{K}(t)\|\|u\|.$$

When $C\|\mathcal{K}_0(t) - \mathcal{K}(t)\| < \frac{1}{2}$, it follows that $\|u\| \leq 2C\|\mathcal{K}(t)u\|$, which proves (1.16).

Let $\mathcal{L}(t)$ be the operator defined as $\mathcal{K}(t)$ with $K(t, x, \xi)$ replaced by $\lambda(t, x, \xi)K(t, x, \xi)$. It is a pseudodifferential operator of order 1, and (1.11) states that

$$\mathrm{Re}(i\mathcal{L}(t)u, \mathcal{K}(t)u) \leq C\|u\|^2.$$

This can be written in the form

$$i((\mathcal{K}(t)^*\mathcal{L}(t) - \mathcal{L}(t)^*\mathcal{K}(t))u, u) \leq 2C\|u\|^2.$$

The principal symbol of the pseudodifferential operator in the left-hand side is

$$K(t, x, \xi)^*\lambda(t, x, \xi)K(t, x, \xi) - (\lambda(t, x, \xi)K(t, x, \xi))^*K(t, x, \xi) = 0$$

since $\lambda(t, x, \xi)$ is Hermitian, so the operator is of order 0 which completes the proof of (1.11). More explicitly, if $\lambda(t)$ is the operator defined as $\mathcal{K}(t)$ but with $K(t, x, \xi)$ replaced by $\lambda(t, x, \xi)$, then the preceding argument consists in writing

$$(2.8) \quad \mathrm{Re}\, i\mathcal{L}(t)u, \mathcal{K}(t)u) = \mathrm{Re}\, i\big((\mathcal{L}(t) - \lambda(t)\mathcal{K}(t))u, \mathcal{K}(t)u\big) + \mathrm{Re}\, i(\lambda(t)\mathcal{K}(t)u, \mathcal{K}(t)u)$$

$$= \mathrm{Re}\, i\big((\mathcal{L}(t) - \lambda(t)\mathcal{K}(t))u, \mathcal{K}(t)u\big) + \tfrac{i}{2}\big((\lambda(t) - \lambda(t)^*)\mathcal{K}(t)u, \mathcal{K}(t)u\big)$$

and using the boundedness of $\mathcal{L}(t) - \lambda(t)\mathcal{K}(t)$ and $\lambda(t) - \lambda(t)^*$.

3. The original proofs. In Section 2 we proved the estimates (1.11) – (1.16) by studying the operators in $L^2(\mathbf{R}^n/l\mathbf{Z}^n)$ obtained by passing from $\{P_\xi^{(j)}(t)\} \in l^2(\mathbf{Z}_l^n)$ to the corresponding Fourier series. However, Petrowsky worked entirely in the Fourier coefficient space $l^2(\mathbf{Z}_l^n)$. This requires that the Fourier coefficients $\widehat{\Psi u}(\xi)$ of Ψu, defined by (2.1), are expressed in terms of the Fourier coefficients of u,

$$u(y) = \sum_{\eta \in \mathbf{Z}_l^n} e^{i\langle y, \eta \rangle} \hat{u}(\eta).$$

(Thus we have changed the notation for Fourier coefficients now.) When $\xi \in \mathbf{Z}_l^n$ we have

$$\widehat{\Psi u}(\xi) = \frac{1}{l^n} \int_Q e^{-i\langle y, \xi \rangle} \psi(y, \xi) u(y) \, dy$$

$$= \sum_{\eta \in \mathbf{Z}_l^n} \frac{1}{l^n} \int_Q e^{-i\langle y, \xi - \eta \rangle} \psi(y, \xi) \hat{u}(\eta) \, dy = \sum_{\eta \in \mathbf{Z}_l^n} \hat{\psi}(\xi - \eta, \xi) \hat{u}(\eta),$$

where

$$\hat{\psi}(\theta, \xi) = \frac{1}{l^n} \int \psi(y, \xi) e^{-i\langle y, \theta \rangle} \, dy, \quad \theta \in \mathbf{Z}_l^n,$$

are the Fourier coefficients of $x \mapsto \psi(x, \xi)$. If $\psi \in S_{1,1}^m$ then

$$|\hat{\psi}(\theta, \xi)| \le C_\nu (1 + |\theta|)^{-\nu} (1 + |\xi|)^m$$

for every ν. (This estimate only requires that $|\partial_y^\beta \psi(y, \xi)| \le C(1 + |\xi|)^m$ when $|\beta| \le \nu$.) When $m = 0$ and $\nu > n$ it follows that

$$(3.1) \qquad \Big(\sum_{\xi \in \mathbf{Z}_l^n} |\widehat{\Psi u}(\xi)|^2 \Big)^{\frac{1}{2}} \le C_\nu' \Big(\sum_{\xi \in \mathbf{Z}_l^n} |\hat{u}(\xi)|^2 \Big)^{\frac{1}{2}}, \quad C_\nu' = C_\nu \sum_{\nu \in \mathbf{Z}_l^n} (1 + |\eta|)^{-\nu},$$

by a classical lemma of Schur, or just application of the triangle inequality. This simple argument yields (1.12) – (1.16).

The estimate (1.11) is harder. In the proof given in Section 2 we used the rules of pseudodifferential calculus giving the principal symbol of an adjoint and of a product. We must do something similar when working in the Fourier representation.

To simplify notation we suppress the variable t from now on which is legitimate since our hypotheses are uniform with respect to t. Writing

$$(3.2) \qquad \hat{\lambda}(\eta, \xi) = \frac{1}{l^n} \int_Q e^{-i\langle x, \eta \rangle} \lambda(x, \xi) \, dx, \quad \xi \in \mathbf{Z}_l^n,$$

$$(3.3) \qquad c(\eta, \xi) = \frac{1}{l^n} \int_Q e^{-i\langle x, \eta \rangle} K(x, \xi) u(x) \, dx, \quad \xi \in \mathbf{Z}_l^n,$$

for the Fourier coefficients of $x \mapsto \lambda(x, \xi)$ and $x \mapsto K(x, \xi)u(x)$, we have $c(\xi, \xi) = c_\xi$ with the notation (1.6), and by (1.8)

$$(3.4) \qquad P_\xi^{(1)} = i \sum_{\eta \in \mathbf{Z}_l^n} \hat{\lambda}(\xi - \eta, \xi) c(\eta, \xi), \quad \xi \in \mathbf{Z}_l^n.$$

Hence (1.11) is equivalent to

$$i \sum_{\xi, \eta \in \mathbf{Z}_l^n} (\hat{\lambda}(\xi - \eta, \xi) c(\eta, \xi), c(\xi, \xi)) - i \sum_{\xi, \eta \in \mathbf{Z}_l^n} (c(\xi, \xi), \hat{\lambda}(\xi - \eta, \xi) c(\eta, \xi)) \le 2C \sum_{\xi \in \mathbf{Z}_l^n} |\hat{u}(\xi)|^2.$$

Since λ is Hermitian symmetric we have

$$\hat{\lambda}(\eta,\xi)^* = \frac{1}{l^n} \int_Q e^{i\langle x,\eta\rangle} \lambda(x,\xi)\, dx = \hat{\lambda}(-\eta,\xi).$$

Interchanging ξ and η in the second sum we can therefore rewrite the estimate in the form

$$(3.5)\quad i \sum_{\xi,\eta\in\mathbf{Z}_l^n} \left((\hat{\lambda}(\xi-\eta,\xi)c(\eta,\xi), c(\xi,\xi)) - (\hat{\lambda}(\xi-\eta,\eta)c(\eta,\eta), c(\xi,\eta)) \right) \leq 2C \sum_{\xi\in\mathbf{Z}_l^n} |\hat{u}(\xi)|^2.$$

Formula (18) in [P1] (cf $(18)_2$ in [P2]) splits the left-hand side into three terms

$$i \sum_{\xi,\eta\in\mathbf{Z}_l^n} \left((\hat{\lambda}(\xi-\eta,\xi)c(\eta,\xi), c(\xi,\xi)) - (\hat{\lambda}(\xi-\eta,\eta)c(\eta,\eta), c(\xi,\eta)) \right)$$

$$(3.6)\qquad = i \sum_{\xi,\eta\in\mathbf{Z}_l^n} \left((\hat{\lambda}(\xi-\eta,\xi) - \hat{\lambda}(\xi-\eta,\eta))c(\eta,\xi), c(\xi,\xi) \right)$$

$$(3.7)\qquad + i \sum_{\xi,\eta\in\mathbf{Z}_l^n} \left(\hat{\lambda}(\xi-\eta,\eta)c(\eta,\xi), c(\xi,\xi) - c(\xi,\eta) \right)$$

$$(3.8)\qquad + i \sum_{\xi,\eta\in\mathbf{Z}_l^n} \left(\hat{\lambda}(\xi-\eta,\eta)(c(\eta,\xi) - c(\eta,\eta)), c(\xi,\eta) \right).$$

(Actually (3.6) is the sum of the first two sums in (18) of [P1], so (3.7) and (3.8) are called the third and fourth sums there.)

With the notation

$$(3.9)\qquad \widehat{K}(\eta,\xi) = \frac{1}{l^n} \int_Q e^{-i\langle x,\eta\rangle} K(x,\xi)\, dx,$$

for the Fourier coefficients of $x \mapsto K(x,\xi)$, we have

$$(3.10)\qquad c(\eta,\xi) = \sum_{\theta\in\mathbf{Z}_l^n} \widehat{K}(\eta-\theta,\xi)\hat{u}(\theta).$$

To estimate (3.6) we observe that for any N

$$(3.11)\qquad |\hat{\lambda}(\xi-\eta,\xi) - \hat{\lambda}(\xi-\eta,\eta)| \leq C_N(1+|\xi-\eta|)^{-N}, \quad \xi,\eta\in\mathbf{Z}_l^n.$$

The estimate is obvious when $\xi=0$ or $\eta=0$, and if $\xi\neq 0, \eta\neq 0$ it follows from the mean value theorem since $|\partial\hat{\lambda}(\xi-\eta,\theta)/\partial\theta| \leq C_N(1+|\xi-\eta|)^{-N-1}$ when $\theta\neq 0$. Hence (3.6) can be estimated by

$$C_N' \sum_{\xi,\eta,\theta\in\mathbf{Z}_l^n} (1+|\xi-\eta|)^{-N}(1+|\eta-\theta|)^{-N}|\hat{u}(\theta)||c(\xi,\xi)|$$

$$\leq C_N'' \sum_{\xi,\theta\in\mathbf{Z}_l^n} (1+|\xi-\theta|)^{-N}|\hat{u}(\theta)||c(\xi,\xi)| \leq C_N''' \left(\sum_{\theta\in\mathbf{Z}_l^n} |\hat{u}(\theta)|^2 \right)^{\frac{1}{2}} \left(\sum_{\xi\in\mathbf{Z}_l^n} |c_\xi|^2 \right)^{\frac{1}{2}},$$

if $N > n$. Here we have used that

$$\sum_{\eta\in\mathbf{Z}_l^n} (1+|\xi-\eta|)^{-N}(1+|\eta-\theta|)^{-N} \leq 2(1+|\xi-\theta|/2)^{-N} \sum_{\eta\in\mathbf{Z}_l^n} (1+|\eta|)^{-N},$$

which follows since either $|\xi - \eta| \geq |\xi - \theta|/2$ or $|\eta - \theta| \geq |\xi - \theta|/2$ in each term. Thus (3.6) can be estimated by $C \sum_{\theta \in \mathbf{Z}_l^n} |\hat{u}(\theta)|^2$.

To estimate (3.7) we use that

$$(3.12) \qquad c(\xi, \xi) - c(\xi, \eta) = \sum_{\theta \in \mathbf{Z}_l^n} (\widehat{K}(\xi - \theta, \xi) - \widehat{K}(\xi - \theta, \eta)) \hat{u}(\theta), \quad \xi, \eta \in \mathbf{Z}_l^n,$$

and that the mean value theorem gives the estimate

$$(3.13) \quad |\widehat{K}(\xi - \theta, \xi) - \widehat{K}(\xi - \theta, \eta)| \leq C_N (1 + |\xi - \theta|)^{-N} |\xi - \eta|/(1 + |\eta|), \quad \xi, \eta \in \mathbf{Z}_l^n.$$

Using (3.10), (3.12) and (3.13) we can therefore estimate (3.7) by

$$C_N' \sum_{\xi, \eta, \theta', \theta'' \in \mathbf{Z}_l^n} \frac{1 + |\eta|}{(1 + |\xi - \eta|)^{N+1}} \frac{|\hat{u}(\theta')|}{(1 + |\eta - \theta'|)^N} \frac{|\xi - \eta| |\hat{u}(\theta'')|}{(1 + |\xi - \theta''|)^N (1 + |\eta|)}.$$

When $N > n$ we obtain as before by summing first over ξ, then over η, that (3.7) can be estimated by $C \sum_{\theta \in \mathbf{Z}_l^n} |\hat{u}(\theta)|^2$. The estimate of (3.8) is quite similar, for interchanging ξ and η in (3.12), (3.13) we estimate (3.8) by

$$C_N' \sum_{\xi, \eta, \theta', \theta'' \in \mathbf{Z}_l^n} \frac{1 + |\eta|}{(1 + |\xi - \eta|)^{N+1}} \frac{|\xi - \eta| |\hat{u}(\theta')|}{(1 + |\eta - \theta'|)^N (1 + |\eta|)} \frac{|\hat{u}(\theta'')|}{(1 + |\xi - \theta''|)^N}.$$

This is the same sum as before which completes the proof. Note that we have only used that the coefficients A_k in (1.1) are in C^{n+2}.

The decomposition (3.6) – (3.8) is not quite the same as the one in (2.8), which would correspond to

$$i \sum_{\xi, \eta \in \mathbf{Z}_l^n} \left((\hat{\lambda}(\xi - \eta, \xi) c(\eta, \xi), c(\xi, \xi)) - (\hat{\lambda}(\xi - \eta, \eta) c(\eta, \eta), c(\xi, \eta)) \right)$$

$$= i \sum_{\xi, \eta \in \mathbf{Z}_l^n} (\hat{\lambda}(\xi - \eta, \xi)(c(\eta, \xi) - c(\eta, \eta)), c(\xi, \xi))$$

$$- i \sum_{\xi, \eta \in \mathbf{Z}_l^n} (\hat{\lambda}(\xi - \eta, \eta) c(\eta, \eta), c(\xi, \eta) - c(\xi, \xi))$$

$$+ i \sum_{\xi, \eta \in \mathbf{Z}_l^n} ((\hat{\lambda}(\xi - \eta, \xi) - \hat{\lambda}(\xi - \eta, \eta)) c(\eta, \eta), c(\xi, \xi)).$$

The first two sums in the right-hand side are complex conjugates so it suffices to estimate the first and the third, which can be done along the same lines as the estimates above of (3.6) – (3.8). Thus Petrowsky's proof is closely related to the pseudodifferential operator approach but it is not identical.

References

[H2] L. Hörmander, *The analysis of linear partial differential operators I*, Springer Verlag, 1983.

[H2] ———, *The analysis of linear partial differential operators III*, Springer Verlag, 1985.

[P1] I, G. Petrowsky, *Über das Cauchysche Problem für Systeme von partiellen Differentialgleichungen*, Mat. Sb. **2(44)** (1937), 815–870.

[P2] ———, *Some remarks on my papers on the problem of Cauchy*, Mat. Sb. **39(81)** (1956), 267–272.

PROOF OF THE EXISTENCE OF FUNDAMENTAL
SOLUTIONS AND OF SOME INEQUALITIES

LARS HÖRMANDER

Let $P(D)$ be a polynomial with constant coefficients. We shall construct a fundamental solution of $P(D)$ and study its properties. The existence of fundamental solutions has been established previously by Malgrange (ref. [20] in [D]) and by Ehrenpreis ([B], [C]), but our proof is different and more elementary.

We assume that the coordinates are chosen so that in the development

$$(1) \qquad P(\xi) = a_n \xi_1^n + a_{n-1} \xi_1^{n-1} + \cdots + a_0$$

the highest coefficient a_n is independent of $\xi_2, \ldots, \xi_\nu$. We may even assume that $a_n = 1$. To obtain a fundamental solution we shall give a sense to the purely formal expression

$$E * u(x) = (2\pi)^{-\nu/2} \int \frac{\hat{u}(\xi)}{P(\xi)} e^{i\langle x, \xi \rangle} \, d\xi$$

when $u \in C_0^\infty(\mathbf{R}^\nu)$. In other words, we have to choose a suitable integration path in the complex domain.

Denote by $t_1, \ldots, t_n$ the zeros of P as a function of ξ_1 and let $\tau(\xi_2, \ldots, \xi_\nu)$ be the smallest of the numbers τ between $-\varepsilon$ and ε for which

$$\min_{k=1,\ldots,n} |\tau - \operatorname{Im} t_k|$$

is maximal. The minimum is then at least equal to ε/n, hence

$$(2) \qquad |\xi_1 + i\tau - t_k| \geq \varepsilon/n$$

for every k and all real $\xi_1, \xi_2, \ldots, \xi_\nu$. The function $\tau(\xi_2, \ldots, \xi_\nu)$ is not necessarily continuous but is always measurable and bounded by ε.

We now set

$$(3) \qquad E(\check{u}) = (2\pi)^{-\nu/2} \int_{-\infty}^{+\infty} d\xi_2 \ldots d\xi_\nu \int_{-\infty}^{+\infty} \frac{\hat{u}(\xi + i\tau e)}{P(\xi + i\tau e)} \, d\xi_1, \quad u \in C_0^\infty,$$

where $e = (1, 0, \ldots, 0)$. Since τ is a bounded and measurable function and $|P(\xi + i\tau e)| \geq (\varepsilon/n)^n$ for every real ξ in virtue of (2), the integral exists and (3) defines a distribution for which

$$(4) \quad E * u(x) = (2\pi)^{-\nu/2} \int e^{i(x^2 \xi_2 + \cdots + x^\nu \xi_\nu)} \, d\xi_2 \ldots d\xi_\nu \int \frac{\hat{u}(\xi_1 + i\tau, \xi_2, \ldots)}{P(\xi_1 + i\tau, \xi_2, \ldots)} e^{ix^1(\xi_1 + i\tau)} \, d\xi_1.$$

From this formula it follows that $E * (P(D)u) = u$ when $u \in C^\infty$. For if we replace u by $P(D)u$ in (4), the denominator disappears and after we have moved the integration path back to the real axis we obtain

$$(E * P(D)u)(x) = (2\pi)^{-\nu/2} \int \hat{u}(\xi) e^{i\langle x, \xi \rangle} \, d\xi = u(x).$$

© Springer International Publishing AG, part of Springer Nature 2018

L. Hörmander, *Unpublished Manuscripts*,

https://doi.org/10.1007/978-3-319-69850-2_5

THEOREM 1. *The formula (3) defines a fundamental solution for which, if α is a sequence of indices containing only the index 1,*[6]

$$(5) \qquad \int |P^{(\alpha)}(D)E * u|^2 e^{-2\varepsilon|x^1|}\, dx \le C^2 \int |u|^2 e^{2\varepsilon|x^1|}\, dx, \quad u \in C_0^\infty.$$

PROOF. First note that it follows from (4) that

$$(6) \quad P^{(\alpha)}(D)E * u = (2\pi)^{-(\nu-1)/2} \int e^{i(x^2\xi_2 + \cdots + x^\nu \xi_\nu)} e^{-x^1\tau} F_\alpha(x^1, \xi_2, \ldots, \xi_\nu)\, d\xi_2 \ldots d\xi_\nu,$$

where

$$(7) \quad F_\alpha(x^1, \xi_2, \ldots, \xi_\nu) = (2\pi)^{-1/2} \int \frac{P^{(\alpha)}(\xi_1 + i\tau, \xi_2, \ldots, \xi_\nu)}{P(\xi_1 + i\tau, \xi_2, \ldots, \xi_\nu)} \hat{u}(\xi_1 + i\tau, \xi_2, \ldots) e^{ix^1\xi_1}\, d\xi_1.$$

It follows from (2) and the formula for the logarithmic derivative of a polynomial that with a constant C depending only on ε and n we have $|P^{(\alpha)}/P| \le C$ for all the arguments figuring in the integral (7). Using this inequality and Parseval's formula we obtain

$$(8) \qquad \int |F_\alpha(x^1, \xi_2, \ldots, \xi_\nu)|^2\, dx^1 \le C^2 \int |\hat{u}(\xi_1 + i\tau, \xi_2, \ldots, \xi_\nu)|^2\, d\xi_1.$$

Noting that $-x^1\tau \le \varepsilon|x^1|$ and using Parseval's formula again, we also obtain

$$(9) \qquad \int |P^{(\alpha)}(D)E * u|^2\, dx^2 \ldots dx^\nu \le e^{2\varepsilon|x^1|} \int |F_\alpha(x^1, \xi_2, \ldots, \xi_\nu)|^2\, d\xi_2 \ldots d\xi_\nu$$

for fixed x^1. If we divide this inequality by $e^{2\varepsilon|x^1|}$, integrate with respect to x^1 and use the inequality (8), we finally get

$$(10) \quad \int |P^{(\alpha)}(D)E * u|^2 e^{-2\varepsilon|x^1|}\, dx \le \int |F_\alpha(x^1, \xi_2, \ldots, \xi_\nu)|^2\, dx^1\, d\xi_2 \ldots d\xi_\nu$$

$$\le C^2 \int |\hat{u}(\xi_1 + i\tau, \xi_2, \ldots, \xi_\nu)|^2\, d\xi_1\, d\xi_2 \ldots d\xi_\nu.$$

In order to estimate the last integral by means of u we introduce the "partial Fourier transform"

$$U(x^1, \xi_2, \ldots, \xi_\nu) = (2\pi)^{-(\nu-1)/2} \int u(x^1, \ldots, x^\nu) e^{-i(x^2\xi_2 + \cdots + x^\nu \xi_\nu)}\, dx^2 \ldots dx^\nu.$$

Parseval's formula gives

$$\int |\hat{u}(\xi_1 + i\tau, \xi_2, \ldots, \xi_\nu)|^2\, d\xi_1 = \int |U(x^1, \xi_2, \ldots, \xi_\nu)|^2 e^{2x^1\tau}\, dx^1$$

$$\le \int |U(x^1, \xi_2, \ldots, \xi_\nu)|^2 e^{2\varepsilon|x^1|}\, dx^1,$$

$$\int |U(x^1, \xi_2, \ldots, \xi_\nu)|^2\, d\xi_2 \ldots d\xi_\nu = \int |u(x^1, \ldots, x^\nu)|^2\, dx^2 \ldots dx^\nu,$$

---

[6]This result is due to Malgrange (Ref. [21] in [D]) when α is the void sequence.

from which it follows that

$$(11) \quad \int |\hat{u}(\xi_1 + i\tau, \xi_2, \ldots, \xi_\nu)|^2 \, d\xi_1 \, d\xi_2 \, \ldots \, d\xi_\nu \leq \int |u(x^1, \ldots, x^\nu)|^2 e^{2\varepsilon|x^1|} \, dx^1 \ldots dx^\nu.$$

Combining (10) and (11) we have proved the inequality (5). — Note that the magnitude of the constant only depends on n, a_n and ε.

Replacing u by $P(D)u$ in (5) we obtain

$$\int |P^{(\alpha)}(D)u|^2 e^{-2\varepsilon|x^1|} \, dx \leq C^2 \int |P(D)u|^2 e^{2\varepsilon|x^1|} \, dx, \quad u \in C_0^\infty,$$

when α only contains the index 1. This suggests the following theorem, which also generalizes Lemma 2.7 in [D].

THEOREM 2. *Let λ be a fixed vector in $\mathbf{R}_\nu$ and set $P_\lambda = \sum_1^\nu \lambda_k \partial P/\partial \xi_k$, where P is a polynomial of degree n in the direction λ. Then, for arbitrary fixed $\eta \in \mathbf{R}_\nu$,*

$$(12) \quad \int |P_\lambda(D)u|^2 e^{2\langle x,\eta\rangle} \, dx \leq n^2 \int |P(D)u|^2 e^{2\langle x,\eta\rangle} \cosh(2\langle x,\lambda\rangle) \, dx, \quad u \in C_0^\infty.$$

For the proof we need a lemma concerning analytic functions. We first recall a lemma which is essentially due to Malgrange (see [D], Lemma 2.1), of which we give another new proof, suggested by L. Gårding.

LEMMA 1. *If $g(z)$ is an analytic function of a complex variable z for $|z| \leq 1$, and $r(z)$ is a polynomial with highest coefficient A, then*

$$(13) \quad |Ag(0)|^2 \leq (2\pi)^{-1} \int_0^{2\pi} |g(e^{i\theta})r(e^{i\theta})|^2 \, d\theta.$$

PROOF. Let n be the degree of r and set $R(z) = z^n \bar{r}(1/z)$. This is a polynomial such that $R(0) = \bar{A}$ and $R(e^{i\theta}) = e^{in\theta}\overline{r(e^{i\theta})}$ so that $|R(e^{i\theta})| = |r(e^{i\theta})|$. Hence

$$(2\pi)^{-1} \int |g(e^{i\theta})r(e^{i\theta})|^2 \, d\theta = (2\pi)^{-1} \int |g(e^{i\theta})R(e^{i\theta})|^2 \, d\theta \geq |g(0)R(0)|^2 = |g(0)A|^2.$$

We next prove the lemma which is essential for us here:

LEMMA 2. *Under the assumptions of Lemma 1 we have, if n is the degree of r*

$$(14) \quad |g(0)r^{(k)}(0)| \leq \frac{n!}{(n-k)!} \left((2\pi)^{-1} \int_0^{2\pi} |g(e^{i\theta})r(e^{i\theta})|^2 \, d\theta \right)^{\frac{1}{2}}.$$

PROOF. We may suppose that $r(z) = \prod_{j=1}^n (z - z_j)$, since (14) is homogeneous in r. Application of the previous lemma to the polynomial $\prod_{j=1}^k (z - z_j)$ and the analytic function $g(z) \prod_{j=k+1}^n (z - z_j)$ gives

$$|g(0)| \prod_{k+1}^n |z_j| \leq \left((2\pi)^{-1} \int_0^{2\pi} |g(e^{i\theta})r(e^{i\theta})|^2 \, d\theta \right)^{\frac{1}{2}}$$

and similarly for the product of any $n - k$ of the numbers $|z_j|$. Hence $P^{(k)}(0)$ being the sum of $n!/(n-k)!$ terms of the same kind as $\prod_{k+1}^n z_j$, (14) follows.

The estimate given by (14) is the best possible which is valid for all g and r as is seen by taking $r(z) = (1+z)^n$ and $f(z) = (1+\varepsilon+z)^{-n}$ with a positive ε, and then letting ε tend to 0.

PROOF OF THEOREM 2. Application of Lemma 2 to the analytic function $\hat{u}(\xi+i\eta+\lambda z)$ and the polynomial $P(\xi+i\eta+\lambda z)$ gives, if $u \in C_0^\infty$,

$$|\hat{u}(\xi+i\eta)P_\lambda(\xi+i\eta)|^2 \le \frac{n^2}{2\pi}\int |\hat{u}(\xi+i\eta+\lambda e^{i\theta})P(\xi+i\eta+\lambda e^{i\theta})|^2\, d\theta.$$

Integrating this inequality with respect to the real variable ξ, we get

$$\int |\hat{u}(\xi+i\eta)P_\lambda(\xi+i\eta)|^2\, d\xi \le \frac{n^2}{2\pi}\int d\theta \int |\hat{u}(\xi+i(\eta+\lambda\sin\theta))P(\xi+i(\eta+\lambda\sin\theta))|^2\, d\xi,$$

which according to Parseval's formula is equivalent to

$$(15) \qquad \int |P_\lambda(D)u|^2 e^{2\langle x,\eta\rangle}\, dx \le \frac{n^2}{2\pi}\int d\theta \int |P(D)u|^2 e^{2\langle x,\eta\rangle} e^{2\langle x,\lambda\rangle\sin\theta}\, dx.$$

Now we have for real t

$$(2\pi)^{-1}\int_0^{2\pi} e^{t\sin\theta}\, d\theta = \frac{2}{\pi}\int_0^{\frac{\pi}{2}} \cosh(t\sin\theta)\, d\theta \le \cosh t,$$

and hence (12) follows if we integrate first with respect to θ in the right hand side of (15).

It would no doubt be possible to prove Theorem 2 by energy integral arguments but the proof given now is at least equally simple.

The result of Theorem 3 can easily be iterated. Thus let Λ be a set of vectors $\lambda_1,\ldots,\lambda_k$ in $\mathbf{R}_\nu$. Denoting the degree of P by n and the polynomial obtained by differentiating $P(\xi)$ sucessively in the directions $\lambda_1,\ldots,\lambda_k$ by P_Λ, we obtain

$$(16) \qquad \int |P_\Lambda(D)u|^2 e^{2\langle x,\eta\rangle}\, dx \le \left(\frac{n!}{(n-k)!}\right)^2 \int |P(D)u|^2 e^{2\langle x,\eta\rangle} \prod_{j=1}^k \cosh(2\langle x,\lambda_j\rangle)\, dx.$$

This inequality generalizes Lemma 2.8 in [D].

PROBLEM. It is unknown to the author whether there always exists a fundamental solution E such that $P^{(\alpha)}(D)E * u$ is locally square integrable for every square integrable function u with compact support and all α. The answer is positive if the set

$$\{\eta; \eta \text{ is real and } P(\xi+i\eta) \ne 0 \text{ when } \xi \text{ is real and } |\xi| > C\}$$

has interior points for some value of C, hence in particular if P is normally hyperbolic or of local type. A positive answer in the general case would have many interesting consequences.

References

[B] L. Ehrenpreis, *Solution of some problems of division I*, Amer. J. Math. **76** (1954), 883–903.

[C] ———, *Solution of some problems of division II*, Amer. J. Math. **77** (1955), 286–292.

[D] L. Hörmander, *On the theory of general partial differential operators*, Acta Math. **94** (1955), 161–248.

INEQUALITIES BETWEEN NORMAL AND TANGENTIAL
DERIVATIVES OF HARMONIC FUNCTIONS

LARS HÖRMANDER

For functions which are harmonic in a sphere, Višik [9] has proved an estimate of the normal derivative in terms of the tangential ones,

$$(1) \qquad \int_S (\partial u/\partial r)^2 \, dS \leq \int_S (\operatorname{grad}_S u, \operatorname{grad}_S u) \, dS.$$

Here $\operatorname{grad}_S u$ denotes the component of $\operatorname{grad} u$ along the surface S. Višik proved (1) by integrating $\int r^2 \partial u/\partial r \Delta u \, dx$ by parts. The inequality was later on used in [10] as an essential tool in the study of general boundary problems for elliptic second order equations.

On the other hand, using the theory of singular integral operators, Mihlin [5] proved that, for functions u which are harmonic in the half-space $x^m > 0$ and tend to zero at infinity sufficiently rapidly, we have

$$(2) \qquad \int_{x^m=0} \{(\partial u/\partial x^1)^2 + \cdots + (\partial u/\partial x^{m-1})^2\} \, dx^1 \ldots dx^{m-1}$$

$$\leq C \int_{x^m=0} (\partial u/\partial x^m)^2 \, dx^1 \ldots dx^{m-1},$$

where C is a positive constant. It is very easy to prove this result in another way. For integrating the identity

$$2\frac{\partial u}{\partial x^m}\Delta u = \frac{\partial}{\partial x^m}\left(\left(\frac{\partial u}{\partial x^m}\right)^2 - \left(\frac{\partial u}{\partial x^1}\right)^2 - \cdots - \left(\frac{\partial u}{\partial x^{m-1}}\right)^2\right)$$
$$+ 2\frac{\partial}{\partial x^1}\left(\frac{\partial u}{\partial x^1}\frac{\partial u}{\partial x^m}\right) + \cdots + 2\frac{\partial}{\partial x^{m-1}}\left(\frac{\partial u}{\partial x^{m-1}}\frac{\partial u}{\partial x^m}\right)$$

which only differs formally from a well-known identity for hyperbolic operators due to Friedrichs and Lewy [2], we obtain, since $\Delta u = 0$,

$$(3) \qquad \int_{x^m=0} \{(\partial u/\partial x^1)^2 + \cdots + (\partial u/\partial x^{m-1})^2\} \, dx^1 \ldots dx^{m-1}$$

$$= \int_{x^m=0} (\partial u/\partial x^m)^2 \, dx^1 \ldots dx^{m-1},$$

which contains (2).

Višik and Mihlin both remark without proof that their results are valid also for simply connected domains and any homogeneous second order elliptic differential operator for

© Springer International Publishing AG, part of Springer Nature 2018

L. Hörmander, *Unpublished Manuscripts*,

https://doi.org/10.1007/978-3-319-69850-2_6

which Dirichlet's problem is solvable instead of the Laplace operator. Višik states, however, that the proof is less elementary in the general case.

Using the energy integral method in the form given by Hörmander [3], we shall give a simple proof of the inequalities announced by Višik and Mihlin, also for domains which are not simply connected. By means of the inequality we then show how the study of the distribution of the eigenvalues of a classical vibration problem with the parameter in the boundary condition can be reduced to the study of the vibration problem when the boundary is considered as a membrane. Since very precise results concerning the membrane problems have been published recently (Avakumovič [1], Lewitan [4]), we can thus in important cases improve the results of Sandgren [8]. Our approach also seems to be new, although Payne and Weinberger [7] have applied similar methods to elliptic equations in other connections.

Let M be a Riemannian manifold with a sufficiently differentiable positive definite metric tensor G_{ik}. Let R be a domain with compact closure in M and a sufficiently differentiable boundary S. We may of course consider S as a closed Riemannian manifold of $n-1$ dimensions. The Riemannian elements of volume in R and S will be denoted by dR and dS, respectively. For twice continuously differentiable functions U, defined in R, we define the Beltrami operator in R by the formula

$$(4) \qquad \Delta_R u = -\frac{1}{\sqrt{G}}\frac{\partial}{\partial x^i}\left(\sqrt{G}\,G^{ik}\frac{\partial U}{\partial x^k}\right).$$

Similarly we define the Beltrami operator Δ_S in S by means of the Riemannian geometry in S.

Let u be a sufficiently differentiable function in S. As is well known, there exists one and only one solution U of the equation $\Delta_R u = 0$ which equals u on the boundary S. The solution U is sufficiently differentiable in $\overline{R}$. We now set

$$(5) \qquad \mathcal{P}u = -\frac{dU}{dn},$$

where n is the interior unit normal. Thus $\mathcal{P}$ is the operation of passing from the boundary values of a harmonic function to its exterior normal derivative. We shall prove that $\mathcal{P}$ has roughly the same metric character as a tangential differentiation.

In virtue of Green's formula we have

$$(6) \quad (\mathcal{P}u, u) = -\int_S U\frac{dU}{dn}\,dS = \int_R ((\operatorname{grad}_R U, \operatorname{grad}_R U) - U\Delta_R U)\,dR$$
$$= \int_R (\operatorname{grad}_R U, \operatorname{grad}_R U)\,dR,$$

where $(\mathcal{P}u, u)$ is the scalar product of $\mathcal{P}u$ and u in the real Hilbert space $L^2(S)$, defined by the measure dS. Furthermore, Green's formula gives

$$(7) \quad (\mathcal{P}u, v) - (u, \mathcal{P}v) = -\int_S \left(\frac{dU}{dn}V - U\frac{dV}{dn}\right)dS = -\int_R (V\Delta_R U - U\Delta_R V)\,dR = 0.$$

Hence the operator $\mathcal{P}$ is positive and symmetric.

We now introduce the energy tensor of Hörmander [3],

$$(8) \qquad T_{ik} = 2\frac{\partial U}{\partial x^i}\frac{\partial U}{\partial x^k} - G_{ik}(\operatorname{grad}_R U, \operatorname{grad}_R U).$$

If f^i is a continuously differentiable contravariant vector field and we denote by $f_{i,k}$ the covariant derivative of f_i with respect to x^k, we obtain ([3], formula (4))

$$
(9) \qquad \operatorname{div}(f^i T_i^k) = \frac{1}{\sqrt{G}}\frac{\partial}{\partial x^k}(\sqrt{G}f^i T_i^k) = -2f^i \frac{\partial U}{\partial x^i}\Delta_R U + T^{ik}f_{i,k}.
$$

When $\Delta_R U = 0$ it thus follows from Green's formula that

$$
(10) \qquad -\int_S T_{ik}f^i n^k\, dS = \int_R T^{ik} f_{i,k}\, dR.
$$

Using (8) we can write the integrand of the left hand side in the form

$$
(11) \qquad T_{ik}f^i n^k = 2(\operatorname{grad}_R U, f)(\operatorname{grad}_R U, n) - (f, n)(\operatorname{grad}_R U, \operatorname{grad}_R U).
$$

Since S is twice continuously differentiable we can find a continuously differentiable vector field f such that $f = -n$ on S. Recalling that u denotes the boundary values of U, we then obtain

$$
-T_{ik}f^i n^k = 2\Big(\frac{dU}{dn}\Big)^2 - \Big((\operatorname{grad}_S U, \operatorname{grad}_S U) + \Big(\frac{dU}{dn}\Big)^2\Big) = (\mathcal{P}u)^2 - (\operatorname{grad}_S u, \operatorname{grad}_S u).
$$

Hence we can write the formula (10) as follows

$$
(12) \qquad (\mathcal{P}u, \mathcal{P}u) - (\Delta_S u, u) = \int_R T^{ik} f_{i,k}\, dR.
$$

EXAMPLE. For the Laplace operator in a sphere with radius r we set $f = x/r$. Then $T^{ik}f_{i,k} = (2 - n)r^{-1}\sum(\partial u/\partial x^k)^2$, and it follows from (12) and (6) that

$$
(13) \qquad (\mathcal{P}u, \mathcal{P}u) + \frac{1}{r}(n - 2)(\mathcal{P}u, u) = (\Delta_S u, u),
$$

which contains Višik's inequality (1).

We now return to the general case. Since the integrand $T^{ik}f_{i,k}$ in (12) is a quadratic form in $\operatorname{grad} U$ with continuous coefficients, there is a constant K such that for every U

$$
|T^{ik}f_{i,k}| \le K(\operatorname{grad}_R U, \operatorname{grad}_R U)
$$

in the whole of R. Using this estimate and the identity (6) in (12), we have proved the following theorem.

THEOREM 1. *There exists a constant K such that*

$$
(14) \qquad |(\mathcal{P}u, \mathcal{P}u) - (\Delta_S u, u)| \le K(\mathcal{P}u, u),
$$

for all sufficiently differentiable functions u, defined in S.

Since the operators Δ_S and $\mathcal{P}$ are positive and symmetric, they have positive and symmetric closures which we shall denote by Δ_S and P. Since S is a closed manifold, the closed operator Δ_S is in fact self-adjoint. This is well known but we give the proof. It is sufficient to prove that the range of $\Delta_S + I$ is the whole of $L^2(S)$, and since the positivity of Δ_S shows that

$$
\|(\Delta_S + I)u\| \ge \|u\|,
$$

we need only prove that the range of $\Delta_S + I$ is dense in $L^2(S)$. Let v be orthogonal to this range so that

$$
\int_S (\Delta_S u + u)v\, dS = 0
$$

for all sufficiently differentiable functions u. This means that v is a weak and hence, after correction on a null set, a strong solution of the equation $\Delta_S v + v = 0$. In virtue of the maximum principle, a solution of this equation can neither have a positive maximum nor a negative minimum so that $v = 0$. Hence the assertion follows.

THEOREM 2. *The closed operator P is positive and self-adjoint. Its domain is identical with the domain of $\sqrt{\Delta_S}$, and for u in this common domain we have*

$$\left| \|Pu\| - \|\sqrt{\Delta_S}u\| \right| \leq K\|u\|. \tag{15}$$

Furthermore, the spectrum of P is discrete, and the eigenfunctions are sufficiently differentiable.

PROOF. If u is sufficiently differentiable, it follows from (14) that

$$(\Delta_S u, u) \leq (\mathcal{P}u, \mathcal{P}u) + K(\mathcal{P}u, u) \leq \|(\mathcal{P} + \tfrac{1}{2}K\,\mathrm{Id})u\|^2.$$

We also obtain

$$\|(\mathcal{P} - \tfrac{1}{2}K\,\mathrm{Id})u\|^2 = (\mathcal{P}u, \mathcal{P}u) - K(\mathcal{P}u, u) + \tfrac{1}{4}K^2(u, u)$$
$$\leq (\Delta_S u, u) + \tfrac{1}{4}K^2(u, u) \leq \|(\sqrt{\Delta_S} + \tfrac{1}{2}K\,\mathrm{Id})u\|^2.$$

These two inequalities may also be written

$$\|\sqrt{\Delta_S}u\| \leq \|\mathcal{P}u\| + \tfrac{1}{2}K\|u\|, \quad \|\mathcal{P}u\| \leq \|\sqrt{\Delta_S}u\| + K\|u\|,$$

which proves (15) for sufficiently differentiable functions u. Let u_n be a sequence of such functions, defined in S. Application of (15) to the differences $u = u_n - u_m$ shows that the sequences u_n and $\mathcal{P}u_n$ converge if and only if the sequences u_n and $\sqrt{\Delta_S}u_n$ converge. Hence, in virtue of the definition of the closure of an operator, the domains of P and $\sqrt{\Delta_S}$ are equal, and (11) evidently remains valid in this common domain.

We shall now prove that the closed, positive, symmetric operator P is self-adjoint. This will follow if we prove that $(P + I)^{-1}$ is densely defined, for we can repeat the arguments given in connection with Δ_S above. But this follows from the classical fact that the third boundary problem

$$\Delta_R U = 0, \quad -\frac{dU}{dn} + U = \varphi \quad \text{on } S$$

has a sufficiently differentiable solution, if φ is sufficiently differentiable in S.

To prove that the spectrum is discrete, we have to show that the set

$$\{u; u \in \mathcal{D}_P, \|u\| \leq 1, \|Pu\| \leq 1\}$$

is pre-compact. Since this set is contained in the set

$$\{u; u \in \mathcal{D}_{\sqrt{\Delta_S}}, \|u\| \leq 1, \|\sqrt{\Delta_S}u\| \leq 1 + K\},$$

our assertion follows from the well-known fact that the spectrum of Δ_S is discrete.

References

1. V.G. Avakumovič, *Über die Eigenfunktionen auf geschlossenen Riemannschen Manningfaltigkeiten*, Math. Z. **65** (1956), 327–344.
2. K. Friedrichs and H. Lewy, *Über die Eindeutigkeit und das Abhängigkeitsgebiet der Lösungen beim Anfangswertproblem linearer hyperbolischer Differentiagleichungen*, Math. Ann. **98** (1928), 299–326.
3. L. Hörmander, *Uniqueness theorems and estimates for normally hyperbolic partial differential equations of the second order*, C.R. 12e Congr. Math. Scand. Lund (1953), 105–115.

4. B. M. Lewitan, *On the asymptotic behavior of the spectral function and on expansion in eigenfunctions of a self-adjoint differential equation of second order II*, Izv. Akad. Nauk SSSR Ser. Mat. **19** (1955), 33–58.

5. S. G. Mihlin, *On an inequality for the boundary values of harmonic functions*, Uspehi Matem. Nauk **6:6** (1951), 158–159, (Russian).

7. L. E. Payne and H. F. Weinberger, *New bounds in harmonic and biharmonic problems*, J. of Math. and Physics **33** (1955), 291–307.

8. L. Sandgren, *A vibration problem*, Comm. Sém. Math. Univ. Lund **19** (1955), 1–84.

9. M. I. Višik, *On an inequality for the boundary values of harmonic f8unctions in a sphere*, Uspehi Matem. Nauk **6:2** (1951), 165–166, (Russian).

10. _____, *On general boundary problems for elliptic differential equations*, Trudy Moskov. Mat. Obšč. **1** (1952), 187–246, (Russian).

THE DIRICHLET PROBLEM IN A MAXIMAL OPEN SET

Lars Hörmander

Let Ω be an open subset of $\mathbf{R}^n$, and let m be a positive integer. In [1] a necessary and sufficient condition was given for the completion $\dot{H}_m$ of $C_0^\infty(\Omega)$ with respect to the norm

$$(1) \qquad \|u\|_m = \left((2\pi)^{-n} \int |\xi|^{2m} |\hat{u}(\xi)|^2 \, d\xi \right)^{\frac{1}{2}} \quad u \in C_0^\infty(\mathbf{R}^n),$$

to be a subspace of $\mathcal{D}'(\Omega)$. It is then in fact a subspace of $\mathcal{S}'(\mathbf{R}^n)$, and the Dirichlet problem for $(-\Delta)^m$ can be solved in Ω.

The result was as follows: When $n \geq 2m+1$ there is no restriction on Ω; when n is odd and $< 2m$ the condition is that the complement is not empty; when $n = 2m$ the condition is that the complement is not $n/2$ polar; and when n is even $< 2m$ the complement must not be $n/2$ polar and contained in a proper affine subspace of $\mathbf{R}^n$. Thus Ω is a maximal admissible set if n is odd $< 2m$ and the complement is a single point, or n is even $< 2m$ and the complement consists of the vertices of a simplex of dimension n in $\mathbf{R}^n$. The purpose here is to discuss these cases explicitly.

We begin with some generalities. The map

$$(2) \qquad C_0^\infty(\Omega) \ni u \mapsto |\xi|^m \hat{u} = F_m u$$

extends by continuity to an isometry F_m from $\dot{H}_m$ into $L^2(\mathbf{R}^n, d\xi/(2\pi)^n)$, and $g \in L^2$ is orthogonal to the range if and only if

$$(|\xi|^m \hat{u}, g) = (\hat{u}, |\xi|^m g) = 0, \quad u \in C_0^\infty(\Omega).$$

If $T \in \mathcal{S}'$ is the inverse Fourier transform of $|\xi|^m g$, this means that $\operatorname{supp} T \subset \complement\Omega$. Thus *the orthogonal complement of $F_m \dot{H}_m$ consists of all $g \in L^2$ such that $|\xi|^m g(\xi) = \widehat{T}(\xi)$ for some $T \in \mathcal{S}'$ with $\operatorname{supp} T \subset \complement\Omega$.*

By Lemme 3 in [1] there is a continuous embedding $\dot{H}_m \subset \mathcal{D}'(\Omega)$ if and only if for every $\psi \in C_0^\infty(\Omega)$ there exists some $T_\psi \in \mathcal{S}'(\mathbf{R}^n)$ with support in $\complement\Omega$ such that $\hat{\psi} - \widehat{T}_\psi \in |\xi|^m L^2$. This is then also true for every $\psi \in \mathcal{S}(\mathbf{R}^n)$, and

$$(3) \qquad (u, \psi) = (2\pi)^{-n}(\hat{u}, \hat{\psi}) = (2\pi)^{-n}(F_m u, (\hat{\psi} - \widehat{T}_\psi)/|\xi|^m), \quad u \in \dot{H}_m, \quad \psi \in \mathcal{S}(\mathbf{R}^n).$$

In fact, this is obvious when $u \in C_0^\infty(\Omega)$ and follows by continuity in general. As observed a moment ago, the right-hand side is independent of the choice of T_ψ. The various conditions on Ω listed above were obtained by examining when it is possible to find T_ψ with the required properties.

© Springer International Publishing AG, part of Springer Nature 2018
L. Hörmander, *Unpublished Manuscripts*,
https://doi.org/10.1007/978-3-319-69850-2_7

When $\dot{H}_m \subset \mathcal{D}'(\Omega)$ and $f \in C_0^\infty(\Omega)$, then the continuous antilinear map $\dot{H}_m \ni \psi \mapsto (f, \psi)$ is the scalar product in $\dot{H}_m$ with an element $u \in \dot{H}_m$ which by definition is the solution of the Dirichlet problem. Thus

$$(4) \qquad (u, \psi)_m = (f, \psi)_0, \quad \psi \in C_0^\infty(\Omega),$$

which means that

$$\int |\xi|^m (F_m u)(\xi) \overline{\hat{\psi}(\xi)} \, d\xi = \int \hat{f}(\xi) \overline{\hat{\psi}(\xi)} \, d\xi = \int \frac{\hat{f}(\xi) - \widehat{T}_f(\xi)}{|\xi|^m} \overline{|\xi|^m \hat{\psi}(\xi)} \, d\xi, \quad \psi \in C_0^\infty(\Omega).$$

The functions $F_m \psi$ are dense in the range of F_m. If $\widehat{T}_f$ is chosen so that $(\hat{f}(\xi) - \widehat{T}_f(\xi))/|\xi|^m$ is in the range, which by the characterization of the orthogonal complement above means precisely that the L^2 norm is minimal, then it follows that

$$(F_m u)(\xi) = (\hat{f}(\xi) - \widehat{T}_f(\xi))/|\xi|^m.$$

Using (3) we obtain

$$(5) \qquad (u, \psi) = (2\pi)^{-n} \int |\xi|^{-2m} (\hat{f}(\xi) - \widehat{T}_f(\xi)) \overline{(\hat{\psi}(\xi) - \widehat{T}_\psi(\xi))}, \quad f, \psi \in C_0^\infty(\Omega),$$

where it should be understood that $\widehat{T}_f$ is the minimizer. We shall use this formula below to compute Green's function explicitly in some cases.

The odd dimensional case. Assume now that $n = 2m - 2k - 1$ where k is a non-negative integer, and that $\Omega = \mathbf{R}^n \setminus \{0\}$. If $\psi \in \mathcal{S}(\mathbf{R}^n)$ then the distribution T_ψ must be of the form $t(D)\delta_0$ since it has support at the origin, and

$$\int |\hat{\psi}(\xi) - t(\xi)|^2 |\xi|^{-2m} \, d\xi < \infty.$$

Since $2m = n + 2k + 1$, integrability at infinity means that the degree of the polynomial t is at most equal to k, and integrability at the origin means that $\psi(\xi) - t(\xi) = O(|\xi|^{k+1})$ when $\xi \to 0$. Hence (see [1, p. 262]) t is the Taylor polynomial of ψ of degree k,

$$\widehat{T}_\psi(\xi) = \sum_{|\alpha| \le k} \hat{\psi}^{(\alpha)}(0) \xi^\alpha / \alpha!, \quad T_\psi = \sum_{|\alpha| \le k} \int x^\alpha \psi(x) \, dx / \alpha! (-\partial)^\alpha \delta_0,$$

and if u is the solution of $(-\Delta)^k u = f \in C_0^\infty(\Omega)$ with Dirichlet boundary condition, then (5) gives

$$(6) \qquad (u, \psi) = (2\pi)^{-n} \int |\xi|^{-2m} (\hat{f}(\xi) - \widehat{T}_f(\xi)) \overline{(\hat{\psi}(\xi) - \widehat{T}_\psi(\xi))} \, d\xi.$$

We shall make this explicit by using that

$$E(x) = c_{n,m} |x|^{2m-n} = c_{n,m} |x|^{2k+1}$$

for an appropriate constant $c_{n,m}$ is a fundamental solution of $(-\Delta)^m$, thus $\widehat{E}$ is a homogeneous distribution of degree $-2m$ such that $|\xi|^{2m} \widehat{E}(\xi) = 1$. It is legitimate to replace

$|\xi|^{-2m}$ by the distribution $\widehat{E}$ in (6). For the proof we choose $\chi \in C_0^\infty(\mathbf{R}^n)$ equal to 1 in a neighborhood of 0 and note that the right-hand side of (6) is equal to the limit as $\varepsilon \to 0$ of the same integral with another factor $(1 - \chi(\xi/\varepsilon))$ in the integrand, and there we may replace $|\xi|^{-2m}$ by $\widehat{E}(\xi)$. What we must prove is therefore that $\widehat{E}(\chi_\varepsilon g) \to 0$ when $\varepsilon \to 0$ if $\chi_\varepsilon(\xi) = \chi(\xi/\varepsilon)$ and $g \in C^\infty$ vanishes of order $2k+2$ at the origin. Taylor's formula gives that $\chi(\xi)g(\varepsilon\xi) = \varepsilon^{2k+2} \sum_{|\alpha|=2k+2} \chi_{\varepsilon\alpha}(\xi)$ where $\chi_{\varepsilon\alpha}$ is bounded in C_0^∞ as $\varepsilon \to 0$. Now

$$\widehat{E}(\chi_\varepsilon g) = \varepsilon^{2k+2} \sum_{|\alpha|=2k+2} E(\varepsilon^n \hat{\chi}_{\varepsilon\alpha}(\varepsilon\cdot)) = c_{n,m}\varepsilon \int |x|^{2k+1} \hat{\chi}_{\varepsilon\alpha}(x)\, dx$$

is $O(\varepsilon)$ which proves the claim. Having replaced $|\xi|^{-2m}$ by $\widehat{E}$ in the right-hand side of (6) we note that it is equal to the limit of the modification obtained by multiplying $\widehat{T}_f$ and $\widehat{T}_\psi$ by $\hat{\varphi}(\delta\varepsilon)$ where $\varphi \in C_0^\infty$ and $\hat{\varphi}(0) = 1$. We can then use Fourier's inversion formula and conclude that

$$(7) \qquad (u, \psi) = \lim_{\delta \to 0}(E * (f - \varphi^\delta * T_f), \psi - \varphi^\delta * T_\psi),$$

where $\varphi^\delta = \delta^{-n}\varphi(\cdot/\delta)$. When $\delta \to 0$ it follows that $u(x) = \int G(x,y)f(y)\,dy$ where Green's function G is given by

$$(8) \qquad G(x,y) = E(x-y) - \sum_{|\alpha|\leq k} ((-y)^\alpha E^{(\alpha)}(x) + (-x)^\alpha E^{(\alpha)}(y))/\alpha!.$$

Note that $G(x,y) = O(|x|^{k+1})$ as $x \to 0$.

The even dimensional case. Assume now that $n = 2m - 2k$ where k is a positive integer, and that $\Omega = \mathbf{R}^n \setminus \{x^0, \ldots, x^n\}$ where $x^0, \ldots, x^n$ are affinely independent. If $\psi \in \mathcal{S}(\mathbf{R}^n)$ then the distribution T_ψ must be of the form $\sum_0^n t_j(D)\delta_{x^j}$, and

$$\int |\hat{\psi}(\xi) - \sum_0^n t_j(\xi)e^{-i\langle x^j,\xi\rangle}|^2 |\xi|^{-2m}\, d\xi < \infty.$$

Since $2m = n + 2k$ the convergence at infinity implies that the polynomials t_j have degree $\leq k-1$, and convergence at the origin means that $\hat{\psi}(\xi) - \sum_0^n t_j(\xi)e^{-i\langle x^j,\xi\rangle} = O(|\xi|^{k+1})$ as $\xi \to 0$. If $k = 1$ then t_j must be constants with

$$(9) \qquad \sum_0^n t_j = \hat{\psi}(0) = \int \psi(x)\, dx, \quad \sum_0^n t_j x^j = i\hat{\psi}'(0) = \int x\psi(x)\, dx.$$

These $n+1$ equations are linearly independent, for if $a + \langle x^j, b\rangle = 0$, $j = 0, \ldots, n$, then $a = 0$ and $b = 0$ since all x^j do not lie in an affine hyperplane. Let $\lambda_0, \ldots, \lambda_n$ be the barycentric coordinates with respect to the points $x^0, \ldots, x^n$, that is, λ_j is affine linear and $\lambda_j(x^k) = \delta_{jk}$, $j, k = 0, \ldots, n$. Then

$$\sum_0^n \lambda_j(x) = 1, \quad \sum_0^n \lambda_j(x)x^j = x, \quad x \in \mathbf{R}^n,$$

for this is true when $x = x^k$, $k = 0, \dots, n$. The unique solution of (8) is therefore given by

$$t_j = \int \lambda_j(x)\psi(x)\,dx, \quad j = 0, \dots, n, \quad \text{that is,}$$

(10)
$$T_\psi = \sum_0^n \Big(\int \lambda_j(x)\psi(x)\,dx \Big)\delta_{x^j}.$$

When $k > 1$ it is still possible to find $t_0, \dots, t_n$ of degree $< k$ so that

$$\hat{\psi}(\xi) - \sum_0^n t_j(\xi)e^{-i\langle x^j,\xi\rangle} = O(|\xi|^{k+1}).$$

(See [1, pp. 267–268].) In fact, by induction it suffices to prove that given a homogeneous polynomial $t(\xi)$ of degree $k > 1$ one can find $t_j(\xi)$ of degree $k - 1$ such that

$$\sum_0^n t_j(\xi)e^{-i\langle x^j,\xi\rangle} = t(\xi) + O(|\xi|^{k+1}).$$

To do so it suffices to note that we can find c_{kj} such that

$$\sum_{j=0}^n c_{kj}e^{-i\langle x^j,\xi\rangle} = \xi_k + O(|\xi|^2),$$

multiply by suitable homogeneous polynomials of degree $k - 1$ and sum. However, this shows that T_ψ will not be uniquely determined when $k > 1$. For that reason we assume that $k = 1$ in what follows.

To evaluate (5) we note that there is a constant $c_{n,m}$ such that

$$E(x) = c_{n,m}|x|^{2m-n}\log|x| = c_{n,m}|x|^2\log|x|$$

is a fundamental solution of $(-\Delta)^m$. Thus $\widehat{E}(\xi) = |\xi|^{-2m}$ in the complement of the origin where it is a finite part. We shall first prove that $|\xi|^{-2m}$ can be replaced by $\widehat{E}$ in (5). As in the odd dimensional case above we must prove that $\widehat{E}(\chi_\varepsilon g) \to 0$ when $\varepsilon \to 0$ if $\chi_\varepsilon(\xi) = \chi(\xi/\varepsilon)$ and $\chi \in C_0^\infty$, $g \in C^\infty$ vanishes of order 4 at the origin. As there we write $\chi(\xi)g(\varepsilon\xi) = \varepsilon^4 \sum_{|\alpha|=4} \chi_{\varepsilon\alpha}(\xi)$ where $\chi_{\varepsilon\alpha}$ is bounded in C_0^∞ as $\varepsilon \to 0$. We have again

$$\widehat{E}(\chi_\varepsilon g) = \varepsilon^4 \sum_{|\alpha|=4} E(\varepsilon^n\hat{\chi}_{\varepsilon\alpha}(\varepsilon\cdot)) = c_{n,m}\varepsilon^2 \int |x|^2 \log|x/\varepsilon|\,\hat{\chi}_{\varepsilon\alpha}(x)\,dx \to 0,$$

and with the same notation as before it follows that (7) is valid. Hence the Green function is given by

(11) $$G(x,y) = E(x - y) - \sum_{j=0}^n E(x - x^j)\lambda_j(y) - \sum_{j=0}^n E(y - x^j)\lambda_j(x)$$

$$+ \sum_{j,k=0}^n E(x^j - x^k)\lambda_j(x)\lambda_k(y).$$

Note that (11) does not change if $E(x)$ is replaced by $E(x) + c|x|^2$, for

$$|x - y|^2 - \sum_{j=0}^{n}(|x - x^j|^2\lambda_j(y) + |y - x^j|^2\lambda_j(x)) + \sum_{j,k=0}^{n} |x^j - x^k|^2\lambda_j(x)\lambda_k(y)$$

$$= -2\langle x, y\rangle + 2\sum_{j=0}^{n}(\langle x, x^j\rangle\lambda_j(y) + \langle y, x^j\rangle\lambda_j(x)) - 2\sum_{j,k=0}^{n} \langle x^j, x^k\rangle\lambda_j(x)\lambda_k(y)$$

$$= -2\langle x, y\rangle + 4\langle x, y\rangle - 2\langle x, y\rangle = 0.$$

When $k > 1$ it is not as easy to give an explicit formula for the extremal T_f to be used in (7), and the explicit computation of the Green function is therefore much harder.

References

[1] L. Hörmander and J.-L. Lions, *Sur la complétion par rapport à une intégrale de Dirichlet*, Math. Scand. **4** (1956), 259–270.

THE DIFFUSION APPROXIMATION
IN NEUTRON TRANSPORT THEORY

LARS HÖRMANDER

Introduction. The purpose of this paper is to make a rigorous investigation of the asymptotics of the solutions of the transport equation (Peierls' equation; see Davison [1])

$$(1) \qquad \varrho(x) = \frac{c}{4\pi l} \int_V \varrho(x') \frac{e^{-|x-x'|/l}}{|x-x'|^2} \, dx' + l \int_V S(x') \frac{e^{-|x-x'|/l}}{|x-x'|^2} \, dx', \quad x \in V,$$

when l, the mean free path of the neutrons in the homogeneous medium V, converges to 0. The function S is a positive source function and c denotes the expected number of neutrons formed at a collision. One assumes that c depends on l according to the formula

$$(2) \qquad c = 1 + \mu l^2/3$$

where μ is a constant.

We first recall the usual terminology (see Davison [1]). The equation (1) is said to be subcritical if the largest eigenvalue of the integral equation (1) is < 1. By the elementary theorems on integral equations with symmetrical kernel one can then solve (1) by iteration for an arbitrary $S \in L^2$, and it is easy to see that this is also possible for an arbitrary measure instead of $S \, dx$. Hence there exists for every positive S a positive solution of (1) in the subcritical case. On the other hand, if there is a positive solution of (1) for some positive S, then the equation (1) is subcritical. For then the Neumann series for (1) must converge, which gives the desired result at once. This fact is the reason for the physical relevance of whether the equation (1) is subcritical or not; that it is subcritical is equivalent to the existence of a stationary neutron distribution when there is a time independent source function, such as the spontaneous fission.

Let λ denote the smallest eigenvalue for the Dirichlet problem for the Laplace operator in V and assume that $\mu < \lambda$. We shall then prove that (1) is subcritical if l is sufficiently small, and that the solution of (1) when $l \to 0$ converges to the solution of the Dirichlet problem for the equation

$$(3) \qquad \Delta u + \mu u + 12\pi S = 0$$

in the weak topology of measures in $\overline{V}$, that is, so that the integral of ϱ over an arbitrary subdomain of V converges to the integral of u over the same domain. This is a weakened form of the diffusion approximation (see Davison [1], Chap. 8). It states more precisely that the solution of (1) in the first approximation is equal to the solution of (3) with vanishing Dirichlet data on the parallel surface at distance 0.7104 times l outside the boundary of V. The author has not yet succeeded to prove or disprove that this is correct.

© Springer International Publishing AG, part of Springer Nature 2018
L. Hörmander, *Unpublished Manuscripts*,
https://doi.org/10.1007/978-3-319-69850-2_8

It should be observed that the equation (1) is the equation for neutron transport only if V is convex. Otherwise it does not have physical significance and the exponent $|x - x'|$ should be replaced by the part of $|x - x'|$ passing through V. Then one no longer obtains the limiting problem (3) with an infinitesimal boundary condition. We shall not in this paper investigate the boundary conditions in this important case.

An a priori inequality. We assume throughout this section that the inequality $\mu < \lambda$ is fulfilled. Under the hypothesis that we know that (1) has a positive solution ϱ for the positive source function S — which will later be proved rigorously for sufficiently small l — we shall deduce an estimate of $\int \varrho \, dx$ over V by means of $\int S \, dx$. We make the following assumption on the domain V:

A) The upper bound for the lowest eigenvalue of the Dirichlet problem for the Laplace operator in domains Ω containing $\overline{V}$ is equal to λ.

This is a regularity condition on the boundary ∂V of V; it is for example well known that condition A) is fulfilled if the boundary is once continuously differentiable or convex. The proof is then an easy consequence of the existence of a contraction of a neighborhod of $\overline{V}$ into V which has Lipschitz continuous inverse with Lipschitz constant close to 1, if one applies the variational definition of the lowest eigenvalue.

Thus we can choose a domain Ω containing $\overline{V}$ so that the lowest eigenvalue λ' of the Dirichlet problem in Ω is $> \mu$. Let Φ be the corresponding eigenfunction. By well-known theorems $\Phi > 0$ in Ω. Now we multiply (1) by Φ and integrate over V. This gives

$$(4) \quad \int \varrho(x)\Phi(x) \, dx = c \iint_{V \times V} \varrho(x)\Phi(x')K_l(x - x') \, dx \, dx'$$
$$+ 4\pi l^2 \iint_{V \times V} S(x)\Phi(x')K_l(x - x') \, dx \, dx'$$

where we have introduced the notation

$$(5) \qquad\qquad K_l(y) = \frac{1}{4\pi l} \frac{e^{-|y|/l}}{|y|^2}.$$

To estimate the right-hand side of (4) we note first that since

$$(6) \qquad\qquad \int_{\mathbf{R}^3} K_l(y) \, dy = 1$$

we have

$$(7) \qquad 4\pi l^2 \iint_{V \times V} S(x)\Phi(x')K_l(x - x') \, dx \, dx' \leq 4\pi l^2 \max_V \Phi \int_V S(x) \, dx.$$

The estimate of the first integral in the right-hand side of (4) requires greater care. Let Ω_1 be a domain which contains $\overline{V}$ but is so small that $\overline{\Omega}_1 \subset \Omega$. Then Φ is analytic in $\overline{\Omega}_1$ and the complement of Ω_1 has a positive distance δ to V. Now we observe first that the integral in question increases if the domain of integration for x' is extended to Ω_1. Then we introduce

$$(8) \qquad\qquad \Phi(x') = T(x'; x) + R(x'; x)$$

where

(9)
$$T(x'; x) = \sum_{|\alpha| \leq 2} \frac{\Phi^{(\alpha)}(x)}{\alpha!} (x' - x)^\alpha$$

is the Taylor expansion of second order with center at x and R is the remainder term. When x and x' belong to Ω_1 then

(10)
$$|R(x'; x)| \leq C|x' - x|^3.$$

We can now write

$$\int_V \Phi(x') K_l(x - x') \, dx' \leq \int_{\Omega_1} (T(x'; x) + R(x'; x)) K_l(x - x') \, dx'$$

$$= \int_{\mathbf{R}^3} T(x'; x) K_l(x - x') \, dx' - \int_{\complement\Omega_1} T(x'; x) K_l(x - x') \, dx' + \int_{\Omega_1} R(x'; x) K_l(x - x') \, dx'.$$

The integral over $\mathbf{R}^3$ is easy to calculate by means of (6) and

(11)
$$\int y_j K_l(y) \, dy = 0, \quad \int y_j y_k K_l(y) \, dy = \frac{2l^2}{3} \delta_{jk}$$

where δ is Kronecker's delta. Moreover we obtain using (10)

(12)
$$\left| \int_{\Omega_1} R(x'; x) K_l(x - x') \, dx' \right| \leq C \int_{\mathbf{R}^3} |x - x'|^3 K_l(x - x') \, dx' = C_1 l^3.$$

Since $T(x'; x) \leq C(1 + |x' - x|^2)$ when $x \in V$ we obtain if we note that $|x - x'| > \delta$ when $x' \in \complement\Omega_1$

(13)
$$\left| \int_{\complement\Omega_1} T(x'; x) K_l(x - x') \, dx' \right| \leq C \int_{|y| > \delta} (1 + |y|^2) K_l(y) \, dy \leq C_2 (1 + l^2) e^{-\delta/l}.$$

The inequalities (11), (12), (13) give now

(14)
$$\int_V \Phi(x') K_l(x - x') \, dx' \leq \Phi(x) + \frac{l^2}{3} \Delta\Phi(x) + Cl^3$$

when $l \to 0$. After inserting (14) and (7) in (4), rearranging the terms and dividing by l^2, we obtain

(15)
$$\int_V \varrho(x)(-\mu\Phi/3 - c\Delta\Phi/3 - Cl) \, dx \leq 4\pi \max \Phi \int_V S(x) \, dx.$$

Now Φ satisfies the equation $\Delta\Phi + \lambda'\Phi = 0$, so the coefficient of ϱ in (15) becomes $(c\lambda' - \mu)\Phi/3 - Cl$. Since $c \to 1$ when $l \to 0$ and $\lambda' > \mu$, and since Φ is bounded above and below in V, we obtain at once from (15) the following inequality

LEMMA 1. *For every $\varepsilon > 0$ there exist constants $l_0 > 0$ and C such that if ϱ is a positive solution of (1) with positive source function S and $\mu < \lambda - \varepsilon$ then*

(16)
$$\int_V \varrho \, dx \leq C \int_V S \, dx,$$

if $l < l_0$.

The asymptotic condition for subcriticality. Using Lemma 1 it is now easy to prove the following theorem.

THEOREM 1. *If $\mu < \lambda$ then the equation* (1) *is subcritical when $l < l_0$.*

PROOF. The eigenvalues of the integral operator in (1) are for fixed l continuous functions of the parameter μ, which proves in particular that the set of those μ for which the equation is subcritical is open. Obviously it contains all negative μ, which follows by taking Fourier transforms in (1) or even simpler by using a classical criterion due to Carleman. We shall now also prove that the set of those $\mu < \lambda - \varepsilon$ for which the equation is subcritical is closed. For if we take a positive source function S and a sequence $\mu_n \to \mu_0$ for which the equation is subcritical, then the equation (1) with $\mu = \mu_n$ has a positive solution ϱ_n which therefore satisfies (16). By Helly's selection theorem one can choose a subsequence such that ϱ_n converges weakly as a measure in $\overline{V}$, but then it follows at once from (1) that ϱ_n actually converges uniformly in $\overline{V}$ to a function ϱ_0, and in the limit we see that ϱ_0 solves (1) with $\mu = \mu_0$. Hence the equation is subcritical then, which proves the assertion. Since a set which is both open and closed in the interval $(-\infty, \lambda - \varepsilon)$ consists of the entire interval if it is not empty, the proof is complete.

As a consequence of Lemma 1 and Theorem 1 we also obtain the following result:

THEOREM 2. *For every $\mu < \lambda - \varepsilon$ the equation* (1) *has for $l \leq l_0$ one and only one solution for fixed S, and it satisfies the inequality*

$$(17) \qquad \int_V |\varrho|\, dx \leq C \int_V |S|\, dx.$$

The limit of the solution of (1). As in the preceding section we assume that $\mu < \lambda$. By (17) the solutions of (1) with $l \leq l_0$ form for fixed S a compact family of mesures in $\overline{V}$. We shall prove the following:

THEOREM 3. *If the boundary ∂V is of class $C^{2+\varepsilon}$ for some $\varepsilon > 0$ then the solution of* (1) *converges weakly as a measure in $\overline{V}$ to the solution of the differential equation*

$$(18) \qquad \Delta u + \mu u + 12\pi S = 0$$

with Dirichlet's boundary condition.

REMARK. The regularity assumption on the boundary can undoubtedly be weakened.

PROOF. The statement means, if we denote the solution of (1) by ϱ_l, that

$$(19) \qquad \int_V \varrho_l(x) f(x)\, dx \to \int_V u(x) f(x)\, dx, \quad l \to 0,$$

for every continuous function f in $\overline{V}$. By (17) it even suffices to consider only a dense set of such f. We shall therefore prove (19) under the hypothesis that $f \in C^\varepsilon$. Let Ψ be the solution of the Dirichlet problem

$$(20) \qquad \Delta\Psi + \mu\Psi + f = 0, \quad \Psi = 0 \text{ on } \partial V.$$

According to Schauder [2] Ψ is in $C^{2+\varepsilon}$ and we extend its definition to a neighborhood of $\overline{V}$ so that it remains in this class. With a constant $\alpha > 0$ we shall consider (4) for $\Phi = \Psi + \alpha$. Since $\Phi > 0$ on the boundary one can estimate the first integral on the right-hand side of

(4) in the same way as before with the exception that the exponent in (10) becomes $2 + \varepsilon$, and we obtain using (17)

$$(21) \qquad \int \varrho_l(x)(-\Delta\Psi - \mu\Psi)\, dx \le 12\pi \iint S(x)\Psi(x')K_l(x - x')\, dx' + C\alpha + C_\alpha l^\varepsilon.$$

From (5) and the positivity of K_l it follows at once that the double integral converges to $\int S(x)\Psi(x)\, dx$ when $l \to 0$. Hence we obtain from (21) and (20)

$$(22) \qquad \overline{\lim_{l \to 0}} \int \varrho_l f\, dx \le 12\pi \int S\Psi\, dx + C\alpha$$

for every $\alpha > 0$, and therefore

$$(23) \qquad \overline{\lim_{l \to 0}} \int_V \varrho_l f\, dx \le 12\pi \int_V S\Psi\, dx.$$

If we apply this to $-f$ instead we obtain

$$(24) \qquad \overline{\lim_{l \to 0}} \int_V \varrho_l(-f)\, dx \le 12\pi \int_V S(-\Psi)\, dx,$$

hence

$$(25) \qquad \underline{\lim_{l \to 0}} \int_V \varrho_l f\, dx \ge 12\pi \int_V S\Psi\, dx.$$

Taken together (23) and (25) prove that

$$(26) \qquad \lim_{l \to 0} \int_V \varrho_l f\, dx = 12\pi \int_V S\Psi\, dx.$$

According to (18)

$$12\pi \int_V S\Psi\, dx = \int_V (-\Delta u - \mu u)\Psi\, dx$$

and since u and Ψ vanish on the boundary we obtain from Green's formula

$$(27) \qquad 12\pi \int_V S\Psi\, dx = \int_V u(-\Delta\Psi - \mu\Psi)\, dx = \int_V uf\, dx.$$

But (19) follows from from (27) and (26), so the proof is complete.

A not yet finished study of the first approximation of the solution of (1) has given the following more precise form of the convergence of the solutions of the equation (1):

THEOREM 4. *If $d(x)$ denotes the distance from x to ∂V then*

$$(28) \qquad \int_V |\varrho_l(x) - u(x)|d(x)\, dx = O\big(l \log \tfrac{1}{l}\big) \quad \text{when } l \to 0.$$

Because of the preliminary nature of this result we omit the proof.

References

[1] B. Davison, *Neutron transport theory*, Oxford, 1957.

[2] J. Schauder, *Ueber lineare elliptische Differentialgleichungen zweiter Ordnung*, Math. Z. **38** (1934), 257–282.

[3] ———, *Numerische Abschätzungen in elliptischen linearen Differentialgleichungen*, Studia Math. **5** (1934), 34–42.

SOME TAUBERIAN THEOREMS OF LEWITAN

LARS HÖRMANDER

In this lecture[7] I am going to present some theorems from Fourier analysis due to Lewitan and give various extensions of them. Finally, if the time allows, I shall also show how these theorems enter in Lewitan's study of the asymptotics of the eigenfunctions for second order elliptic differential equations.

The prototype for Lewitan's "Tauberian theorems" — if they can be called that — is a classical inequality of Bohr:

If f is a function such that $|f(x) - f(y)| \leq C|x - y|$, $x, y \in \mathbf{R}$, and f has no spectrum in the interval $(-\Lambda, \Lambda)$, then

$$|f(x)| \leq C/4\Lambda.$$

The meaning of the notion of spectrum must be made precise. I interpret it in the sense of Schwartz, so the meaning is that

$$\int f\hat{\varphi}\, dx = 0$$

if $\hat{\varphi}(x) = \int_{-\infty}^{+\infty} e^{-2\pi i x \xi} \varphi\, d\xi$ is the Fourier transform of a function $\varphi \in C^\infty$ with support in $(-\Lambda, \Lambda)$. This definition is clearly meaningful if $\int_{-\infty}^{+\infty} |f(x)|(1 + x^2)^{-n}\, dx < \infty$ for some n, that is, if f is temperate.

Strictly speaking this inequality should perhaps be called the Bohr-Lewitan inequality since Lewitan was the first who proved it for functions which are not almost periodic, in a paper in 1937.

The generalizations made by Lewitan in the study of eigenfunction expansions can be written in the following form:

If f has locally bounded variation, $\int_a^{a+1} |df| \leq C(1 + |a|)^k$ where k is a nonnegative integer, and f has no spectrum in $(-\Lambda, \Lambda)$, then there is a constant K such that if $0 < \Lambda < 1$

$$|f(x)| \leq KC\Lambda^{-k-1}(1 + |x|)^k.$$

Moreover, if $k = 0$ and we also assume that

$$\lim_{x \to \infty} f(x + y) - f(x) = 0$$

it follows that

$$\lim_{x \to \infty} f(x) = 0.$$

---

[7]Lecture in Gothenburg at the Swedish Mathematical Society November 11, 1958.

Lewitan proved these results by repeating his old proof of Bohr's inequality. However, we shall see that one can obtain much stronger results by using Bohr's inequality directly, and a further improvement by using a onesided variant of Bohr's inequality which I proved in Math. Scand. **2**; these new variants are even better suited for the applications to eigenfunction expansions.

Onesided Bohr inequality. *If f is temperate and has no spectrum in $(-\Lambda, \Lambda)$ and $f(x) - f(y) \leq C(x - y)$ when $x \geq y$, then*

$$|f(x)| \leq C/2\Lambda.$$

A less precise inequality is easily proved as follows, using the proofs of Nagy-Strauss and Esseen for Bohr's inequality.

For arbitrary n there exists a function $\varphi = O((1 + |x|)^{-n})$ such that

$$\varphi \geq 0, \quad \hat{\varphi}(\xi) = 1/2\pi i\xi \text{ in a neighborhood of } |\xi| \geq 1.$$

Such a function can be constructed in the following way: Take a function $\chi \in C_0^\infty(-1, 1)$ which $= 1$ in a neighborhood of 0, and set $\hat{\varphi}_1(\xi) = (1 - \chi(\xi))/2\pi i\xi$. Then φ_1 differs from $\operatorname{sgn} \xi/2$ only by the Fourier transform of a distribution with compact support, hence an analytic function, and furthermore $(2\pi ix)^n \varphi_1$ is the Fourier transform of the nth derivative of $(1 - \chi(\xi))/2\pi i\xi$, which is in L^1, and hence bounded. The only flaw is that φ_1 is not positive. Hoever, we can attain that by adding a suitable multiple of the function

$$\left(\frac{\sin^2 \pi x/2n + \sinh^2 1}{(\pi x/2n)^2 + 1}\right)^n$$

which has its spectrum in $[-1/2, 1/2]$.

Note that $\varphi(\Lambda x)$ and $-\varphi(-\Lambda x)$ have the analogous property with conditions on the Fourier transform in $(-\Lambda, \Lambda)$. We can therefore if we have taken φ sufficiently rapidly decreasing write

$$f(x) = \int \varphi(\Lambda y) df(x - y) = -\int \varphi(-\Lambda y) df(x - y).$$

This gives at once the preceding theorem apart from a constant in the right-hand side, for the hypotheses imply that $df \leq C\, dx$ and we obtain

$$|f(x)| \leq C \int \varphi(\Lambda y)\, dy = C\Lambda^{-1} \int \varphi\, dy.$$

If we use the trivial inequality $(1 + |x - y|)^\alpha \leq (1 + |x|)^\alpha (1 + |y|)^\alpha \ (\alpha > 0)$ we obtain in the same way:

THEOREM. *If f is of locally bounded variation,*

$$df(x) \leq C(1 + |x|)^\alpha\, dx \quad (\alpha \geq 0)$$

and f is temperate with no spectrum in $(-\Lambda, \Lambda)$, then with a constant K_α

$$|f(x)| \leq K_\alpha C(1 + |x|)^\alpha \left(1 + \frac{1}{\Lambda}\right)^\alpha / \Lambda.$$

If in addition

$$df(x) \le o(1)(1 + |x|)^\alpha \, dx$$

where $o(1) \to 0$ *when* $x \to +\infty$, *it follows that*

$$|f(x)| = o((1 + |x|)^\alpha), \quad x \to +\infty.$$

Analogous theorems can of course be stated for other weight functions. We shall now see how formal manipulations of this theorem can give Lewitan's theorem with an improvement where we have replaced the condition on the total variation by a condition of Schmidt's type.

THEOREM. *Let f be real, measurable and temperate and have no spectrum in* $(-\Lambda, \Lambda)$. *If for some y*

$$(1) \qquad\qquad f(x + y) - f(x) \le C(1 + |x|)^\alpha, \quad -\infty < x < \infty,$$

it follows that

$$(2) \qquad\qquad \left| \int_x^{x+y} f(t) \, dt \right| \le K_\alpha C(1 + |x|)^\alpha \left(1 + \frac{1}{\Lambda}\right)^\alpha / \Lambda.$$

If we can put $o(1)$ in the right-hand side of (1) when $x \to +\infty$ we can also do so in (2).

PROOF. The function

$$G(x) = \int_x^{x+y} f(t) \, dt$$

is the convolution of f with the characteristic function of $(-y, 0)$ so it has no spectrum in $(-\Lambda, \Lambda)$. It is absolutely continuous and $G'(x) \le C(1 + |x|)^\alpha$ almost everywhere. Hence the theorem follows from the preceding one.

THEOREM. *If (1) is valid for all y between 0 and 1, with C independent of y too, then it follows that*

$$(3) \qquad\qquad |f(x)| \le K_\alpha C(2 + |x|)^\alpha \left(1 + \frac{1}{\Lambda}\right)^{\alpha+1}.$$

If in (1) we are allowed to put $o(1)$ on the right-hand side we can put $o(1)$ in the right-hand side of (3) also.

PROOF. By hypothesis we have

$$\int_{x-1}^x f(t) \, dt \ge f(x) - C \int_{x-1}^x (1 + |t|)^\alpha \, dt, \qquad \int_x^{x+1} f(t) \, dt \le f(x) + C \int_x^{x+1} (1 + |t|)^\alpha \, dt$$

whence the statement follows by estimating the integrals in the right-hand side and using the preceding theorem. The same reasoning, used with Lebesgue's theorem on dominated convergence, gives the o-theorem.

I shall now sketch how Lewitan reduces the study of the asymptotic behavior of eigenfunctions to these theorems. Let us for the sake of formal simplicity consider the very simplest case, the eigenvalue problem

$$\varphi'' + (\lambda - q)\varphi = 0$$

on a finite interval, with selfadjoint boundary conditions. I also assume that all eigenvalues are positive and denote them by μ_k^2 ($\mu_k > 0$); the eigenfunctions are denoted by φ_k. The functions $\varphi_k(x) \cos \mu_k t$ satisfy the wave equation

$$\partial^2 u / \partial x^2 - q(x)u - \partial^2 u / \partial t^2 = 0,$$

the eigenvalue problem stems from a separation of variables in this equation. Furthermore the normal derivative of our solution vanishes when $t = 0$. If we therefore introduce the time derivative of the Riemann function $w(x, t, y)$ when $t \geq 0$ (note the translation invariance in the t direction) we can write

$$\varphi_k(x) \cos \mu_k t = \tfrac{1}{2}(\varphi_k(x+t) + \varphi_k(x-t)) + \tfrac{1}{2} \int_{|x-y| \leq |t|} w(x, |t|, y) \varphi_k(y) \, dy,$$

provided that $|t|$ is smaller than the distance ε from x to the end points. If we now choose a function g which vanishes outside $(-\varepsilon, \varepsilon)$ we obtain by multiplication and integration

$$\varphi_k(x)(\hat{g}(\mu_k) + \hat{g}(-\mu_k)) = \int (g(t-x) + g(x-t))\varphi_k(t) \, dt + \iint\limits_{|x-y| \leq |t|} w(x, |t|, y)\varphi_k(y)g(t) \, dt \, dy.$$

Multiplying by $\varphi_k(y)$ and summing we obtain, since the functions φ_k form a complete orthonormal system, formally — and using basic estimates for the eigenfunctions also rigorously if g is sufficiently differentiable — that

$$\sum_1^\infty \varphi_k(x)\varphi_k(y)(\hat{g}(\mu_k) + \hat{g}(-\mu_k)) = g(y-x) + g(x-y) + \int_{|x-y| \leq |t|} w(x, |t|, y)g(t) \, dt.$$

Write

$$\theta(x, y, \mu) = \sum_{\mu_k \leq \mu} \varphi_k(x)\varphi_k(y) \quad \text{when } \mu > 0, \ \theta(x, y, \mu) = -\theta(x, y, -\mu).$$

Then the formula above says that

$$\int \hat{g}(\mu)d\theta(x, y, \mu) = g(y-x) + g(x-y) + \int_{|x-y| \leq |t|} w(x, |t|, y)g(t) \, dt,$$

that is, the Fourier transform of $d\theta(x, y, \mu)$ in the sense of Schwartz consists in the interval $(-\varepsilon, \varepsilon)$ of the point masses at $x - y$ and $y - x$ and the function $w(x, |t|, y)$ when $|t| \geq |x - y|$, 0 otherwise. Let $d\theta^*$ correspond to the same problem with $q = 0$ in the infinite interval; this is obtained formally by putting $w = 0$ in the formulas above. Thus we obtain $d\theta^*(x, y, \mu) = \frac{1}{\pi} \cos(y-x)\mu$. Take $y = x$ for the sake of simplicity. If we write $d\phi = d\theta^* - d\theta$ and note that $w(x, |t|, x) = O(|t|^c)$ with $c > 0$ even if we just have $q \in L^p$ for some $p > 1$, it follows that the Fourier transform of $d\phi$ is in $(-\varepsilon, \varepsilon)$ a function H with $H/t \in L^1$, and

$$\phi(\mu + a) - \phi(\mu) < a/\pi, \quad a > 0.$$

The Fourier transform is defined since $d\phi(\mu)$ is temperate. Let $\alpha(\mu)$ be the inverse Fourier transform of H/it. Then $\alpha(\mu)$ is bounded and tends to 0 at infinity by the Riemann-Lebesgue theorem. Thus we see that the function

$$F(\mu) = \phi(\mu) - \alpha(\mu)$$

has no spectrum in $(-\varepsilon, \varepsilon)$ and satisfies the condition

$$F(\mu + a) - F(\mu) \leq C \quad \text{when } 0 < a < 1,$$

by the inequality satisfied by ϕ and the bound for α. By the theorems above it follows now that F is bounded, hence ϕ is bounded which was to be proved. (This argument does not exactly follow that of Lewitan. Since he does not have onesided Tauberian conditions as above he must extract from the problem additional estimates to get a bound for the total variation of F in intervals of length 1, which makes the arguments somewhat more complicated.)

SOME VARIANTS OF AN INEQUALITY DUE TO H. BOHR

Lars Hörmander

In his study of the asymptotic properties of spectral functions belonging to differential operators, Lewitan has given some Tauberian theorems related to an inequality of Bohr (Lewitan [2], [3], [4]). His proofs depend on estimates of certain trigonometrical polynomials, which he had previously introduced in proving Bohr's inequality for nonperiodic functions (Lewitan [5]). We shall here give simple proofs of his results in a sharper form by using Bohr's inequality directly, in the form given by Hörmander [1].

THEOREM 1. *Let f be a real measurable function, defined for* $-\infty < x < \infty$ *such that* $|f(x)|(1 + x^2)^{-n}$ *is integrable for some n and which has no spectrum in the sense of L. Schwartz in the interval* $(-\Lambda, \Lambda)$. *Then, if for some y,*

$$(1) \qquad\qquad f(x + y) - f(x) \leq C, \quad -\infty < x < \infty,$$

it follows that

$$(2) \qquad\qquad \left| \int_x^{x+y} f(t)\, dt \right| \leq \frac{C}{2\Lambda}, \quad -\infty < x < \infty.$$

PROOF. Let F be the function without spectrum in $(-\Lambda, \Lambda)$ such that $F' = f$. Then the function

$$G(x) = F(x + y) - F(x) = \int_x^{x+y} f(t)\, dt$$

has no spectrum there either. G is absolutely continuous, and according to (1) we have almost everywhere

$$G'(x) = f(x + y) - f(x) \leq C.$$

Hence Bohr's inequality ([1], p. 42) shows that

$$|G(x)| \leq \frac{C}{2\Lambda},$$

which proves (2).

THEOREM 2. *If f satisfies the assumptions of the preceding theorem for every y with $0 \le y \le 1$, it follows that*

$$(3) \qquad\qquad |f(x)| \le C\left(1 + \frac{1}{2\Lambda}\right).$$

PROOF. In virtue of the assumptions we have

$$\int_{x-1}^{x} f(t)\, dt \ge f(x) - C, \qquad \int_{x}^{x+1} f(t)\, dt \le f(x) + C$$

so that, in virtue of Theorem 1,

$$|f(x)| \le C + \frac{C}{2\Lambda}.$$

REMARK. Lewitan [2] proved essentially the same result under the assumption that instead of (1) we know that f is of locally bounded variation and $\int_{x}^{x+1} |df| \le C$.

THEOREM 3. *If f satisfies the assumptions of Theorem 1 and, furthermore,*

$$(4) \qquad\qquad \varlimsup_{x \to \infty} (f(x + y) - f(x)) \le 0$$

then it follows that

$$(5) \qquad\qquad \int_{x}^{x+y} f(t)\, dt \to 0 \quad \text{when } x \to \infty.$$

PROOF. With the notations of the proof of Theorem 1 we have to prove that $G(x) \to 0$ when $x \to \infty$. Let $H(x)$ be the function without spectrum in $(-\Lambda, \Lambda)$ such that $H'(x) = G(x)$. Theorem 1 and Bohr's inequality show that

$$|G(x)| = |H'(x)| \le \frac{C}{2\Lambda}, \quad |H(x)| \le \frac{C}{8\Lambda^2}.$$

We shall first prove that $H(x) \to 0$ when $x \to \infty$. To do so it is sufficient to prove that if $x_n \to \infty$ and $H(x_n) \to a$ we have $a = 0$. We may also assume that the functions $H(x + x_n)$ converge locally uniformly to a continuous function $H_0(x)$, for since these functions are uniformly bounded and uniformly continuous we could otherwise obtain this property by passing to a suitable subsequence. Being the limit in S' of functions with no spectrum in $(-\Lambda, \Lambda)$ the function H_0 has no spectrum there either. Furthermore, H_0 is concave since, for arbitrary x and $h > 0$,

$$H_0(x + h) + H_0(x - h) - 2H_0(x) = \lim_{n \to \infty} \int_{x-h}^{x} (G(t + h + x_n) - G(t + x_n))\, dt,$$

and our assumption that $\varlimsup_{x \to \infty} G'(x) \le 0$ implies that the limit is ≤ 0. Thus $H_0(x)$ is bounded and concave, hence equal to the constant a, and since H_0 has no spectrum in $(-\Lambda, \Lambda)$ we obtain $a = 0$.

It now easily follows that also $G(x) \to 0$ when $x \to \infty$. In fact,

$$G(x) = \int_{0}^{1} (G(x) - G(x - t))\, dt + H(x) - H(x - 1),$$

$$-G(x) = \int_{0}^{1} (G(x + t) - G(x))\, dt + H(x) - H(x + 1),$$

and since $\varlimsup_{x \to \infty} G'(x) \le 0$ it follows from these two identities that $\varlimsup_{x \to \infty} G(x) \le 0$ and that $\varlimsup_{x \to \infty} (-G(x)) \le 0$, hence $\lim_{x \to \infty} G(x) = 0$.

THEOREM 4. *If f satisfies the assumptions of Theorem 2, and (4) is valid for every fixed y, $0 \le y \le 1$, it follows that $f(x) \to 0$ when $x \to \infty$.*

PROOF. Theorem 2 shows that f is uniformly bounded. Hence, using Lebesgue's theorem and (4) in the identity

$$f(x) = \int_0^1 (f(x) - f(x-t))\, dt + \int_{x-1}^x f(t)\, dt$$

and noting that the last integral tends to zero in virtue of Theorem 3, we obtain that $\overline{\lim}_{x\to\infty} f(x) \le 0$. Similarly, the identity

$$-f(x) = \int_0^1 (f(x+t) - f(x))\, dt - \int_x^{x+1} f(t)\, dt$$

shows that $\overline{\lim}_{x\to\infty}(-f(x)) \le 0$. Hence $\lim_{x\to\infty} f(x) = 0$.

REMARK. Theorem 4 generalizes a result given by Lewitan [4]. It is obvious that a similar result holds when $x \to -\infty$.

COROLLARY 1. *Let f be a locally integrable measurable function such that $\int_x^{x+1} |f(t)|\, dt \to 0$ when $|x| \to \infty$. Then, if ξ does not belong to the spectrum of f, the limit*

$$\lim_{\substack{x_1 \to -\infty \\ x_2 \to +\infty}} \int_{x_1}^{x_2} f(x) e^{-2\pi i x \xi}\, dx$$

exists and equals 0.

PROOF. By assumption the function $f(x)e^{-2\pi i x \xi}$ has no spectrum in the neighborhood of the origin. Thus this function is the derivative of a function $F_\xi(x)$ with no spectrum in an interval $(-\Lambda, \Lambda)$, and the previous theorem (applied to the real and imaginary parts) shows that $F_\xi(x) \to 0$ when $x \to \pm\infty$. This proves the assertion.

References

[1] L. Hörmander, *A new proof and a generalization of an inequality of Bohr*, Math. Scand. **2** (1954), 33–45.

[2] B. M. Lewitan, *On the asymptotic behaviour of the spectral function of a self-adjoint differential euation of the second order*, Izv. Akad. Nauk SSSR, Ser. Mat. **16** (1952), 325–352.

[3] ———, *On a special Tauberian theorem*, Izv. Akad. Nauk SSSR, Ser. Mat. **17** (1953), 269–284.

[4] ——— *On the asymptotic behaviour of the spectral function and the eigenfunction expansion of self-adjoint differential equations of the second order*, Izv. Akad. Nauk SSSR, Ser. Mat. **19** (1955), 33–58.

[5] ———, *Über eine Verallgemeinerung der Ungleichungen von S. Bernstein und H. Bohr*, Doklady Akad. Nauk SSSR **15** (1937), 169–172.

ELECTROMAGNETIC WAVE PROPAGATION OVER GROUND WITH SMALL INHOMOGENEITIES IN THE ELECTRICAL CONSTANTS

Lars Hörmander

1. Introduction. The purpose of these pages is to study how the vertical electrical component of an electromagnetic wave propagating over a homogeneous earth with plane boundary will be affected by small local variations in its electrical constants. This problem has been frequently studied in the literature (see Bremmer [1] and the references there). We shall here follow a somewhat different path, by solving the variational equation for the first variation of the field. This may have the advantage that the results are applicable also to inhomogeneities of small depth, provided that the electrical properties are not changed too much. On the other hand, the results are perhaps less suitable for numerical calculations and less appropriate for large variations of the electrical constants as for example in a salt water lake.

The solution of our problem is essentially contained in Sommerfeld's formulas for the field generated by a vertical or horizontal dipole located in air above a homogeneous earth. The difference is really just that we are interested in a fictitious transmitter placed in the earth instead. However, we shall give a complete deduction of this result which will be the main contents of this manuscript, and aim at giving them a more symmetrical form. (Compare Sommerfeld [2], [3] and Weyl [4].)

2. Maxwell's equations in a homogeneous space. For the monochromatic case with frequency ω, that is, where all field variables vary with t as $e^{-i\omega t}$, the equations become (cf. Bremmer [1], p. 518)

$$(2.1) \qquad \operatorname{rot} E - \frac{i\omega\mu}{c} H = 0$$

$$(2.2) \qquad \operatorname{rot} H + \frac{i\omega}{c}\varepsilon_{\text{eff}} E = \frac{4\pi}{c} I$$

where $I e^{-i\omega t}$ is the forced current with frequency π and

$$(2.3) \qquad \varepsilon_{\text{eff}} = \varepsilon + \frac{4\pi\sigma}{\omega} i.$$

(μ is the permeability, ε the dielectric constant, and σ the electrical conductivity.) In particular we have $\operatorname{Re}\varepsilon_{\text{eff}} > 0$ and $\operatorname{Im}\varepsilon_{\text{eff}} \geq 0$. In what follows we shall assume that $\operatorname{Im}\varepsilon_{\text{eff}} > 0$, that is, that $\sigma \neq 0$. This has the advantage that Sommerfeld's radiation condition is replaced by the simple requirement that the field must not grow exponentially at infinity.

We shall first derive the fundamental solution (the "Hertz solution"), that is, study (2.2) when $I = I_0\delta$ (δ = the Dirac measure at the origin). The solution shall be temperate

© Springer International Publishing AG, part of Springer Nature 2018

L. Hörmander, *Unpublished Manuscripts*,

https://doi.org/10.1007/978-3-319-69850-2_10

so we can apply the Fourier transformation in all the variables to (2.1) and (2.2) which gives

(2.1)′
$$i\xi \times \hat{E} - \frac{i\omega\mu}{c}\hat{H} = 0$$

(2.2)′
$$i\xi \times \hat{H} + \frac{i\omega}{c}\varepsilon_{\text{eff}}\hat{E} = \frac{4\pi}{c}I_0,$$

where $\hat{E}$ and $\hat{H}$ are the Fourier transforms of E and H respectively. Now take the vector product of (2.2)′ with ξ from the left and use (2.1)′. This gives

$$i(\xi \times (\xi \times \hat{H})) + i\frac{\omega^2 \mu\varepsilon_{\text{eff}}}{c^2}\hat{H} = \frac{4\pi}{c}\xi \times I_0.$$

Now set

(2.4)
$$k^2 = \omega^2 \mu\varepsilon_{\text{eff}}/c^2,$$

which is a number in the first quadrant. Since (2.1)′ gives that

(2.5)
$$(\xi, \hat{H}) = 0$$

we now obtain

$$i(\xi \times (\xi \times \hat{H})) = -i(\xi, \xi)\hat{H},$$

whence

(2.6)
$$\hat{H} = \frac{4\pi i}{c}\xi \times I_0((\xi, \xi) - k^2)^{-1}.$$

(Note that the right-hand side is well defined since $\operatorname{Im} k^2 > 0$. The case where $\sigma = 0$ will later on be considered as the limit $\sigma \to +0$, which will give a unique sense to (2.6) also in that case.) From (2.2)′ we now obtain

(2.7)
$$\varepsilon_{\text{eff}}\hat{E} = -\frac{4\pi i}{\omega}\xi \times (\xi \times I_0)((\xi, \xi) - k^2)^{-1} - \frac{4\pi i}{\omega}I_0.$$

To calculate E and H we now only have to study the inverse Fourier transform

(2.8)
$$V(x) = (2\pi)^{-3} \int \frac{e^{i\langle x,\xi\rangle}}{(\xi, \xi) - k^2} \, d\xi,$$

which we shall calculate in several ways. Note that (2.6) means that

(2.6)′
$$H = \frac{4\pi}{c}\operatorname{rot}(VI_0),$$

and that (2.7) implies that

(2.7)′
$$\varepsilon_{\text{eff}}E = \frac{4\pi i}{\omega}\operatorname{rot}\operatorname{rot}(VI_0) - \frac{4\pi i}{\omega}I_0\delta.$$

To calculate (2.8) we note first that it is obvious that V is a function of $|x|$ only. It is therefore sufficient to consider for $x_1 > 0$

$$V(x_1, 0, 0) = (2\pi)^{-3} \int \frac{e^{ix_1\xi_1}}{(\xi, \xi) - k^2} \, d\xi = (2\pi)^{-2} \int_0^\infty \int_{-\infty}^\infty \frac{e^{itx_1}}{t^2 + r^2 - k^2} r \, dr \, dt.$$

We integrate first with respect to t. Set $z = \sqrt{k^2 - r^2}$ where the square root is defined as a number in the first quadrant, just as k. In the upper half plane e^{itx_1} is ≤ 1 in absolute value, and the integrand has its only pole at the point $t = z$, with residue $e^{izx_1}/2z$. Hence we obtain

$$V(x_1, 0, 0) = \frac{i}{2\pi} \int_0^\infty e^{izx_1} r \, dr / 2z.$$

Now we have $zdz = -rdr$, which gives

$$V(x_1, 0, 0) = -\frac{i}{4\pi} \int e^{izx_1} \, dz,$$

where z describes a contour from k to $+i\infty$. Hence we obtain

$$V(x_1, 0, 0) = e^{ikx_1}/4\pi x_1,$$

which gives

(2.9)
$$V(x) = \frac{e^{ik|x|}}{4\pi|x|}.$$

Here we can let $\sigma \to 0$ without difficulty, if we wish, and (2.9) remains valid also in the case of conductivity 0. (The formula (2.9) is of course the Hertz potential, the fundamental solution of the operator $\Delta + k^2$.) Inserting (2.9) in (2.6)' and (2.7)' now gives the usual Hertz solution.

We shall now derive an alternative expression for the Hertz potential which was used by Weyl to study the wave reflexion against a homogeneous half space (see Weyl [4]). The idea is that in (2.8) we only calculate the Fourier transform with respect to one variable, for example ξ_3; then the Fourier transform of V with respect to x_1 and x_2 for fixed x_3 is left under the integral sign, which is advantageous if one wants to introduce an inhomogeneity bounded by a plane $x_3 =$ constant. Set $x' = (x_1, x_2, 0)$ and $\xi' = (\xi_1, \xi_2, 0)$. The integral with respect to ξ_3 in (2.8) is calculated as before, which gives

(2.10)
$$V(x) = \frac{i}{8\pi^2} \int \frac{\exp(i(x', \xi') + ix_3\sqrt{k^2 - (\xi', \xi')})}{\sqrt{k^2 - (\xi', \xi')}} \, d\xi',$$

where the square root again shall be chosen in the first quadrant. Set $\xi_3 = \sqrt{k^2 - (\xi', \xi')}$, that is, ξ_3 is determined by the conditions

(2.11)
$$(\xi, \xi) = k^2, \quad \mathrm{Im}\, \xi_3 \geq 0, \quad \mathrm{Re}\, \xi_3 \geq 0.$$

Then we can write (2.10) in the form

(2.10)'
$$V(x) = \frac{i}{8\pi^2} \int e^{i(x, \xi)} \xi_3^{-1} \, d\xi', \quad x_3 > 0.$$

Thus in the half plane $x_3 > 0$ we have a decomposition of V into purely harmonic solutions of the equation $\Delta + k^2 = 0$, and a corresponding decomposition of H and E into harmonic solutions of Maxwell's equations

$$(2.12) \qquad H = -\frac{1}{2\pi c}\int \xi \times I_0 e^{i(x,\xi)}d\xi'/\xi_3; \quad x_3 > 0;$$

$$(2.13) \qquad \varepsilon_{\text{eff}}E = \frac{1}{2\pi\omega}\int \xi \times (\xi \times I_0) e^{i(x,\xi)}\, d\xi'/\xi_3; \quad x_3 > 0.$$

Formally one can consider $d\xi'/\xi_3$ as the surface element on the sphere with radius k, divided by k. This is the interpretation of Weyl. Note, however, that the integration to a large extent takes place over a complex sphere even if k is real. The contributions when $(\xi',\xi') > k^2$ then correspond to rapidly decaying oscillations.

3. Reflection against a homogeneous half space. We assume now that each of the half spaces $x_3 > 0$ and $x_3 < 0$ is homogeneous with electrical constants μ^+, μ^- and so on. With a source placed as before at a point in the half space $x_3 < 0$ we shall calculate the resulting field. In doing so we require that the solution shall be bounded at infinity and, during the derivation, that the conductivities σ^+ and σ^- are both $\neq 0$. The problem is thus to find for the field which is determined by (2.12), (2.13) a "reflected" additional field in the half space $x_3 < 0$ and a refracted field in the half space $x_3 > 0$ so that the discontinuity conditions are fulfilled. In view of the decomposition (2.12), (2.13) it is sufficient to consider a plane wave

$$H = \xi \times A e^{i(x,\xi)}; \quad (\xi,\xi) = k_-^2, \quad \operatorname{Im}\xi_3 > 0$$

$$E = -\frac{c}{\omega\varepsilon_{\text{eff}}}\xi \times (\xi \times A)e^{i(x,\xi)}$$

in the half space $x_3 < 0$. We make an analogous "ansatz" for the reflected wave $\check{H}$, $\check{E}$ but with ξ replaced by $\check{\xi} = (\xi_1,\xi_2,-\xi_3)$ and for the refracted wave H^*, E^* with ξ replaced by $\xi^* = (\xi_1,\xi_2,\xi_3^*)$ where $\operatorname{Im}\xi_3^* > 0$ and $(\xi^*,\xi^*) = k_+^2$;[8] the vector A is replaced in these waves by $\check{A}$ and A^* respectively. Since H and E are not changed if a multiple of ξ is added to A, we may assume that

$$(3.1) \qquad (A,\xi) = (\check{A},\check{\xi}) = (A^*,\xi^*) = 0,$$

and the expression for E is then simplified to

$$E = \frac{ck_-^2}{\omega\varepsilon_{\text{eff}}}A e^{i(x,\xi)} = \frac{\omega\mu^-}{c}A e^{i(x,\xi)}$$

and similarly for $\check{E}$ and E^*. If we set $N = (0,0,1)$ then the continuity of the tangential components of E gives that

$$(3.2) \qquad \mu^-(A + \check{A}) = \mu^+ A^* + \lambda N$$

for some λ. Let $a = (A,N)$ and define $\check{a}$, a^* similarly. Then we obtain from the continuity of the normal component of $\varepsilon_{\text{eff}}E$ that

$$(3.3) \qquad k_-^2(a + \check{a}) = k_+^2 a^*.$$

---

[8]The motivation for this ansatz is that the refracted wave shall be bounded in the half plane $x_3 > 0$ and purely harmonic in the boundary variables.

The corresponding relations for the magnetic field become

$$(3.4) \qquad \xi \times A + \breve{\xi} \times \breve{A} = \xi^* \times A^* + \nu N$$

for some ν, and a corresponding equation for the normal components. However, this equation follows at once from (3.2) if we take the vector prodeuct of both sides with $\xi' = (\xi_1, \xi_2, 0)$ and note that $\xi - \xi'$, $\breve{\xi} - \xi'$ and $\xi^* - \xi'$ are all proportional to N. In the same way (3.3) and (3.4) follow using (3.2). For a given A satisfying (3.1) the equations (3.1) – (3.4) for $\breve{A}$ and A^* therefore contain at most 6 linearly independent equations for 6 unknowns. If we prove that they have at most one solution it will follow that they do have a solution so that a result obtained by formal calculations must in fact satisfy all the equations.

Note now that (3.1) implies that

$$(3.1)' \qquad (A, \xi') + a\xi_3 = (\breve{A}, \xi') - \breve{a}\xi_3 = (A^+, \xi') + a^*\xi_3^* = 0.$$

If we take the scalar product of (3.2) with ξ' we thus obtain the equation

$$(3.5) \qquad \mu^-(-a\xi_3 + \breve{a}\xi_3) + \mu^+ a^*\xi_3^* = 0.$$

If this equation is combined with (3.3) we obtain

$$(3.6) \qquad 2a = (k_+^2/k_-^2 + \mu^+ \xi_3^*/\mu^- \xi_3)a^*$$

$$(3.7) \qquad 2\breve{a} = (k_+^2/k_-^2 - \mu^+ \xi_3^*/\mu^- \xi_3)a^*.$$

Note that the parenthesis in the right-hand side of (3.6) is $\neq 0$ since both terms have a positive real part. Hence the equations determine a^* and $\breve{a}$ uniquely in terms of a, and this also determines the constant λ in (3.2),

$$(3.8) \qquad \lambda = \mu^-(a + \breve{a}) - \mu^+ a^* = (\mu^- k_+^2/k_-^2 - \mu^+)a^*.$$

Since we are not interested in explicit expressions for the remaining components of $\breve{A}$ and A^* it suffices as pointed out above to prove the uniqueness of the solution of the equations (3.1) – (3.4). Thus suppose for a moment that $A = 0$; then it follows from (3.6) – (3.8) that $\breve{a} = a^* = \lambda = 0$. By (3.1) $\breve{A}$ and A^* are thus orthogonal to ξ' and to N, so $\xi' \times \breve{A}$ and $\xi' \times A^*$ have the direction N. The tangential component of the equation (3.4) is therefore $(\xi_3 \breve{A} + \xi_3^* A^*) \times N = 0$, hence $\xi_3 \breve{A} + \xi_3^* A^* = 0$. Combined with the equation (3.2) which gives $\mu^- \breve{A} - \mu^+ A^* = 0$ it follows now that $\breve{A} = A^* = 0$, for $\mu^- \xi_3^* + \mu^+ \xi_3$ has a positive real part.

Summing up we see that the refracted and the reflected waves are uniquely determined and that the x_3 component E_3 of the refracted wave is given by

$$(3.9) \qquad E_3 = \frac{\omega\mu^+}{c} a^* e^{i(x,\xi^*)},$$

where a^* is determined by (3.6). We shall now use this to study the field generated by a source I^0 at a point x^0 with $x_3^0 < 0$. The resulting field shall be bounded at infinity, so the Fourier transform in the variables x_1 and x_2 shows that it must be given by replacing x by $x - x^0$ and k by k_- in (2.12), adding to every harmonic component a reflected wave in the half space $x_3 < 0$ and replacing it with a refracted wave in the half space $x_3 > 0$.

Since these shall be bounded they are uniquely determined by the discussion above, and the refracted wave E^* in the half space $x_3 > 0$ gets the vertical component

$$(3.10) \qquad E_3^* = -\frac{\omega\mu^+}{\pi c^2} \int \frac{I_3^0 - (I^0, \xi)\frac{\xi_3}{k_-^2}}{\frac{k_+^2}{k_-^2} + \frac{\mu^+\xi_3^*}{\mu^-\xi_3}} e^{i((x,\xi^*)-(x^0,\xi))} d\xi'/\xi_3.$$

(We have here applied (3.9) to every Fourier component of (2.12), with

$$A = -(2\pi c)^{-1}(I^0 - \xi(I^0,\xi)/(\xi,\xi)),$$

in order to satisfy the normalization (3.1).) If we have a source with continuous density $I(y)$ in the half space $y_3 < 0$ we obtain by integration of (3.10)

$$(3.11) \qquad E_3^* = -\frac{\omega\mu^-}{\pi c^2} \int (\hat{I}_3(\xi) - (\hat{I}(\xi),\xi)\xi_3/k_-^2)(\mu^-k_+^2/\mu^+k_-^2 + \xi_3^*/\xi_3)^{-1}e^{i(x,\xi^*)} d\xi'/\xi_3,$$

where $\hat{I}$ is the Fourier transform of I.

4. The perturbation equation. Let us again consider a solution of Maxwell's equations, in a medium with variable electrical properties,

$$\text{rot } E - \frac{i\omega\mu}{c}H = 0$$
$$\text{rot } H + \frac{i\omega}{c}\varepsilon_{\text{eff}}E = \frac{4\pi}{c}I.$$

We assume at first that μ and ε_{eff} are continuous; the limiting case of surfaces of discontinuity for these quantities implies the usual discontinuity relations. Let us now make an infinitesimal variation $\tilde{\varepsilon}_{\text{eff}}$ of ε_{eff} while we keep I, ω and μ unchanged. If $E + \tilde{E}$, $H + \tilde{H}$ is the perturbed solution of Maxwell's equations it follows by differentiation of them — assuming that the perturbation is small and that the derivative exists — that

$$\text{rot } \tilde{E} - \frac{i\omega\mu}{c}\tilde{H} = 0$$
$$\text{rot } \tilde{H} + \frac{i\omega}{c}\varepsilon_{\text{eff}}\tilde{E} = -\frac{i\omega}{c}\tilde{\varepsilon}_{\text{eff}}E = \frac{4\pi}{c}\frac{-i\omega}{4\pi}\tilde{\varepsilon}_{\text{eff}}E.$$

Hence $\tilde{E}$, $\tilde{H}$ are equal to the field generated by currents with the density $(-i\omega/4\pi)\tilde{\varepsilon}_{\text{eff}}E$. The boundary condition is still boundedness at infinity, and if we again accept surfaces of discontinuity for the electrical constants we obtain there the usual discontinuity conditions. If we prescribe E and H, and the unperturbed medium as in Section 3 consists of two homogeneous half spaces which are only perturbed in the half space $x_3 < 0$, then the perturbation of the field in the half space $x_3 > 0$ can now be obtained from (3.11). To study the variations of the field which emanate from a distant dipole in the air it should be sufficient to make as ansatz for the unperturbed fields E and H an almost tangentially incident plane wave in the half space $x_3 > 0$, superimposed on the reflected wave and in the half space $x_3 < 0$ replaced by the refracted plane wave. These are determined according to Section 3, which thus in principle gives an answer to the question about the perturbation of the vertical component of the electrical field. The numerical difficulties in exploiting these formulas are very large, however, unless one is not at a distance from the perturbations of so many wave lengths that the known asymptotic formulas for the solution (3.10) can be exploited. (See Weyl [4], Bremmer [1], pp. 520–531.)

References

[1] H. Bremmer, *Propagation of electromagnetic waves*, Handbuch der Physik **16**(1958), pp. 423–639.

[2] A. Sommerfeld, *Über die Ausbreitung der Wellen in der drahtlosen Telegraphie*, Ann. d. Phys. **28** (1909), 665–736.

[3] _____, *Über die Ausbreitung der Wellen in der drahtlosen Telegraphie*, Ann. d. Phys. **81** (1926), 1135–1153.

[4] H. Weyl, *Ausbreitung elektromagnetischer Wellen über einem ebenen Leiter*, Ann. d. Phys. **60** (1919), 481–500.

CLASSES OF INFINITELY DIFFERENTIABLE FUNCTIONS

JAN BOMAN AND LARS HÖRMANDER

1. Definitions and general properties.

It is well known that the set of analytic functions of a real variable can be characterized by the growth of the derivatives with the order of differentiation. A real or complex valued function u defined in a closed bounded interval $I \subset \mathbf{R}$ is said to be analytic if there exists an analytic function u_1 defined in a *complex* neighborhood of I such that $u_1 = u$ on I. That u is analytic in I is equivalent to the existence of constants C_1 and C_2 such that

$$\sup_{x \in I} |u^{(k)}(x)| \leq C_1 C_2^k k!, \quad k = 0, 1, 2, \dots.$$

We shall study other classes of functions which are defined by the growth of the derivatives with the order of differentiation. Denote by $C^\infty(I)$ the set of functions which are infinitely differentiable in a closed subinterval I of $\mathbf{R}$ which may be equal to $\mathbf{R}$. Let M be an arbitrary sequence of positive numbers M_k and set

$$C_M(I) = \{u \mid u \in C^\infty(I), \exists C_1, C_2 \text{ such that } \sup_{x \in I} |u^{(k)}(x)| \leq C_1 C_2^k M_k, \ k = 0, 1, 2, \dots \}.$$

EXAMPLE. Take $M_k^{(\alpha)} = (k!)^\alpha$ where $\alpha \geq 1$. $C_{M^{(1)}}$ is equal to the class of analytic functions, and for $\alpha > 1$ new classes of functions are obtained which occur in the theory of partial differential equations. For example, all solutions of the heat equation are elements of $C_{M^{(2)}}$. As examples of functions in $C_{M^{(\alpha)}}$ we list the function f defined by

$$f(x) = \begin{cases} 0, & \text{for } x \leq 0, \\ e^{-x^{-a}}, & \text{for } x > 0, \end{cases} \quad (a > 0).$$

One can prove that $f \in C_{M^{(\alpha)}}$ if and only if $\alpha \geq 1 + a^{-1}$.

Different sequences can define the same class. If $I = \mathbf{R}$ every class can be defined with a logarithmically convex sequence M. In fact, by an inequality of Kolmogorov it is true that if $F_k = \sup_{x \in \mathbf{R}} |f^{(k)}(x)|$, where $f(x)$ is an arbitrary function in $C^\infty(\mathbf{R})$, then

(1.1) $$F_k \leq 2 F_m^{\frac{n-k}{n-m}} F_n^{\frac{k-m}{n-m}}, \quad n < k < m.$$

This inequality implies that F_k is "almost" logarithmically convex, for if the factor 2 did not occur in the right-hand side, then (1.1) would precisely mean logarithmic convexity. Using (1.1) it is easy to prove that $C_M(\mathbf{R}) = C_{\overline{M}}(\mathbf{R})$, where $\overline{M} = (\overline{M}_k)$ is the largest logarithmically convex minorant sequence of M. (See Bang, Om Quasi-Analytiske Funktioner, page 13).

© Springer International Publishing AG, part of Springer Nature 2018
L. Hörmander, *Unpublished Manuscripts*,
https://doi.org/10.1007/978-3-319-69850-2_11

Note that in general it is not true that $C_M(I) = C_{\overline{M}}(I)$ when I is a finite interval. Consider for example the class $C_M(I)$ where $M_k = e^{-k^2}$ for all k. In this case $\overline{M}_k = 0$ for $k > 0$. Thus every non-constant polynomial is in $C_M(I)$ without belonging to $C_{\overline{M}}(I)$.

However, also when I is a finite interval we can define all important classes using logarithmically convex sequences. We shall therefore in what follows only use logarithmically convex sequences M. We also make the convention that $M_0 = 1$, which is no restriction of the generality, and since M_k/M_1^k is increasing and defines the same class as M_k (see below), we can assume that M_k is increasing.

The question which sequences define the same class is completely solved for logarithmically convex sequences by the following theorem.

THEOREM 1.1. *Let I be an arbitrary interval and L and M two logarithmically convex sequences. Then*

$$(1.2) \qquad\qquad C_L(I) \subset C_M(I),$$

if and only if there is a constant C such that

$$(1.3) \qquad\qquad L_k \leq C^k M_k \quad \text{for every } k.$$

For the proof of Theorem 1.1 we need the following lemma.

LEMMA 1.1. *Let M be a logarithmically convex sequence and set*

$$h(r) = \sup_k r^k/M_k, \quad r > 0.$$

Then

$$M_k = \sup_{r>0} r^k/h(r), \quad k = 0, 1, 2, \dots.$$

PROOF. We shall actually prove that for an arbitrary positive sequence M

$$(A) \qquad\qquad M_k \geq \sup_{r>0} r^k/h(r), \quad k = 0, 1, 2, \dots,$$

and that if M is logarithmically convex then in addition

$$(B) \qquad\qquad M_k \leq \sup_{r>0} r^k/h(r), \quad k = 0, 1, 2, \dots.$$

(A) For arbitrary k and $r > 0$ we have

$$h(r) \geq r^k/M_k, \quad \text{that is,} \quad M_k \geq r^k/h(r),$$

which gives (A).

(B) if M is logarithmically convex then we have for every s in the interval $M_k/M_{k-1} \leq s \leq M_{k+1}/M_k$

$$M_k s^{j-k} \leq M_j \quad \text{for every } j, \quad \text{that is,} \quad s^j/M_j \leq s^k/M_k \quad \text{for every } j,$$

whence $h(s) \leq s^k/M_k$. Hence

$$M_k \leq s^k/h(s) \leq \sup_{r>0} r^k/h(r),$$

which proves (B).

PROOF OF THEOREM 1.1. Assume first that $L_k \leq C^k M_k$ for every k. If $f \in C_L(I)$, that is, $\sup_{x \in I} |f^{(k)}(x)| \leq C_1 C_2^k L_k$ for suitable constants C_1, C_2, then $\sup_{x \in I} |f^{(k)}(x)| \leq C_1 (CC_2)^k M_k$, that is, $f \in C_M(I)$. This proves that $C_L(I) \subset C_M(I)$.

To prove the other part of the theorem we shall consider the set

$$B = \{u \mid \exists C \text{ such that } \sup_{x \in I} |u^{(k)}(x)| \leq CL_k \text{ for all } k\}.$$

In B we introduce the norm

$$\|u\|_B = \sup_{x \in I, k} |u^{(k)}(x)|/L_k.$$

B is a complete normed space, that is, a Banach space. Set for $j = 1, 2, \ldots$

$$V_j = \{u \mid u \in B, \sup_{x \in I} |u^{(k)}(x)| \leq j^{k+1} M_k \text{ for all } k\}.$$

$V_1, V_2, \ldots$ are closed subsets of the space B. Our hypothesis that $C_L(I) \subset C_M(I)$ means that $\cup_{j=1}^{\infty} V_j = B$. By Baire's theorem (Bourbaki, Topologie générale Chap. IX, p. 75) some set V_j must have an interior point. Since all sets $V_1, V_2, \ldots$ are symmetrical with respect to the origin and convex, the origin must be an interior point of V_j. This means that there is some $\delta > 0$ such that

(1.4) $$\|u\|_B \leq \delta \implies u \in V_j.$$

If we set $h_L(\xi) = \sup_k \xi^k / L_k$ for $\xi > 0$ and take

$$u(x) = \delta (h_L(\xi))^{-1} e^{ix\xi},$$

then $\|u\|_B = \delta$ whence by (1.4)

$$\delta (h_L(\xi))^{-1} \xi^k \leq j^{k+1} M_k \text{ for all } \xi > 0 \text{ and } k \geq 0.$$

If we now take the supremum over all $\xi > 0$ and use Lemma 1.1, we obtain

$$L_k \leq \delta^{-1} j^{k+1} M_k \text{ for all } k \geq 0.$$

Since $L_0 = M_0 = 1$, (1.3) follows if $C \geq j^2/\delta$ and $C \geq j$. Theorem 1.1 is thereby proved.

COROLLARY 1.1. *The smallest class containing $C_L(I)$ and $C_M(I)$ is $C_{\sup(L,M)}(I)$.*

THEOREM 1.2. *If M and L are two logarithmically convex sequences, I an arbitrary interval, $f \in C_L(I)$ and $g \in C_M(I)$, then $fg \in C_{\sup(L,M)}(I)$.*

PROOF. We use the notation $D^k = (d/dx)^k$. The hypothesis means that

(1.5) $$\sup_{x \in I} |D^k f(x)| \leq C_1 C_2^k L_k, \quad k = 0, 1, 2, \ldots,$$

(1.6) and $$\sup_{x \in I} |D^k g(x)| \leq C_1 C_2^k M_k, \quad k = 0, 1, 2, \ldots,$$

for suitable constants C_1 and C_2. Leibnitz' formula reads:

$$(1.7) \qquad D^k(fg) = \sum_{j=0}^{k} \binom{k}{j} D^j f D^{k-j} g.$$

From (1.5), (1.6), and (1.7) we obtain:

$$(1.8) \qquad \sup_{x \in I} |D^k(fg)| \le \sum_{j=0}^{k} \binom{k}{j} C_1 C_2^j L_j C_1 C_2^{k-j} M_{k-j}, \quad k = 0, 1, 2, \ldots.$$

Since L and M are logarithmically convex sequences and $L_0 = M_0 = 1$ we have

$$L_j \le L_0^{(k-j)/k} L_k^{j/k} = L_k^{j/k}, \quad 0 < j < k,$$

and correspondingly for M_j. Now (1.8) becomes:

$$\sup_{x \in I} |D^k(fg)| \le C_1^2 C_2^k \sum_{j=0}^{k} \binom{k}{j} (L_k^{1/k})^j (M_k^{1/k})^{k-j} = C_1^2 C_2^k (L_k^{1/k} + M_k^{1/k})^k$$

$$\le C_1^2 (2C_2)^k \sup(L_k, M_k), \quad k = 0, 1, 2, \ldots.$$

Theorem 1.2 is thereby proved.

If f and g belong to the same class $C_M(I)$, where M is logarithmically convex, the theorem shows that $fg \in C_M(I)$. Hence we have:

COROLLARY 1.2. *If M is a logarithmically convex sequence and I an arbitrary interval, then $C_M(I)$ is a ring.*

2. The Denjoy-Carleman theorem.

We shall now treat the unique continuation problem for a class. Differently stated: for which classes $C_M(I)$ is a function in $C_M(I)$ uniquely determined by all its derivatives at a point?

In what follows we shall say that a function $u \in C^\infty(I)$ vanishes of infinite order at a point $x \in I$ if $u^{(k)}(x) = 0$ for every k.

DEFINITION. The class $C_M(I)$ is said to be quasianalytic if every $u \in C_M(I)$ vanishing of infinite order at some point in I must vanish identically in I.

It is well known that the analytic class is quasianalytic in this sense. From the examples at the beginning of Section 1 it follows directly that the class $C_{M^{(\alpha)}}$ is not quasianalytic if $\alpha > 1$. The problem to characterize the sequences M for which $C_M(I)$ is quasianalytic was posed by Hadamard already in 1912. The problem was completely solved by the Denjoy-Carleman theorem which we shall now prove for classes defined by logarithmically convex sequences M. The first complete proof of this theorem has been given by Carleman (Fonctions quasi analytiques (1926)). The sufficiency of condition (b) for quasianalyticity had been proved earlier by Denjoy (1921).

THEOREM 2.1 (THE DENJOY-CARLEMAN THEOREM). *Let I be a finite or infinite subinterval of $\mathbf{R}$, and M a logarithmically convex sequence of positive numbers. Then the class $C_M(I)$ is quasianalytic if and only if the following equivalent conditions are fulfilled:*

(a)
$$\int_{-\infty}^{\infty} \log\left(\sum_{k=0}^{\infty} \frac{|\xi|^k}{M_k}\right) \frac{d\xi}{1+\xi^2} \quad \text{is divergent}$$

(b)
$$\sum_{k=0}^{\infty} \frac{1}{M_k^{1/k}} \quad \text{is divergent}$$

(c)
$$\sum_{k=1}^{\infty} \frac{M_{k-1}}{M_k} \quad \text{is divergent.}$$

Note that the conditions for $C_M(I)$ to be quasianalytic depend only on the squence M and not on the interval I.

There are plenty of quasianalytic classes which are strictly larger than the quasianalytic class. Such a class is obtained by taking $M_0 = M_1 = 1$, $M_k = (k \log k)^k$ for $k = 2, 3, \ldots$. That the class defined by this sequence is strictly larger than the analytic class follows from Theorem 1.1.

Since the proof of Theorem 2.1 depends on the Fourier transformation we need to study conditions on the Fourier transform

$$\hat{f}(\xi) = \int_{-\infty}^{\infty} f(x)e^{-ix\xi}\,dx$$

of a function f in order that f shall belong to $C_M(\mathbf{R})$. Set

(2.1)
$$q_M(\xi) = \sum_{k=0}^{\infty} \frac{|\xi|^k}{M_k}.$$

Assume first that this series has a finite radius of convergence. Then $\underline{\lim}\, M_k^{1/k} = C < \infty$. But by the logarithmic convexity $M_k^{1/k}$ is increasing so it follows that $M_k^{1/k} \leq C$, or $M_k \leq C^k$ for every k, which proves that $C_M(I)$ is contained in the analytic class, and $C_M(I)$ is therefore obviously quasianalytic. In the proof of Theorem 2.1 we may therefore assume that the series (2.1) converges for all ξ.

We denote by $C_0^\infty(\mathbf{R})$ the set of functions in $C^\infty(\mathbf{R})$ which have compact support, that is, vanish outside a finite subinterval of $\mathbf{R}$.

LEMMA 2.1. *If $f \in C_0^\infty(\mathbf{R}) \cap C_M(\mathbf{R})$ then there are constants $a > 0$ and C such that $|\hat{f}(\xi)q_M(a\xi)| \leq C$ for all ξ.*

PROOF. We start from the identity

$$\xi^k \hat{f}(\xi) = \int_{-\infty}^{\infty} (D/i)^k f(x)e^{-ix\xi}\,dx.$$

Since f has compact support and $f \in C_M(\mathbf{R})$ this gives $|\xi|^k|\hat{f}(\xi)| \leq C_1 C_2^k M_k$, whence with $a = 1/2C_2$

$$q_M(a\xi)|\hat{f}(\xi)| = \left(\sum_{k=0}^{\infty} \frac{|a\xi|^k}{M_k}\right)|\hat{f}(\xi)| \leq C_1 \sum_k 2^{-k} = 2C_1,$$

which proves the lemma.

LEMMA 2.2. *If $|f(x)q_M(ax)(1+x^2)| \leq C$ where $a > 0$ and C are constants, then $\hat{f} \in C_M(\mathbf{R})$.*

PROOF. By the formula

$$(iD)^k \hat{f}(\xi) = \int_{-\infty}^{\infty} f(x)x^k e^{-ix\xi}\, d\xi$$

it follows from the hypotheses in the lemma that

$$\sup_{\xi \in \mathbf{R}} |D^k \hat{f}(\xi)| \leq C \int_{-\infty}^{\infty} \frac{|x|^k\, dx}{q_M(ax)(1+x^2)} \leq Ca^{-k}M_k \int_{-\infty}^{\infty} \frac{dx}{1+x^2}$$

$$\leq C_1 a^{-k} M_k, \quad \text{for } k = 0, 1, 2, \ldots,$$

that is, $\hat{f} \in C_M(\mathbf{R})$.

For the proof of Theorem 2.1 we also need the following two lemmas which give a characterization of the boundary values of a function which is analytic in a half plane.

LEMMA 2.3. *Let ϕ be a positive continuously differentiable function on $\mathbf{R}$ such that $\phi(\xi)(1+\xi^2)$ is bounded and*

$$\int_{-\infty}^{\infty} \frac{\log \phi(\xi)}{1+\xi^2}\, d\xi > -\infty.$$

Then there is a function F defined in the half plane $\operatorname{Im} \zeta \geq 0$ such that F is analytic when $\operatorname{Im} \zeta > 0$, continuous when $\operatorname{Im} \zeta \geq 0$, $|F(\xi)| = \phi(\xi)$ for real ξ, and $|F(\zeta)| \leq C(1+|\zeta|^2)^{-1}$ for $\operatorname{Im} \zeta \geq 0$.

LEMMA 2.4. *If F is analytic for $\operatorname{Im} \zeta > 0$, continuous and bounded for $\operatorname{Im} \zeta \geq 0$, then*

$$\int_{-\infty}^{\infty} \frac{\log |F(\xi)|}{1+\xi^2}\, d\xi > -\infty.$$

PROOF OF LEMMA 2.3. Let ϕ be a function which fulfills the hypotheses of Lemma 2.3. We shall construct a function F which has all the properties stated in Lemma 2.3 and is different from zero everywhere. For such an F we have

$$\log F(\zeta) = \log |F(\zeta)| + i \arg F(\zeta).$$

Here $\log F(\zeta)$ is analytic, and hence the functions $u(\xi, \eta) = \log |F(\zeta)|$ and $v(\xi, \eta) = \arg F(\zeta)$ are conjugate harmonic functions of ξ and η, where $\zeta = \xi + i\eta$. For real ξ we must have

(2.2) $$\log |F(\xi)| = u(\xi, 0) = \log \phi(\xi).$$

A harmonic function u satisfying this condition can be determined by means of Poisson's formula. Set

$$u(\xi, \eta) = \frac{1}{\pi} \int_{-\infty}^{\infty} \frac{\eta \log \phi(\xi')\, d\xi'}{(\xi - \xi')^2 + \eta^2}$$

for $\eta > 0$, and let $u(\xi, \eta)$ be defined by (2.2) for $\eta = 0$. Then u is a harmonic function for $\eta > 0$. One can easily prove that u becomes continuously differentiable for $\eta \geq 0$ since

ϕ is continuously differentiable. From the Cauchy-Riemann equations it follows that the conjugate harmonic function v becomes continuous for $\eta \geq 0$. Hence the function

$$F(\zeta) = e^{u(\xi,\eta)+iv(\xi,\eta)}$$

is analytic for $\operatorname{Im} \zeta > 0$ and continuous for $\operatorname{Im} \zeta \geq 0$. It remains to derive an estimate for F. Since $\phi(\xi) \leq C(1+\xi^2)^{-1}$ we have

$$(2.3) \qquad u(\xi,\eta) \leq \frac{1}{\pi} \int_{-\infty}^{\infty} \frac{\eta \log(C(1+\xi'^2)^{-1})\, d\xi'}{(\xi-\xi')^2 + \eta^2} = w(\xi,\eta),$$

for $\eta > 0$. Here $w(\xi,\eta)$ is the Poisson integral of $\log(C(1+\xi^2)^{-1})$, so w is a harmonic function in the upper half space with the boundary values $w(\xi,0) = \log(C(1+\xi^2)^{-1})$ for real ξ. Such a function is $\log(C|\zeta+i|^{-2})$, where $\zeta = \xi + i\eta$. Using a suitable uniqueness theorem for the Dirichlet problem in a half plane we thus obtain

$$w(\xi,\eta) = \log(C|\zeta+i|^{-2}), \quad \eta > 0,$$

which is also easily obtained by residue calculus. Hence we have $u(\xi,\eta) \leq \log(C|\zeta+i|^{-2})$, $\eta > 0$, by (2.3), which gives

$$|F(\zeta)| = e^{u(\xi,\eta)} \leq C|\zeta+i|^{-2} \leq C(1+|\zeta|^2)^{-1}.$$

PROOF OF LEMMA 2.4. Let F be a function satisfying the hypotheses of Lemma 2.4. The function $\zeta = i(1-w)(1+w)^{-1}$ maps the disc $|w| < 1$ on the half plane $\operatorname{Im} \zeta > 0$. Denote by G the function $w \mapsto F(\zeta(w))$. $G(w)$ is analytic for $|w| < 1$ and continuous for $|w| \leq 1$ except for $w = -1$. We shall prove that

$$(2.4) \qquad \int_0^{2\pi} \big| \log|G(e^{i\theta})| \big|\, d\theta < \infty.$$

Let us assume for a moment that (2.4) holds. Then we can again introduce ξ as integration variable in (2.4), which changes this integral after an easy computation to

$$\int_{-\infty}^{\infty} \frac{|\log|F(\xi)||}{1+\xi^2}\, d\xi < \infty.$$

It only remains to prove (2.4). To do so we shall use Jensen's formula.

   Jensen's formula. Let $g(z)$ be analytic for $|z| < R$. Assume that $g(0) \neq 0$, and let $z_1, z_2, \ldots$ be the zeros of $g(z)$ situated in the disc $|z| < R$, repeated as many times as the multiplicity and ordered so that $|z_1| \leq |z_2| \leq |z_3| \ldots$. Then we have, if $|z_n| \leq r \leq |z_{n+1}|$,

$$\log \frac{r^n |g(0)|}{|z_1|\,|z_2| \ldots |z_n|} = \frac{1}{2\pi} \int_0^{2\pi} \log|g(re^{i\theta})|\, d\theta.$$

If $G(0) \neq 0$ we can apply the theorem to $g = G$ and obtain

$$\int_0^{2\pi} \log|G(re^{i\theta})|\, d\theta \geq 2\pi \log|G(0)|, \quad r < 1.$$

If G has a zero with multiplicity m for $w = 0$, we apply Jensen's formula instead to the function $G(w)/w^m$ and obtain

$$\int_0^{2\pi} \log \left| \frac{G(re^{i\theta})}{r^m e^{im\theta}} \right| d\theta \geq 2\pi \log(|G^{(m)}(0)/m!|).$$

Hence

$$(2.5) \qquad \int_0^{2\pi} \log |G(re^{i\theta})| \, d\theta \geq 2\pi m \log r + 2\pi \log(G^{(m)}(0)/m!) \geq A,$$

where A is independent of r for $\frac{1}{2} < r < 1$. By hypothesis $|F(\zeta)|$ is bounded. It is no restriction to assume that $|F(\zeta)| \leq 1$. Then also $|G(re^{i\theta})| \leq 1$ and hence $\log |G(re^{i\theta})| \leq 0$, so (2.5) can be written

$$\int_0^{2\pi} \big| \log |G(re^{i\theta})| \big| \leq -A, \quad \tfrac{1}{2} < r < 1.$$

When $r \to 1$ then $G(re^{i\theta})$ tends to $G(e^{i\theta})$ almost everywhere (for all θ except for $\theta = \pi$). By Fatous's lemma we therefore obtain

$$\int_0^{2\pi} \big| \log |G(e^{i\theta})| \big| \, d\theta \leq \lim_{r \to 1} \int_0^{2\pi} \big| \log |G(re^{i\theta})| \big| \, d\theta \leq -A.$$

Lemma 2.4 is thereby proved.

The interest in this connection of functions analytic and bounded in a half plane is shown by Lemmas 2.5 and 2.6.

LEMMA 2.5. *If $f \in C_0^\infty(\mathbf{R})$ and $f(x) = 0$ for $x \geq 0$, then the Fourier-Laplace transform*

$$\hat{f}(\zeta) = \int_{-\infty}^{\infty} f(x) e^{-ix\zeta} \, dx$$

is analytic and bounded when $\operatorname{Im} \zeta \geq 0$, that is, $\hat{f}$ satisfies the hypotheses of Lemma 2.4.

PROOF. $\hat{f}$ is analytic in the entire complex plane. That $\hat{f}$ is bounded when $\operatorname{Im} \zeta \geq 0$ follows from the fact that $e^{x\eta} \leq 1$ if $f(x) \neq 0$ and $\eta \geq 0$.

LEMMA 2.6. *If F is such a function as was constructed in Lemma 2.3 then $\widehat{F}(x) = 0$ for $x < 0$ (here $\widehat{F}$ denotes the Fourier transform of the restriction of F to the real axis).*

PROOF. By the definition of $\widehat{F}$

$$(2.6) \qquad \widehat{F}(x) = \int_{-\infty}^{\infty} F(\xi) e^{-ix\xi} \, d\xi.$$

By hypothesis F satisfies an estimate

$$(2.7) \qquad |F(\zeta)| \leq C(1 + |\zeta|^2)^{-1}, \quad \operatorname{Im} \zeta \geq 0.$$

Since F is analytic for $\operatorname{Im} \zeta \geq 0$ and since for arbitrary fixed x and $\eta > 0$ the integral

$$\left| \int_0^\eta F(\xi + i\eta_1) e^{-ix(\xi + i\eta_1)} \, d\eta_1 \right| \leq e^{|x\eta|} \int_0^\eta |F(\xi + i\eta_1)| \, d\eta_1$$

tends to zero when $\xi \to \infty$ in virtue of the estimate (2.7), we can move the integration path in (2.6) and obtain the formula

$$\widehat{F}(x) = \int_{-\infty}^{\infty} F(\xi + i\eta)e^{-ix(\xi + i\eta)}\, d\xi = e^{x\eta} \int_{-\infty}^{\infty} F(\xi + i\eta)e^{-ix\xi}\, d\xi \quad \text{when } \eta > 0.$$

Using (2.7) again we obtain $|\widehat{F}(x)| \leq Ce^{x\eta}$ where C is a constant which does not depend on η. Letting η tend to plus infinity for fixed $x < 0$ gives $\widehat{F}(x) = 0$.

PROOF OF THEOREM 2.1. I: (a) implies that $C_M(I)$ is quasianalytic. We assume that $C_M(I)$ is not quasianalytic and shall prove the negation of (a), that is, that

$$\int_{-\infty}^{\infty} (1 + \xi^2)^{-1} \log q_M(\xi)\, d\xi < \infty.$$

We prove first that if $C_M(I)$ is not quasianalytic then there is a function $f \in C_0^\infty(\mathbf{R}) \cap C_M(\mathbf{R})$ which does not vanish identically. It is no restriction to assume that I is the interval $-1 \leq x \leq 1$. If $C_M(I)$ is not quasianalytic there exists a function $g \in C_M(I)$ and a point $x_0 \in I$ such that $g^{(k)}(x_0) = 0$ for all k, and g is not identically zero. After a possible change of variables we may assume that $g(0) \neq 0$. Form the function $g_1(x) = g(x)g(-x)$, $-1 \leq x \leq 1$. Since $C_M(I)$ is closed under multiplication (Corollary 1.1) we have $g_1 \in C_M(I)$. Moreover, $g_1(0) \neq 0$ and $g_1^{(k)}(x_0) = g_1^{(k)}(-x_0) = 0$ for all k. Define a function f by

$$f(x) = \begin{cases} g_1(x), & |x| \leq |x_0| \\ 0, & |x| > |x_0|. \end{cases}$$

Then $f \in C_0^\infty(\mathbf{R}) \cap C_M(\mathbf{R})$ and f does not vanish identically, for $f(0) \neq 0$.

By a translation of the function constructed above we obtain a function f such that $f \in C_0^\infty(\mathbf{R}) \cap C_M(\mathbf{R})$, $f(x) = 0$ for $x \geq 0$ and $f \not\equiv 0$. This function satisfies the hypotheses of Lemma 2.5. It follows from this that the Fourier transform $\hat{f}$ satisfies the hypotheses of Lemma 2.4. Hence

(2.8)
$$\int_{-\infty}^{\infty} \frac{\log |\hat{f}(\xi)|}{1 + \xi^2}\, d\xi > -\infty.$$

But since $f \in C_0^\infty(\mathbf{R}) \cap C_M(\mathbf{R})$ there exist (by Lemma 2.1) constants $a > 0$ and C such that $|q_M(a\xi)\hat{f}(\xi)| \leq C$, thus

(2.9)
$$\log q_M(a\xi) \leq C - \log |\hat{f}(\xi)|.$$

From (2.8) and (2.9) it follows that

$$\int_{-\infty}^{\infty} \frac{\log q_M(a\xi)}{1 + \xi^2}\, d\xi < \infty.$$

The negation of (a) follows now by the change of variables $a\xi \to \xi$ and a trivial estimate.

II: If $C_M(I)$ is quasianalytic then (a) is valid. Assume that (a) is not fulfilled, that is, that $\int_{-\infty}^{\infty} (1 + \xi^2)^{-1} \log q_M(\xi)\, d\xi < \infty$. It must be proved that $C_M(I)$ is not quasianalytic. Take $\phi(\xi) = ((1 + \xi^2)q_M(\xi))^{-1}$. The function ϕ satisfies all hypotheses in Lemma 2.3. Hence there exists a function F such that F is analytic for $\text{Im}\,\zeta > 0$, continuous for

Im $\zeta \geq 0$, $|F(\xi)| = \phi(\xi)$ for real ξ, and $|F(\zeta)| \leq C(1+|\zeta|^2)^{-1}$ for Im $\zeta \geq 0$. But according to Lemma 2.6 these properties of F imply that $\widehat{F}(x) = 0$ for $x < 0$. Since $|F(\xi)| = \phi(\xi) \not\equiv 0$ we have $\widehat{F} \not\equiv 0$. Furthermore it follows from our choice of ϕ and Lemma 2.2 that $\widehat{F} \in C_M(\mathbf{R})$. The existence of a function $\widehat{F}$ with the stated properties shows that the class $C_M(\mathbf{R})$ is not quasianalytic. That $C_M(I)$ is not quasianalytic for any interval I is seen by making a suitable translation of $\widehat{F}$ and taking the restriction to I.

III: (b) $\implies$ (a). Assume that the integral (a) is convergent. Put $a_k = M_k^{1/k}$. When $\xi \geq ea_k$ then
$$\log q_M(\xi) \geq \log(\xi^k/M_k) \geq \log(e^k) = k,$$
and hence
$$\int_{ea_k}^{e\eta} \xi^{-2} \log q_M(\xi)\, d\xi \geq k \int_{ea_k}^{e\eta} \xi^{-2}\, d\xi = e^{-1}k(a_k^{-1} - \eta^{-1}).$$

From this we obtain
$$\sum_1^N a_k^{-1} = \sum_1^N k(a_k^{-1} - a_{k+1}^{-1}) + N a_{N+1}^{-1}$$
$$\leq e \sum_1^N \int_{ea_k}^{ea_{k+1}} \xi^{-2} \log q_M(\xi)\, d\xi + e \int_{ea_{N+1}}^{\infty} \xi^{-2} \log q_M(\xi)\, d\xi$$
$$\leq e \int_{a_0}^{\infty} \xi^{-2} \log q_M(\xi)\, d\xi,$$

which proves that $\sum a_k^{-1}$ is convergent.

IV: (a) $\implies$ (b). We assume that $\sum M_k^{-1/k} = \sum a_k^{-1}$ is convergent and shall prove that the integral (a) is convergent. Since partial integration gives

(2.10)
$$\int_1^x \xi^{-2} \log q_M(\xi)\, d\xi \leq \int_1^x \xi^{-2} U(\xi)\, d\xi,$$

where $U(\xi) = \left(\sum_{k=1}^{\infty} k(\xi/a_k)^k\right) / \left(\sum_{k=0}^{\infty} (\xi/a_k)^k\right)$, it is sufficient to prove the convergence of $\int_1^{\infty} \xi^{-2} U(\xi)\, d\xi$. Let κ be a positive number such that $\sum_{k=1}^{\infty} k\kappa^k < 1$. Then, if $0 < \xi < \kappa a_n$,
$$\sum_{k=n}^{\infty} k(\xi/a_k)^k \leq \sum_{k=1}^{\infty} k\kappa^k < 1 < n, \quad \text{and}$$
$$\sum_{k=1}^{n-1} k(\xi/a_k)^k \leq n \sum_{k=1}^{n-1} (\xi/a_k)^k < n \sum_{k=1}^{\infty} (\xi/a_k)^k.$$

By adding the preceding formulas one obtains
$$\sum_{k=1}^{\infty} k(\xi/a_k)^k < n \sum_{k=0}^{\infty} (\xi/a_k)^k, \quad \text{for } \xi \leq \kappa a_n,$$

that is, $U(\xi) < n$ for $\xi \leq \kappa a_n$. Using this inequality in the interval $\kappa a_{n-1} \leq \xi \leq \kappa a_n$ one obtains
$$\int_{\kappa a_0}^{\kappa a_N} \xi^{-2} U(\xi)\, d\xi \leq \sum_{n=1}^N n \int_{\kappa a_{n-1}}^{\kappa a_n} \xi^{-2}\, d\xi = \sum_{n=1}^N \kappa^{-1} n(a_{n-1}^{-1} - a_n^{-1})$$
$$= \kappa^{-1}\left(\sum_{n=0}^{N-1} a_n^{-1} - N a_N^{-1}\right) \leq \kappa^{-1} \sum_{n=0}^{N-1} a_n^{-1}.$$

Since $a_N \to \infty$ when $N \to \infty$ it follows that

$$\int_{\kappa a_0}^{\infty} \xi^{-2} U(\xi)\, d\xi \leq \kappa^{-1} \sum_{n=0}^{\infty} a_n^{-1},$$

which completes the proof.

V: (c) $\implies$ (b). Since $M_0 = 1$ and M is logarithmically convex, we have $M_{k-1} \leq M_k^{(k-1)/k}$ or $M_{k-1}/M_k \leq M_k^{-1/k}$ for all k, which proves that (c) $\implies$ (b).

To prove that (b) $\implies$ (c) we need the following lemma.

LEMMA 2.7 (CARLEMAN'S INEQUALITY). [9] *If $\sum b_n$ is a convergent series with non-negative terms which are not all equal to 0, then*

$$\sum_{k=1}^{\infty} (b_1 b_2 \cdots b_k)^{1/k} < e \sum_{k=1}^{\infty} b_k.$$

PROOF. Let $c_1, c_2, \ldots$ be positive constants which will be chosen later. By the inequality between arithmetic and geometric means

$$\sum_{k=1}^{\infty} (b_1 b_2 \cdots b_k)^{1/k} \leq \sum_{k=1}^{\infty} \left(\frac{c_1 b_1\, c_2 b_2 \cdots c_k b_k}{c_1 c_2 \cdots c_k} \right)^{1/k}$$

$$\leq \sum_{k=1}^{\infty} (c_1 c_2 \cdots c_k)^{-1/k} k^{-1} \sum_{n \leq k} c_n b_n = \sum_{n=1}^{\infty} b_n c_n \sum_{k \geq n} k^{-1} (c_1 c_2 \cdots c_k)^{-1/k}.$$

We choose $c_n = (n+1)^n / n^{n-1}$ which gives

$$(c_1 c_2 \cdots c_k)^{-1/k} = 1/(k+1) \quad \text{and}$$

$$\sum_{k \geq n} k^{-1} (c_1 c_2 \cdots c_k)^{-1/k} = \sum_{k \geq n} (1/k(k+1)) = 1/n.$$

Hence we obtain

$$\sum_{k=1}^{\infty} (b_1 b_2 \cdots b_k)^{1/k} \leq \sum_{k=1}^{\infty} b_k c_k / k \leq \sum_{k=1}^{\infty} (1 + (1/k))^k b_k < e \sum_{k=1}^{\infty} b_k.$$

VI: (b) $\implies$ (c). Setting $b_k = M_{k-1}/M_k$ in Carleman's inequality gives $(b_1 b_2 \cdots b_k)^{1/k} = M_k^{-1/k}$, since $M_0 = 1$, and one obtains

$$\sum_{k=0}^{\infty} M_k^{-1/k} < e \sum_{k=1}^{\infty} M_{k-1}/M_k,$$

which proves that (b) $\implies$ (c). Theorem 2.1 is thereby completely proved.

---

[9]See Hardy-Littlewood-Pólya: Inequalities.

THEOREM 2.2. [10] *Let M^0 be logarithmically convex and $\sum (M_k^0)^{-1/k}$ be divergent. The intersection of all classes $C_M(I)$ such that*

(1) M/M^0 *is logarithmically convex and*

(2) $C_M(I)$ *is not quasianalytic*

is equal to the class $C_{M^*}(I)$ where

$$(2.11) \qquad (M_k^*)^{1/k} = (M_k^0)^{1/k} \sum_0^k (M_n^0)^{-1/n}, \quad k = 0, 1, \dots.$$

COROLLARY. *The intersection of all non-quasianalytic classes $C_M(I)$ with logarithmically convex M is the analytic class.*

PROOF OF THEOREM 2.2. We prove first that if M is an arbitrary sequence such that (1) and (2) are valid, then $C_M(I) \supset C_{M^*}(I)$. To do so it is sufficient to prove that

$$(2.12) \qquad \sup (M_k^*/M_k)^{1/k} < \infty,$$

if M/M^0 is logarithmically convex and $\sum M_k^{-1/k}$ is convergent. Since the logarithmic convexity of M/M^0 implies that $(M_k/M_k^0)^{1/k}$ is increasing we obtain

$$\sum_0^n M_k^{-1/k} = \sum_0^n (M_k^0)^{-1/k} (M_k^0/M_k)^{1/k} \geq (M_n^0/M_n)^{1/n} \sum_0^n (M_k^0)^{-1/k} = (M_n^*/M_n)^{1/n},$$

which proves (2.12) and therefore one inclusion in Theorem 2.2.

To prove the other inclusion we shall prove that for an arbitrary $f \notin C_{M^*}(I)$ one can find a sequence M such that M/M^0 is logarithmically convex, the class $C_M(I)$ is not quasianalytic, and $f \notin C_M(I)$. We shall choose M so that

$$(2.13) \qquad \log M_{k_j} = \log M_{k_j}^* + jk_j, \quad j = 1, 2, \dots$$

and

$$\log(M_k/M_k^0) \quad \text{is linear for } k_{j-1} \leq k \leq k_j, \ j = 1, 2, \dots, \quad \text{that is,}$$

$$\log(M_k/M_k^0) = \log(M_{k_{j-1}}/M_{k_{j-1}}^0) + (k - k_{j-1})\kappa_j, \ k_{j-1} \leq k \leq k_j, \quad \text{where}$$

$$(2.14) \qquad \kappa_j = \frac{\log(M_{k_j}^*/M_{k_j}^0) + jk_j - \log(M_{k_{j-1}}^*/M_{k_{j-1}}^0) - (j-1)k_{j-1}}{k_j - k_{j-1}}.$$

Suppose that $k_1, \dots, k_{j-1}$ have already been chosen. Set $F_k = \sup |f^{(k)}(x)|$. If we can prove that k_j can be chosen so that

(i) $$\kappa_j \geq \kappa_{j-1}$$

(ii) $$\sum_{k_{j-1}}^{k_j} M_{k-1}/M_k \leq 2e^{-j}$$

(iii) $$(F_{k_j}/M_{k_j})^{1/k_j} \geq e^j,$$

---

[10]See Bang, Om Quasi-Analytiske Funktioner $(M_n^0 = 1)$, and Rudin, Division in algebras of C^∞-functions, MRC Techn. Rep. Nov. 1961, $(M_n^0 = n!)$.

then we are finished. We enter the expression (2.11) for M_k^* into (2.14) and estimate κ_j when $k_j \to \infty$. If B denotes a constant depending only on $k_1, \ldots, k_{j-1}$, then we can write

$$\kappa_j = \frac{k_j \log \sum_1^{k_j} (M_k^0)^{-1/k} + jk_j - B}{k_j - k_{j-1}}, \quad \text{which gives}$$

(2.15) $$\kappa_j = j + \log \sum_1^{k_j} (M_k^0)^{-1/k} + o(1), \quad \text{when } k_j \to \infty.$$

For $k_{j-1} \leq k \leq k_j$ we have $(M_{k-1}/M_k) = e^{-\kappa_j}(M_{k-1}^0/M_k^0) \leq e^{-\kappa_j}(M_k^0)^{-1/k}$, since M^0 is logarithmically convex. Hence we obtain

$$\sum_{k_{j-1}}^{k_j} (M_{k-1}/M_k) \leq e^{-\kappa_j} \sum_{k_{j-1}}^{k_j} (M_k^0)^{-1/k}$$

and using (2.15)

(2.16) $$\log \sum_{k_{j-1}}^{k_j} (M_{k-1}/M_k) \leq -\kappa_j + \log \sum_{k_{j-1}}^{k_j} (M_k^0)^{-1/k} = -j + o(1), \quad \text{when } k_j \to \infty.$$

From (2.15) it follows, since $\sum (M_k^0)^{-1/k}$ is divergent, that $\kappa_j \to \infty$ when $k_j \to \infty$. In virtue of this fact and (2.16) one sees that there is a constant N such that (i) and (ii) are valid if $k_j > N$.

It remains to prove that (iii) is fulfilled for a suitable choice of k_j. The formula (2.13) can be written

(2.17) $$(M_{k_j}/M_{k_j}^*)^{1/k_j} = e^j, \quad j = 1, 2, \ldots.$$

Our hypothesis that $f \notin C_M^*(I)$ means that $\overline{\lim}(F_k/M_k^*)^{1/k} = +\infty$. Hence we can for arbitrary N choose $k_j > N$ such that

(2.18) $$(F_{k_j}/M_{k_j}^*)^{1/k_j} > e^{2j}.$$

(iii) follows now from (2.17) and (2.18).

REMARK. One proves easily that $\sum (M_k^*)^{-1/k} = \infty$.

3. Interpolation in classes of infinitely differentiable functions.

Let I be a closed interval $\subset \mathbf{R}$. If $f \in C^\infty(I)$ and x is a point in I we shall denote the sequence $f^{(k)}(x)$, $k = 0, 1, \ldots$ by $d_x f$ and call $d_x f$ the function element of f at the point x. In analogy with the definition of the class $C_M(I)$ we denote by c_M the set of all sequences $a = (a_0, a_1, \ldots)$ for which there exist constants C_1 and C_2 such that

(3.1) $$|a_n| \leq C_1 C_2^n M_n, \quad n = 0, 1, \ldots.$$

With this notation our main task is to examine when it is possible for every $a \in c_M$ to find $f \in C_M(I)$, for some neighborhood I of x, such that $d_x f = a$. First we prove a classical theorem of Borel.

THEOREM 3.1. *For every sequence $a = (a_0, a_1, \dots)$ there exists a function $f \in C_0^\infty(\mathbf{R})$ such that $d_0 f = a$.*

PROOF. Choose a function $\varphi \in C_0^\infty(\mathbf{R})$ which is equal to 1 in a neighborhood of the origin, and set with real constants b_n such that $b_n \geq 1$ and $b_n \geq |a_n|$

$$(3.2) \qquad f(x) = \sum_0^\infty x^n a_n \varphi(b_n x)/n!.$$

If $\varphi(x) = 0$ when $|x| \geq A$, then it is certain that also $f(x) = 0$ when $|x| \geq A$. Now we claim that the series (3.2) and all the formally differentiated series

$$f^{(k)}(x) = \sum_{j=0}^k \sum_{n=k-j}^\infty \binom{k}{j} \frac{x^{n+j-k}}{(n+j-k)!} a_n b_n^j \varphi^{(j)}(b_n x),$$

are uniformly convergent also when $|x| \leq A$. This follows at once from the fact that

$$|x^{j+1} a_n b_n^j \varphi^{(j)}(b_n x)| \leq \sup |t^{j+1} \varphi^{(j)}(t)|.$$

Hence (3.2) defines a function $f \in C_0^\infty(\mathbf{R})$, and it is trivial that $d_0 f = a$.

REMARK. Whitney (Trans. Amer. Math. Soc. **36**(1934), 63–89) has generalized this theorem to the case where one prescribes the function element of f at every point in a closed subset of $\mathbf{R}^n$, provided of course that certain compatibility conditions are fulfilled.

The problem of finding a function in a given class with prescribed function element at a point has a completely different character depending on whether the class is quasianalytic or not, that is, whether the extension is unique or not. We shall first treat the simple quasianalytic case.

THEOREM 3.2. *Assume that M is logarithmically convex and that the class $C_M(\mathbf{R})$ is quasianalytic. In order that to every $a \in c_L$ there shall exist a neighborhood I of the origin and a function $f \in C_M(I)$ such that $d_0 f = a$ it is necessary and sufficient that for some constant K*

$$(3.3) \qquad L_n/n! \leq K^{n+1} M_p/p! \quad when \ p \leq n.$$

PROOF. a) *The sufficiency.* Since $|a_n| \leq C_1 C_2^n L_n$ and (3.3) implies that $L_n \leq K^{n+1} n!$, the power series

$$(3.4) \qquad f(x) = \sum_0^\infty x^n a_n/n!$$

converges when $|x| < 1/C_2 K$, and it is clear that $d_0 f = a$. Now

$$|f^{(k)}(x)| \leq \sum_k^\infty |x^{n-k} a_n/(n-k)!| \leq \sum_k^\infty |x|^{n-k} C_1 C_2^n L_n/(n-k)!$$

$$\leq \sum_k^\infty |x|^{n-k} C_1 C_2^n K^{n+1} M_k \binom{n}{k}.$$

Since $\binom{n}{k} \leq 2^n$ we obtain the estimate

$$|f^{(k)}(x)| \leq C_1 C_2^k K^{k+1} 2^k (1 - 2C_2|x|K)^{-1} M_k$$

if $|x| < 1/2C_2K$. Hence f belongs to the class C_M in a neighborhood of the origin.

b) *The necessity.* Let B be the Banach space of all $a \in c_M$ for which the norm $\|a\| = \sup |a_n|/L_n < \infty$. Let I_j be the interval $[-1/j, 1/j]$ and denote by F_j the set of all $a \in B$ such that one can find $f \in C_M(I_j)$ such that $d_0 f = a$ and

$$(3.5) \qquad |f^{(p)}(x)| \leq j^{p+1} M_p; \quad x \in I_j, \ p = 0, 1, 2, \ldots .$$

Since the set of all functions satisfying (3.5) is a compact subset of $C^\infty(I_j)$ (Ascoli's theorem), F_j is a closed subset of B. We have $\cup_1^\infty F_j = B$, for by hypothesis every element in B can be extended in the class C_M in some neighborhood of the origin. By Baire's theorem we can therefore choose j so that F_j has an interior point, and since F_j is convex and symmetric, the origin is an interior point (see also the proof of Theorem 1.1). Hence there exists a positive number δ such that every element $a \in B$ for which $|a_n| \leq \delta L_n$, $n = 0, 1, \ldots$, can be extended to a function in I_j satisfying (3.5).

So far we have not used that the class is quasianalytic. To do so we observe now that for $f(x) = x^n L_n \delta / n!$ we have $f^{(j)}(0) = 0$ if $j \neq n$ and $f^{(n)}(0) = L_n \delta$. Hence $\|d_0 f\| = \delta$, and since f is the only extension of $d_0 f$ in the quasianalytic class C_M it follows that f must satisfy (3.5), which for $x = 1/j$ gives

$$(1/j)^{n-p} L_n \delta / (n-p)! \leq j^{p+1} M_p, \quad p \leq n.$$

Hence we obtain

$$L_n / n! \leq \delta^{-1} j^{n+1} M_p / p!$$

if we use that $1 \leq \binom{n}{p} = n!/p!(n-p)!$. This proves the theorem.

In the non-quasianalytic case the problem of extending arbitrary function elements is harder and has not been treated until quite recently, simultaneously and independently by Carleson (Math. Scand. **9**(1961), 197–206), Ehrenpreis (unpublished) and Mityagin (Soviet Mathematics **2**(1961), 594–597). All three have obtained the same result; Ehrenpreis is also said to have proved a converse. Mityagin has in addition sketched a more general result analogous to Whitney's extension theorem.

Before we formulate any result we note that if $f \in C_M(I)$, where I is a neighborhood of the origin and C_M is not quasianalytic, then one can find $f_1 \in C_M(\mathbf{R})$ such that $d_0 f_1 = d_0 f$. For if $g \in C_0^\infty(I) \cap C_M(\mathbf{R})$ and $g = 1$ in a neighborhood of the origin we can take $f_1 = gf$. Hence we need not consider local extensions as in Theorem 3.2.

THEOREM 3.3. *Let M be a logarithmically convex sequence such that C_M is not quasianalytic, and set*

$$(3.6) \qquad h_M(\xi) = \sup_n |\xi|^n / M_n, \quad \log H(r) = \frac{1}{\pi} \int_{-\infty}^{\infty} \frac{|r|}{r^2 + t^2} \log h_M(t) \, dt.$$

(The existence of the integral follows from Theorem 2.1.) Define further

$$(3.7) \qquad \widetilde{M}_n = \sup r^n / H(r).$$

If L_n is a positive sequence for which

$$(3.8) \qquad L_n \leq C^n \widetilde{M}_n$$

one can for every $a \in c_L$ find a function $f \in C_M(\mathbf{R})$ such that $d_0 f = a$.

Before the proof we give a lemma which shows the role of the function $H(r)$, which obviously is the restriction to the imaginary axis of the Poisson integral of $\log h_M$.

LEMMA 3.1. *The Poisson integral of* $\log h_M$,

$$U(x, y) = \frac{|y|}{\pi} \int_{-\infty}^{\infty} \frac{1}{(x - t)^2 + y^2} \log h_M(t)\, dt,$$

has on the circle $x^2 + y^2 = r^2$ *its maximum when* $x = 0$, *and this maximum is thus* $\log H(r)$.

PROOF. If the polar coordinates in the upper half plane are denoted by r, θ where $0 < \theta < \pi$, then the map $(x, y) \mapsto (\varrho, \theta)$ where $\varrho = \log r$ is conformal and maps the upper half plane on the parallel strip $0 < \theta < \pi$. In the new coordinates U is therefore still harmonic and is the Poisson integral with boundary values $\log h_M(e^\varrho)$ when $\theta = 0$ and $\theta = \pi$. Now

$$\log h_M(e^\varrho) = \sup(\varrho n - \log M_n)$$

which shows that $\log h_M(e^\varrho)$ is a convex function of ϱ. Furthermore the Poisson kernel for the strip $0 < \theta < \pi$ is obviously invariant for translations along the strip. For fixed θ the Poisson integral is thus a *convolution* of a convex and a positive function, hence convex. But a harmonic function of ϱ and θ which is convex as a function of ϱ is automatically a concave function of θ. Since the Poisson integral is symmetrical around $\theta = \pi/2$ it must therefore have its maximum for fixed ϱ when $\theta = \pi/2$. This proves the lemma.

The proof of the lemma shows at the same time that $H(r) \geq h_M(r)$, hence that $\widetilde{M}_n \leq M_n$ (see Lemma 1.1). However, for many important examples the classes $C_{\widetilde{M}}$ and C_M are identical; in particular this is true when $M_n = n!^\alpha$ where $\alpha > 1$. It is also easy to give examples where $C_M \neq C_{\widetilde{M}}$; this is true if $\log h_M(t)$ behaves at infinity as $t/(\log t)^2$.

As a final preparation before the proof we note that for every polynomial $p(z)$

$$(3.9) \qquad \sup_{z \in \mathbf{C}} |p(z)|/H(|z|) \leq \sup_{x \in \mathbf{R}} |p(x)|/h_M(x).$$

In fact, according to the maximum principle the function $|p(z)|e^{-U(x,y)}$ assumes its maximum on the real axis, for it tends to 0 at infinity and its logarithm is subharmonic in each half plane. Since $U(x, y) \leq H(|z|)$ this implies the inequality (3.9).

Using (3.7) and Cauchy's inequality we can estimate the coefficients of the polynomial p; for every $r > 0$ we have

$$|p^{(j)}(0)/j!| \leq r^{-j} \sup_{|z|=r} |p(z)| \leq r^{-j} H(r) \sup_{x \in \mathbf{R}} |p(x)|/h_M(x).$$

If we take the infimum over r we obtain by the definition of $\widetilde{M}_n$ that

$$(3.10) \qquad |p^{(j)}(0)/j!| \leq \widetilde{M}_j^{-1} \sup_{x \in \mathbf{R}} |p(x)|/h_M(x), \quad j = 0, 1, 2, \ldots.$$

PROOF OF THEOREM 3.3. It suffices to construct a measure $d\mu$ on $\mathbf{R}$ which satisfies the condition

$$(3.11) \qquad \int_{-\infty}^{\infty} h_M(\xi)|d\mu(\xi)| < \infty$$

and has the moments

$$(3.12) \qquad \int_{-\infty}^{\infty} \xi^n d\mu(\xi) = a_n/i^n.$$

For if f is the Fourier transform of μ,

$$f(x) = \int e^{ix\xi} d\mu(\xi),$$

then (3.11) implies that $f \in C_M$ (see Lemma 2.2), and it follows from (3.12) that $d_0 f = a$. Now the measures satisfying (3.11) are nothing else than the continuous linear functionals on the space B of continuous functions φ for which $\varphi(\xi)/h_M(\xi) \to 0$ as $\xi \to \infty$, normed with $\sup |\varphi|/h_M$. The condition (3.12) means that if $p(z) = \sum b_n z^n$ is a polynomial then

$$\int_{-\infty}^{\infty} p(\xi) d\mu(\xi) = \sum_0^{\infty} a_n b_n / i^n = \sum_0^{\infty} a_n p^{(n)}(0) / i^n n!.$$

What must be done is thus to prove that the linear functional

$$p \mapsto \sum_0^{\infty} a_n p^{(n)}(0) / i^n n!$$

which is defined for all polynomials p can be extended to a continuous linear functional on the space B. By Hahn-Banach this is possible if for some constant C we have

$$(3.13) \qquad \left| \sum_0^{\infty} a_n p^{(n)}(0) / i^n n! \right| \leq C \sup |p(\xi)|/h_M(\xi),$$

which by (3.10) is true provided that

$$(3.14) \qquad \sum_0^{\infty} |a_n|/\widetilde{M}_n \leq C.$$

We have now proved that every a satisfying (3.14) can be extended in the class C_M. But if $a \in C_L$ it follows from (3.6) for suitable constants C_1 and C_2 that

$$|a_n| \leq C_1 C_2^n \widetilde{M}_n.$$

The condition (3.14) is thus fulfilled with a_n replaced by $a_n/(2C_2)^n$. We can therefore find $g \in C_M$ such that

$$g^{(n)}(0) = a_n/(2C_2)^n, \quad n = 0, 1, \ldots.$$

But this implies that $f(x) = g(2C_2 x)$ is in C_M and that $d_0 f = a$. The proof is complete.

REMARK. Carleson's proof differs formally from the one above by working with L_2 norms which allows him to use a projection argument (orthogonal polynomials) instead of the Hahn-Banach theorem. Apart from that the three quoted proofs are essentially identical.

4. Division in classes of infinitely differentiable functions.

By Corollary 1.2 the class $C_M(I)$ is closed for addition and multiplication if M is a logarithmically convex sequence. If $f \in C_M$ it follows therefore that $F(f) \in C_M$ for every polynomial F. We shall now study the composite function $F(f)$ when F is not a polynomial. In particular we shall investigate if $1/f$ must belong to C_M when $f \in C_M$ and $1/f$ is bounded. Because of the great technical advantages we shall then consider complex valued functions f throughout. This does not affect the results since division by a complex valued function according to the identity $1/f = \bar{f}/f\bar{f}$ can be reduced to division by a real valued function. The following results are essentially all due to Rudin, Division in algebras of infinitely differentiable functions, MRC Techn. Summ. Rep. 273, November 1961.

In the statement of the results it is convenient to use the notation

$$A_n = (M_n/n!)^{1/n}.$$

First we prove a positive result.

THEOREM 4.1. *Assume that the sequence A_n is almost increasing in the sense that there is a constant K such that*

$$(4.1) \qquad A_s \leq K A_n, \quad s \leq n.$$

If $f \in C_M(I)$ and $\inf |f| = c > 0$ it follows then that $1/f \in C_M(I)$. More precisely, if

$$(4.2) \qquad \sup_I |D^n f| \leq C_1 C_2^n M_n, \quad n = 0, 1, \dots,$$

and we set $B_1 = 2/c$, $B_2 = K C_2(1 + 2C_1 c^{-1})$, it follows that

$$(4.3) \qquad \sup_I |D^n(1/f)| \leq B_1 B_2^n M_n, \quad n = 0, 1, \dots.$$

PROOF. Let x_0 be an arbitrary point in I and form the Taylor expansion of f of order n at x_0,

$$f_n(x) = \sum_0^n f^{(j)}(x_0)(x - x_0)^j / j!.$$

Then $D^n(1/f) = D^n(1/f_n)$ at x_0. On a circle in the complex plane with radius r and center at x_0 we have according to (4.2) and (4.1)

$$|f_n(z)| \geq c - C_1 \sum_1^n (C_2 A_j r)^j \geq c - C_1 \sum_1^n (C_2 K A_n r)^j \geq c/2 = B_1^{-1}$$

if $C_2 K A_n r = \varepsilon$ where $C_1 \varepsilon/(1 - \varepsilon) = c/2$. Hence by Cauchy's inequality

$$|D^n(1/f_n)(x_0)| \leq n! r^{-n} B_1 = B_1 B_2^n M_n$$

where the last equality follows since $C_2 K/\varepsilon = B_2$. This proves (4.3).

It is also easy to prove a more general version of Theorem 4.1.

THEOREM 4.2. *Let A_n be almost increasing. Then all analytic functions operate on C_M, that is, if $f \in C_M(I)$ and F is analytic in a neighborhood of the closure of the range of f, then it follows that $F(f) \in C_M(I)$.*

PROOF. We can find an open neighborhood of the closure of the range of f such that its boundary Γ consists of rectifiable Jordan curves and F is analytic in the closed set bounded by Γ. Then we have

$$F(z) = (2\pi i)^{-1} \int_\Gamma F(\zeta)/(\zeta - z) \, d\zeta$$

if z belongs to the range of f, hence

$$(4.4) \qquad F(f(x)) = (2\pi i)^{-1} \int_\Gamma F(\zeta)/(\zeta - f(x)) \, d\zeta.$$

Now we can find a positive number c such that

$$|\zeta - f(x)| > c \quad \text{when } x \in I \text{ and } \zeta \in \Gamma.$$

Furthermore we can find constants C_1 and C_2 such that

$$\sup_I |D^n(f - \zeta)| \le C_1 C_2^n M_n; \quad n = 0, 1, \ldots; \ \zeta \in \Gamma.$$

It suffices to choose C_1 and C_2 so that this is true for $\zeta = 0$ and then possibly increase C_1 so that the inequality is valid for every $\zeta \in \Gamma$ when $n = 0$. From Theorem 4.1 it follows now that (4.3) is valid with f replaced by $(\zeta - f)$ for every $\zeta \in \Gamma$; the constants B_1 and B_2 do not depend on ζ. Hence differentiation of (4.4) under the integral sign gives

$$(4.5) \qquad \sup_I |D^n F(f)| \le (2\pi)^{-1} \int_\Gamma |F(\zeta)| \, |d\zeta| \, B_1 B_2^n M_n.$$

This proves the theorem.

We shall now prepare for the proof of a converse of Theorems 4.1 and 4.2.

LEMMA 4.1. *Let F be an infinitely differentiable function in the circle $|z| \le 1$. Assume further that*

$$(4.6) \qquad f \in C_M(\mathbf{R}), \ \sup|f| \le 1 \implies F(f) \in C_M(\mathbf{R}).$$

Then there is a constant C such that

$$(4.7) \ f \in \dot{B}, \ \sup|D^n f| < M_n, \ n = 0, 1, \cdots \implies \sup|D^n F(f)| \le C^{n+1} M_n, \ n = 0, 1, \ldots.$$

Here $\dot{B}$ denotes the class of functions $f \in C^\infty(\mathbf{R})$ such that $f^{(n)}(x) \to 0$ as $x \to \infty$, for every n. By the Riemann-Lebesgue lemma the Fourier transform of an arbitrary integrable function with compact support is in the class $\dot{B}$, for example. Note that the result has the same form as if one could have used the closed graph theorem, which of course is not possible since the map $f \mapsto F(f)$ is not linear.

PROOF OF THE LEMMA. Assume that (4.7) is not valid. Then we can for every integer $j > 0$ choose a function $f_j \in \dot{B}$ such that

$$(4.8) \qquad \sup|f_j^{(n)}| < M_n, \quad n = 0, 1, \ldots$$

whereas for some integer n_j

$$(4.9) \qquad \sup|D^n F(f_j)| > j^{n+1} M_n \quad \text{when } n = n_j.$$

Since we can make an arbitrary translation of f_j without affecting (4.8), we may assume that

$$(4.9)' \qquad |D^n F(f_j)(0)| > j^{n+1} M_n \quad \text{when } n = n_j.$$

Now we set

$$(4.10) \qquad g_k(x) = \sum_{j=1}^k f_j(x - x_j),$$

where x_j are constants which we shall determine successively so that for every k

$$(4.11) \qquad \sup |g_k^{(n)}| < 2^n M_n, \quad n = 0, 1, 2, \ldots,$$

$$(4.12) \qquad |(D^{n_j} F(g_k))(x_j)| > j^{n_j+1} M_{n_j}, \quad j = 1, \ldots, k.$$

If $k = 1$ then (4.11) and (4.12) follow from (4.8) and (4.9)$'$ for any choice of x_1. Assuming now that $x_1, \ldots, x_{k-1}$ have already been chosen we shall prove that (4.11) and (4.12) are valid if x_k is chosen large enough. First we note that (4.8) implies that

$$\sup |g_k^{(n)}| < k M_n,$$

so (4.11) is valid if $2^n \geq k$. Since g_{k-1} and f_k belong to $\dot{B}$ and $\sup |D^n g_{k-1}| < 2^n M_n$, $\sup |D^n f_k| < M_n \leq 2^n M_n$, it follows that (4.11) is also valid when $2^n < k$ if we choose $|x_k|$ large enough. For sufficiently large $|x_k|$ (4.12) is also a consequence of the inductive hypothesis if $j < k$ and of (4.9) if $j = k$.

(4.11) implies in particular that the sequence g_k is equicontinuous. We can therefore choose a subsequence $g_{k'}$ which converges uniformly on every finite interval to a function g. Since the derivatives of $g_{k'}$ also are equicontinuous it follows at once that g is infinitely differentiable and that $D^n g_{k'}$ converges uniformly to $D^n g$ on every finite interval. Passage to the limit in (4.11) and (4.12) gives

$$(4.11)' \qquad \sup |g^{(n)}| \leq 2^n M_n, \quad n = 0, 1, \ldots,$$

$$(4.12)' \qquad |(D^{n_j} F(g))(x_j)| \geq j^{n_j+1} M_{n_j}, \quad j = 1, 2, \ldots,$$

which contradicts (4.6) and proves the lemma.

LEMMA 4.2. *If in addition to the hypotheses of Lemma 4.1 we assume that*

$$F(z) = \sum_0^\infty a_k z^k$$

where the series is absolutely convergent when $|z| < 1$ then

$$(4.13) \qquad |a_k| \sup |D^n(f^k)| \leq C^{n+1} M_n; \quad k = 0, 1, \ldots; \; n = 0, 1, \ldots,$$

for all $f \in \dot{B}$ satisfying the inequalities

$$(4.14) \qquad \sup |f^{(n)}| \leq M_n, \quad n = 0, 1, \ldots.$$

PROOF. We can apply (4.7) to ζf for every complex number ζ with $|\zeta| < 1$, and obtain then

$$\left| \sum_0^\infty a_k D^n(f^k) \zeta^k \right| \leq C^{n+1} M_n, \quad |\zeta| < 1.$$

Hence (4.13) follows from Cauchy's inequalities for the coefficients of a power series.

We can now prove a converse of Theorems 4.1 and 4.2.

THEOREM 4.3. *Assume that there exists a complex number λ such that*

$$(4.15) \qquad f \in C_M(\mathbf{R}), \ |f| \leq 1 \implies (\lambda - f)^{-1} \in C_M(\mathbf{R}).$$

Then the sequence A_n is almost increasing.

PROOF. If we use the notation

$$q_M(\xi) = \sup |\xi|^n / M_n$$

as before, then (4.14) is fulfilled if we set $f(x) = \hat{\varphi}(x)$ where

$$(4.16) \qquad \int q_M(\xi)|\varphi(\xi)|\, d\xi \leq 1,$$

for we have (compare Lemma 2.2)

$$|f^{(n)}| \leq \int |\xi|^n |\varphi(\xi)|\, d\xi \leq M_n \int q_M(\xi)|\varphi(\xi)|\, d\xi \leq M_n.$$

Furthermore it follows from the Riemann-Lebesgue lemma that $f \in \dot{B}$. Hence the estimate (4.13) is valid for f. If we now let φ converge to $q_M(\xi)^{-1}$ times the Dirac measure at ξ, it follows that

$$(4.17) \qquad |a_k| \sup |D^n e^{ikx\xi}/q_M(\xi)^k| \leq C^{n+1} M_n,$$

where $a_k = \lambda^{-k-1}$ is the coefficient of z^k in the Taylor expansion of $(\lambda - z)^{-1}$. (It is clear that $|\lambda| > 1$ if (4.15) is valid, for otherwise we could take $f = \lambda$.) Hence it follows from (4.17) that

$$(4.18) \qquad (k|\xi|)^n / q_M(\xi)^k \leq C^{n+1} |\lambda|^{k+1} M_n.$$

In particular we can choose n as a multiple of k in (4.18), that is, put $n = ks$. Then we obtain

$$k^{ks}(|\xi|^s/q_M(\xi))^k \leq C^{ks+1}|\lambda|^{k+1} M_{ks} \leq (C|\lambda|)^{ks+1} M_{ks}.$$

If we take the maximum with respect to ξ in the left-hand side and use Lemma 1.1, we obtain if we afterwards raise both sides to the power $1/ks$,

$$(4.19) \qquad s^{-1} M_s^{1/s} \leq K_1(ks)^{-1} M_{ks}^{1/ks},$$

where K_1 is a constant. If n and s are arbitrary integers with $s \leq n$ we can choose an integer k such that $n/2 \leq ks \leq n$ and obtain using (4.18) and that $M_n^{1/n}$ is an increasing sequence

$$s^{-1} M_s^{1/s} \leq K_1(ks)^{-1} M_{ks}^{1/ks} \leq 2K_1 n^{-1} M_n^{1/n}.$$

Since Stirling's formula proves that $(s!)^{1/s}/s \to 1$ as $s \to \infty$, this implies the inequality (4.1). The proof is complete.

REMARK. One also proves easily (see Rudin's paper) that if A_n is not almost increasing then one can choose $f \in C_M(\mathbf{R})$ so that $(f - \lambda)^{-1}$ is not in $C_M(\mathbf{R})$ for any complex λ and so that there even exist entire functions F such that $F(f) \notin C_M(\mathbf{R})$. However, Rudin only proved Theorem 4.3 for non-quasianalytic classes.

THEOREM 4.4. *The intersection of all non-quasianalytic classes $C_M(I)$, where M is logarithmically convex and the sequence A_n is almost increasing, is equal to the quasianalytic class corresponding to the sequence $(n \log n)^n$.*

PROOF. Set $M_n^0 = n^n$. By Stirling's formula (4.1) is equivalent to

$$(4.20) \qquad\qquad (M_s/M_s^0)^{1/s} \leq K(M_n/M_n^0)^{1/n}, \quad s \leq n,$$

for some other constant K. Since (4.20) is satisfied if the sequence M_n/M_n^0 is logarithmically convex, it follows from Theorem 2.2 that the intersection of the classes in Theorem 4.4 is contained in the class C_{M^*} defined in Theorem 2.2. But from the first part of the proof of Theorem 2.2, where actually only (4.20) would have been required, it follows also that (4.20) implies that $C_M \supset C_{M^*}$. This proves the theorem, for $M_n^{*1/n}$ is asymptotically equal to $n \log n$.

APPROXIMATION ON TOTALLY REAL MANIFOLDS

LARS HÖRMANDER

1. Introduction. Twenty years ago John Wermer and the author [4] proved a theorem on approximation by analytic functions on a totally real submanifold of $\mathbf{C}^n$. One part of the argument was quite precise and assumed only that the manifold was of class C^1. However, another part using L^2 estimates required higher differentiability assumptions, depending on the dimension n. This flaw was later removed by other authors using integral formulas and hyperfunction theory. (See [1], [2] and the references in the latter.) When writing down an elementary introduction to hyperfunction theory [3, Chapter 9], the author noticed that it would actually have been sufficient to use Weierstrass' original proof of his approximation theorem, by convolution with a Gaussian. In this note we shall present this very elementary argument. Section 4 also gives a simple proof of a theorem of Valentine [5] used in [4].

2. Local approximation. Let us first recall that a C^1 submanifold Σ of $\mathbf{C}^n$ is said to be totally real if it has no complex tangent, that is, if

$$(2.1) \qquad\qquad T \cap iT = \{0\}$$

when T is the tangent plane of Σ at some point z_0. This implies that the dimension of Σ is at most n. A real basis for T is then linearly independent over $\mathbf{C}$, so it can be extended to a complex basis. With suitable complex coordinates we may therefore assume that $z_0 = 0$ and that

$$T = \{(x_1, \ldots, x_\sigma, 0, \ldots, 0); x_j \in \mathbf{R}\}.$$

Thus

$$\Sigma \ni x + iy \mapsto (x_1, \ldots, x_\sigma) = x'$$

has bijective differential at 0. Hence there is an open neighborhood ω of 0 in $\mathbf{R}^\sigma$ such that a neighborhood $\tilde{\omega}$ of 0 in Σ is the range of $\omega \ni t \mapsto \zeta(t)$ where $\zeta \in C^1(\omega)$ is bijective and

$$\operatorname{Re}\zeta_j(t) = t_j, \quad \operatorname{Im}\zeta_j'(0) = 0 \text{ if } j \le \sigma; \quad \zeta_j'(0) = 0 \text{ if } j > \sigma.$$

If $\Sigma \in C^r$ then $\zeta \in C^r$.

When $t, s \in \omega$ we have with the notation $z^2 = \langle z, z \rangle$, $z \in \mathbf{C}^n$,

$$\operatorname{Re}(\zeta(t) - \zeta(s))^2 = (t - s)^2 + \sum_{\sigma+1}^n (\operatorname{Re}(\zeta_j(t) - \zeta_j(s)))^2 - \sum_1^n (\operatorname{Im}(\zeta_j(t) - \zeta_j(s)))^2.$$

Replacing ω by a smaller neighborhood of 0 if necessary we obtain

$$(2.2) \qquad\qquad \operatorname{Re}(\zeta(t) - \zeta(s))^2 \ge |t - s|^2/2, \qquad t, x \in \omega.$$

© Springer International Publishing AG, part of Springer Nature 2018

L. Hörmander, *Unpublished Manuscripts*,

https://doi.org/10.1007/978-3-319-69850-2_12

Let K be a compact subset of ω and let $u \in C_0^{r-1}(K)$. We shall approximate u, or rather the function $u \circ \zeta^{-1}$ on $\tilde{\omega} \subset \Sigma$ by the Gaussian convolution

$$(2.3) \qquad u_\varepsilon(z) = \varepsilon^{-\sigma} \int e^{-(z-\zeta(t))^2/\varepsilon^2} a(t) u(t)\, dt,$$

which is of course an entire analytic function. To motivate the choice of a we note that if $\theta : \mathbf{R}^\sigma \to \mathbf{C}^n$ is a linear function such that $\operatorname{Re}\theta(t)^2 \geq |t|^2/2$, $t \in \mathbf{R}^\sigma$, then

$$\int e^{-\theta(t)^2}\, dt = \pi^{\sigma/2}(\det A_\theta)^{-1/2}$$

where A_θ is the symmetric matrix defined by $\langle A_\theta t, t\rangle = \theta(t)^2$. (See e.g. [4, (3.4.1)''] where a discussion of the definition of the square root is also given.) We define

$$(2.4) \qquad a(t) = \pi^{-\sigma/2}(\det A_{\zeta'(t)})^{1/2}.$$

With $z = \zeta(s)$, $s \in \omega$, and $t = s + \varepsilon\tau$ we can write

$$u_\varepsilon(\zeta(s)) = \int e^{-\Phi(s,\tau,\varepsilon)} a(s + \varepsilon\tau) u(s + \varepsilon\tau)\, d\tau$$

$$\Phi(s,\tau,\varepsilon) = (\zeta(s+\varepsilon\tau) - \zeta(s))^2/\varepsilon^2 = \left(\int_0^1 \zeta'(s+\varepsilon\delta\tau)\tau\, d\delta\right)^2.$$

Φ is in C^{r-1} as a function of (s,τ) and as such depends continuously on ε; $\Phi(s,\tau,0) = (\zeta'(s)\tau)^2$. Hence it follows that the derivatives of $u_\varepsilon(\zeta(s))$ with respect to s of order $\leq r-1$ are continuous functions of $\varepsilon \in [0,1)$, and

$$u_0(\zeta(s)) = \int e^{-(\zeta'(s)\tau)^2} a(s) u(s)\, d\tau = u(s)$$

by (2.4). This means that

$$u_\varepsilon \to u \circ \zeta^{-1} \text{ in } C^{r-1}(\tilde{\omega}) \text{ when } \varepsilon \to 0.$$

Let K_0 be a compact subset of ω with $K_0 \cap K = \emptyset$. Then we can choose $c > 0$ and a neighborhood Ω of $\zeta(K_0)$ such that

$$\operatorname{Re}(z - \zeta(t))^2 \geq c, \quad \text{if } z \in \Omega \text{ and } t \in K,$$

for $\operatorname{Re}(\zeta(s) - \zeta(t))^2 \geq 2c > 0$ by (2.2) if $s \in K_0$. Hence u_ε converges uniformly to 0 in Ω as $\varepsilon \to 0$. We have proved

THEOREM 2.1. *Every point in a totally real C^r manifold $\Sigma \subset \mathbf{C}^n$, $r \geq 1$, has a neighborhood ω in Σ with the following property. If K, K_0 are disjoint compact subsets of ω one can find a neighborhood Ω_0 of K_0 in $\mathbf{C}^n$ such that for every $u \in C_0^{r-1}(K)$ there exist entire functions u_ε converging to u in $C^{r-1}(\omega)$ and converging uniformly to 0 in Ω_0, as $\varepsilon \to 0$.*

3. Global approximation. Keeping the notation of Theorem 2.1 we can choose $\psi \in C_0^\infty(\mathbf{C}^n)$ equal to 1 in a neighborhood of K such that

$$\Sigma \cap \operatorname{supp} \psi \subset\subset \omega, \qquad K_0 = \Sigma \cap \operatorname{supp} \psi' \subset \omega \setminus K.$$

Then $\bar{\partial}(\psi u_\varepsilon) \to 0$ in $C^\infty(\Omega)$ if Ω is a sufficiently small neighborhood of Σ in $\mathbf{C}^n$. In fact, it suffices to choose Ω so small that $\Omega \cap \operatorname{supp} \psi' \subset \Omega_0$ where Ω_0 is defined as in Theorem 2.1. This leads to the following global approximation theorem:

THEOREM 3.1. *Let K be a compact subset of a totally real C^r submanifold Σ of $\mathbf{C}^n$, and let $u \in C^{r-1}(K)$. Then one can choose a neighborhood Ω of K in $\mathbf{C}^n$ and functions u_ε analytic in Ω such that $u_\varepsilon \to u$ in $C^{r-1}(K)$ as $\varepsilon \to 0$.*

PROOF. We can cover K by a finite number of open subsets ω_j of Σ for which the conclusions of Theorem 2.1 hold. By means of a subordinate partition of unity we split u into a sum $\sum u_j$ where $u_j \in C_0^\infty(K_j)$. Choose corresponding cutoff functions ψ_j as in the discussion preceding the statement. Then we obtain functions $\psi_j u_{j\varepsilon}$ converging to u_j in $C_0^{r-1}(\Sigma)$ such that

$$\bar{\partial}(\psi_j u_{j\varepsilon}) \to 0 \quad \text{in } C^\infty(\Omega)$$

if Ω is a sufficiently small neighborhood of K. Now K has a fundamental system of strictly pseudoconvex neighborhoods. (See [4, Theorem 3.1] and its proof.) We can choose such a neighborhood Ω for which

$$f_\varepsilon = \sum \bar{\partial} \psi_j \, u_{j\varepsilon} \to 0 \quad \text{in } C^\infty(\overline{\Omega}).$$

Then solution of the $\bar{\partial}$ Neumann problem yields v_ε with $\bar{\partial} v_\varepsilon = f_\varepsilon$ in Ω converging to 0 in $C^\infty(\overline{\Omega})$. The functions

$$u_\varepsilon = \sum \psi_j u_{j\varepsilon} - v_\varepsilon$$

have the required property.

4. Valentine's extension theorem. The purpose of this note is to write down a completely elementary proof of the following theorem of Valentine [5] used in [4]:

THEOREM 4.1. *If E is a subset of a real separable Hilbert space H_1 and $f : E \to H_2$ is a map to another real Hilbert space H_2 satisfying the Lipschitz condition*

$$(4.1) \qquad \|f(x_1) - f(x_2)\|_2 \le \|x_1 - x_2\|_1, \quad x_1, x_2 \in E,$$

then f can be extended to a map $H_1 \to H_2$ with the same property.

PROOF. We shall first assume that H_2 *is finite dimensional.* It suffices to prove that for every $\xi \in H_1 \setminus E$ one can find $\eta \in H_2$ such that

$$(4.2) \qquad \|\eta - f(x)\|_2 \le \|\xi - x\|_1, \quad x \in E,$$

for if this is possible we can extend f to a dense set by successively adding the points in a dense sequence, and then f extends by continuity to the whole of H_1. The inequality (4.2) means that the balls

$$B_x = \{\eta \in H_2; \|\eta - f(x)\|_2 \le \|\xi - x\|_1\}$$

with $x \in E$ shall have a point in common. That is true if any finite number of them have a point in common, so we may assume in the proof that E is finite. Set

$$M = \min_{\eta \in H_2} \max_{x \in E} \|\eta - f(x)\|_2 / \|\xi - x\|_1;$$

we have to prove that $M \leq 1$. The definition of M implies that the balls

$$B_{x,M} = \{\eta \in H_2; \|\eta - f(x)\|_2 \leq M\|\xi - x\|_1\}$$

with $x \in E$ have at least one point in common. If they had two points in common then the interior of the interval between them would be in the interior of every $B_{x,M}$ which contradicts the extremal property of M. Let η be the unique point in $\cap_{x \in E} B_{x,M}$, and let E_0 be the set of points $x \in E$ such that $\|\eta - f(x)\|_2 = M\|\xi - x\|_1$, which is not empty. If one could find $y \in H_2 \setminus \{0\}$ such that $\langle y, \eta - f(x) \rangle < 0$ for $x \in E_0$, then $\|\eta + \varepsilon y - f(x)\|_2 < M\|\xi - x\|_1$, $x \in E$, if $\varepsilon > 0$ is small enough, which is a contradiction. Hence there is no such y which means that the origin is in the convex hull of $\{\eta - f(x); x \in E_0\}$. Thus we can find $\lambda_x \geq 0$ when $x \in E_0$ with sum 1 such that $\sum_{x \in E_0} \lambda_x (\eta - f(x)) = 0$.

Now note that by the hypothesis (4.1) we have

$$\|\eta - f(x_1)\|_2^2 + \|\eta - f(x_2)\|_2^2 - 2\langle \eta - f(x_1), \eta - f(x_2) \rangle_2$$
$$\leq \|\xi - x_1\|_1^2 + \|\xi - x_2\|_1^2 - 2\langle \xi - x_1, \xi - x_2 \rangle_1, \quad x_1, x_2 \in E.$$

If we take $x_1, x_2 \in E_0$, multiply by $\lambda(x_1)\lambda(x_2)$ and sum, it follows that

$$(M^2 - 1) \sum_{x \in E_0} \lambda_x \|\xi - x\|^2 + \left\| \sum_{x \in E_0} \lambda_x (\xi - x) \right\|^2 \leq 0,$$

which implies that $M^2 \leq 1$.

The proof is now complete when H_2 is finite dimensional. If not, it is no restriction to assume that H_2 is separable since the range of f is contained in a separable subspace. Thus we may assume that $H_2 = l^2$. Let P_N be the projection $H_2 \to \mathbf{R}^N$ on the first N coordinates. Then $P_N \circ f$ satisfies (4.1) so we can find $\eta \in \mathbf{R}^N$ with the property (4.2); set

$$M_N = \{\eta \in \mathbf{R}^N; \|\eta - P_N f(x)\|_2 \leq \|\xi - x\|_1, x \in E\}$$

which is thus a closed convex non-empty set for every N. If $\nu > N$ then $P_N M_\nu \subset M_N$ and $P_N M_\nu$ decreases when ν increases. Let $\widetilde{M}_N$ be the limit as $\nu \to \infty$. Then $P_N \widetilde{M}_\nu = \widetilde{M}_N$ when $\nu > N$, so it follows that we can find $\eta_N \in \widetilde{M}_N$ with $P_N \eta_\nu = \eta_N$ for every $\nu > N$. The norms of the elements η_ν are bounded so they converge as $\nu \to \infty$ to an element $\eta \in H_2$ satisfying (4.2). The proof is complete.

REMARK. The proof of Valentine uses Helly's theorem to reduce the number of points in E to $\dim H_2 + 1$, when this is finite, then the theorem of Knaster, Kuratowski and Mazurkiewicz to get a proof of (4.2) by contradiction. To extend to infinite dimensions he then uses the Smulian theorem. Both these sophisticated tools are thus superfluous.

References

[1] Bo Berndtsson, *Integral kernels and approximation on totally real submanifold of* $\mathbf{C}^n$, Report Dept. of Math. Gothenburg (1979), no. 1, 8 pp..

[2] F. Reese Harvey and R. O. Wells, Jr, *Holomorphic approximation and hyperfunction theory on a* C^1 *totally real submanifold of a complex manifold*, Math. Ann. **197** (1972), 287–318.

[3] L. Hörmander, *The analysis of linear partial differential operators I. Distribution theory*, Springer Verlag, Berlin, Heidelberg, New York, Tokyo, 1983.

[4] L. Hörmander and J. Wermer, *Uniform approximation on compact sets in* $\mathbf{C}^n$, Math. Scand. **23** (1968), 5–21.

[5] F. A. Valentine, *A Lipschitz condition preserving extensions of a vector function*, Amer. J. Math. **67** (1945), 83–93.

THE WORK OF CARLESON AND HOFFMAN ON
BOUNDED ANALYTIC FUNCTIONS IN THE DISC

Lars Hörmander

References

[1] L. Carleson, *An interpolation problem for bounded analytic functions*, Amer. J. Math. **80** (1958).
[2] ———, *Interpolation by bounded analytic functions and the corona problem*, Ann. of Math. **76** (1962).
[3] K. Hoffman, *Bounded analytic functions and Gleason parts*, Ann. of Math. **86** (1967) (to appear).

Notations:[11] D = open unit disc, $H^\infty(D)$ = Banach algebra of analytic functions in D with $\|f\| = \sup |f(z)| < \infty$.

THEOREM [1]. *Let $z_j \in D$. Then the map*

$$H^\infty \ni f \to \{f(z_j)\} \in l^\infty$$

is onto if and only if

(1) $$\prod_{j;j \neq k} |z_j - z_k|/|1 - z_j \bar{z}_k| \geq c > 0.$$

(Note that if $\|f\| \leq 1$ and $f(z_j) = 0$ when $j \neq k$, then $f(z_k)$ is at most equal to the product.)

Equivalent condition: The Blaschke product

$$B(z) = \prod((z - z_j)/(1 - \bar{z}_j z))(-\bar{z}_j)/|z_j|$$

converges (that is, $\sum(1 - |z_j|) < \infty$) and

(1)′ $$(1 - |z_j|^2)|B'(z_j)| \geq c.$$

(Note that $(1 - |z|^2)|f'(z)| \leq \|f\|$ so we have a bound both ways.)

DEFINITION. The sequence z_j is called interpolating if (1), (1)′ hold.

Let M be the maximal ideal space = space of continuous multiplicative linear functionals with weak topology, so that for $f \in H^\infty$ the Gelfand transform $\hat{f}(m) = m(f)$ is continuous on M. This is a compact Hausdorff space. If $m \in M$ and $m(z) = w$ belongs to D, then $m(f) = f(w)$, for $f(z) - f(w) = (z - w)g(z)$, $g \in H^\infty$, hence $m(f - f(w)) = 0$. So M can be thought of as the unit disc with a fringe attached to the boundary. If w is a boundary point there are of course many m with $m(z) = w$, for we can take $m(f) = \lim f(z_i)$ where z_i is an ultrafilter in D converging to w. This is not only an example, for

---

[11]Lecture in the Princeton Current Literature Seminar

© Springer International Publishing AG, part of Springer Nature 2018
L. Hörmander, *Unpublished Manuscripts*,
https://doi.org/10.1007/978-3-319-69850-2_13

THEOREM [2]. *D is dense in M.*

If $z_j \in D$ we define $\varrho(z_1, z_2) = |z_1 - z_2|/|1 - \bar{z}_1 z_2|$ =pseudohyperbolic distance. It lies between 0 and 1 with the latter value excluded. The hyperbolic distance $r(z_1, z_2) = \frac{1}{2} \log(1+\varrho)/(1-\varrho)$ satisfies the triangle inequality. Let us recall Schwarz' lemma: If $|f| \leq 1$ then $\varrho(f(z_1), f(z_2)) \leq \varrho(z_1, z_2)$ (we exclude the constant function or define $\varrho(z, z) = 0$ when z is on the boundary of D). Hence an extension of ϱ to all of $M \times M$ is given by

$$\varrho(m_1, m_2) = \sup_{\|f\| \leq 1} \varrho(\hat{f}(m_1), \hat{f}(m_2)) = \sup_{\substack{\hat{f}(m_2)=0 \\ \|f\| \leq 1}} |\hat{f}(m_1)|.$$

We obtain $\varrho(m, z) = 1$ when $z \in D$ if $m \notin D$. *Gleason parts* are equivalence classes for the equivalence relation $m_1 \sim m_2$ if $\varrho(m_1, m_2) < 1$. (Note that $r(m_1, m_2)$ satisfies the triangle inequality though it may take the value $+\infty$.)

Let A be a connected analytic space. A map $A \xrightarrow{\phi} M$ is called analytic if $\hat{f} \circ \phi$ is analytic on A for every $f \in H^\infty$. Then $\phi(A)$ belongs to one Gleason part for if an analytic function on A vanishes at one point $a_1 \in A$ and is ≤ 1 everywhere, it cannot come close to 1 at any point $a_2 \in A$ by the maximum principle. — We shall see that it suffices to take $A = D$; all other analytic maps factor through maps into D.

The analytic maps $D \to M$ form a closed, thus compact, subset of the space M^D of all maps $D \to M$. Set $L_\alpha(z) = (z + \alpha)/(1 + \bar{\alpha}z)$, $\alpha \in D$.

THEOREM [3]. *When $\alpha \to m$ it follows that L_α converges to an analytic map L_m from D to M. The map L_m maps D onto $P(m)$ and is one-to-one unless $P(m) = \{m\}$ so that it must map all of D to m. If G is the union of the closures of all interpolating sequences then*

$$G = \bigcup_{P(m) \neq \{m\}} P(m) = \bigcup \text{ range of all non-trivial analytic maps } D \to M;$$

thus G is open.

We shall first discuss the proof of Theorem [3], assuming Theorem [2]. The main point is a study of Blaschke products, for $f \in H^\infty$ can be written $f = Bh$ where B is a Blaschke product, $h \in H^\infty$ has no zeros and $\|h\| = \|f\|$; we have

LEMMA 1. *If $h \neq 0$ in D and $\hat{h}(m) = 0$ it follows that $\hat{h} = 0$ in $P(m)$.*

PROOF. We can write $h = g^n$ and obtain if $\|h\| = \|g\| = 1$ that

$$\varrho(m, m_1) \geq |\hat{g}(m_1)| = |\hat{h}(m_1)|^{1/n} \to 1 \quad \text{if } \hat{h}(m_1) \neq 0.$$

If $f = Bh$ and $\hat{f}(m) = 0$ it follows that $\hat{B}(m) = 0$ unless $\hat{f}$ vanishes identically in $P(m)$. If $\hat{f}(m_1) \neq 0$ it follows that $\hat{B}(m_1) \neq 0$. For this reason one can concentrate on studying Blaschke products.

The Blaschke product B (or rather $\hat{B}$) is said to have a zero order at least k at m if $B = B_1 \cdots B_k$ with $\hat{B}_j(m) = 0$ for $j = 1, \ldots, k$. The proof of Lemma 1 also gives

LEMMA 2. *If $\hat{B}$ vanishes of infinite order at m it vanishes in $P(m)$.*

Notation. If B is a Blaschke product with zeros z_j we set

$$\delta(B) = \inf(1 - |z_j|^2)|B'(z_j)|.$$

Thus $\delta(B) \leq 1$, and $\delta(B) > 0$ means that $\{z_j\}$ is an interpolation sequence.

FACTORIZATION LEMMA I. *Every B can be written $B = B_1 B_2$ where $(1-|z|^2)|B_1'(z)| \geq |B_2(z)|$ when $B_1(z) = 0$ and vice versa; hence*

$$\delta(B_j) \geq \delta(B)^{\frac{1}{2}}, \quad j = 1, 2.$$

FACTORIZATION LEMMA II. *Let $0 < \delta < 1$, $0 < r < 1$. Then one can find constants a and b so that every Blaschke product B with no zero when $|z| < r$ has a factorization $B = B_1 B_2$ where*

$$a|B_1(z)|^{1/b} \leq |B_2(z)| \leq |B_1(z)|^b/a$$

when z has pseudohyperbolic distance at least δ from the zeros of B.

MAIN LEMMA. *Let B be a Blaschke product and let $m \in M$ be a zero of $\hat{B}$. Then either $\hat{B}$ has a zero of infinite order at m or else m belongs to the closure of an interpolating subsequence of the zero sequence of B.*

PROOF. We may assume that $\hat{B}$ has a zero of order 1 only at m and have to prove the last alternative of the statement. Let $K_\delta(B)$ be the set of points at pseudohyperbolic distance at least δ from all zeros of B. If for some $\delta > 0$ we have $m \in$ closure of $K_\delta(B)$, then factorization lemma II gives $B = B_1 B_2$ with $\hat{B}_1(m) = \hat{B}_2(m) = 0$ for if we approach m within K_δ we know that B tends to 0, hence the two factors must both tend to 0. This is a contradiction, so m belongs to the closure of the δ-pseudohyperbolic neighborhood of the zero sequence for any $\delta > 0$. Now factor $B = B_1 B_2$ according to factorization lemma I. Both factors do not vanish at m; assume $\hat{B}_2(m) \neq 0$. In the δ-pseudohyperbolic neighborhood of any point with $|\hat{B}_2(z)| < |\hat{B}_2(m)|/4$ we have $|B_2| < |\hat{B}_2(m)|/4 + 2\delta < |\hat{B}_2(m)|/2$ if $\delta < |\hat{B}_2(m)|/8$. Hence, for all small δ, m belongs to the closure of the δ-pseudohyperbolic neighborhood of the set T of zeros of B_1 where $|\hat{B}_2(z)| > |\hat{B}_2(m)|/4$. By the conditions in factorization lemma I this is an interpolating sequence. We claim that m is in the closure of T. Indeed, consider a neighborhood of m defined by

$$|\hat{f}_j| < 1, \quad j = 1, 2, \ldots, k, \quad \text{where } \hat{f}_j(m) = 0.$$

For any $\delta > 0$ we can find z, z' with $z' \in T$ and $\varrho(z', z) < \delta$ so that $|f_j(z)| < 1/2$, $j = 1, \ldots, k$; it follows then that $|f_j(z')| < 1/2 + 2\delta \|f_j\|$. If $\delta < 1/(4\|f_j\|)$ we conclude that $|f_j(z')| < 1$; hence m is in the closure of T.

COROLLARY. *If $P(m) \neq \{m\}$, then m belongs to the closure of an interpolating sequence.*

We shall now prove the existence of the limit of L_α when $\alpha \to m$. This is obvious when $P(m) = m$, for every limit must then be equal to the map of D to m. Hence it remains to consider the case when $P(m) \neq \{m\}$ and we can then take an interpolating sequence S with $m \in \bar{S}$. First note the identity $(F \in H^\infty)$

$$(F \circ L_\alpha)'(0) = (1 - |\alpha|^2) F'(\alpha).$$

If we take for F the Blaschke product B of S and take $\alpha \in S$, the modulus of the right hand side is bounded from below. Hence no limit L of L_α when $\alpha \in S$ can be constant. Now if L_1, L_2 are two different limits, then $L_1(0)$ and $L_2(0)$ are different. Indeed, choose disjoint neighborhoods U_1 and U_2 of L_1 and L_2 in M^D and set

$$T_j = \{z \in S; L_z \in U_j\}.$$

These sequences are disjoint, and since $L_z(0) = z$ we have $L_j(0) \in \bar{T}_j$. Now S is an interpolating sequence so we can find $f \in H^\infty$ equal to j in T_j, $j = 1, 2$. Hence $\hat{f}(L_j(0)) = j$, so $L_1(0) \neq L_2(0)$. It follows that there is a unique limit L of L_α when $\alpha \in S$ such that $L(0) = m$. We claim that $L_\alpha \to L$ when $\alpha \to m$ even if α is not restricted to the sequence S. Indeed, we may assume for any $\varepsilon > 0$ that $|B(\alpha)| < \varepsilon$. Since S is interpolating it is easy to show that for small ε this set is contained in the union of the disjoint $C\varepsilon$-pseudohyperbolic neighborhoods of the points of S, where C is a suitable constant. Hence when $|B(z)| < \varepsilon$ we have a unique projection π from z to S, mapping z to the nearest point in S. If $\|f\| \leq 1$ we have

$$|\hat{f} \circ L_\alpha(\zeta) - \hat{f} \circ L_{\pi\alpha}(\zeta)| \leq 2\varrho(L_\alpha(\zeta), L_{\pi\alpha}(\zeta))$$
$$= 2\varrho(\zeta, L_\alpha^{-1} L_{\pi\alpha}(\zeta)) \to 0 \quad \text{for fixed } \zeta \text{ when } \varepsilon \to 0.$$

Hence any limit of L_α when $\alpha \to m$ is also a limit of $L_{\pi\alpha}$, and since the limit must map 0 to m, it is equal to the one found in the beginning of the proof.

By factorization lemma I we can find factors $B_0 = B$, B_1, B_2, ... of B such that m is in the closure of the zeros of each of them and $\delta(B_{j+1}) \geq \delta(B_j)^{\frac{1}{2}}$, hence $\delta(B_j) \to 1$ as $j \to \infty$. Since

$$|(\hat{B}_j \circ L_m)'(0)| \geq \delta(B_j) \to 1, \quad j \to \infty,$$

and since $|\hat{B}_j \circ L_m| \leq 1$, it follows if we normalize the argument of B_j properly that $\hat{B}_j \circ L_m(z) \to z$ as $j \to \infty$. Hence L_m is 1–1. We also obtain $\varrho(L_m(z_1), L_m(z_2)) \geq \varrho(z_1, z_2)$. Now by Schwarz' lemma $\varrho(\hat{f} \circ L_m)(z_1), (\hat{f} \circ L_m)(z_2)) \leq \varrho(z_1, z_2)\|f\|$ when $z \in D$, hence when $z \in M$, so the opposite inequality holds too.

It remains to show that L_m maps D onto $P(m)$. Let $\varrho(m, m_1) < 1$ and choose ε with $\varrho(m, m_1) < \varepsilon < 1$. One can then find δ so that for every Blaschke product A with $\delta(A) \geq \delta$ the set where $|A(z)| < \varepsilon$ consists of disjoint islands around the zeros, with pseudohyperbolic distance from the zeros bounded away from 1; these are mapped by A onto the disc with radius ε. Replacing B by a suitable B_j we may assume that $\delta(B) \geq \delta$. Since $\hat{B}(m) = 0$ we have $|\hat{B}(m_1)| \leq \varrho(m, m_1) < \varepsilon$ so m_1 belongs to the closure of the set of points where $|B(z)| < \varepsilon$. Now let T be a subsequence of S such that $m \in \bar{T}$ and L_z is close to L_m when $z \in T$; we obtain T by intersecting S with a small neighborhood of m. Then $\delta(B_T) \geq \delta(B) \geq \delta$. Since $\hat{B}_T(m) = 0$ we obtain $|\hat{B}_T(m_1)| \leq \varrho(m, m_1) < \varepsilon$ and conclude that m_1 belongs to the closure of the set where $|B_T(z)| < \varepsilon$. Hence there exists a limit L' of L_z for $z \in T$ such that $L'(\zeta) = m_1$ for some ζ with $|\zeta|$ bounded away from 1 by a number depending only on δ. It follows that m_1 belongs to the closure of the range of $L_m(\zeta)$ when ζ belongs to this disc; however, this is a compact set so we conclude finally that m_1 is in the range of L_m. The proof is complete.

We shall now discuss the proofs of Theorems [1] and [2]. First note that Theorem [2] can be formulated as follows:

THEOREM [2]'. *If $f_1, \ldots, f_N \in H^\infty$ and $|f_1(z)| + \cdots + |f_N(z)| \geq \delta > 0$ for some constant δ and all $z \in D$, then there exist $g_1, \ldots, g_N \in H^\infty$ such that*

$$g_1 f_1 + \cdots + g_N f_N = 1.$$

The equivalence with Theorem [2] is proved as follows. Let $m \in M$. If m is not in the closure of D, then there exists a neighborhood of m which does not intersect D, that is, one can find $f_1, \ldots, f_N \in H^\infty$ such that $\hat{f}_j(m) = 0$ but the set defined by $|\hat{f}_1| < 1$, ...,

$|\hat{f}_N| < 1$ does not meet D, hence $|f_1(z)| + \cdots + f_N(z)| \geq 1$ for all $z \in D$. From Theorem [2]' we obtain $g_j \in H^\infty$ such that $\sum_1^N \hat{g}_j \hat{f}_j = 1$, which gives a contradiction at the point m. On the other hand, if D is dense in M and $|f_1(z)| + \cdots + |f_N(z)| \geq \delta$ when $z \in D$, then this inequality extends to M and it follows that the ideal generated by $f_1, \ldots, f_N$ is not contained in any maximal ideal so it is equal to H^∞, hence Theorem [2] is valid.

When proving Theorems [1] and [2]' we shall modify Carleson's arguments slightly and emphasize the solution of the inhomogeneous equation

(3) $$\partial u / \partial \bar{z} = \mu \quad \text{in } D.$$

Here μ is assumed to be a measure with finite total mass, and the equation shall be understood in the sense of distribution theory. If ϕ is an integrable function on ∂D we shall say that $u = \phi$ on ∂D if there exists a distribution U with support in $\bar{D}$ such that $U = u$ in D and

(4) $$\partial U / \partial \bar{z} = \mu - \phi \, dz / 2i,$$

where $\phi \, dz$ is of course a measure with support on ∂D and μ is extended so that there is no mass in the complement of D. If $u = 0$ then (4) implies that $U = 0$, for $\partial U / \partial \bar{z}$ would otherwise be a distribution with support on ∂D with positive transversal order. Hence u determines both μ, ϕ and U uniquely, so it is legitimate to say that ϕ is the boundary value of u when (4) is valid.

If $u \in H^p$ for some $p \geq 1$, then ϕ coincides with the boundary value in the usual sense. Conversely, if u is analytic and has boundary values belonging to $L^p(\partial D)$ in the sense of (4), then $u \in H^p$, $p \geq 1$. If $f \in H^\infty$ and u is a solution of (3) with boundary value ϕ, then fu satisfies (3) with μ replaced by $f\mu$ and has boundary values $f\phi$. This is obvious when f is analytic in a neighborhood of $\bar{D}$ and follows in general if we first consider $f(rz)$ with $r < 1$ and then let $r \uparrow 1$, noting that the solution $U \in \mathcal{E}'(\bar{D})$ of the equation $\partial U / \partial \bar{z} = F$ is a continuous function of $F \in \mathcal{E}'(\bar{D})$ when it exists.

The existence of a solution of (4) with support in $\bar{D}$ means precisely that the right hand side is orthogonal to all (entire) analytic functions. Thus (3) has a solution with boundary values ϕ if and only if for entire analytic f

$$\int_D f \, d\mu - (2i)^{-1} \int_{\partial D} \phi(z) f(z) \, dz = 0.$$

In view of the Hahn-Banach theorem there exists a solution with boundary values $\leq C$ if and only if for entire analytic f

$$\left| \int_D f \, d\mu \right| \leq C \int_{\partial D} |f(z)| \, dz / 2.$$

A sufficient condition for this is given by the following lemma of Carleson [1], [2].

LEMMA 3. *There is a constant C such that*

(5) $$\int_D |v(z)|^p \, |d\mu(z)| \leq CM \int_{\partial D} |v|^p \, |dz|, \quad v \in H^p(D), \ p > 0,$$

for every measure μ in D such that

(6) $$|\mu|\{\zeta; |\zeta - z| < r\} \leq Mr; \quad z \in \partial D, \ r > 0.$$

Conversely, if (5) is valid for some CM then (6) is valid for some M.

It suffices to prove (5) for positive harmonic v when $p = 2$, for $|v(z)|^{p/2}$ is a subharmonic function; it is then closely related to the Hardy-Littlewood maximal theorem. (An extension to several complex variables will be published by the speaker in Math. Scand.)

From Lemma 3 we obtain

LEMMA 4. *If* (6) *is valid, it follows that the equation* $\partial u/\partial \bar{z} = \mu$ *has a solution with boundary values bounded by* CM.

PROOF OF THEOREM [1]. Let z_j be an interpolating sequence and let w_j be a sequence of complex numbers with $|w_j| \leq 1$. If u is a bounded analytic function with $u(z_j) = w_j$, $j = 1, 2, \ldots$ and if we set $v = u/B$, where B is the Blaschke product of the sequence, then

$$\partial v/\partial \bar{z} = \sum (u(z_j)/B'(z_j))\pi\delta(z_j) = \pi \sum (w_j/B'(z_j))\delta(z_j) = f.$$

Conversely, if we can prove that this equation has a solution with bounded boundary values, then $u = Bv$ will be analytic, have bounded boundary values and satisfy the interpolation problem. In view of (1)$'$, we have

$$|w_j/B'(z_j)| \leq (1 - |z_j|^2)$$

and it follows from the standard relations between convergence of products and sums that

$$\sum_{|z_j - z| < r} (1 - |z_j|^2) \leq Cr, \quad z \in \partial D,$$

if one notes that $|B(z)|$ is bounded from below outside the disjoint ε-pseudohyperbolic neighborhoods of the points z_j. Theorem I now follows from Lemma 4.

The main step in the proof of Theorem [2] is contained in

LEMMA 5. *Let* $f_1, \ldots, f_N \in H^\infty$ *and assume that*

$$|f_1(z)| + \cdots + |f_N(z)| \geq \delta > 0, \quad z \in \partial D.$$

For sufficiently small $\varepsilon > 0$ *one can then find a partition of unity* ϕ_j *subordinate to the covering of* D *by the open sets* $D_j = \{z; |f_j(z)| > \varepsilon\}$ *such that* $\partial\phi_j/\partial\bar{z}$, *defined in the sense of distribution theory, is a measure which satisfies* (6) *for all* j *and some* M.

PROOF OF THEOREM [2]$'$ MODULO LEMMA 5. Since $1 = \sum f_j\phi_j/f_j$ we obtain a solution to our problem by setting

$$g_j = \phi_j/f_j + \sum_1^N a_{jk}f_k$$

provided that a_{jk} can be chosen so that a_{jk} has bounded boundary values and

1° $a_{jk} = -a_{kj}$
2° g_j is analytic, that is,

$$\sum_1^N f_k \partial a_{jk}/\partial\bar{z} + \frac{1}{f_j}\partial\phi_j/\partial\bar{z} = 0.$$

Since $\sum \partial\phi_k/\partial\bar{z} = 0$ the condition 2° is fulfilled if

$$\partial a_{jk}/\partial\bar{z} = ((\bar{f}_j/f_k)\partial\phi_k/\partial\bar{z} - (\bar{f}_k/f_j)\partial\phi_j/\partial\bar{z})/\sum_m |f_m|^2.$$

The right-hand side is skew symmetric in j and k and satisfies (6). Hence it follows from Lemma 4 that one can find a_{jk} satisfying these equations and antisymmetric in j and k so that the boundary values are bounded functions. This proves Theorem [2]$'$.

Now the set of bounded functions ϕ with $\partial\phi/\partial\bar{z}$ satisfying (6) forms a ring, so the standard technique for constructing partitions of unity can be applied to derive Lemma 5 from

LEMMA 6. *There exists a constant k such that if $0 < \varepsilon < \frac{1}{2}$ and $f \in H^\infty$, $\|f\| \leq 1$, one can find ϕ with $0 \leq \phi \leq 1$, $\phi(z) = 0$ when $|f(z)| < \varepsilon^k$, $\phi(z) = 1$ when $|f(z)| > \varepsilon$, so that $\partial\phi/\partial\bar{z}$ satisfies (6).*

PROOF. By the Schwarz lemma the pseudo-hyperbolic distance between the two sets $\{z; |f(z)| > \varepsilon\}$ and $\{z; |f(z)| < \varepsilon^k\}$ is at least $\frac{1}{2}(\varepsilon - \varepsilon^k) > 0$. (There is nothing to prove unless these two sets are non-empty.) Let $\chi(t)$ be a Lipschitz continuous function which is 1 when $t = 0$ and 0 when $t > \frac{1}{2}(\varepsilon - \varepsilon^k)$. Then $\psi(z) = \chi(\delta(z))$ where $\delta(z)$ is the pseudo-hyperbolic distance from z to $\{z; |f(z)| > \varepsilon\}$ satisfies the conditions

$$
\psi(z) = 0 \ \text{ when } \ |f(z)| < \varepsilon^k, \quad \psi(z) = 1 \ \text{ when } \ |f(z)| > \varepsilon,
$$
(7)
$$
|\operatorname{grad}\psi(z)| \leq C/d(z),
$$

where $d(z) = 1 - |z|$ is the boundary distance from z. Obviously (7) does not suffice to give the estimate (6) for $\partial\psi/\partial\bar{z}$ but we do get such control of $\int |\partial\psi/\partial\bar{z}|$ if the integral is taken over a set such that $d(z)$ never changes by more than a fixed factor on its intersection with any normal to ∂D. The function ϕ will contain ψ as an essential ingredient: in particular we can take $\phi = \psi$ on any compact subset of D and restrict attention to what happens near the boundary. We need the following result, valid for any dimension if D is replaced by any set with a C^2 boundary.

LEMMA 7. *Let π denote the projection from D to ∂D. For every $a > 0$ there is a constant C_a such that for every subharmonic function $u \leq 0$ in D*

$$
\sigma(\{\pi\zeta; |\zeta - z| \leq ad(z), \ u(\zeta) \leq tu(z)\}) \leq C_a d(z)/t,
$$

where σ denotes arc length on ∂D.

This is a simple consequence of the Hardy-Littlewood maximal theorem. (Carleson uses closely related estimates of harmonic measures due to Beurling and Hall.)

By a standard arc on ∂D we shall mean an arc obtained by bisecting the range $0 \leq \theta \leq 2\pi$ for the argument one or more times; if I is a standard arc with length $2\pi 2^{-N}$ we let $D(I)$ be the set of all z with $\arg z \in I$ and $d(z) \leq 2^{-N}$ and call $D(I)$ the standard sector corresponding to I. Note that standard arcs (or sectors) either have no inner points in common or else one contains the other. Since every arc is contained in the union of two standard arcs of at most twice the length, it is clear that (6) holds for $\partial\phi/\partial\bar{z}$ if for every standard arc I we have

(6)′
$$
\int_{D(I)} |\partial\phi/\partial\bar{z}| \, d\lambda \leq C\sigma(I).
$$

We shall define ϕ successively in decreasing standard sectors $D(I)$. Let us consider the situation when we have already handled the construction outside a certain $D(I_0)$. We distinguish two cases:

1° $|f(z)| < \varepsilon$ in the upper half of $D(I_0)$, that is, $|f(z)| < \varepsilon$ outside the two sectors corresponding to the two halves of I_0. We are then allowed to set $\phi = 0$ in the upper half of $D(I)$. Continuing this choice as far as possible, we let $\mathcal{I}$ be the set of maximal standard arcs $I \subset I_0$ such that $|f(z)| \geq \varepsilon$ for some z in the upper half of $D(I)$. In

$$
D'(I_0) = D(I_0) \setminus \bigcup_{I \in \mathcal{I}} D(I)
$$

we then have $|f(z)| < \varepsilon$ so we are allowed to choose $\phi = 0$ there. Define $\phi_{I_0} = 0$ everywhere for reasons of symmetry with the next case.

2° Suppose that $|f(z)| \geq \varepsilon$ for some z in the upper half of $D(I_0)$. If k is sufficiently large it follows by applying Lemma 7 to $u = \log|f|$ that if $E = \{\pi\zeta; \zeta \in D(I_0), |f(\zeta)| < \varepsilon^{k/2}\}$, then

$$(8) \qquad\qquad \sigma(E) < \sigma(I_0)/2.$$

Let $\mathcal{I}$ be the set of maximal standard arcs contained in E; thus

$$(9) \qquad\qquad \sum_{I\in\mathcal{I}} \sigma(I) \leq \sigma(I_0)/2.$$

We define $D'(I_0)$ as above. If $z \in D'(I_0)$ and $|f(z)| < \varepsilon^k$ so that we are forced to have $\phi(z) = 0$, then it follows by the Schwarz lemma for a certain constant $\gamma > 0$ (depending on ε) that an interval of length $4\gamma d(z)$ around πz belongs to E. Hence we can find $I \in \mathcal{I}$ so that $\pi z \in I$ and $\sigma(I) \geq \gamma d(z)$; on the other hand we have $d(z) \geq \sigma(I)/2\pi$ since $z \notin D(I)$. Note that the ratio between these two bounds for $d(z)$ is bounded by a fixed number $2\pi/\gamma$. Now set

$$\phi_{I_0}(t) = 0 \quad\text{outside}\quad D'(I_0)$$
$$= 1 \quad\text{in}\quad D'(I_0) \quad\text{except that for}\quad I \in \mathcal{I} \quad\text{we define}\quad \phi_{I_0} = \psi$$
$$\text{in}\quad \{z; \pi z \in I, \sigma(I)/2\pi \leq d(z) \leq \sigma(I)/\gamma\}.$$

Then ϕ_{I_0} takes the values required for ϕ in $D'(I_0)$; we have

$$(10) \qquad\qquad \int_{D(I')} |\partial\phi_{I_0}(z)/\partial\bar{z}|\, d\lambda \leq C\min(\sigma(I'), \sigma(I_0))$$

if I' is any standard arc. Indeed, this follows from the properties of ψ pointed out initially and the fact that $\sum_{I\in\mathcal{I}, I\subset I'} \sigma(I) \leq \min(\sigma(I'), \sigma(I_0))$.

Starting with $\mathcal{I}^0$, the set composed of the two standard arcs with $\sigma(I) = \pi$, we can with these procedures choose successive generations of standard arcs $\mathcal{I}^k$; the passage from $\mathcal{I}^k$ to $\mathcal{I}^{k+1}$ is obtained by applying the procedure 1° or 2° to each arc in $\mathcal{I}^k$. At least every other step is of type 2°, so the sum of the lengths of the arcs in $\mathcal{I}^k$ falls off exponentially.

We set $\phi = \sup_{I\in\mathcal{I}} \phi_I$ where $\mathcal{I} = \cup\mathcal{I}^j$. Then $0 \leq \phi \leq 1$ and $\phi = 0$ or $\phi = 1$ where required. Let I be a standard arc. Clearly $\partial\phi/\partial\bar{z}$ can at most contain a measure of linear density 1 supported by the boundary of $D(I)$, so it suffices for us to consider the contribution to $(6)'$ inside $D(I)$. If $I \subset I' \in \mathcal{I}$ and $\phi_{I'}$ is not identically 0 in the interior of $D(I)$, then I is not contained in any of the arcs of next generation constructed inside I'; clearly this defines I' uniquely. In view of (9) the sum of the lengths of the arcs belonging to $\mathcal{I}$ which are contained in I is at most equal to $2\sigma(I)$. Outside of the boundaries of the corresponding sectors, whose total length is bounded by $C\sigma(I)$, we have $\phi = \sum_{J\in\mathcal{I}} \phi_J$, and $(6)'$ therefore follows from (10). This completes the proof of Lemma 6; hence we have proved Theorem [2]'.

SOME PROBLEMS CONCERNING LINEAR PARTIAL DIFFERENTIAL EQUATIONS

LARS HÖRMANDER

As support for future work of myself and of students I plan to collect here a number of problems related more or less to the material in my Springer book.[12]

1. Existence of linear right inverse. (Problem from Trèves.) When does there exist a map $E : C^\infty(\mathbf{R}^n) \to C^\infty(\mathbf{R}^n)$ which is linear and continuous and has the property that $P(D)E =$identity. Trèves conjectures that this is possible only in the hyperbolic case. — The property means that for any bounded O there exists a bounded O' such that if $f = 0$ in O' the equation $P(D)u = f$ has a solution with $u = 0$ in O. In particular, there is a fundamental solution vanishing in $O + t$ for any large enough translation t. This is an equivalent property.

2. Existence of solutions for real analytic right hand side. (Raised by Trèves and also by Ehrenpreis.) Let f be real analytic in $\mathbf{R}^n$. (Analogous problem in open sets.) Does there always exist a real analytic solution u of $P(D)u = f$? We know of course that one can be found in any non-quasianalytic class of functions. I am doubtful in general.

3. Properties of the comparison relation. Since this has turned out to fit surprisingly well to all sorts of problems it seems motivated to explore its properties further. For example, what can one say about the topological structure of the sets

$$\{Q; Q \prec P\} \quad \text{and} \quad \{Q; P \prec Q \prec P\}$$

for given P. When P is elliptic the second question reduces to one asked by Gelfand in his paper on the index problem.

4. Singular supports of solutions. The general problem is to decide when a set can be the singular support of a solution of a differential equation. Present work should give a number of such results. Applications should be made to the study of geometric properties of strong P-convexity. In particular, it should be decided if, as seems likely, strong P-convexity and P-convexity is the same thing when $n = 2$.

5. Admissible lower order terms in a hyperbolic operator. The improved understanding of zeros of a polynomial given by my paper on the characteristic Cauchy problem may perhaps at last prove the conjecture.

---

[12]This was written in Princeton, probably during the early Spring 1968 when the return to Sweden was close. Comments on later developments have been added at the end.

© Springer International Publishing AG, part of Springer Nature 2018
L. Hörmander, *Unpublished Manuscripts*,
https://doi.org/10.1007/978-3-319-69850-2_14

6. Supports of fundamental solutions. Let K be a convex cone with vertex at the origin. When does $P(D)$ have a fundamental solution with support in K. So far the results are complete if K is a proper cone (Gårding) or a half space (my evolution operators). For intermediate cases an analogue of Petrowsky's results has been given by Gindikin. These should be refined in analogy with my paper on evolution operators.

7. Regularity of fundamental solutions. It should be decided for evolution operators and more generally for the operators encountered in 6) whether the fundamental solution with support in K can be chosen in $B^{\mathrm{loc}}_{\infty, \tilde{P}}$. Can one find it with exponential growth? Possible relations between the exponential growth and the geometric facts reflected by the constant A_1 in my paper. What are the properties of the tempered fundamental solution of an operator satisfying Petrowsky's condition? (Know it is proper for Schrödinger equation.)

8. Global uniqueness theorems. Let $P(D)$ be a differential operator, let $\langle x, N \rangle = 0$ be a non-characteristic plane; assume that P_m is not hyperbolic with respect to N. Let K be a convex closed set in $\langle x, N \rangle = 0$. When does there exist a solution of $P(D)u = 0$ (in a possibly onesided neighborhood of the plane) with Cauchy data vanishing outside K, $u \neq 0$. We know by John's theorem that K cannot be compact. Furthermore, we know that if $\langle x, \xi \rangle = C$ is a supporting plane of K, then John-Holmgren shows that if $P_m(s\xi + tN)$ is elliptic in s, t, then $u = 0$ identically. We must therefore require at least one real zero for each ξ. For the wave equation and a timelike plane this means that there must not be any timelike supporting line. Conversely, if that is so we conclude that K contains a hyperbola, intersection of a light cone with vertex perhaps at some point outside the plane. The fundamental solution at that point then has the required property. — By superposition we find in this case that every K having no timelike supporting line can be exactly equal to the support of the Cauchy data. — If K is a half plane the condition that there exists at least one characteristic mentioned above is of course also sufficient (the existence of null solutions).

9. Solvability. There is an obvious problem in extending the results of Chapter VI in my book. Trèves is working on it now but I should return to it later.

10. Pseudodifferential operators related to operators of principal type. The constructions used by Courant, Lax, Gårding, Leray, Ludwig and others to determine the singularities of the fundamental solution of the hyperbolic equations should be refined to a systematic technique playing a role for operators of principal type analogous to pseudodifferential operators in the elliptic case. The construction of fundamental solutions in that setup should of course not be purely algebraic in terms of the symbol but reduce to solution of first order differential equations. The constructions should yield fundamental solutions with singularities along "one half" of the characteristic conoid and thereby give back the results on continuation of singularities in Chapter VIII of my book.

11. The uniqueness of the Cauchy problem. After the counterexample by Plis the interest has died out here. However, there are very many problems left.

a) Investigate strong uniqueness. At this time it seems to be known only for second order operators though I have a feeling that I heard that Agmon has proved it for the square of the Laplacean plus lower order terms.

b) Study the construction of counterexamples more systematically. Plis had some conjectures in his lecture at Stockholm but they did not seem right from an invariance point of view.

c) Led by b) one should improve the uniqueness theorems in my book.

d) Investigate the influence of real coefficients. This should allow more multiplicities.

12. Non-elliptic boundary problems. Beyond my paper in the Annals there are the interesting notes of Egorov and Kondratev. These should be put in a general context and properties of the index should be studied.

13. The exactness of the Spencer sequence. In connection with 12 one should again try to do something about the problems left open in my Annals paper and apply the result to the Spencer sequence. The use of weight functions should also be explored.

14. Universal hypoellipticity. My paper on the second order case suggests the following problem: Which are the non-commutative polynomials $P(X_1, \ldots, X_r)$ such that the differential operator obtained by replacing X_j by first order operators, whose Lie algebra (generated) has full rank at every point, must always be hypoelliptic. Perhaps one should develop a functional calculus based on such operators.

15. Boundary problems for non-elliptic operators. In my thesis the existence of abstract boundary problems was proved, that is, the existence of boundedly invertible operators $\widehat{P}$ with $P_0 \subset \widehat{P} \subset P$. The next question originally intended has hardly been touched apart from a small part of Thomées thesis, but the techniques developed for the non-characteristic Cauchy problem together with some unpublished manuscripts of mine on the two-dimensional case should give more. The problem is to find for given differential boundary conditions on a plane piece of the boundary when the analogue is true if P is restricted by these boundary conditions and P_0 is extended to allow Cauchy data with compact support in the plane piece of boundary satisfying the boundary conditions.

16. Approximation theorem for boundary problems. In connection with my paper on the approximation of solutions of convolution equations I have solved this problem originally intended for Kiselman's licentiate thesis, apart from the precise conditions to be put on the set where the approximation is to be made. This should be completed to a full real analogue of his results in the complex domain.

17. Analytic continuation of solutions of elliptic boundary problems. Let O be a convex subset of a half space and consider solutions of an elliptic differential equation with elliptic boundary conditions on the part of the boundary which lies in the boundary of the half space. Which is the largest set in the real (complex) domain to which the solutions can all be extended? This seems closely related to the methods being developed for 16.

18. Continuation of solutions of boundary problems. Problem 17 has a meaning also in the non-elliptic case. Attention should be paid to papers by Hans Lewy and his students (Filippenko?).

19. Mixed problems for hyperbolic equations. It seems that one is far from a complete understanding of the conditions for correctness. See e.g. the article by Agmon in the Paris symposium 1962.

20. Asymptotic properties of eigenfunctions. It should be possible to improve all results to $\sigma < 1$. Perhaps one should change the present attitude where one first constructs a fundamental solution approximately and then the corresponding approximation to the spectral function by inverting the Stieltjes transform. It may be better to guess directly what the approximation to the spectral function is going to be and define the parametrix as the Stieltjes transform. The solutions of the Hamilton-Jacobi equations will of course have to play a fundamental role.

21. L^p theory of the $\bar{\partial}$ operator. Since the present results all come by an integration by parts they are restricted to L^2, and it does not seem very likely that they remain valid for all L^p classes. However, it would lead to many consequences if they did so the question should be examined carefully.

22. Division of distributions. Let f be a real valued function in Ω. We wish to define a distribution $1/f$ as a limit of $1/(f + i\varepsilon)$ when $\varepsilon \to 0$. (Note that this occurs for the case $f(\xi) = |\xi|^2 - 1$ in scattering theory.) For which f does the limit exist? Let $u \geq 0$ have compact support and be 1 in a certain compact set $K \subset \Omega$. Then we have

$$\text{Im} \int -u/(f + i\varepsilon)\, dx = \varepsilon \int u/(f^2 + \varepsilon^2)\, dx \geq \int_K \varepsilon/(f^2 + \varepsilon^2)\, dx.$$

Let $\mu(t) = m\{x; x \in K; |f(x)| \leq t\}$. Then $\mu(t) =$ constant$= m(K)$ for large t and we obtain

$$\int \varepsilon/(f^2 + \varepsilon^2)\, dx = \varepsilon \int_0^\infty d\mu(t)/(t^2 + \varepsilon^2) = 2\varepsilon \int_0^\infty \mu(t) t\, dt/(t^2 + \varepsilon^2)^2 \leq C$$

if the desired limit exists. Hence

(1) $$\mu(t) \leq 2Ct$$

as follows if we restrict the integration to (t, ∞). Conversely, this implies that

$$\int \varepsilon/(f^2 + \varepsilon^2)\, dx \leq 2C\varepsilon \int t d(-1/(t^2 + \varepsilon^2)) = 2C\varepsilon \int_0^\infty dt/(t^2 + \varepsilon^2) = \pi C.$$

Hence (1) is a necessary and sufficient condition for $\text{Im}(f + i\varepsilon)^{-1}$ to remain bounded in $\mathcal{D}'(\Omega)$. Problem: What is needed in addition to (1) to imply the existence of the limit. Something is needed in general. For if $f = x_1^2 + x_2^2$ then $\lim \text{Im}\, 1/(f + i\varepsilon)$ exists but the positivity of f shows that the limit of $1/(f + i\varepsilon)$ does not exist.

The condition (1) implies in particular that the zeros of f are of order $\leq n =$ dimension and that n^{th} order zeros are isolated.

23. Asymptotic behavior of Fourier-type integrals. Let f be real valued. Then Ono has asked for the conditions in order that

$$\int u(x) e^{i\xi f(x)}\, dx$$

shall belong to L^1 for all $u \in C_0^\infty$ (or $\mathcal{S}$, which should amount to the same). More generally, Weil has asked the following question: Let $f \in C^\infty(\Omega, T^*)$ where $\Omega \subset X = \mathbf{R}^n$ and $T = \mathbf{R}^m$. What it the condition in order that

$$\int u(x) \exp(2\pi i \langle t, f \rangle)\, dx \in L^1(T), \quad u \in C_0^\infty(\Omega).$$

This should be closely related to 22, generalized to a question about boundary values at any point on the real axis. There are simple necessary conditions, such as $n \geq m$ and that for $n = m$ it is necessary and sufficient that the Jacobian is nonsingular at every point in Ω. Surjectivity of f' is always sufficient. However, the question seems difficult in general.

24. Unique continuation. (Cf. 11 above.) Try to construct a solution of the Tricomi equation + lower order terms which are 0 in the elliptic half plane but not in the hyperbolic one. Note that Chapter VIII in my book shows that the opposite cannot occur.

25. Unique continuation. Examine estimates such as in Chapter VIII where τ is replaced by $1/\delta$ where $\operatorname{supp} u \subset \Omega_\delta = \{x; |x| < \delta\}$. Should lead to results containing those of Aronszajn, Cordes, Heinz.

26. Nonexistence theorems. Try to prove global nonexistence when there is a closed bicharacteristic. Investigate for possible nonexistence equations for which there is no half plane into which all "turning" bicharacteristics point. Example: product of Tricomi equations for three half planes + lower order terms.

27. Domination. P, Q constant coefficients. We say that P dominates Q locally if to every $\varepsilon > 0$ there exists a domain Ω and a positive measure μ with $\mu(\Omega) \neq 0$ such that

$$\int |Qu|^2 \, d\mu \le \varepsilon^2 \int |Pu|^2 \, d\mu, \quad u \in C_0^\infty(\Omega).$$

By regularisation one sees easily that it is always possible to take the Lebesgue measure in the definition. This domination then turns out to be equivalent to the one in my book, that is,

$$\sup_\xi |Q(\xi)|^2 / \tilde{P}(\xi, t)^2 \to 0, \quad t \to \infty.$$

Now consider global domination where we require μ to be *fixed*. Again we can choose μ with continuous density — if we shrink Ω slightly. However, the answer does not seem in any way obvious. The question is of course suggested by Trèves' thesis and an answer leads to existence theorems for perturbed equations in large sets. Sufficient conditions were given in my first paper in Math. Scand. on the Cauchy problem.

28. Multipliers on Fourier transforms. The first problem which should be solved is whether the characteristic function of a bounded open set with smooth boundary — say a ball — is a multiplier on FL^p when $|1/p - 1/2| < 1/2n$.

29. Asymptotic properties of eigenfunctions, second order case. In the case of a sphere with radius R, in n dimensions, the explicit formulas give if my calculations are correct

$$\left| e(x, x; \lambda) - (2\pi)^{-n} \int_{p(x,\xi)<\lambda} d\xi \right| \le \frac{n^2 + 2n}{4R^2} e(x, x; \lambda) \lambda^{-2/n} (1 + o(1)).$$

Is it possible that this holds at any point x on a compact Riemannian manifold such that at the corresponding point on the universal covering the exponential map is diffeomorphic up to radius πR? ($n \ge 1$.)

Comments added in November 1997

1. Negative results were quickly obtained by Trèves' student Cohoon. Very complete results have been obtained by R. Meise, B. A. Taylor and D. Vogt in *Continuous linear right inverses for partial differential operators with constant coefficients and Phragmén-Lindelöf conditions*, Lecture notes in pure and applied math. **150**(1994), 357–389. It is curious that there is a close relation to [P58].

2. The first counterexamples were given by DeGiorgi, Cattabriga and Piccinini, and I gave a general solution in [P58].

3. In [B5], Theorem 10.4.7, I proved that the set of Q with the same strength as a given P is a domain of holomorphy, but the topological properties have probably not been explored. It may be an uninteresting and unrewarding task.

4. A great deal of progress was made in [P45], [P47] particularly in the constant coefficient case. The last question was answered affirmatively by Corollary 7.3 in [P47].

5. The conjecture on the lower order terms was proved by Leif Svensson in his thesis *Necessary and sufficient conditions for the hyperbolicity of polynomials with hyperbolic principal part*, Ark. Mat. **8**(1968), 145–162, just before I arrived in Lund. A somewhat simpler presentation is given in Section 12.4 of [B5].

6. I assigned this problem to my student Arne Enqvist. Partial results were given in his thesis *On fundamental solutions supported by a convex cone*, Ark. Mat. **12**(1962), 1–40, and it seems very hard to solve it completely.

7, 8. I don't know of any progress on these problems.

9. Solvability has of course been studied extensively for operators of principal type; a survey is given in [P108]. However, the reference to Chapter VI indicates operators of constant strength. In that case my student Gudrun Gudmundsdottir obtained results on global solvability in her thesis *Global properties of differential operators of constant strength*, Ark. Mat. **15**(1977), 169–198. They are contained in Chapter XIII of [B5].

10. This is the program for the development of Fourier integral operators and the goals set in the problem have been more than fulfilled.

11. This problem was taken up systematically in [P61] and numerous later papers such as S. Alinhac, *Non-unicité du problème de Cauchy*, Ann. of Math. **117**(1983), 77–108. In particular it has turned out that strong uniqueness is very rarely true except for second order equations. In that case [P81] is a fairly general result, included in Chapter XVII of [B5].

12. The thesis of my student Johannes Sjöstrand, *Operators of principal type with interior boundary conditions*, Acta Math. **130**(1973), 1–51, grew out of this problem. The oblique derivative problem itself was studied in great generality in joint work of Melin and Sjöstrand as well as by many others.

13. As far as I know there are no general results yet.

14. If the problem has not been solved as stated there is anyway an enormous related literature such as Anders Melin, *Parametrix constructions for right invariant differential operators on nilpotent groups*, Ann. Glob. Analysis and Geometry **1**(1983), 79–130, and the literature quoted there.

15, 16, 17, 18. I know no further work on these problems except Kiselman's paper *Prolongement des solutions d'une équation aux dérivées partielles à coefficients constants*, Bull. Soc. Math. France **97**(1969), 329–356. There he determined the largest convex set to which all solutions of a differential equations with constant coefficients in a given convex set can be continued. This can also be obtained from the Ehrenpreis "fundamental principle". However, no boundary conditions are present in this result.

19. There has been a great deal of progress on this problem. In the case of constant coefficients and a flat boundary it is reported in Section 12.9 of [B5], and for operators with variable coefficients there are results of Kreiss, Sakamoto and others quoted in the references there.

20. This problem was solved in [P44] toward the end of the Spring term 1968, and since [P42] is referred to in problem 5, it follows that the list of problems was made in the early Spring 1968.

21. Such results have come out of the integral formulas of Henkin, Ramirez and others including my student Øvrelid. There are also numerous results with Hölder classes and so on.

22. This problem was given to Anders Melin who produced a number of discouraging examples which suggested that there is no nice solution. The problem was therefore dropped and replaced by what became Melin's inequality.

23. The literature contains of course many cases of this problem but I do not know any general answer.

24. My manuscript carries a handwritten notation that this is not possible, but I do not remember the motivation.

25, 26, 27. I have no further information on these questions.

28. Charles Fefferman proved in *The multiplier problem for the ball*, Ann. of Math. **94**(1971), 330–336, that the answer is negative for every dimension $n > 1$. However, when $n = 2$ Carleson and Sjölin proved that it suffices to take the Riesz mean of any positive order, and this was simplified in [P57]. For higher dimensions only less precise results are known.

29. This problem was undoubtedly motivated by the good Landau remainder term for constant coefficient operators on the torus discussed at the end of [P36]. The search for a second term or a replacement for it took another direction. I suppose that the answer to the question raised here is negative but have no certain evidence. See also Chapter XXIX in [B7].

Altogether, many of these problems led to fruitful research. They were in particular useful both for me and some of my students.

REMARKS ON CONVEXITY WITH RESPECT
TO OPERATORS OF REAL PRINCIPAL TYPE

LARS HÖRMANDER

0. Introduction. [13] *P*-convexity with respect to singular supports has been completely characterized in geometrical terms when P is an operator of principal type. (See [2, section 1.4].) However, no satisfactory geometrical interpretation of P-convexity (with respect to supports) is known even when P is of real principal type as we shall assume throughout this paper. Malgrange [4] (see also [1, section 3.7]) has proved that if an open set $X \subset \mathbf{R}^n$ is P-convex it follows that at every smooth characteristic point on the boundary ∂X the normal curvature of ∂X in the direction of the corresponding bicharacteristic is non-negative. On the other hand, it was proved in [1] that X is P-convex if all such curvatures are positive. For the case where the bicharacteristics are tangents of higher order additional results were given by Treves [5] (see also Zachmanoglou [6]).

The improvement of Holmgren's uniqueness theorem given in [3] leads to considerably weaker sufficient conditions for P-convexity which we discuss in section 1. They require only that cylinders contained in X cannot have a boundary point in ∂X with characteristic tangent plane and the corresponding bicharacteristic along the generator if the ends do not meet ∂X. In order to examine to what extent this is also a necessary condition we give in sections 2 and 3 a rather complete description of the conditions for uniqueness of the Cauchy problem across a convex cylinder. The conditions connect the length of the cylinder to the curvature. As an application we obtain necessary conditions for P-convexity and also an example given in section 4 which shows that the sufficient conditions of section 1 are not necessary. [14]

1. A sufficient condition for P-convexity. In this section we shall give an improved version of Theorem 1.3.6 in [2]. Let $P(D)$ be a differential operator of real principal type and let X be an open set in $\mathbf{R}^n$. One of the equivalent definitions of P-convexity for X (with respect to supports) is that

$$(1.1) \qquad d(\operatorname{supp} P(-D)v, \complement X) = d(\operatorname{supp} v, \complement X), \quad v \in \mathcal{E}'(X).$$

Here d denotes the distance with respect to say the Euclidean norm.

Suppose now that X is *not* P-convex. Then we can find $v \in \mathcal{E}'(X)$ such that $\operatorname{supp} v$ contains points with distance $< a = d(\operatorname{supp} P(-D)v, \complement X)$ to $\complement X$. Let $x_0 \in \operatorname{supp} v$ have minimal distance δ to $\complement X$ and choose $y_0 \in \complement X$ with $|x_0 - y_0| = \delta$. Then $v = 0$ in $B = \{x; |x - y_0| < \delta\}$ so it follows from Holmgren's uniqueness theorem that the tangent plane B_{x_0} at x_0 is characteristic. Theorem 8.2 in [3] implies that if I is the largest interval

---

[13]This paper was mainly written while the author enjoyed the hospitality of IHES in Bures.

[14]Section 4 has been omitted since it contained a serious error caused by a missing term.

© Springer International Publishing AG, part of Springer Nature 2018
L. Hörmander, *Unpublished Manuscripts*,
https://doi.org/10.1007/978-3-319-69850-2_15

on the corresponding bicharacteristic such that $x_0 \in I$ and $d(x, \complement X) < a$ for $x \in I$, then $I \subset \operatorname{supp} v$. Hence the distance from I to $\complement X$ is at least equal to δ and I is a finite open interval. Let Y be the union of the open discs with radius δ and center in I lying in planes orthogonal to I. Then $Y \subset X$ and y_0 is on the cylindrical part of ∂Y while the two discs in the boundary of Y are at distance at least $a - \delta$ from $\complement X$. Note that the tangent plane of Y at y_0 is parallel to the tangent plane of B at x_0 so it is characteristic and the axis of Y has the direction of the corresponding bicharacteristic L through y_0. Since Y can be replaced by a smaller cylinder Y_1 of the same length containing y_0 such that $\overline{Y}_1 \cap \complement X \subset L$, we have proved

THEOREM 1.1. *X is P-convex if X contains no open finitely truncated circular cylinder Y for which*

   a) *the tangent plane of ∂Y at some $y \in \partial X \cap \partial Y$ is characteristic and the corresponding bicharacteristic L has the direction of the axis of Y.*
   b) *$\overline{Y} \cap \complement X \subset L$ and does not meet the flat ends of $\overline{Y}$.*

On the other hand, the existence of such a cylinder Y is clearly ruled out by (1.1) if there is some $v \in \mathcal{E}'$ with $y \in \operatorname{supp} v \subset \overline{Y}$ and $\operatorname{supp} P(-D)v \subset X$. In fact, we can then apply the condition (1.1) to v after a small translation in the direction of the normal of Y at y which gives a contradiction.

2. Construction of local null solutions. Consider a cylinder defined by

$$(2.1) \qquad x_{n-1} \geq F(x')/2, \quad x' = (x_1, \dots, x_{n-2}),$$

where F is a positive definite quadratic form. Let $P(D)$ be a differential operator with real principal part p such that $p(N) = 0$ and $p'(N) = (0, \dots, 0, \gamma) \neq 0$ if $N = (0, \dots, 0, -1, 0)$ is the normal of the cylinder at the x_n axis. Thus the tangent plane is characteristic there and the x_n axis is the corresponding bicharacteristic. We wish to construct a solution of the equation $P(D)u = 0$ in a neighborhood of a given interval

$$J = \{(0, x_n); a \leq x_n \leq b\}$$

on the x_n axis such that the support belongs to the parabolic cylinder (2.1). The first step is to construct a characteristic surface containing J on which (2.1) is fulfilled.

By the implicit function theorem there is a conic neighborhood of N in the surface $p(\xi) = 0$ which is defined by an equation

$$\xi_n = \tau(\xi_1, \dots, \xi_{n-1})$$

where τ is homogeneous of degree 1 near $N_0 = (0, \dots, 0, -1)$ and $d\tau = 0$ at N_0. The corresponding bicharacteristic direction is

$$(\partial \tau / \partial \xi_1, \dots, \partial \tau / \partial \xi_{n-1}, -1).$$

Let $A(x')$ be a quadratic form such that $A - F$ is positive definite, and let $a \leq c \leq b$. Using the Hamilton-Jacobi theory we shall construct a characteristic surface through the manifold of dimension $n - 2$

$$x_{n-1} = A(x')/2, \quad x_n = c.$$

To do so we have to draw the bicharacteristic with direction $(\partial\tau/\partial\xi_1,\ldots,\partial\tau/\partial\xi_{n-1},-1)$ through $(y, A(y)/2, c)$ where $(\xi', \xi_{n-1}) = (Ay, -1)$. Here A is the symmetric matrix corresponding to the form A. Thus the surface must be the image of the map

$$(2.2) \qquad (y, t) \mapsto (y + t\partial\tau/\partial\xi', A(y)/2 + t\partial\tau/\partial\xi_{n-1}, c - t), \quad c - t \in [a, b].$$

A characteristic surface with the required properties is obtained near J if this map has an injective differential in $\{0\} \times [c - b, c - a]$ and if in a neighborhood of this set

$$(2.3) \qquad A(y) + 2t\partial\tau/\partial\xi_{n-1} \geq F(y + t\partial\tau/\partial\xi').$$

For $y = 0$ we have $(\xi', \xi_{n-1}) = (0, -1)$ and at this point we have by hypothesis $\partial\tau/\partial\xi_j = 0$, $j < n$. By Euler's relations for homogeneous functions we also have $\partial^2\tau/\partial\xi_{n-1}\partial\xi_j = 0$, and $\partial^3\tau/\partial\xi_{n-1}\partial\xi_j\partial\xi_k = \partial^2\tau/\partial\xi_j\partial\xi_k$ there. Let T be the symmetric matrix given by the second derivatives $(\partial^2\tau(0, -1)/\partial\xi_j\partial\xi_k); j, k = 1, \ldots, n - 2$. Then we have

$$2t\partial\tau/\partial\xi_{n-1} = t(TAy, Ay) + O(|y|^3),$$
$$y + t\partial\tau/\partial\xi' = y + tTAy + O(|y|^2).$$

It follows that (2.2) has an injective differential at $\{0\} \times [c - b, c - a]$ if and only if $I + tTA$ is non-singular for $t \in [c - b, c - a]$. The condition (2.3) is fulfilled if

$$(Ay, y) + t(TAy, Ay) - F(y + tTAy, y + tTAy) > 0 \quad \text{when } y \neq 0, \ t \in [c - b, c - a].$$

In particular we see again that A must be positive definite. Writing $z = (I + tTA)y$ we have $Ay = (A^{-1} + tT)^{-1}z$, and our conditions reduce to

(i) $\qquad\qquad A^{-1} + tT$ is non-singular for $t \in [c - b, c - a]$,

(ii) $\qquad ((A^{-1} + tT)^{-1}z, z) - (Fz, z) > 0$ if $z \neq 0$ and $t \in [c - b, c - a]$.

These are equivalent to

(iii) $\qquad\qquad 0 < A^{-1} + tT < F^{-1}, \quad t \in [c - b, c - a],$

where by $B < C$ we mean that $C - B$ is positive definite. It follows from (iii) that

(iv) $\qquad\qquad -F^{-1} < (b - a)T < F^{-1}.$

Conversely, if this condition is fulfilled we can find A satisfying (iii). This is obvious when $n - 2 = 1$ and a reduction to this case is immediately obtained if T and F^{-1} are diagonalized simultaneously.

Another way of stating (iv) is

(v) $\qquad\qquad F^{-1} + tT$ is non-singular when $|t| \leq b - a.$

The preceding argument can be used to give the geometrical interpretation which occurs in the following

PROPOSITION 2.1. *Assume that the bicharacteristics corresponding to characteristic tangent planes of $\{(x', F(x')/2, 0)\}$ close to the plane $x_{n-1} = 0$ have no focal point with $|x_n| \leq b - a$. Then there is an analytic characteristic surface for P containing $J = \{(0, x_n); a \leq x_n \leq b\}$ on which (2.1) is valid.*

The proof shows that the condition is essentially necessary also. This will also follow from Theorems 2.2 and 3.1.

We can write the constructed characteristic surface in the form

$$x_{n-1} = f(x', x_n)$$

where f is analytic and $f(x', x_n) = O(|x'|^2)$ for x' near 0 and x_n near $[a, b]$. Following Malgrange [4] (see also the proof of Theorem 5.2.1 in [1]) we shall construct a null solution with support containing J by finding an analytic solution of the Goursat problem

$$P(D)u = 0, \ u = O((x_{n-1} - f(x', x_n))^m), \ u(x', x_{n-1}, c) = (x_{n-1} - f(x', c))^m$$

where $c = (a + b)/2$. The null solution is then obtained by changing u to 0 when $x_{n-1} < f(x', x_n)$. The only point which is not obvious is that the solution exists in a neighborhood of J. To see this we introduce new coordinates by writing

$$x' = \varepsilon y', \ x_{n-1} - f(x', x_n) = \varepsilon^2 y_{n-1}, \ x_n = y_n$$

and we let $v = u/\varepsilon^{2m}$ considered as a function of the y variables. Then we obtain the boundary conditions

$$v = O(y_{n-1}^m), \quad v = y_{n-1}^m \text{ when } y_n = c.$$

The differential equation for v is obtained by noting that

$$\partial/\partial x' = \varepsilon^{-1}\partial/\partial y' - \varepsilon^{-2}\partial f/\partial x'\partial/\partial y_{n-1}, \ \partial/\partial x_{n-1} = \varepsilon^{-2}\partial/\partial y_{n-1},$$

$$\partial/\partial x_n = \partial/\partial y_n - \varepsilon^{-2}\partial f/\partial x_n\partial/\partial y_{n-1}.$$

Here $\varepsilon^{-1}\partial f(\varepsilon y', y_n)/\partial x'$ is bounded and $\varepsilon^{-2}\partial f(\varepsilon y', y_n)/\partial y_n \to g(y', y_n)$ as $\varepsilon \to 0$, uniformly in a neighborhood of any compact set in $\mathbf{C}^n$ with $|y_n - c| \leq (b - a)/2$ and y_n real. The function g is a quadratic form in y'. If we recall that $P(\xi) = \xi_{n-1}^{m-1}(A\xi_n + B)+$ terms of lower order in ξ_{n-1}, $A \neq 0$, it follows that the equation for v after multiplication by ε^{-2} converges to

$$D_{n-1}^{m-1}(AD_n + ig(y', y_n) + B)v = 0.$$

The substitution $v = w \exp(\int_c^{y_n}(g(y', t) - iB)\, dt/A)$ reduces it to the form $D_{n-1}^{m-1}D_n w = 0$. Inspection of the proof of Theorem 5.1.1 in [1] now shows that for small ε our Goursat problem has a solution in a neighborhood of the interval $y' = 0$, $y_n \in [a, b]$. Hence we have proved

THEOREM 2.2. *Under the hypotheses of Proposition 2.1 one can find a distribution u with $P(D)u = 0$ in a neighborhood of J, such that $J \subset \text{supp}\, u$ and (2.1) is valid in $\text{supp}\, u$.*

COROLLARY 2.3. *Assume that the open set $X \subset \mathbf{R}^n$ contains an open convex cylinder Y with C^2 cross section for which in addition to conditions a) and b) in Theorem 1.1 we have*

    c) *the bicharacteristics corresponding to characteristic tangent planes at either one of the flat ends which are close to the tangent plane at y have no focal point in $\overline{Y}$.*

Then X is not P-convex.

In section 4 we shall see that the condition c) cannot be relaxed and that a complete characterization of P-convex sets X cannot be made in terms of the cylinders contained in X.

3. A uniqueness theorem. In this section we shall prove that if the condition (iv) in section 2 is not even fulfilled with inequality in the wide sense, then there is a uniqueness theorem instead of Theorem 2.2.

THEOREM 3.1. *Let u be a solution in a neighborhood of the interval $J = \{(0, x_n); a \leq x_n \leq b\}$ of the equation of real principal type $P(D)u = 0$ and assume that (2.1) is valid in supp u with a positive definite F. Then $u = 0$ in a neighborhood of J unless the plane $x_{n-1} = 0$ is characteristic, the x_n axis has the corresponding bicharacteristic direction and the bicharacteristics corresponding to characteristic tangent planes of $\{(x', F(x')/2, 0)\}$ close to the plane $x_{n-1} = 0$ have no focal point with $|x_n| < b - a$.*

PROOF. By Holmgren's uniqueness theorem and Theorem 8.2 in [3] we only have to show that $u = 0$ near *some point* in J when the plane $x_{n-1} = 0$ is characteristic and the x_n axis has the corresponding bicharacteristic direction, but the bicharacteristics in the statement of the theorem have a focal point with $|x_n| < b - a$. This means with the notations used in section 2 that (iv) is not even valid with inequality in the wide sense. We can choose the coordinates $x_1, \ldots, x_{n-2}$ so that both F and T have diagonal form and $(b - a)|T_{11}| > F_{11}^{-1}$. Replacing x_n by $-x_n$ changes the sign of T so we may even assume that

$$(3.1) \qquad\qquad (b - a)T_{11}F_{11} > 1.$$

We shall use the same arguments as in the proof of Theorem 5.3.2 in [1]. Thus we shall construct a one parameter family of non-characteristic surfaces which start from the set where $x_{n-1} < F(x')/2$ and sweep over some point of J while staying in a compact set where $P(D)u = 0$. Since everything happens near J it is convenient to enlarge a neighborhood by introducing new variables

$$x' = \varepsilon y', \ x_{n-1} = \varepsilon^2 y_{n-1}, \ x_n = y_n.$$

In the dual space this gives the substitution

$$\xi' = \eta'/\varepsilon, \ \xi_{n-1} = \eta_{n-1}/\varepsilon^2 \text{ and } \xi_n = \eta_n.$$

The condition (2.1) remains the same in the new variables and the equation $\xi_n = \tau(\xi', \xi_{n-1})$ becomes $\eta_n = \tau(\eta'/\varepsilon, \eta_{n-1}/\varepsilon^2) = -\varepsilon^{-2}\eta_{n-1}\tau(-\varepsilon\eta'/\eta_{n-1}, -1) = -T(\eta', \eta')/2\eta_{n-1} + O(\varepsilon)$. It is therefore sufficient to carry out the program for the characteristic form $2\eta_n\eta_{n-1} + T(\eta', \eta')$ and then go back to the variables x with a sufficiently small ε.

Consider the surfaces S_t:

$$y_{n-1} = F_{11}(y_1 - 1/2) + t + g(y_n), \quad a \leq y_n \leq b,$$

where g will be chosen later. We have $y_{n-1} < F(y')/2$ if

$$t + g(y_n) < F_{11}(y_1 - 1)^2/2 + \sum_{2}^{n-2} F_{jj}y_j^2/2.$$

This is true at the boundary and at infinity in S_t if $t \leq 0$ and $g(a) < 0$, $g(b) < 0$. If $t + \sup g < 0$ it is true on all of S_t. The surfaces S_t are non-characteristic if $F_{11}^2 T_{11} - 2g'(y_n) > 0$. These conditions on g imply that $g(y_n) < F_{11}^2 T_{11}(y_n - a)/2$. On the other hand, all such values can actually be attained. Hence we can sweep over the set of all y with $a \leq y_n \leq b$ and

$$y_{n-1} < F_{11}(y_1 - 1/2) + F_{11}^2 T_{11}(y_n - a)/2.$$

The condition is fulfilled by some point on J if

$$0 < -F_{11}/2 + F_{11}^2 T_{11}(b - a)/2$$

which is precisely the condition (3.1). The theorem is proved.

Theorem 3.1 can of course be used to improve Theorem 1.1. This leads to a rather complicated statement so we shall content ourselves with constructing in section 4 an example of a P-convex set which does not satisfy the conditions in Theorem 1.1.

References

[1] L. Hörmander, *Linear partial differential operators*, Springer-Verlag, Berlin-Göttingen-Heidelberg, 1963.

[2] _____, *On the existence and regularity of solutions of linear pseudo-differential equations*, Ens. Math. **17** (1971), 99–163.

[3] _____ Uniqueness theorems and wave front sets for solutions of linear differential equations with analytic coefficients, Comm. Pure Appl. Math. **24** (1971), 671–704.

[4] B. Malgrange, *Sur les ouverts convexes par rapport à un opérateur différentiel*, C. R. Acad. Sci. Paris **254** (1962), 614–615.

[5] F. Treves, *Linear partial differential equations with constant coefficients*, Gordon and Breach, New York, 1966.

[6] E.C. Zachmanoglou, *Non-uniqueness of the Cauchy problem for linear partial differential equations with variable coefficients*, Arch. Rat. Mech. Anal. **27** (1968), 373–384.

FOURIER MULTIPLIERS WITH SMALL NORM

Lars Hörmander

By $M_p(\mathbf{R}^n)$ or M_p for short we denote the space of multipliers on Fourier transforms of L^p functions in $\mathbf{R}^n$, that is, the space of functions m such that

$$S \ni u \mapsto \mathcal{F}^{-1}(m\mathcal{F}u)$$

is bounded for the L^p norm if $\mathcal{F}$ is the Fourier transformation. For basic properties of multipliers we refer to Hörmander [1]. In particular, $M_2 = L^\infty$ and for every p

$$(1) \qquad \|m\|_{M_p} \geq \|m\|_{M_2} = \|m\|_{L^\infty}, \quad \text{if } m \in M_p \subset M_2.$$

Shapiro [2, 3] has called attention to the interesting properties of multipliers for which there is equality in (1). When $p = 1$ the elements of M_p are the Fourier transforms of measures $d\mu$, and the norm is the total mass of $d\mu$. If we strengthen equality in (1) to

$$(2) \qquad m(0) = \|m\|_{M_1},$$

which is reasonable since m is then continuous, the condition (2) means that $d\mu$ is a positive measure, hence that m is a positive definite function. If $m(\xi) - m(0) = o(|\xi|^2)$ the second moments of the measure are 0 which means that m is a constant. Thus $m(\xi) - \|m\|_{M_1}$ can only vanish to the second order. For any $p \neq 2$ it was proved by Shapiro that $m(\xi) - \|m\|_{M_p}$ can at most vanish to the third order. The following theorem improves his result. Since $M_p = M_{p'}$ when $1/p + 1/p' = 1$, we assume that $1 \leq p < 2$.

THEOREM 1. *Let $m \in M_p(\mathbf{R}^n)$ and assume that*

$$(3) \qquad m(\xi) - \|m\|_{M_p} = O(|\xi|^a).$$

If $1 \leq p < 2$ and $a > p + 1$ it follows that m is a constant.

We shall begin the proof by reducing it to the case $n = 1$ and fairly smooth functions m.

LEMMA 2. *Let G be the group of $n \times n$ matrices with $T^*T = TT^* = c(T)I$, where I is the identity matrix, and choose $\chi \in C_0^\infty(G)$ with $0 \leq \chi$, $\int \chi(T)\,dT = 1$ where dT is the Haar measure. If m satisfies the hypotheses in Theorem 1, the same is true of*

$$m_\chi(\xi) = \int m(T\xi)\chi(T)\,dT.$$

© Springer International Publishing AG, part of Springer Nature 2018
L. Hörmander, *Unpublished Manuscripts*,
https://doi.org/10.1007/978-3-319-69850-2_16

In addition $m_\chi \in C^\infty(\mathbf{R}^n \setminus \{0\})$ and

(4) $$|D_\xi^\alpha m_\chi(\xi)| < C_\alpha |\xi|^{-|\alpha|} |\xi|^a (1 + |\xi|)^{-a}, \quad \alpha \neq 0.$$

PROOF. Assume $\|m\|_{M_p} = 1$. Since $\xi \mapsto m(T\xi)$ is also a multiplier with norm 1 ([1, Theorem 1.13]) it is clear that $\|m_\chi\|_{M_p} \leq 1$. But $m_\chi(\xi) - 1 = O(|\xi|^a)$, $\xi \to 0$, so $\|m_\chi\|_{M_p} \geq 1$ by (1). Hence $\|m_\chi\|_{M_p} = 1$. If X is in the Lie algebra of G then

$$m_\chi(e^{tX}\xi) = \int m(Te^{tX}\xi)\chi(T)\, dT = 1 + \int (m(T\xi) - 1)\chi(Te^{-tX})\, dT.$$

All derivatives with respect to t are therefore bounded by $C|\xi|^a(1+|\xi|)^{-a}$ when $t = 0$. Now X can be chosen so that $X\xi$ is any one of the vector fields $\sum \xi_j \partial/\partial \xi_j$ or $\xi_j \partial/\partial \xi_k - \xi_k \partial/\partial \xi_j$. Since

$$\sum_j \xi_j |\xi|^{-1}(\xi_j/\partial\xi_k - \xi_k\partial/\partial\xi_j) + \xi_k|\xi|^{-1}\sum \xi_j \partial/\partial\xi_j = |\xi|\partial/\partial\xi_k,$$

the same conclusion is valid for any product of operators $|\xi|\partial/\partial\xi_k$ applied to m_χ. This completes the proof.

If m_χ is a constant for every choice of χ, then m is itself a constant. It is therefore sufficient to prove Theorem 1 when m satisfies (4). Since m is then continuous, the function

$$\mathbf{R} \ni t \mapsto m(t\xi)$$

is for fixed $\xi \in \mathbf{R}^n$ an element of $M_p(\mathbf{R})$ of norm 1. (This degenerate case of Theorem 1.13 in [1] follows since the unit ball in M_p is closed in $\mathcal{D}'$.) If the theorem is known when $n = 1$ we conclude that $m(t\xi) = m(0)$ for all t and ξ so it follows for arbitrary n. From now on we therefore assume that $n = 1$, that m satisfies (4) as well as (3), and that $\|m\|_{M_p} = 1$. Choose $M \in \mathcal{S}'$ so that $\widehat{M} = m$.

Let f and g be functions in $\mathcal{S}$ such that $\hat{f} \in C_0^\infty(\mathbf{R})$ and $\hat{g} \in C_0^\infty(\mathbf{R} \setminus 0)$. In addition we require f to have a simple zero at 0 and we set $f_\varepsilon(x) = \varepsilon^{-\gamma} f(\varepsilon x)$ where for reasons which will appear later

(5) $$1 < \gamma < a - p.$$

(Recall that $a - p > 1$ by hypothesis.) The proof of the theorem will follow when $\varepsilon \to 0$ in the inequality

(6) $$\|M * f_\varepsilon + M * g\|_p^p \leq \|f_\varepsilon + g\|_p^p.$$

It is then important that $M * f_\varepsilon$ is close to f_ε in virtue of (3), (4), and we shall expand both sides using the inequality

(7) $$\left||z + w|^p - |z|^p - p\operatorname{Re}|z|^p w/z\right| \leq C|w|^p; \quad z, w \in \mathbf{C}, \ 1 \leq p \leq 2.$$

For reasons of homogeneity it is sufficient to verify (7) when $z = 1$. For small $|w|$ it is then a consequence of the expansion

$$|1 + w|^p = (1 + w + \bar{w} + w\bar{w})^{p/2} = 1 + p\operatorname{Re} w + O(|w|^2)$$

and it is obviously valid for $|w|$ bounded away from 0.

Writing $M * f_\varepsilon = f_\varepsilon + (M * f_\varepsilon - f_\varepsilon)$ we obtain from (6) and (7) after assembling the interesting first order terms

$$(8) \quad \operatorname{Re} \int |f_\varepsilon|^p f_\varepsilon^{-1} (M * g - g) \, dx$$

$$\leq C \int \left(|f_\varepsilon|^{p-1} |M * f_\varepsilon - f_\varepsilon| + |M * f_\varepsilon - f_\varepsilon|^p + |M * g|^p + |g|^p \right) dx.$$

Here $M * g$ and g are in $\mathcal{S}$ so the last two terms are integrable. The Fourier transform of $M * f_\varepsilon - f_\varepsilon$ is

$$\varepsilon^{-\gamma-1}(m(\xi) - 1)\hat{f}(\xi/\varepsilon) = \varepsilon^{-\gamma-1+a}\hat{h}_\varepsilon(\xi/\varepsilon)$$

where

$$\hat{h}_\varepsilon(\xi) = \hat{f}(\xi)(m(\varepsilon\xi) - 1)\varepsilon^{-a}.$$

Since $\hat{h}_\varepsilon$ has fixed support and is bounded in C^2 by (3), (4) where $a \geq 2$, we conclude that

$$|h_\varepsilon(x)| < C/(1 + x^2).$$

Recalling that $(M * f_\varepsilon - f_\varepsilon)(x) = \varepsilon^{-\gamma+a} h_\varepsilon(\varepsilon x)$ we conclude that the first two terms on the right hand side of (8) can be estimated by

$$(9) \quad\quad\quad C\varepsilon^{-p\gamma+a} \int (1 + |\varepsilon x|)^{-2} \, dx = C\varepsilon^{-p\gamma+a-1}.$$

It remains to study the left hand side of (8). Set $G = M * g - g$, which is a function in $\mathcal{S}$ with Fourier transform $(m - 1)\hat{g}$. We can write

$$\int |f_\varepsilon|^p f_\varepsilon^{-1}(M * g - g) \, dx = \varepsilon^{-\gamma(p-1)} \int |f(\varepsilon x)|^p / f(\varepsilon x) \, G(x) \, dx.$$

Since $|f(\varepsilon x)| < C\varepsilon|x|$ because $f(0) = 0$, we can estimate the integrand in modulus by $C\varepsilon^{p-1}|x|^{p-1}|G(x)|$, and $|x|^{p-1}|G(x)|$ is integrable. Hence

$$(10) \quad \varepsilon^{(\gamma-1)(p-1)} \int |f_\varepsilon|^p f_\varepsilon^{-1}(M * g - g) \, dx \to |f'(0)|^p / f'(0) \int |x|^p x^{-1} G(x) \, dx.$$

(Note that since all moments of G vanish, the left hand side would decrease faster than any power of ε if $f(0)$ were different from 0.) Now (5) gives

$$(5)' \quad\quad (\gamma - 1)(p - 1) > 0, \quad (\gamma - 1)(p - 1) - p\gamma + a - 1 = a - p - \gamma > 0,$$

if $p > 1$ as we may well assume, so (8), (9) and (10) imply when $\varepsilon \to 0$

$$(11) \quad\quad\quad\quad \operatorname{Re} \int |x|^p x^{-1} G(x) \, dx \leq 0.$$

The Fourier transform of $|x|^p/x$ is for $\xi \neq 0$ a constant times $|\xi|^{-p} \operatorname{sgn} \xi$, and $\hat{G} = (m - 1)\hat{g}$. Since $\hat{g}$ is any function in $C_0^\infty(\mathbf{R} \setminus 0)$ we can now conclude that $m = 1$ by passing to Fourier transforms in (11). Theorem 1 is proved.

We do not know if the condition on a in Theorem 1 can be weakened, but we shall make a few remarks on this question. Assume that m is an element of norm 1 in M_p such that for some positively homogeneous h of degree a

$$(12) \qquad m(\xi) = 1 + h(\xi) + o(|\xi|^a), \quad \xi \to 0.$$

Since $m(\xi n^{-1/a})^n \to e^{h(\xi)}$ boundedly we obtain $\|e^h\|_{M_p} \leq 1$, hence $\|e^h\|_{M_p} = 1$ because $h(0) = 0$. Using the homogeneity we conclude that $\|e^{th}\|_{M_p} = 1$ for every $t > 0$, so the Fourier multipliers e^{th} define a contraction semigroup in L^p. Writing $H = \mathcal{F}^{-1} h \mathcal{F}$ for the infinitesimal generator and differentiating the inequality

$$\|e^{tH} f\|_p{}^p \leq \|f\|_p{}^p, \quad f \in \mathcal{S}, \ t > 0,$$

we obtain

$$(13) \qquad \mathrm{Re} \int |f|^p f^{-1} H f \, dx \leq 0, \quad f \in \mathcal{S}.$$

Conversely, (13) implies that $\|e^{th}\|_{M_p} = 1$ when $t > 0$.

It is easy to show that if (13) is valid for some h not identically 0, then (13) is valid for $h(\xi) = -|\xi|^a$ or $h(\xi) = i\xi|\xi|^{a-1}$. Theorem 1 shows that there is no such h if $a > p + 1$. The basic properties of multipliers show that if (13) holds for one p then it is also true for all other values between p and the conjugate exponent p'. When h is hermitian symmetric the operator H is real and it is then easily shown that (13) is valid for complex valued f if it is valid in the real valued case. When $2 < a < 3$ and $h(\xi) = -|\xi|^a$ the inequality (13) can then be written

$$\int |f|^p / f \, |D|^a f \, dx = (p-1) \int |f|^{p-2} f' \, |D|^{a-2} f' \, dx \geq 0$$

or if we use a well known expression for the scalar product in H_s spaces

$$(13)' \qquad \int \left(|f(x)|^{p-2} f'(x) - |f(y)|^{p-2} f'(y) \right) \left(f'(x) - f'(y) \right) |x - y|^{1-a} \, dx \, dy \geq 0,$$

where $f \in \mathcal{S}$ is assumed to be real valued. Unfortunately we do not know when (13)$'$ is fulfilled.

References

[1] L. Hörmander, *Estimates for translation invariant operators in L^p spaces*, Acta Math. **104** (1960), 93–140.

[2] H. S. Shapiro, *Fourier multiplirs whose multiplier norm is an attained value*, Linear Operators and Approximation, Conference held at Oberwolfach August 14–22, 1971, Birkhäuser Verlag, 1972, pp. 338–347.

[3] ———, *Regularity properties of the element of closest approximation*, Trans. Amer. Math. Soc. (to appear).

THE COSINE THEOREM ON A SURFACE
AND THE NOTION OF CURVATURE

Lars Hörmander[15]

0. Introduction. Every student of differential geometry learns that the fundamental work of Gauss [2] on curved surfaces was influenced by his interest in geodesy. However, the nature of this influence is seldom spelled out. This is unfortunate, for the work of Gauss seems extremely natural if one knows a theorem of Legendre on spherical triangles which was familiar to contemporary geodesists. Legendre's theorem can still be found in elementary text books on geodesy, but as far as I know it does not occur in any text book on differential geometry. There is no doubt that Gauss knew it for he devotes a large part of [2] to proving a better and more general result. (Cf. Dombrowski [6, p. 114]; in this paper there is also a complete translation of Gauss [2].) Whether Gauss was actually motivated by the theorem of Legendre must of course remain a matter of speculation.

In section 2 we recall basic facts on spherical triangles. In particular we prove the spherical cosine theorem and the theorem of Legendre which is an approximate version of it. The extension to geodesic triangles on surfaces in $\mathbb{R}^2$ is discussed in section 3. Asking for the higher dimensional analogue of Legendre's theorem leads us to the Riemann curvature tensor in section 4. Some passages in Riemann's paper [4] indicate that he may have had such arguments in mind.

In order to emphasize the geometrical ideas we have often used geometrical intuition in the text rather than rigorous proofs. The serious reader can find the missing technical details in Appendix A. In Appendix B finally we derive a higher order approximation to the cosine theorem on a curved two dimensional surface which improves that of Gauss and might even be new.

2. Spherical triangles. Let us consider a sphere with center at the origin in $\mathbb{R}^3$ and radius R. We parametrize it with the longitude θ and the latitude φ which it is convenient to measure from the north pole. The cartesian coordinates are then

$$(R\sin\varphi\cos\theta, R\sin\varphi\sin\theta, R\cos\varphi)$$

so the square of the element of arc is

$$ds^2 = dx^2 + dy^2 + dz^2 = R^2(d\varphi^2 + \sin^2\varphi d\theta^2).$$

It follows at once that the shortest distance along the sphere to the north pole is $R\varphi$ and that it is attained only for the curve where θ is constant. Thus the shortest distance between two points on the sphere is always attained along the great circle through them,

---

[15]This is an expanded version of a lecture at the Swedish Mathematical Society March 17, 1979

that is, the intersection of the sphere and a plane through the two points and the center. This is uniquely determined when the points are not antipodes, and we shall choose the arc of the great circle which is $< \pi R$ then.

Let us now consider a spherical triangle, that is, a domain bounded by three such great circle arcs. We call their lengths a, b, c and the opposite angles α, β, γ. It is convenient to put the north pole at the corner where the angle is α and let the xz plane go through that where the angle is γ. Then the coordinates of the vertices are

$$(0, 0, R), \quad (R \sin b/R, 0, R \cos b/R), (R \sin c/R \cos \alpha, R \sin c/R \sin \alpha, R \cos c/R),$$

and the scalar product of the last two vectors divided by R^2 is

$$(2.1) \qquad \cos a/R = \cos b/R \cos c/R + \sin b/R \sin c/R \cos \alpha.$$

This is the spherical cosine theorem which in another form was known already to Albattani (850(?)–929). (See [5, p. 138].) Spherical trigonometry has even older roots in astronomy; it was developed before plane trigonometry (see [1, p. 421]).

For triangles on the earth which can be measured with classical geodetic methods the quotients a/R, b/R, c/R are all small, so in such contexts it is natural to use a series expansion in (2.1). If we replace $\cos x$ by $1 - x^2/2$ and $\sin x$ by x except in the factor $\cos \alpha$ and discard terms of order four or higher, we obtain the cosine theorem

$$(2.2) \qquad a^2 = b^2 + c^2 - 2bc \cos \alpha$$

valid for triangles in the plane. The next approximation

$$\cos x = 1 - x^2/2 + x^4/24 - \dots, \quad \sin x = x - x^3/6 + \dots$$

gives in the same way when terms of order 6 are neglected

$$a^2 = b^2 + c^2 - 2bc \cos \alpha + (a^4 - b^4 - c^4 - 6b^2c^2)/12R^2 + bc(b^2 + c^2)/3R^2 \cos \alpha + \dots.$$

Approximating a^2 on the right by (2.2) we obtain

$$(2.3) \qquad a^2 = b^2 + c^2 - 2bc(\cos \alpha + (bc \sin^2 \alpha)/6R^2) + \dots.$$

Let α_e be the angle replacing α in the Euclidean triangle with the sides a, b, c. Then (2.2) is valid with α replaced by $\alpha_e = \alpha - \delta$, that is,

$$a^2 = b^2 + c^2 - 2bc(\cos \alpha + \delta \sin \alpha + O(\delta^2)).$$

Comparison with (2.3) gives when b and c are small but $\sin \alpha$ is not

$$(2.4) \qquad \delta = \alpha - \alpha_e = (bc \sin \alpha)/6R^2 + \dots = (bc \sin \alpha_e)/6R^2 + \dots = T/3R^2 + \dots$$

where dots indicate terms of order four or higher and T is the area of the Euclidean triangle with sides a, b, c. (The existence of such a triangle is clear since the triangle inequality is also valid for the geodesic distance.)

THEOREM 2.1 (LEGENDRE [3]). *For a spherical triangle with small sides the angles are approximately equal to the angles of the Euclidean triangle with the same sides increased by its area divided by $3R^2$. The error is of order four in a/R, b/R, c/R when the angles of the triangles are not small.*

Legendre published this theorem in 1787 in connection with a geodetical survey which he directed, but it had been used earlier in geodetic practice (see [5, p. 170]). An easy consequence is

THEOREM 2.2 (GIRARD). *For a spherical triangle the area divided by R^2 is equal to the spherical excess $\alpha + \beta + \gamma - \pi$.*

PROOF. If we divide the spherical triangle into four by geodesics connecting the midpoints of the sides, then the spherical excess of the whole triangle is equal to the sum of those of the four smaller ones. In fact, the sum of all the angles in the small triangles is equal to the sum of the angles in the large one plus 3π, the sum of the angles at the new vertices, and this is compensated by the subtraction of π for each triangle. Now repeat the construction until the triangles are very small. For small triangles Legendre's theorem shows that the excess minus the area divided by R^2 is small compared to the area. Summing up we see that the difference between the spherical excess of the original triangle and its area divided by R^2 is arbitrarily small, which proves the theorem.

Girard's proof was published in 1629 but the result seems to have been known already to Regiomontanus (1436–1476) (see [5, p. 128]). An elementary proof using no infinitesimal arguments was given by Cavalieri in 1632.

The interesting point in Legendre's theorem is that for small triangles the spherical excess is about equally distributed on the three vertices. However, this is no longer true if one includes one more term in the calculations above, for this gives

$$(2.5) \qquad \delta = \alpha - \alpha_e = S/3R^2 + S(b^2 + c^2 - 2a^2)/180R^4 + \cdots$$

where dots now indicate terms of sixth order and S is the area of the spherical triangle. Since

$$(\alpha - \alpha_e) - (\beta - \beta_e) = S(b^2 - a^2)/60R^4 + \cdots$$

this means that a larger correction must be made opposite a shorter side.

The approximation (2.5) is given by Gauss in section 27 of his fundamental paper [2]. We shall give it a more general form in Appendix B. However, our immediate goal is to show how one is led to the main results of [2] by trying to extend Legendre's theorem to general surfaces.

3. Geodesic triangles on surfaces. Let us now consider an infinitely differentiable surface in $\mathbb{R}^3$. It can be described by giving the Cartesian coordinates as functions of two parameters $u = (u_1, u_2)$. The square of the element of arc $ds^2 = dx^2 + dy^2 + dz^2$ then becomes

$$ds^2 = \sum_{j,k=1}^{2} g_{jk}(u)du_j du_k,$$

where the matrix (g_{jk}) is positive definite. A curve on the surface is defined by giving u as a function of t, and its length is

$$\int \left(\sum_{j,k=1}^{2} g_{jk}(u(t))du_j/dt\, du_k/dt \right)^{1/2} dt.$$

As a substitute for the great circles on the sphere we shall use *geodesic* lines, that is, curves which give the shortest path along the surface between any two of its nearby points. This extremal condition is easily converted to a second order differential equation which shows that for any point x_0 on the surface and tangent direction t_0 at x_0 there is a unique geodesic through x_0 with tangent t_0. Moreover, one can choose the local coordinates $u = (u_1, u_2)$ so that $u = 0$ corresponds to x_0 and the straight lines through 0 in the u plane correspond to geodesics on the surface with the same arc length. If $r > 0$ is small, then the circle

$$|u| = (u_1^2 + u_2^2)^{1/2} = r$$

in the u plane corresponds to a curve C_r on the surface which is orthogonal to the geodesics through x_0. In fact, otherwise we could obtain a curve from x_0 to C_r of length $< r$ by cutting across from a geodesic to C_r where they meet at an angle $< \pi/2$. (For a rigorous proof of these facts see Appendix A.) Thus

$$\sum_{j,k=1}^{2} g_{jk}(u) u_j v_k = 0 \quad \text{if} \quad \sum_{1}^{2} u_j v_j = 0,$$

$$\sum_{j,k=1}^{2} g_{jk}(u) u_j u_k = \sum_{1}^{2} u_j^2.$$

We can sum up these two conditions in one

(3.1) $$\sum_{k=1}^{2} g_{jk}(u) u_k = u_j, \quad j = 1, 2.$$

Conversely, assume that we have local coordinates for which (3.1) is valid. Writing $u = (r \cos\theta, r \sin\theta)$ and using the conditions which led to (3.1) we obtain

$$ds^2 = dr^2 + (g_{11} \sin^2\theta - 2g_{12} \sin\theta \cos\theta + g_{22} \cos^2\theta) r^2 d\theta^2 \geq dr^2.$$

Here we have equality only for curves with $d\theta = 0$. As in the proof that great circle arcs on a sphere are geodesics we conclude that the curves $t \to tu$ are geodesics with length equal to the Euclidean length. Coordinate systems with this property, or equivalently satisfying (3.1), are called *geodesic (normal) coordinates*. We have seen that there is one centered at any point on the surface, and it is uniquely determined apart from an orthogonal transformation. At the same time we have found that there is a unique geodesic connecting nearby points.

To give an example we return to the sphere. Then it is clear that polar normal coordinates are given by the longitude θ and $r = R\varphi$. The element of arc on the sphere is

$$ds^2 = dr^2 + R^2 \sin^2(r/R) d\theta^2 = dr^2 + r^2 d\theta^2 + (R^2/r^2 \sin^2(r/R) - 1) r^2 d\theta^2.$$

If we pass to the rectangular coordinates $u_1 = r \cos\theta$, $u_2 = r \sin\theta$, then

(3.2) $$ds^2 = du_1^2 + du_2^2 - (1/3R^2 - 2r^2/45R^4 + \ldots)(u_1 du_2 - u_2 du_1)^2.$$

As we shall soon see, the coefficient $-1/3R^2$ here is closely related to Legendre's theorem.

Now consider geodesic normal coordinates u on an arbitrary surface, thus assume that (3.1) is fulfilled. If we replace u by tu and divide by t, it follows when $t \to 0$ that

$$\sum_k g_{jk}(0)u_k = u_j,$$

so $(g_{jk}(0))$ is the unit matrix. If we set $h_{jk}(u) = g_{jk}(u) - g_{jk}(0)$ then

(3.1)'
$$\sum_k h_{jk}(u)u_k = 0.$$

Thus the quadratic form $\sum h_{jk}(u)v_jv_k$ is proportional to $(u_1v_2 - u_2v_1)^2$, so for $u \neq 0$ we have

$$\sum h_{jk}(u)v_jv_k = \lambda(u)(u_1v_2 - u_2v_1)^2.$$

In particular, $h_{22}(u) = \lambda(u)u_1^2$ so h_{22} vanishes of second order when $u_1 = 0$ and $u_2 \neq 0$, hence also when $u_2 = 0$. By Taylor's formula this proves that $\lambda(u)$ is a C^∞ function at 0 also. Thus (3.1) is equivalent to

(3.3)
$$ds^2 = du_1^2 + du_2^2 + \lambda(u)(u_1du_2 - u_2du_1)^2$$

which is very similar to (3.2).

DEFINITION 3.1. $K = -3\lambda(0)$ is called the Gauss curvature or total curvature of the surface at the center x_0 of the coordinate system.

Note that the geodesic coordinates are uniquely determined apart from a rotation which does not affect this value of λ.

The preceding definition of the curvature is only applicable at the center of a geodesic normal coordinate system. Assume more generally that we have an element of arc

$$\sum_{j,k=1}^{2} g_{jk}(u)du_jdu_k, \quad g_{jk}(u) - \delta_{jk} = O(|u|^2)$$

where (δ_{jk}) is the identity matrix. Then we must find geodesic coordinates U at 0, that is, functions $U_j(u)$ such that for some function $\lambda(U)$

$$\sum_{j,k=1}^{2} g_{jk}(u)du_jdu_k = dU_1^2 + dU_2^2 + \lambda(U)(U_1dU_2 - U_2dU_1)^2,$$

When $U = 0$ this means that the matrix $\partial U_j/\partial u_k$ is orthogonal, and since geodesic coordinates remain geodesic if they are rotated, we may assume that this is the identity matrix. Thus

$$U_j = u_j + V_j(u) + W_j(u) + \cdots$$

where V_j and W_j are homogeneous of second and third order and the error vanishes of order four. Identification of first order terms gives

$$du_1dV_1 + du_2dV_2 = 0, \quad \text{that is,}$$
$$\partial V_1/\partial u_1 = \partial V_1/\partial u_2 + \partial V_2/\partial u_1 = \partial V_2/\partial u_2 = 0.$$

Hence $\partial^2 V_1/\partial u_2^2 = -\partial^2 V_2/\partial u_1 \partial u_2 = 0$; similarly $\partial^2 V_2/\partial u_1^2 = 0$ so $V_1 = V_2 = 0$. The equality of second order terms gives, if h_{jk} is the second order term in the Taylor expansion of g_{jk}

$$h_{11} = 2\partial W_1/\partial u_1 + \lambda(0)u_2^2, \quad 2h_{12} = 2\partial W_1/\partial u_2 + 2\partial W_2/\partial u_1 - 2\lambda(0)u_1 u_2,$$
$$h_{22} = 2\partial W_2/\partial u_2 + \lambda(0)u_1^2.$$

We can eliminate W by applying the differential operators $\partial^2/\partial u_2^2$, $-\partial^2/\partial u_1 \partial u_2$ and $\partial^2/\partial u_1^2$ and adding. This gives

$$6\lambda(0) = \partial^2 h_{11}/\partial u_2^2 - 2\partial^2 h_{12}/\partial u_1 \partial u_2 + \partial^2 h_{22}/\partial u_1^2.$$

so the curvature K at 0 is given by

(3.4) $$K = \partial^2 g_{12}/\partial u_1 \partial u_2 - (\partial^2 g_{11}/\partial u_2^2 + \partial^2 g_{22}/\partial u_1^2)/2.$$

As an example let us compute the curvature of a surface $z = f(x,y)$ in $\mathbb{R}^3$ which is tangent to the xy plane at 0, that is, $f = df = 0$ at 0. Then

$$ds^2 = dx^2 + dy^2 + df^2; \quad g_{11} = (\partial f/\partial x)^2 + 1, \; g_{12} = \partial f/\partial x \partial f/\partial y, \; g_{22} = (\partial f/\partial y)^2 + 1$$

with the parameters $u_1 = x$ and $u_2 = y$, so at the origin we have

$$K = (\partial^2 f/\partial x \partial y)^2 + \partial^2 f/\partial x^2 \partial^2 f/\partial y^2 - 2(\partial^2 f/\partial x \partial y)^2 = \partial^2 f/\partial x^2 \partial^2 f/\partial y^2 - (\partial^2 f/\partial x \partial y)^2.$$

If for example $f(x,y) = ax^2/2 + by^2/2 +$ terms of higher order, then $K = ab$ is the product of the two principal curvatures a and b at 0. That this can be computed in terms of the metric of the surface is one of the main results of Gauss [2] (the "theorema egregium").

Our discussion of small spherical triangles is easily adapted to geodesic triangles, domains bounded by three geodesics. We choose a geodesic coordinate system centered at the vertex with angle α. The other vertices are at u and v (in the coordinate plane) and the lengths of the corresponding sides are equal to their Euclidean lengths $|u|$ and $|v|$. To compute the geodesic distance a between u and v we first observe that

$$a^2 = \int_0^1 \sum g_{jk}(U) dU_j/dt\, dU_k/dt\, dt$$

if $U(0) = u$, $U(1) = v$ and $t \to U(t)$ is the geodesic with parameter proportional to the arc length. Indeed, the integrand is then equal to a^2. In Appendix A we shall show that $U(t) - u - t(v - u)$ is of third order in u and v, just because there are no first order terms in (3.3). Since the determinant $|u + t(v - u), v - u| = |u, v|$, we obtain

(3.5) $$a^2 = |u - v|^2 + \int_0^1 \lambda(u + t(v - u))|u, v|^2 \, dt + O_6 = |u - v|^2 + \lambda((u + v)/2)|u, v|^2 + O_6$$

where O_6 denotes an error of order six in (u, v). Here we have used the analogue in the calculus of variations of the fact that if the minimum of the C^2 function f is attained at x_0, then $f(x) - f(x_0) = O(|x - x_0|^2)$ since $f'(x_0) = 0$. We have also used Taylor expansion of $\lambda(u + t(v - u))$ at $t = 1/2$ for reasons of symmetry. Now (3.5) is exactly similar to (2.3), with $-1/3R^2$ replaced by $\lambda((u + v)/2)$, if we note that $|u, v|^2 = |u|^2 |v|^2 \sin^2 \alpha$, so Legendre's theorem remains valid with (2.4) modified to

(3.6) $$\delta = \alpha - \alpha_e = -\lambda((u + v)/2)T + O_4$$

where α_e is the angle corresponding to α in the Euclidean triangle with the same sides and T is its area. Taking the value of λ at 0 instead we have

(3.6)' $$\delta = \alpha - \alpha_e = KT/3 + O_3$$

where K is the curvature at 0. (Later on we shall return to the more precise approximation (3.6).) The proof of Girard's theorem is now applicable with hardly any change and gives another main result of Gauss [2]:

THEOREM 3.2. *For a geodesic triangle the surface integral of the total curvature is equal to the sum of the angles minus π.*

4. The higher dimensional case. In section 3 we considered only two dimensional surfaces but we shall now discuss the higher dimensional case as indicated by Riemann in his "Habilitationsschrift" [4] in 1854. We assume that near 0 in $\mathbb{R}^n$ we are given a positive definite smooth arc element

$$ds^2 = \sum_{j,k=1}^{n} g_{jk}(u) du_j du_k.$$

Just as in the two dimensional case one can introduce geodesic normal coordinates, and these are characterized by

(4.1)
$$\sum_{k=1}^{n} g_{jk}(u) u_k = u_j, \quad j = 1, \dots, n.$$

We expand g in a Taylor series at 0,

$$g = \mathrm{id} + g^1 + g^2 + \dots$$

where g^j is a homogeneous polynomial in u of degree j. Here $g^1 = 0$ because this is true in the two dimensional case by the equivalence of (3.1) and (3.3). Next g^2 is a quadratic form $G(u; v)$ in v whose coefficients are quadratic forms in u. We can polarize G to a 4-linear form $G(u_1, u_2; v_1, v_2)$ which is symmetric in u_1 and u_2 as well as in v_1 and v_2 and has the property

$$G(u; v) = G(u, u; v, v).$$

The condition (4.1) means that

(4.2)
$$G(u, u; u, v) = 0$$

which can be written more generally

(4.3)
$$G(u_1, u_2; u_3, u_4) + G(u_3, u_1; u_2, u_4) + G(u_2, u_3; u_1, u_4) = 0.$$

From (4.3) or directly from the validity when $n = 2$ we obtain

(4.4)
$$G(u; v) = G(v; u).$$

When u and v are small, the same argument as in the two dimensional case shows that the square of the geodesic distance a between u and v is

$$a^2 = |u - v|^2 + \int_0^1 G(u + t(v - u); v - u) \, dt + O_5 = |u - v|^2 + G(u; v) + O_5.$$

For the geodesic triangle with vertices 0, u and v, we now obtain

(4.5)
$$\delta = \alpha - \alpha_e = -G(u; v)/2|u \wedge v| + O_3$$

for the excess angle at 0. Here $|u \wedge v|$ denotes the Euclidean area of the parallelogram spanned by u and v. The term $G(u; v)$ corresponds to $-K|u, v|^2/3$ in the two dimensional case, so it is natural to interpret $-3G(u;v)/|u \wedge v|^2$ as the curvature in the two plane spanned by these vectors at 0. Note that the quotient only depends on the plane spanned by u and v, for $G(u;v)$ does not change if a multiple of v (resp. u) is added to u (resp. v), and the quotient is homogeneous of degree 0.

The preceding observations are extremely close to Riemann [4, p. 279]:

"Führt man diese Grössen ein, so wird für unendlich kleine Werthe von x das Quadrat des Linienelements $= \sum dx^2$, das Glied der nächsten Ordnung in demselben aber gleich einem homogenen Ausdruck zweiten Grades der $n(n-1)/2$ Grössen $(x_1 dx_2 - x_2 dx_1)$, $(x_1 dx_3 - x_3 dx_1)$, $\ldots$, also eine unendlich kleine Grösse der vierten Dimension, so dass man eine endliche Grösse erhält, wenn man sie durch das Quadrat des unendlich kleinen Dreiecks dividirt, in dessen Eckpunkten die Werthe der Veränderlichen sind $(0, 0, 0, \ldots)$, $(x_1, x_2, x_3, \ldots)$, $(dx_1, dx_2, dx_3, \ldots)$. Diese Grösse behält denselben Werth, so lange die Grössen x und dx in denselben binären Linearformen enthalten sind, oder so lange die beiden kürzesten Linien von den Werthen 0 bis zu den Werthen x und von den Werthen 0 bis zu den Werthen dx in demselben Flächenelement bleiben, und hängt also nur von Ort und Richtung derselben ab. Sie wird offenbar $= 0$, wenn die dargestellte Mannigfaltigkeit eben, d.h. das Quadrat des Linienelements auf $\sum dx^2$ reducirbar ist, und kann daher als das Mass der in diesem Punkte in dieser Flächenrichtung stattfinden Abweichung der Manningfaltigkeit von der Ebenheit angesehen werden. Multiplicirt mit $-3/4$ wird sie der Grösse gleich, welche Herr Geheimer Hofrath Gauss das Krümmingsmass einer Fläche genannt hat."

When $n = 2$ the curvature was described by a single parameter, the Gauss curvature K. Since the dimension of the space of quadratic forms in n variables is $n(n+1)/2$ and that of cubic forms is $n(n+1)(n+2)/6$, the space of forms satisfying (4.2) is easily seen to have the dimension

$$(n(n+1)/2)^2 - n^2(n+1)(n+2)/6 = n^2(n^2-1)/12.$$

Already for $n = 3$ this is equal to 6, for $n = 4$ it is 20 and so on.

The dependence of $G(u;v)$ on the two plane spanned by u, v becomes more clear if one passes to the Riemann-Christoffel curvature tensor, which is the 4-linear form defined by

(4.6) $R(u_1, u_2; u_3, u_4) = 2G(u_1, u_4; u_2, u_3) - 2G(u_1, u_3; u_2, u_4).$

Using (4.3) we obtain

$$-6G(u_1, u_2; u_3, u_4) = R(u_1, u_4; u_2, u_3) + R(u_1, u_3; u_2, u_4),$$
$$R(u, v; u, v) = 2G(u, v; u, v) - 2G(u, u; v, v) = -3G(u, u; v, v).$$

We have a one to one correspondence between 4-linear forms G and R having the symmetry properties:

	$G(u_1, u_2; u_3, u_4)$	$R(u_1, u_2; u_3, u_4)$
symmetric	$u_1 \leftrightarrow u_2,\ u_3 \leftrightarrow u_4$ or $(u_1, u_2) \leftrightarrow (u_3, u_4)$	$(u_1, u_2) \leftrightarrow (u_3, u_4)$
antisymmetric		$u_1 \leftrightarrow u_2$ or $u_3 \leftrightarrow u_4$
	(4.3)	(4.7)

Here (4.7) is the first Bianchi identity

(4.7) $R(u_1, u_2; u_3, u_4) + R(u_3, u_1; u_2, u_4) + R(u_2, u_3; u_1, u_4) = 0.$

The symmetries mean that

$$R(u_1, u_2; u_3, u_4) = Q(u_1 \wedge u_2, u_3 \wedge u_4)$$

where the exterior product (vector product) $u \wedge v$ has $\binom{n}{2}$ components $u_j v_k - u_k v_j$ $(j < k)$ and Q is a symmetric bilinear form on $\mathbb{R}^{\binom{n}{2}}$ with an additional property which expresses the Bianchi identity. It implies that Q, hence R, is uniquely determined by $Q(u \wedge v, u \wedge v) = R(u, v; u, v)$ for this determines $G(u, u; v, v)$ and hence the 4-linear form G. Thus the angle excess $R(u, v; u, v)/6|u \wedge v|$ in "Legendre's theorem" contains complete information about the curvature also in the n dimensional case.

We shall now give the formulas for R in an arbitrary coordinate system. This could be done as in the two dimensional case but it is easier to give the result and verify it afterwards. Thus we introduce the Christoffel symbols of the first kind

$$\Gamma_{ijk} = (\partial g_{ik}/\partial u_j + \partial g_{jk}/\partial u_i - \partial g_{ij}/\partial u_k)/2.$$

We claim that the curvature tensor has the components

$$(4.8) \qquad R_{ijkl} = \partial \Gamma_{ikj}/\partial u_l - \partial \Gamma_{ilj}/\partial u_k + \sum_{r,s}(\Gamma_{ilr}\Gamma_{kjs} - \Gamma_{ikr}\Gamma_{ljs})g^{rs}$$

where (g^{rs}) is the inverse of the matrix (g_{rs}). This is true at the center of a normal coordinate system, for the non-linear terms vanish there and the linear terms

$$(4.8)' \qquad \tfrac{1}{2}(\partial^2 g_{kj}/\partial u_i \partial u_l + \partial^2 g_{il}/\partial u_j \partial u_k - \partial^2 g_{ik}/\partial u_j \partial u_l - \partial^2 g_{lj}/\partial u_i \partial u_k)$$

agree with (4.6). Assume now that v denotes some other coordinates with

$$u = v + O(|v|^2) \quad \text{at } 0$$

where u still denotes normal coordinates. The arc element in terms of the new coordinates is then $\sum \tilde{g}_{ij} dv_i dv_j$ where

$$\tilde{g}_{ij} = \sum g_{rs}(u)\partial u_r/\partial v_i \partial u_s/\partial v_j.$$

Since $dg_{rs} = 0$ at 0 we obtain

$$\partial \tilde{g}_{ij}/\partial v_k = \partial g_{ij}/\partial u_k + \sum_r (\partial^2 u_r/\partial v_i \partial v_k \partial u_r/\partial v_j + \partial u_r/\partial v_i \partial^2 u_r/\partial v_j \partial v_k) + O(|v|^2)$$

so the new Christoffel symbols are

$$\tilde{\Gamma}_{ijk} = \Gamma_{ijk} + \sum_r \partial^2 u_r/\partial v_i \partial v_j \partial u_r/\partial v_k + O(|v|^2).$$

Since $\tilde{g}^{ts} = \delta_{ts}$ we obtain at 0

$$\partial \tilde{\Gamma}_{ikj}/\partial v_l + \sum_{t,s} \tilde{\Gamma}_{ilt}\tilde{\Gamma}_{kjs}\tilde{g}^{ts} = \partial \Gamma_{ikj}/\partial u_l + \partial^3 u_j/\partial v_i \partial v_k \partial v_l$$

$$+ \sum_r (\partial^2 u_r/\partial v_i \partial v_k \partial^2 u_r/\partial v_j \partial v_l + \partial^2 u_r/\partial v_i \partial v_l \partial^2 u_r/\partial v_k \partial v_j).$$

When we interchange k and l and subtract, it follows that $\tilde{R}_{ijkl} = R_{ijkl}$ if $\tilde{R}_{ijkl}$ is defined by (4.8). Hence we may conclude that

$$(4.9) \qquad \tilde{R}_{ijkl} = \sum_{\alpha,\beta,\gamma,\delta} R_{\alpha\beta\gamma\delta}\partial u_\alpha/\partial v_i \partial u_\beta/\partial v_j \partial u_\gamma/\partial v_k \partial u_\delta/\partial v_l$$

if u are normal coordinates and v arbitrary coordinates, for this remains true if we compose with a linear change of coordiantes. Now (4.9) follows for arbitrary changes of coordinates, that is, (4.8) defines invariantly a tensor which agrees with our earlier definition of the curvature tensor.

Appendix A: Geodesics. In section 3 we considered a surface with parameters $u = (u_1, u_2)$ and element of arc defined by

$$ds^2 = \sum_{j,k=1}^{2} g_{jk}(u) du_j du_k.$$

A curve from u^0 to u^1 can be described by a C^∞ function $[0,1] \ni t \to u(t)$ with $u(0) = u^0$, $u(1) = u^1$, and its length is

$$\int_0^1 ds = \int_0^1 \Big(\sum g_{jk}(u) du_j/dt\, du_k/dt\Big)^{1/2} dt \le \Big(\int_0^1 \sum g_{jk}(u) du_j/dt\, du_k/dt\, dt\Big)^{1/2}.$$

Here we have used Schwarz' inequality. There is equality if and only if ds/dt is constant, that is, the parameter t is proportional to the arc length. The infimum of $\int_0^1 ds$ over all curves from u^0 to u^1 is called the geodesic distance from u^0 to u^1. Its square is clearly the infimum of

$$\text{(A.1)} \qquad I = \int_0^1 \sum g_{jk}(u) du_j/dt\, du_k/dt\, dt$$

among all curves from u^0 to u^1.

The curve $t \to u(t)$ is called a geodesic if the arc length between nearby points is equal to their geodesic distance. We choose the parameter proportional to the arc length. Now make a variation of the curve, that is, choose $u(t, \varepsilon)$ depending smoothly on a parameter ε so that $u(t, 0) = u(t)$ for all t and $u(t, \varepsilon) = u(t)$ when $t = 0$ or $t = 1$ and for all t outside a small interval. Then I becomes a function of ε which must be minimal when $\varepsilon = 0$. If $\dot{u}$ denotes the derivative with respect to ε when $\varepsilon = 0$ we obtain

$$0 = \dot{I} = \int_0^1 \sum_{j,k} (\partial g_{jk}/\partial u_i \dot{u}_i du_j/dt\, du_k/dt + 2g_{jk} d\dot{u}_j/dt\, du_k/dt)\, dt.$$

After removing the differentiation on $\dot{u}_j$ by partial integration we conclude that the Euler equations

$$\text{(A.2)} \qquad 2\frac{d}{dt} \sum_k g_{jk}(u) du_k/dt = \sum_{i,k} \partial g_{ik}/\partial u_j du_i/dt\, du_k/dt, \quad j = 1,2$$

must be valid. Conversely, if we just assume $\dot{u}(0) = 0$ we obtain from (A.2)

$$\text{(A.3)} \qquad \dot{I} = 2\sum_{j,k} g_{jk}(u)\dot{u}_j\, du_k/dt \quad \text{for } t = 1.$$

By the basic existence theorems for ordinary differential equations the Euler equations (A.2) have a unique solution for small t with arbitrarily prescribed initial data

$$u = u^0, \; du/dt = v \quad \text{for } t = 0.$$

The solution $u(u^0, t, v)$ is a C^∞ function and depends only on tv rather than on t and v (because (A.2) is independent of t and homogeneous in dt). This means that $U(tv) =$

$u(u^0, t, v)$ is a C^∞ function from a neighborhood of 0 in $\mathbb{R}^2$ to a neighborhood of u^0 with $U(v) = u^0 + v + O(|v|^2)$ as $v \to 0$. By the inverse function theorem we can therefore introduce U as new local coordinates on the surface. The arc length can then be written

$$\sum G_{jk}(U) dU_j/dt \, dU_k/dt.$$

Since the curves $t \to tU$ have a parameter proportional to the arc length we have

$$\int_0^1 \sum G_{jk}(tU) U_j U_k \, dt = \sum G_{jk}(0) U_j U_k.$$

When we vary U it follows from (A.3) that

$$\sum G_{jk}(U) \dot{U}_j U_k = \sum G_{jk}(0) \dot{U}_j U_k.$$

By a linear change of variables we can arrange that $(G_{jk}(0))$ is the unit matrix (completion of squares), and then we have

(A.4)
$$\sum_k G_{jk}(U) U_k = U_j, \quad j = 1, 2.$$

Thus we have proved (3.1). As observed after (3.1) it follows immediately from (A.4) that the curves $t \to tU$ are geodesics with length equal to the Euclidean length. Every solution of (A.2) is therefore a geodesic.

No change is required in the preceding discussion in the case of n dimensions so we have in fact proved (4.1). From now on we consider the n dimensional case and assume that the given local coordinates are geodesic. We want to find the geodesic between two points u and v close to 0, that is, the solution of (A.2) with $u(0) = u$ and $u(1) = v$. We enter in (A.2) the Taylor expansion
$$g = e + G + g^3 + g^4 + \dots$$

of g, where e is the Euclidean metric and g^j is homogeneous of degree j. Write

$$u(t) = u^1(u, v, t) + u^2(u, v, t) + \dots$$

with u^j homogeneous with respect to (u, v) of degree j and

$$u^1(u, v, 0) = u, \ u^1(u, v, 1) = v, \ u^j(u, v, t) = 0 \quad \text{for } t = 0, 1; \ j > 1.$$

First we obtain the equation $d^2 u^1/dt^2 = 0$ which gives the straight line

$$u^1 = u + tz, \quad z = v - u.$$

The next equation is $d^2 u^2/dt^2 = 0$, so $u^2 = 0$ by the boundary conditions. If we multiply (A.2) by arbitrary constants c_j and sum, the first interesting equation is therefore

$$2d^2(u^3, c)/dt^2 + 2dG(u + tz; z, c)/dt = \langle c, \partial/\partial u \rangle G(u + tz; z, z)$$

or equivalently, by (4.3),

$$d^2(u^3, c)/dt^2 = G(u + tz, c; z, z) - 2G(u + tz, z; z, c) = 2G(u, c; z, z).$$

Recalling the boundary conditions we obtain

(A.5) $\qquad u^3 = (t^2 - t)w \quad$ where $\langle w, c \rangle = G(u, c; z, z) \quad$ for all c.

The approximation to the geodesic found now suffices to determine the square of the geodesic distance a between u and v with an error O_8 where again O_k denotes functions vanishing of order k in (u, v). Thus

$$a^2 = \int_0^1 g(u + tz + (t^2 - t)w; z + (2t - 1)w)\, dt + O_8,$$

and we note that (by (4.2))

$$\int_0^1 |z + (2t - 1)w|^2\, dt - |z|^2 = |w|^2/3 = G(u, w; z, z)/3,$$

$$\int_0^1 \left(G(u + tz + (t^2 - t)w; z + (2t - 1)w) - G(u + tz; z) \right) dt$$

$$= \int_0^1 \left(G(u - t^2 w; z + (2t - 1)w) - G(u; z) \right) dt = -2G(u, w; z, z)/3 + O_8$$

since the integral of $(2t - 1)$ is 0 and terms which are quadratic in w are O_8. This means that

$$a^2 = \int_0^1 g(u + tz; z)\, dt - |w|^2/3 + O_7$$

$$= g((u + v)/2; z) + \frac{1}{24} \frac{d^2}{dt^2} g((u + v)/2 + tz; z)_{t=0} - |w|^2/3 + O_7$$

for integrals of odd powers of $(t - 1/2)$ vanish and a fourth order derivative could only contain terms from g^5 or higher.

Now we specialize to the case of two dimensions with the metric (3.3). Then we have $G(u; v) = \lambda_0 |u, v|^2$ so w is determined by

$$\langle w, c \rangle = \lambda_0 |u, z| |c, z|.$$

Hence $w = \lambda_0 |u, z| (z_2, -z_1)$ so $|w|^2 = \lambda_0^2 |u, v|^2 |z|^2$. It follows that

(A.6) $\quad a^2 = |z|^2 + |u, v|^2 \left(\lambda((u + v)/2) + \frac{1}{24} \frac{d^2}{dt^2} \lambda((u + v)/2 + tz)_{t=0} - \lambda_0^2 |z|^2/3 \right) + O_7,$

or if we introduce the Taylor expansion $\lambda = \lambda_0 + \lambda_1 + \lambda_2 + \dots$

(A.7) $\quad a^2 = |z|^2 + |u, v|^2 \left(\lambda_0 + \lambda_1((u + v)/2) + (\lambda_2(u) + \lambda_2(u, v) + \lambda_2(v))/3 \right.$
$$\left. - \lambda_0^2 |z|^2/3 \right) + O_7.$$

Here λ_2 is the polarized form of λ_2 in the middle term. As in the proof of (2.4) and with the same notations as there we have, if $|u, v| > 0$,

$$a^2 = |z|^2 - 2|u||v|(\delta \sin \alpha - \delta^2 \cos \alpha/2) + O_8$$

$$= |z|^2 - 2|u, v| \delta + (u, v)\lambda_0^2 |u, v|^2/4 + O_7$$

by the crude form (3.6)' of Legendre's theorem. Comparison with (A.7) gives us now a refinement of Legendre's theorem

(A.8) $\quad \delta = -\frac{1}{2}|u, v| \left(\lambda_0 + \lambda_1((u + v)/2) + (\lambda_2(u) + \lambda_2(u, v) + \lambda_2(v))/3 \right.$
$$\left. - \lambda_0^2 (u, v)/4 - \lambda_0^2 |z|^2/3 \right) + O_5.$$

Here we have assumed that α is bounded away from 0 and from π.

Appendix B: Higher order approximations. In (A.8) the term λ_0 is the curvature at the vertex 0, but to make (A.8) more appealing one should express the other terms by means of the curvature at some other points. To do so we must compute the curvature of the metric (3.3) when $u \neq 0$. In the two dimensional case the curvature is

$$K = R_{1212}/(g_{11}g_{22} - g_{12}^2)$$

for this is true in geodesic coordinates and the quotient is invariant under coordinate changes. (In view of (4.8)' this contains (3.4).) For the metric (3.3) we have

$$g_{11} = 1 + \lambda u_2^2, \; g_{12} = -u_1 u_2 \lambda, \; g_{22} = 1 + \lambda u_1^2,$$
$$g_{11}g_{22} - g_{12}^2 = 1 + \lambda|u|^2, \; (1 + \lambda|u|^2)g^{jk} = \delta_{jk} + \lambda u_j u_k.$$

In R_{1212} the linear term is (see also (3.4))

$$\partial^2 g_{12}/\partial u_1 \partial u_2 - (\partial^2 g_{11}/\partial u_2^2 + \partial^2 g_{22}/\partial u_1^2)/2$$
$$= -3\lambda - 3\langle u, \partial\lambda/\partial u\rangle - \tfrac{1}{2}\sum u_j u_k \partial^2\lambda/\partial u_j \partial u_k.$$

Since Γ_{111} and Γ_{222} vanish of second order at 0 while $\Gamma_{121} = \lambda u_2 + O_2$ and $\Gamma_{122} = \lambda u_1 + O_2$, the non-linear term in R_{1212} is $\lambda^2|u|^2 + O_3$, hence

(B.1) $$K = -3\lambda - 3\langle u, \partial\lambda/\partial u\rangle - \tfrac{1}{2}\sum u_j u_k \partial^2\lambda/\partial u_j \partial u_k + 4\lambda^2|u|^2 + O_3.$$

A somewhat longer but elementary computation gives

$$K = -(1 + \lambda|u|^2)^{-1}(3\lambda + 3\langle u, \partial\lambda/\partial u\rangle + \tfrac{1}{2}\sum u_j u_k \partial^2\lambda/\partial u_j \partial u_k)$$
$$+ (1 + \lambda|u|^2)^{-2}(\langle u, \partial\lambda/\partial u\rangle + 2\lambda)^2|u|^2/4$$

but we shall not make use of that.

If we introduce the Taylor expansion $\lambda = \lambda_0 + \lambda_1 + \lambda_2 + \ldots$ where λ_j is homogeneous of degree j, then (B.1) gives

(B.1)' $$K = -3\lambda_0 - 6\lambda_1 - 10\lambda_2 + 4\lambda_0^2|u|^2 + O_3.$$

In particular this allows us to interpret the more precise form (3.6) of Legendre's theorem by noting that

$$12\lambda((u + v)/2) = 12\lambda_0 + 6\lambda_1(u) + 6\lambda_1(v) + O_2 = -2K(0) - K(u) - K(v) + O_2.$$

Thus we may formulate (3.6) as follows: If K_α, K_β, K_γ are the curvatures at the vertices with angles α, β, γ, then the angle excess is

(B.2) $$\delta = \alpha - \alpha_e = (2K_\alpha + K_\beta + K_\gamma)T/12 + O_4.$$

This result is also due to Gauss [2]. Note that the excesses at the different vertices will therefore not be the same in third order terms except when the curvature is stationary.

To examine the accuracy of the Gauss approximation we note that (B.1)' gives

$$(2K(0) + K(u) + K(v))/12 = -\lambda_0 - \lambda_1((u+v)/2) - \tfrac{5}{6}(\lambda_2(u) + \lambda_2(v))$$
$$+ \lambda_0^2(|u|^2 + |v|^2)/3 + O_3.$$

Another approximation which agrees in the first two terms is

$$K((u+v)/4)/3 = -\lambda_0 - \lambda_1((u+v)/2) - \tfrac{5}{24}(\lambda_2(u) + \lambda_2(v) + 2\lambda_2(u,v))$$
$$+ \lambda_0^2|u+v|^2/12 + O_3.$$

We combine them with the weights $1/5$ and $4/5$ and set

(B.3) $$\overline{K} = (2K(0) + K(u) + K(v) + 16K((u+v)/4)/20.$$

Then we obtain using (A.8)

(B.4) $$\delta = \tfrac{1}{2}|u,v|\big(\overline{K}/3 + \lambda_0^2(|z|^2/3 + (u,v)/4 - (|u|^2 + |v|^2 + |u+v|^2)/15\big) + O_5.$$

The area T of the Euclidean triangle with sides a, b, c, thus angle $\alpha - \delta$, is

$$\tfrac{1}{2}|b||c|\sin(\alpha - \delta) = \tfrac{1}{2}|b||c|(\sin\alpha - \delta\cos\alpha + O(\delta^2)) = |u,v|/2 + (u,v)\lambda_0|u,v|/4 + O_5$$

by $(3.6)'$, so

$$|u,v|/2 = T(1 - \lambda_0(u,v)/2) + O_5.$$

This gives if we recall that $\overline{K}/3 = -\lambda_0 + O_1$

(B.4)' $$\delta = T\big(\overline{K}/3 + \lambda_0^2(|z|^2/3 + 3(u,v)/4 - (|u|^2 + |v|^2 + |u+v|^2)/15\big) + O_5.$$

Now $b^2 = |u|^2$, $c^2 = |v|^2$ and $a^2 = |u-v|^2 + O_4$ so we obtain finally

(B.5) $$\alpha - \alpha_e = T\big((2K_\alpha + K_\beta + K_\gamma + 16K'_\alpha)/60 + K_\alpha^2(7b^2 + 7c^2 + a^2)/360\big) + O_5$$

where K'_α is the curvature at the midpoint of the geodesic between the vertex with angle α and the midpoint of the opposite side. In fact, this point differs from $(u+v)/4$ by O_3.

To compare with the formula (2.5) of Gauss we shall now compute the Riemannian area S of the geodesic triangle. Since

$$g_{11}g_{22} - g_{12}^2 = 1 + |x|^2\lambda_0 + O_3$$

and the geodesic line is a parabola with error O_4 (see (A.5)) we have

$$S = \iint (1 + |x|^2\lambda_0/2)\, dx - \tfrac{2}{3}|z,-w/4| + O_5$$

with the double integral taken over the Euclidean triangle. Since $|z,w|/6 = -|z|^2|u,v|\lambda_0/6$ and

$$\iint |x|^2\, dx = |u,v| \iint_{0 \le y_j, y_1+y_2 < 1} (y_1^2(u,u) + 2y_1y_2(u,v) + y_2^2(v,v))\, dy$$
$$= |u,v|(|u|^2 + (u,v) + |v|^2)/12$$

we obtain

$$S = \tfrac{1}{2}|u,v|\big(1 + \lambda_0((|u|^2 + (u,v) + |v|^2)/12 - |z|^2/3)\big) + O_5.$$

As in the proof of (B.5) we now obtain

(B.6) $$\alpha - \alpha_e = S\big((2K_\alpha + K_\beta + K_\gamma + 16K'_\alpha)/60 + K_\alpha^2(b^2 + c^2 - 2a^2)/180\big) + O_5.$$

Summing up, we have proved

THEOREM B.1. *Let a geodesic triangle have small sides a, b, c and area S. Denote by T the area of the Euclidean triangle with the same sides and by α resp. α_e the angle opposite a in the geodesic and the Euclidean triangle. Assume α bounded away from 0 and π. Denote by K_α, K_β, K_γ the curvatures at the vertices and by K'_α the curvature at the midpoint of the geodesic between the vertex with angle α and the midpoint of the opposite side. Then (B.5) and (B.6) are valid.*

It is clear that (B.6) contains (2.5) as a special case. Note that (B.5) and (B.6) only contain the curvature at four points although the second order Taylor expansion of λ or K involves six parameters.

References

[1] M. Cantor, *Vorlesungen über Geschichte der Mathematik. Band I*, Teubner Verlag, Leipzig-Berlin, 1906.

[2] D.F. Gauss, *Disquisitiones generales circa superficies curvas*, Comm. soc. reg. scient. Gött. **6** (1823–1827), German translation in Ostwald's Klassiker der exakten Wissenschaften 5(1889), English translation in General investigations of curved surfaces, Raven Press, Hewlett N.Y. 1965, which also contains an earlier version.

[3] A.M. Legendre, *Sur les opérations trigonométriques dont les résultats dépendent de la figure de la terre*, Mém. de l'Acad. de Paris (1787), 352 ff.

[4] B. Riemann, *Ueber die Hypothesen welche der Geometrie zu Grunde liegen*, Werke, ed. H. Weber, 372–387. Teubner Verlag, Leipzig, 1892.

[5] J. Tropfke, *Geschichte der Elementar-Mathematik. Band V*, Walter de Gruyter et Co., Berlin-Leipzig, 1923.

[6] P. Dombrowski, *150 years after Gauss' "disquisitiones generales circa superficies curvas"*, Astérisque **62** (1979), 1–153.

CORRECTION TO MY PAPER
ON SOBOLEV SPACES ASSOCIATED WITH SOME LIE ALGEBRAS

LARS HÖRMANDER

Professor Claude Zuily has kindly called my attention to a serious error in [1]. It occurs in the last paragraph of Section 4, ending with the estimate $(4.6)'$ (where the left-hand side should have been $\|F(u)\|_{s,p}^{\#m,g}$). We shall examine here what the arguments of [1] actually prove. Fortunately it turns out that the applications are not affected at all.

Let g be a slowly varying metric in the open set $X \subset \mathbf{R}^n$, and let m be g continuous. (We keep the notation and definitions of [1].) If $D^\alpha u \in L_{\mathrm{loc}}^p(X)$, $|\alpha| \le s$, we define in analogy with (2.4)

$$\|u\|_{s,p}^{\#m,g} = \sup_{x_1} \sum_{j \le s} \left(\int m(x_1, x')^p |u^{(j)}(x_1, x')|_{(x_1, x')}^p dx' \right)^{1/p},$$

where $|.|_x$ indicates that the norm is taken with respect to the metric g at x. (Formula $(*.*)$ is always a reference to [1].) If ϕ_ν, $\nu = 1, 2, \ldots$, form a standard partition of unity corresponding to the metric, and $u_\nu = \phi_\nu u$, then

$$(1) \qquad \|u_\nu\|_{s,p}^{\#m,g} \le C\|u\|_{s,p}^{\#m,g}.$$

Let $\mu(x)$ be the $n-1$ dimensional Lebesgue measure of

$$\{y; |y - x|_x < 1, y_1 = x_1\}.$$

Then it follows from (4.1), or from the standard Sobolev lemma, that

$$(2) \qquad \|m\mu^{1/p}u\|_\infty \le C_s\|u\|_{s,p}^{\#m,g}$$

when $sp > n - 1$. If m_0 is another g continuous weight, we have from (4.4) an analogue of $(2.3)''$

$$(3) \qquad \|u_\nu\|_{j,ps/j}^{\#m_j,g} \le C(\|u_\nu\|_{s,p}^{\#m,g})^{j/s}(\|u_\nu\|_{s_0,p}^{\#m_0,g})^{(s-j)/s},$$

where $m_j = m^{j/s}(m_0\mu^{1/p})^{(s-j)/s}$. Let $U_\nu = \sum u_\mu$ with summation for all μ such that $\mathrm{supp}\,\phi_\mu \cap \mathrm{supp}\,\phi_\nu \ne \emptyset$. The number of terms is bounded, so we obtain from (3) and (1)

$$(4) \qquad \|U_\nu\|_{j,ps/j}^{\#m_j,g} \le C(\|u\|_{s,p}^{\#m,g})^{j/s}(\|u\|_{s_0,p}^{\#m_0,g})^{(s-j)/s}.$$

© Springer International Publishing AG, part of Springer Nature 2018
L. Hörmander, *Unpublished Manuscripts*,
https://doi.org/10.1007/978-3-319-69850-2_18

Assume that

(5)
$$1 \leq C m_0^p \mu$$

and that $p s_0 > n - 1$. Then (2) gives a bound

$$|U_\nu| \leq M = C \|u\|_{s_0,p}^{\#m_0,g},$$

if the right-hand side is finite, as we assume from now on. We want to estimate $\|F(u)\|_{s,p}^{\#m,g}$ in terms of $\|u\|_{s,p}^{\#m,g}$ and M, assuming that $F \in C^s$ and that $F(0) = 0$. Since $F(t) \leq C(M)|t|$ it is clear that

$$\|F(u)\|_{0,p}^{\#m,g} \leq C_0(M)\|u\|_{0,p}^{\#m,g}$$

so we may assume that $s > 0$. The differential of $F(u)$ of order s consists besides the term $F'(u)u^{(s)}$, which also has an obvious estimate, of the sum over ν of terms with norm at x bounded by a constant times

$$\phi_\nu(x)|U_\nu^{(k_1)}|_x \ldots |U_\nu^{(k_j)}|_x; \quad k_1 + \cdots + k_j = s; \ k_1 > 0, \ldots, k_j > 0; \ j \geq 2.$$

Now

$$\|f_1 \ldots f_j\|_p \leq \prod_1^j \|f_i\|_{sp/k_i},$$

since $\sum k_i/sp = 1/p$, and since

$$\prod_1^j m_{k_i} = m(m_0 \mu^{1/p})^{j-1} \geq m m_0 \mu^{1/p}$$

it follows that

$$\|\phi_\nu|F(u)^{(s)} - F'(u)u^{(s)}|_x\|_{0,p}^{\#m,g} \leq C_1(M) m_0(x_\nu)^{-1} \mu(x_\nu)^{-1/p} \|u\|_{s,p}^{\#m,g},$$

where x_ν is some point in $\operatorname{supp} \phi_\nu$. For a fixed a we see that when $\operatorname{supp} \phi_\nu$ contains some point x_ν with first coordinate a, then $m_0(x_\nu)^{-1}\mu(x_\nu)^{-1/p}$ can be estimated by the integral of $m_0(x)^{-1}\mu(x)^{-1-1/p}$ over $\{x'; |(x', a) - x_\nu|_{x_\nu} \leq c\}$. If c is a sufficiently small positive number there is a bound for the number of overlapping such balls, so we conclude using the triangle inequality that

$$\| |F(u)^{(s)} - F'(u)u^{(s)}|_x \|_{0,p}^{\#m,g} \leq C_2(M)\|u\|_{s,p}^{\#m,g},$$

if

(6)
$$\int m_0(x)^{-1}\mu(x)^{-1-1/p} dx' \leq C_3.$$

Combining this with the trivial estimate of the second term on the left, we obtain

THEOREM 1. *If $ps_0 > n - 1$ and (5), (6) hold, then*

(7)
$$\|F(u)\|_{s,p}^{\#m,g} \leq C(s, p, F, \|u\|_{s_0,p}^{\#m_0,g})\|u\|_{s,p}^{\#m,g}.$$

This result should replace (4.6)' in [1].

We shall now consider more carefully the problem of estimating a product uv when we know

$$A = \|u\|_{s,p}^{\#m,g}\|v\|_{s_0,p}^{\#m_0,g} + \|u\|_{s_0,p}^{\#m_0,g}\|v\|_{s,p}^{\#m,g}.$$

As in the discussion leading to Theorem 1 we obtain from (4.4)

$$m_0(x_\nu)\mu(x_\nu)^{1/p}\|m(x)\phi_\nu(x)|u^{(j)}(x)|_x|v^{(s-j)}(x)|_x\|_p \leq CA.$$

Here we have also used the inequality between geometric and arithmetic means. If (6) is strengthened to

(8)
$$\int m_0(x)^{-1}\mu(x)^{-1-1/p}dx' \leq \kappa(x_1) \leq C_3,$$

we conclude that with L_p norms taken with respect to x' for fixed x_1

$$\||m|u^{(j)}|_x|v^{(s-j)}|_x\|_p \leq C\kappa(x_1)(\|u\|_{s,p}^{\#m,g}\|v\|_{s_0,p}^{\#m_0,g} + \|u\|_{s_0,p}^{\#m_0,g}\|v\|_{s,p}^{\#m,g}).$$

Let us now look at the spaces $W_{s,p}^{\#A^*}$ in [1, Section 5]. After a localization at a non-isotropic direction we identified this space with $W_{s,p}^{\#1,g}$, where $g = |dx|^2/(1 + |x|)^2$. Thus $\mu(x)$ is equivalent to $(1 + |x|)^{(n-1)}$, and $m = m_0 = 1$. The integral in (8) becomes (with l_1 norms of x for the sake of convenience)

$$\int (1 + |x|)^{-(n-1)(1+1/p)}dx' = C(1 + |x_1|)^{-(n-1)/p},$$

which is small at infinity. The other case is (see [1, p. 278])

$$g = |dx_n|^2/(1 + |x|)^2 + |dx''|^2/(1 + |x''|)^2, \quad m(x) = m_0(x) = (1 + |x''|)^{-1/p},$$

where $x'' = (x_1, \ldots, x_{n-1})$. Then $\mu(x)$ is equivalent to $(1 + |x|)(1 + |x''|)^{n-2}$, and the integral in (8) is equivalent to

$$\int (1 + |x''|)^{1/p}((1 + |x'|)(1 + |x''|)^{n-2})^{-1-1/p} dx_2 \ldots dx_n.$$

Integration with respect to x_n gives with $z = (x_2, \ldots, x_{n-1})$ an equivalent integral

$$\int (1 + |x_1| + |z|)^{1/p-(n-2)(1+1/p)}(1 + |z|)^{-1/p}dz$$

$$\leq (1 + |x_1|)^{-(n-2)/p} \int (1 + |z|)^{1/p-(n-2)(1+1/p)}|z|^{-1/p}dz.$$

The integral converges so we obtain the following weakened version of [1, Theorem 5.1]:

THEOREM 2. *Assume that the quadratic form A as well as its restriction to the plane $x_1 = 0$ is non-singular. If $1 \leq p < \infty$ and $(n-1)/p < s_0 \leq s$ then*

$$(9) \qquad \|\mu^{1/p} u\|_\infty \leq C \|u\|_{s_0,p}^{\#A^*}, \quad u \in W_{s_0,p}^{\#A^*},$$

where $\mu(x) = (1 + |x|)^{n-3} \tilde{A}(x)$. For every $F \in C^s$ with $F(0) = 0$, we have

$$(10) \qquad \|F(u)\|_{s,p}^{\#A^*} \leq C(p,F,s, \|u\|_{s_0,p}^{\#A^*}) \|u\|_{s,p}^{\#A^*}, \quad u \in W_{s,p}^{\#A^*}.$$

When $Z_1, \ldots, Z_j$ are among the vector fields (3.1),(3.2), (3.3) in [1] and $j \leq s$, then

$$
(11) \quad
\begin{aligned}
&\|(Z_1 \ldots Z_i u)(Z_{i+1} \ldots Z_j v)\|_p \\
&\qquad \leq C\kappa(x_1)(\|u\|_{s,p}^{\#A^*} \|v\|_{s_0,p}^{\#A^*} + \|u\|_{s_0,p}^{\#A^*} \|v\|_{s,p}^{\#A^*}); \quad u, v \in W_{s,p}^{\#A^*}.
\end{aligned}
$$

Here $\kappa(x_1) = (1 + |x_1|)^{-(n-2)/p}$, and the L^p norm is taken for fixed x_1 with respect to the variables x' only.

Although (11) is weaker than (5.3), Theorem 2 contains everything in Theorem 5.1 of [1] which was actually used later on. The proof given above shows that (11) can be strengthened in non-characteristic directions. It is conceivable that (5.3) is actually valid, but we have not pursued this question since it is not important in the applications envisaged so far.

References

[1] Lars Hörmander, *On Sobolev spaces associated with some Lie algebras*, Current topics in partial differential equations, Kinokuniya Company Ltd, Tokyo, 1985, pp. 261–287.

GÅRDING'S INEQUALITY DURING THREE DECADES

LARS HÖRMANDER[16]

1. The original result. The Gårding inequality was first published in 1953 and was stated as follows in [4]:

THEOREM 1.1. *Let $p_{\alpha\beta}(x)$ be real valued and symmetric in the multi-indices α, β with $|\alpha| = |\beta| = m$, uniformly continuous in $\Omega \subset \mathbf{R}^n$. If*

$$(1.1) \qquad \sum p_{\alpha\beta}(x)\xi^{\alpha+\beta} \geq c|\xi|^{2m} \quad \text{when } x \in \Omega, \ \xi \in \mathbf{R}^n,$$

then there is a constant K such that

$$(1.2) \qquad \sum (p_{\alpha\beta}D^\alpha u, D^\beta u) + K(u, u) \geq 0, \quad u \in C_0^\infty(\Omega).$$

Here $(\ ,\ )$ is the scalar product in L^2 where the norm will be written $\|\ \|$. A stronger statement is given during the proof,

$$(1.2)' \qquad \|u\|_{(m)}^2 = \sum_{|\alpha| \leq m} \|D^\alpha u\|^2 \leq C \sum (p_{\alpha\beta}D^\alpha u, D^\beta u) + K\|u\|^2, \quad u \in C_0^\infty(\Omega),$$

and this stronger statement is essential in the proof. Indeed, in view of the interpolation inequality

$$(1.3) \qquad \|u\|_{(k)} \leq C\|u\|_{(m)}^{k/m}\|u\|_{(0)}^{(m-k)/m}$$

the estimate $(1.2)'$ follows by means of a partition of unity if it is known when u has small support. By the uniform continuity assumed it suffices therefore to prove $(1.2)'$ with coefficients frozen at any point in Ω. However, in the constant coefficient case $(1.2)'$ follows from Parseval's formula which allows one to exploit directly the positivity in (1.1). When $m = 1$ one can of course write $\sum p_{\alpha\beta}(x)\xi^{\alpha+\beta}$ as a sum of squares to prove the estimate. However, this is in general impossible in the higher order case, so the use of Fourier analysis essential. This was one of the essential points in Gårding's paper; the other was the effective use of a partition of unity.

The stability of $(1.2)'$ under small perturbations shows that one may assume that $p_{\alpha\beta}$ and all their derivatives are defined and bounded in the whole of $\mathbf{R}^n$. Integrating by parts and using (1.4) we see that the result is then equivalent to

$$(1.4) \qquad \|u\|_{(m)}^2 \leq C \operatorname{Re}(P(x, D)u, u) + K\|u\|^2, \quad u \in C_0^\infty(\mathbf{R}^n),$$

---

[16]Lecture at the Gårding symposium May 31, 1985.

© Springer International Publishing AG, part of Springer Nature 2018
L. Hörmander, *Unpublished Manuscripts*,
https://doi.org/10.1007/978-3-319-69850-2_19

if

$$P(x, \xi) = \sum_{|\alpha|=2m} p_\alpha(x)\xi^\alpha$$

where p_α and the derivatives are bounded in $\mathbf{R}^n$ and

$$c|\xi|^{2m} \leq \sum_{|\alpha|=2m} p_\alpha(x)\xi^\alpha; \quad x, \xi \in \mathbf{R}^n.$$

Jumping a decade ahead from Gårding's 1953 paper we can generalize and simplify by considering a pseudo-differential operator $P(x, D)$ with $P \in S_{1,0}^{2m}$, that is,

$$|D_x^\beta D_\xi^\alpha P(x, \xi)| \leq C_{\alpha\beta}(1 + |\xi|)^{2m-|\alpha|},$$

which is elliptic in the sense that

$$(1 + |\xi|^2)^m \leq C_0 \operatorname{Re} P(x, \xi) + C_1; \quad (x, \xi) \in \mathbf{R}^n.$$

Then we have for any $t < m$

$$(1.4)' \qquad \|u\|_{(m)}^2 \leq C \operatorname{Re} (P(x, D)u, u) + K_t \|u\|_{(t)}^2; \quad u \in C_0^\infty(\mathbf{R}^n).$$

In fact, $Q = (C_0 \operatorname{Re} P + C_1 - \frac{1}{2}(1 + |\xi|^2)^m)^{\frac{1}{2}} \in S_{1,0}^m$, and

$$\tfrac{1}{2}C_0(P(x, D) + P(x, D)^*) - Q(x, D)^*Q(x, D) - \tfrac{1}{2}(1 - \Delta)^m \in \operatorname{Op} S_{1,0}^{2m-1},$$

which implies that

$$C_0 \operatorname{Re} (P(x, D)u, u) \geq \tfrac{1}{2}((1 - \Delta)^m u, u) - C\|u\|_{(m-\frac{1}{2})}^2 = \tfrac{1}{2}\|u\|_{(m)}^2 - C\|u\|_{(m-\frac{1}{2})}^2,$$

where we have switched to the definition

$$\|u\|_{(s)}^2 = \left((2\pi)^{-n} \int |\hat{u}(\xi)|^2 (1 + |\xi|^2)^s \, d\xi\right)^{\frac{1}{2}}$$

of the Sobolev norms. Using the interpolation inequality

$$(1.3)' \qquad \|u\|_{(\lambda s + (1-\lambda)t)} \leq \|u\|_{(s)}^\lambda \|u\|_{(t)}^{1-\lambda},$$

and the inequality between geometric and arithmetic means, we obtain

$$2C_0 \operatorname{Re} (P(x, D)u, u) \geq \tfrac{1}{2}\|u\|_{(m)}^2 - C_t \|u\|_{(t)}^2, \quad u \in C_0^\infty(\mathbf{R}^n),$$

for any $t < m$. Thus it had become legitimate in the mid 1960's to take the square root of the "symbol" and write an elliptic operator essentially as a square for any order.

The original purpose of Gårding's inequality was to prove the existence of solutions of the Dirichlet problem for elliptic operators of higher order and to study the distribution of the eigenvalues. Soon afterwards Gårding [5] showed how one can use it to prove energy estimates for strictly hyperbolic operators. The later refinements have all been intended for similar purposes. The need for refinements has come from the study of hypoelliptic rather than elliptic operators, and from the study of microlocal energy estimates leading to results on propagation of singularities.

2. The sharp Gårding inequality. We can rephrase Gårding's inequality (1.4)′ as follows: Assume that $P \in S_{1,0}^{2m}$ and that

$$(2.1) \qquad\qquad P(x,\xi) \geq 0.$$

Then there is for every $\varepsilon > 0$ a constant $C_{\varepsilon,t}$ such that

$$(2.2) \qquad \mathrm{Re}\,(P(x,D)u,u) \geq -\varepsilon\|u\|_{(m)}^2 - C_{\varepsilon,t}\|u\|_{(t)}^2, \quad u \in C_0^\infty(\mathbf{R}^n).$$

For the proof we just have to apply the earlier version to $P(x,\xi) + \varepsilon(1 + |\xi|^2)^m$. To my knowledge Nirenberg was the first who raised the question if there is a sharp form of Gårding's inequality where $\varepsilon = 0$ in (2.2) and $t = m - \frac{1}{2}$; the latter condition makes the estimate obvious when $P \in S_{1,0}^{2m-1}$, so it only depends on the principal symbol $\in S_{1,0}^{2m}/S_{1,0}^{2m-1}$. The first proof was given in [6] and many others have been given later on. I shall sketch a proof given in [9] which exploits the convolution to express $P(x,D)$ modulo operators of lower order as a (continuous) sum of "squares". First we choose any even $\varphi \in C_0^\infty(\mathbf{R}^{2n})$ with L^2 norm equal to one and introduce the "square"

$$\psi(x,D) = \varphi(x,D)^*\varphi(x,D).$$

Then ψ is an even function in $\mathcal{S}$, and

$$(2.3) \qquad\qquad \iint \psi(y,\eta)\,dy\,d\eta = 1.$$

The operator $\psi(t(x-y),(D-\eta)/t)$ is unitarily equivalent, hence positive, for any $t > 0$ and $(y,\eta) \in \mathbf{R}^{2n}$; it is concentrated at (y,η) with a spread of the magnitude $\Delta x = 1/t$ and $\Delta\xi = t$.

Choose $\chi \in C_0^\infty(\mathbf{R}^n)$ as a decreasing function of $|\xi|$ with $\chi(\xi) = 1$ when $|\xi| < 1$ and $\chi(\xi) = 0$ when $|\xi| > 2$, and write $P = \sum P_k$,

$$P_0(x,\xi) = \chi(\xi)P(x,\xi); \quad P_k(x,\xi) = (\chi(2^{-k}\xi) - \chi(2^{1-k}\xi))P(x,\xi), \ k > 0.$$

With $t_k = 2^{k/2}$ we have

$$(2.4) \qquad\qquad |D_x^\beta D_\xi^\alpha P_k(x,\xi)| \leq C_{\alpha\beta} t_k^{-2|\alpha|+4m}.$$

Now form the regularization

$$a_k(x,\xi) = \iint P_k(x-y,\xi-\eta)\psi(t_ky,\eta/t_k)\,dy\,d\eta = \iint P_k(y,\eta)\psi(t_k(x-y),(\xi-\eta)/t_k)\,dy\,d\eta.$$

Since $P_k \geq 0$ the second representation shows that $a_k(x,D) \geq 0$, as a sum of squares with positive coefficients. The first form shows that a_k has the same estimates (2.4) as P_k with a slight change of constants. Since $2^{k-1} < |\xi - \eta| < 2^{k+1}$ in the support of the integrand we have $|\eta| > 2^{k-2}$ there if $|\xi| < 2^{k-2}$, and $|\eta| > |\xi|/2$ if $|\xi| > 2^{k+2}$. Hence $|\eta|/t_k > 2^{k/2-2} > |\xi|^{\frac{1}{2}}/2$ and $|\eta|/t_k > |\xi|^{\frac{1}{2}} > 2^{k/2+1}$ respectively in the two cases, so the estimate of a_k is in both cases improved by any negative power of $2 + |\xi|$. To estimate the regularization error

$$b_k(x,\xi) = P_k(x,\xi) - a_k(x,\xi)$$

when $2^{k-2} \leq |\xi| \leq 2^{k+2}$ we apply Taylor's formula with error term of order 2 in the first representation of a_k, recalling (2.3) and that ψ is even which makes first order terms drop out. This gives an estimate for b_k of the form (2.4) but with another factor t_k^{-2} in the right-hand side. Summing we obtain

$$b(x,\xi) = \sum b_k(x,\xi) \in S^{2m-1},$$

and since $P(x,D) - b(x,D) \geq 0$, we have proved:

THEOREM 2.1. *(The sharp Gårding inequality.)* If $0 \leq P \in S^{2m}_{1,0}$ then

$$(2.5) \qquad\qquad \text{Re}\,(P(x,D)u,u) \geq -C\|u\|^2_{(m-\frac{1}{2})}.$$

The preceding proof is quite flexible. First of all it applies in the vector valued case where the result was first established by Lax and Nirenberg [11]. Secondly, if $P \in S^{2m}_{\varrho,\delta}$ for some $\delta < \varrho$, that is,

$$|D^\beta_x D^\alpha_\xi P(x,\xi)| \leq C_{\alpha\beta}(1+|\xi|)^{2m-\varrho|\alpha|+\delta|\beta|}$$

then (2.5) is replaced by

$$(2.5)' \qquad\qquad \text{Re}\,(P(x,D)u,u) \geq -C\|u\|^2_{(m-(\varrho-\delta)/2)}.$$

In the proof one just has to note that P_k has bounded derivatives with respect to the scales $\Delta x = 2^{-k\delta}$ and $\Delta\xi = 2^{k\varrho}$, so we take $t^2_k = \Delta\xi/\Delta x = 2^{k(\varrho+\delta)}$. Since

$$t_k\Delta x = t^{-1}_k\Delta\xi = 2^{k(\varrho-\delta)/2}$$

we gain a factor $2^{k(\varrho-\delta)}$ in the estimate of b_k, and $(2.5)'$ follows.

We shall come back to another extension in Section 5 but end the proof with another remark which will be useful in Section 3. If one uses a Taylor expansion of order 4 when estimating b_k then it follows under the hypotheses of Theorem 2.1 that $P(x,D)-b(x,D) \geq 0$ where $b \in S^{2m-1}_{1,0}$ and

$$(2.6) \qquad |b(x,\xi)| \leq C\Big(\sum_{|\alpha+\beta|\leq 2}(1+|\xi|)^{|\alpha|-1}|P^{(\alpha)}_{(\beta)}(x,\xi)| + (1+|\xi|)^{2m-2}\Big).$$

3. Melin's inequality. The estimate (2.6) already suggests that to refine the sharp Gårding inequality one must focus attention on the second order derivatives of the symbol. We shall assume from now on that P has a "classical" symbol

$$P \sim p_{2m} + p_{2m-1} + \cdots$$

with homogeneous terms and $p_{2m} \geq 0$. Since the original Gårding inequality is applicable at non-characteristic points, one has to examine what happens at the characteristics. There the first order derivatives of p_{2m} vanish and the Hessian is a non-negative quadratic form (invariantly defined in the tangent space of the cotangent bundle). The basic observation leading to an improved estimate is that in the one dimensional case

$$((D^2+x^2)u,u) = \|(D-ix)u\|^2 + \|u\|^2 \geq \|u\|^2, \quad u \in C^\infty_0(\mathbf{R}^n),$$

in spite of the fact that the infimum of the symbol $\xi^2 + x^2$ is 0. (Equality is attained for the Gaussian $e^{-\frac{1}{2}x^2}$.) The n-dimensional version of this argument is as follows. If $Q(x,\xi)$ is a positive semi-definite quadratic form in $\mathbf{R}^{2n}$ then the polarized symmetric bilinear form $Q(x,\xi;y,\eta)$; $(x,\xi),(y,\eta) \in \mathbf{R}^{2n}$; together with the symplectic form $\sigma(x,\xi;y,\eta) = \langle\xi,y\rangle - \langle x,\eta\rangle$ defines a linear map F,

$$Q(x,\xi;y,\eta) = \sigma((x,\xi);F(y,\eta)).$$

The eigenvalues $\neq 0$ are of the form $\pm i\mu$ with $\mu > 0$, and the corresponding generalized eigenvectors are in fact genuine ones. The hermitian form $Q(v, \bar{v})/2$ is positive definite in the span V^+ of the eigenfunctions belonging to eigenvalues on the open positive imaginary axis. Let $v_1, \ldots, v_k$ be a unitary basis there, and let $v_{k+1}, \ldots, v_l$ be an orthogonal basis of the space of real generalized eigenvectors belonging to the eigenvalue 0, modulo the radical of Q. If $L_j(x, \xi) = Q(v_j; x, \xi)$ it follows that

$$(3.1) \qquad (Q(x, D) + Q(x, D)^*)/2 = \sum L_j(x, D)^* L_j(x, D) + \mathrm{Tr}^+ Q.$$

Here $\mathrm{Tr}^+ Q = \sum \mu_j$ where $i\mu_j$ runs through the eigenvalues on the positive imaginary axis, repeated according to their multiplicities. The lower bound of (3.1) becomes $\mathrm{Tr}^+ Q$.

The preceding discussion can be repeated on the symbol level for the operator P provided that the characteristic set is smooth and the numbers k and l remain (locally) constant. After dropping terms corresponding to $\|L_j(x, D)u\|^2$ one can apply the usual sharp Gårding inequality. The result is as follows:

THEOREM 3.1. *Assume that P is a "classical" pseudo-differential operator in $X \subset \mathbf{R}^n$ which is properly supported and self-adjoint; let p_{2m} be the principal symbol and p^s_{2m-1} be the subprincipal symbols. (Thus p_{2m} and p^s_{2m-1} are the leading terms if P is written in the Weyl form.) Assume that*

$$(3.2) \qquad p_{2m}(x, \xi) \geq 0 \quad in \ T^*(X) \setminus \{0\},$$
$$(3.3) \qquad \mathrm{Re}\, p^s_{2m-1}(x, \xi) + \mathrm{Tr}^+ Q_{x,\xi} \geq 0 \quad if \ (x, \xi) \in T^*(X) \setminus 0, \ p_{2m}(x, \xi) = 0.$$

Here $Q_{x,\xi}$ is the Hessian of $\frac{1}{2}p_{2m}$ at (x, ξ). Finally assume that the characteristic set Σ is a C^∞ manifold, with tangent plane equal to the radical of $Q(x, \xi)$ when $p_{2m}(x, \xi) = 0$, and that the symplectic form has constant rank on Σ. For every compact set $K \Subset X$ one can then find a constant C_K such that

$$(3.4) \qquad \mathrm{Re}\,(Pu, u) \geq -C_K \|u\|^2_{(m-1)}, \quad u \in C^\infty_0(K).$$

The conditions (3.2), (3.3) are also necessary for (3.4) to hold.

Even without the smoothness and constant rank conditions one can for every characteristic point subtract a sum such as the one in (3.1) so that the principal symbol of the remaining operator vanishes of fourth order at the chosen characteristic while $\mathrm{Tr}^+ Q$ is added to the subprincipal symbol. Using (2.6) to estimate the remainder one obtains

THEOREM 3.2. *(Melin's inequality.) Assume that P is a "classical" pseudo-differential operator with principal and subprincipal symbols p_{2m} and p^s_{2m-1}. Assume that (3.2) and (3.3) hold. For every compact set $K \Subset X$ and every $\varepsilon > 0$ one can then find $C_{K,\varepsilon}$ such that*

$$(3.5) \qquad \mathrm{Re}\,(Pu, u) \geq -\varepsilon \|u\|^2_{(m-\frac{1}{2})} - C_{K,\varepsilon} \|u\|^2_{(m-1)}, \quad u \in C^\infty_0(K).$$

This was originally proved by Melin [11] by a direct localization argument improving that of [6] and an analysis of estimates for quadratic symbols. Later on Theorem 3.1 was proved in Hörmander [7], and in [9, Section 22.3] the arguments were put together to the proofs outlined above.

The analogy between (2.2) and (3.5) makes it natural to ask if (3.5) is valid with $\varepsilon = 0$. Various results in that direction are known besides Theorem 3.1. In [8] it was proved that if $P \in S^1_{1,0}$ is real valued and, for the total symbol,

$$(3.6) \qquad P(x,\xi) + \tfrac{1}{2}\mathrm{Tr}^+(P''(x,\xi) + Ktg_\xi) \geq 0 \quad \text{if } t \geq 0 \text{ and } P''(x,\xi) + tg_\xi \geq 0,$$

for some constant $K \geq 1$, then

$$(3.7) \qquad \mathrm{Re}\,(P(x,D)u, u) \geq -C\|u\|^2_{(-1/10)}, \quad u \in C^\infty_0(\mathbf{R}^n).$$

Here P'' is the Hessian of P and g_ξ is the metric form $|dx|^2 + |d\xi|^2/(1 + |\xi|^2)$. In particular, (3.6) is of course valid if $P(x,\xi) \geq 0$. A result of that type was already proved in [10] where it was essential for the study of the propagation of singularities. However, it was greatly surpassed by a later theorem of Fefferman and Phong [2] which will be discussed in Section 4.

For systems it seems hard to find an analogue of Theorems 3.1 and 3.2. The difficulty is that when Q is a quadratic form valued in hermitian symmetric matrices and Q^w is the corresponding Weyl operator, it is not easy to determine the lower bound for $(Q^w(x,D)u, u)$. A counterexample showing that the lower bound may be negative even if Q is positive was given in Hörmander [9] when u has two components. For the case

$$Q(x,\xi) = \begin{pmatrix} ax^2 + b\xi^2 & \alpha x\xi \\ \alpha x\xi & cx^2 + d\xi^2 \end{pmatrix} \quad (x,\xi) \in \mathbf{R}^2,$$

where $a, b, c, d \geq 0$ and $ad + bc \neq 0$ it has been proved recently by Sung [13] that $Q(x,D)$ is positive if and only if $(\lambda_1, \lambda_2) \in \Omega$,

$$\lambda_{1,2} = ((ad)^{\frac{1}{2}} - (bc)^{\frac{1}{2}} \pm \alpha)/((ad)^{\frac{1}{2}} + (bc)^{\frac{1}{2}}).$$

Here $\Omega \subset \mathbf{R}^2$ is a convex set inscribed in the square $|\lambda_1| + |\lambda_2| \leq 2$ and tangent to it at $(\pm 1, \pm 1)$ but not identical to it. On the other hand, positivity of the matrix $Q(x,\xi)$ for all (x,ξ) is equivalent to $|\lambda_1| + |\lambda_2| \leq 1$. Unless Ω should have a simpler description than the one given by Sung, it seems impossible to find an explicit analogue of Theorems 3.1 and 3.2.

4. The Fefferman-Phong estimates. In [2] it was proved that if $P \in S^{2m}_{1,0}$ and $P \geq 0$, then

$$(4.1) \qquad \mathrm{Re}\,(P(x,D)u, u) \geq -C\|u\|^2_{(m-1)}, \quad u \in C^\infty_0(\mathbf{R}^n),$$

which of course is a much better result than (3.7) for non-negative symbols. However, one should note that (3.6) allows the symbol to be quite negative occasionally (cf. also Theorem 3.1). The estimate (4.1) is important in many applications, see for example Chapters XXII and XXVIII in [9]. A proof of a general version of (4.1) adapted to the Weyl calculus can be found in [9, Chapter XVIII]. After an appropriate localisation the proof depends on the fact that a non-negative function with bounded fourth derivatives is bounded in a ball if the function and the second derivatives are bounded at the origin. In a neighborhood of a point where the symbol is not small one has essentially the situation in the original Gårding inequality, and in a neighborhood of a point where a second derivative is not small one can use the Morse lemma to write the symbol as the sum of a square and a positive symbol independent of one variable. An induction with respect to the dimension then gives the desired estimates.

For non-negative symbols Fefferman and Phong [3] have also obtained necessary and sufficient conditions for the validity of estimates stronger than (4.1),

$$(4.2) \qquad \text{Re} \, (P(x, D)u, u) \geq c\|u\|^2_{(m-\delta)} - C\|u\|^2_{(m-1)}, \quad u \in C_0^\infty(\mathbf{R}^n),$$

for some $c > 0$ and $\delta \in (0, 1)$. When $m = 1$ the condition is essentially that

$$(4.3) \qquad \max_Q |P \circ \Phi| \geq cM^{2(1-\delta)}$$

if $\Phi = (\varphi, \psi)$ is a canonical map from the unit cube Q to $T^*(\mathbf{R}^n)$ such that $|\psi(0)| = M > 1$ and for some positive $\varepsilon(\delta)$, $L(\delta)$

$$(4.4) \qquad \left| D_{x,\xi}^\alpha \big(\varphi(x, \xi) - \varphi(0, 0), \psi(x, \xi) - \psi(0, 0)/M \big) \right| \leq CM^{-\varepsilon(\delta)}, \quad |\alpha| \leq L(\delta).$$

Unfortunately this geometrical condition is extremely hard to check in spite of its appeal to intuition. One should note that the positivity of the symbol is a very strong condition ruling out applications to results of Melin's type. In a very impenetrable manuscript "Symplectic geometry and positivity of pseudo-differential operators" Fefferman and Phong have announced similar conditions necessary or sufficient for positivity of operators of order < 2. I have not seen any detailed proofs. The gap between necessary and sufficient conditions seems too large to give back even Melin's original result.

5. The sharp Gårding inequality for paradifferential operators. Let us first recall the paraproduct defined by Bony [1]. If $a \in \mathcal{S}'$ and $u \in \mathcal{S}'$ then the paraproduct $\pi(a, u)$ of u by a is the restriction to the diagonal of the inverse Fourier transform of the Fourier transform $\hat{a}(\xi)\hat{u}(\eta)$ of $a \otimes u$, cut off by multiplication with a function $\chi(\xi, \eta)$ such that

(i) $|D_{\xi,\eta}^\alpha \chi(\xi, \eta)| \leq C_\alpha(1 + |\xi| + |\eta|)^{-|\alpha|}$;

(ii) $\chi = 1$ at infinity in a conic neighborhood of $0 \times \mathbf{R}^n$;

(iii) $\chi = 0$ at infinity in a conic neighborhood of $\mathbf{R}^n \times 0$ and of the twisted diagonal $\{(\xi, -\xi), \, \xi \in \mathbf{R}^n\}$.

(The paraproduct depends of course on χ, and to emphasize that one sometimes writes $\pi_\chi(a, u)$ instead.) We can write $\pi(a, u)$ as a pseudo-differential operator $\sigma(x, D)u$ where

$$\hat{\sigma}(\xi, \eta) = \chi(\xi, \eta)\hat{a}(\xi).$$

Thus $\sigma(x, \eta)$ is the regularization of a by the inverse Fourier transform of $\chi(\cdot, \eta)$.

Let C^ϱ be the space of uniformly Hölder continuous functions of order ϱ in $\mathbf{R}^n$, with the norm

$$\|a\|_{\langle \varrho \rangle} = \sum_{|\alpha| \leq k} \sup |D^\alpha a| + \sum_{|\alpha| = k} \sup |D^\alpha a(x) - D^\alpha a(y)|/|x - y|^{\varrho - k}$$

where k is the largest integer $< \varrho$. Then it follows if $a \in C^\varrho$ that

$$\|D_\eta^\alpha \sigma(x, \eta)\|_{\langle \varrho \rangle} \leq C_\alpha(1 + |\eta|)^{-|\alpha|}$$

where the norm is taken with respect to the x variables. Since $|\xi| \leq C(1 + |\eta|)$ in the spectrum of $\sigma(x, \eta)$ as a function of x, it follows from Bernstein's theorem that

$$|D_x^\beta D_\eta^\alpha \sigma(x, \eta)| \leq C_{\alpha\beta}(1 + |\eta|)^{|\beta| - |\alpha| - \varrho}, \quad \text{if } |\beta| > \varrho.$$

If the coefficients of a differential operator are in C^ϱ and one replaces them by para-multiplication, then one obtains a paradifferential operator, that is, a pseudo-differential operator with symbol satisfying the following conditions:

$$(5.1) \qquad \qquad \|D_\eta^\alpha P(x, \eta)\|_{\langle \varrho \rangle} \leq C_\alpha(1 + |\eta|)^{m - |\alpha|};$$

$$(5.2) \qquad \qquad |\xi| \leq k|\xi + 2\eta|, \quad (\xi, \eta) \in \operatorname{supp} \widehat{P}(\xi, \eta).$$

Here k is assumed to belong to $(0, 1)$, and $\widehat{P}(\xi, \eta)$ is the Fourier transform of $P(x, \eta)$ with respect to x. It is easily proved that the adjoint $P(x, D)^*$ is then of the form $Q(x, D)$ where Q satisfies the same conditions and

$$Q(x, \xi) - \sum_{j \leq \varrho} \langle iD_x, D_\xi \rangle^j \overline{P}(x, \xi)/j! \in S_{1,1}^{m-\varrho}.$$

For every $P \in S_{1,1}^m$ satisfying (5.2) the operator $P(x, D)$ is continuous from $H_{(s+m)}$ to $H_{(s)}$ for all s. For any $R \in S_{1,1}^\mu$ we have $R(x, D)P(x, D) = S(x, D)$ where $S \in S_{1,1}^{m+\mu}$ and

$$S(x, \xi) - \sum_{|\alpha| \leq \varrho} R^{(\alpha)}(x, \xi)D_x^\alpha P(x, \xi)/\alpha! \in S_{1,1}^{m+\mu-\varrho}.$$

These facts allow one to use the standard arguments in the calculus of pseudo-differential operators with the limitations imposed by the error terms. The sharp Gårding inequality now takes the following form, which is somewhat stronger than the estimate given by Bony [1]:

THEOREM 5.1. *If P satisfies (5.2) and for some $\varrho \in (0, 2]$*

$$(5.3) \qquad \qquad \|D_\eta^\alpha P(\cdot, \eta)\|_{\langle \varrho \rangle} \leq C_\alpha(1 + |\eta|)^{2m + 2\varrho/(\varrho+2) - |\alpha|},$$

$$(5.4) \qquad \qquad \operatorname{Re} P(x, \xi) \geq -C(1 + |\xi|)^{2m},$$

then it follows that

$$(5.5) \qquad \qquad \operatorname{Re}(P(x, D)u, u) \geq -C'\|u\|_{(m)}^2; \quad u \in C_0^\infty.$$

It is easy to deduce the standard sharp Gårding inequality from this theorem for $\varrho = 2$, applied to a suitable regularization of the symbol.

PROOF OF THEOREM 5.1. The operator

$$(1 + |D|^2)^{-\frac{1}{2}m} P(x, D)(1 + |D|^2)^{-\frac{1}{2}m} - P(x, D)(1 + |D|^2)^{-m}$$

is in $\operatorname{Op} S_{1,1}^{2\varrho/(\varrho+2) - \min(\varrho, 1)}$, for $2\varrho/(\varrho + 2) - 1 = (\varrho - 2)/(\varrho + 2) \leq 0$ and $2\varrho/(\varrho + 2) - \varrho = -\varrho^2/(\varrho + 2) < 0$. Thus the difference is L^2 continuous which reduces the proof to the case $m = 0$. The same is true for

$$\tfrac{1}{2}(P(x, D) + P(x, D)^*) - (\operatorname{Re} P)(x, D)$$

so we may assume that P is real valued. Replacing $P(x, \xi)$ by $P(x, \xi) + C$ we may assume that $P \geq 0$ in what follows.

As in the proof of Theorem 2.1 we write now

$$P = \sum_0^\infty P_k$$

where $2^{k-1} \leq |\xi| \leq 2^{k+1}$ if $P(x, \xi) \neq 0$, $k > 0$. We regularize the terms as before but with

$$t_k = 2^{2k/(\varrho+2)}$$

in order to balance two error terms which will occur below. Note that $t_k \ll 2^k$. As before, the regularization a_k defines a positive operator, and for the error b_k we shall prove

(5.6) $$|D_x^\beta D_\xi^\alpha b_k(x, \xi)| \leq C_{\alpha\beta}'' 2^{k(|\beta|-|\alpha|)};$$

(5.7) $$|D_x^\beta D_\xi^\alpha b_k(x, \xi)| \leq C_{\alpha\beta N}''(2^k + |\xi|)^{-N} \quad \text{if } |\xi| < 2^{k-2} \text{ or } |\xi| > 2^{k+2}.$$

This will prove that $b = \sum b_k$ is the sum of one term which is in $S_{1,1}^0$ and satisfies (5.2) for a slightly larger k, and one term which is in $S_{1,0}^{-\infty}$. (Note that convolution does not increase the spectrum.) Thus the theorem will follow from (5.6), (5.7).

It will suffice to prove these estimates when $\alpha = \beta = 0$, for differentiation of the convolution defining a_k can be placed directly on P, and the bound in the basic estimate

(5.8) $$\|D_x^\alpha D_\xi^\beta P_k(x, \xi)\|_{\langle \varrho \rangle} \leq C_{\alpha\beta}' 2^{k(2\varrho/(\varrho+2)+|\beta|-|\alpha|)}$$

is multiplied by the right factor when P is replaced by a derivative. In the integral

$$a_k(x, \xi) = \iint P_k(y, \eta) \psi(t_k(x - y), (\xi - \eta)/t_k) \, dy \, d\eta$$

we have $2^{k-1} < |\eta| < 2^{k+1}$ when the integrand is not 0, hence $|\xi| < 2^{k-2}$ or $|\xi| > 2^{k+2}$ implies $2^k + |\xi| < C|\xi - \eta|$, and

$$|\xi - \eta|/t_k > (2^k + |\xi|)/(C2^{2k/(\varrho+2)}) > (2^k + |\xi|)^{\varrho/(\varrho+2)}/C.$$

Since $\psi \in \mathcal{S}$ the estimate (5.7) follows at once. To prove (5.6) we write

$$a_k(x, \xi) = \iint \psi(y, \eta) P_k(x - y/t_k, \xi - \eta t_k) \, dy \, d\eta$$

and observe that by Taylor's formula

$$\left| P_k(x - y/t_k, \xi - \eta t_k) - \sum_{\substack{|\alpha+\beta|<2 \\ |\beta|<\varrho}} P_{k(\beta)}^{(\alpha)}(x, \xi)(-y/t_k)^\beta(-\eta t_k)^\alpha/\alpha!\beta! \right| \leq C(1 + |y| + |\eta|)^2.$$

Indeed,

$$\left| P_k(x - y/t_k, \xi - \eta/t_k) - \sum_{|\alpha|<2} P_k^{(\alpha)}(x - y/t_k, \xi)(-\eta t_k)^\alpha \right| \leq C2^{k(2\varrho/(\varrho+2)-2)}t_k^2|\eta|^2 = C|\eta|^2,$$

$$\left| P_k(x - y/t_k, \xi) - \sum_{|\beta|<\varrho} P_{(\beta)}(x, \xi)(-y/t_k)^\beta \right| \leq C2^{k2\varrho/(\varrho+2)}t_k^{-\varrho}|y|^\varrho,$$

$$|P_k^{(\alpha)}(x - y/t_k, \xi) - P_k^{(\alpha)}(x, \xi)|t_k \leq C2^{k(2\varrho/(\varrho+2))}|y/t_k|^{\min(\varrho,1)}t_k 2^{-k} \leq C|y|^{\min(\varrho,1)},$$

if $|\alpha| = 1$. Since ψ is odd the first order terms in the Taylor expansion drop out when we integrate, and this proves (5.6) when $\alpha = \beta = 0$. The proof is complete.

We refer to Bony [1] for the application of Theorem 5.1 to the proof of the basic theorem on propagation of singularities for solutions of non-linear differential equations.

References

[1] J.-M. Bony, *Calcul symbolique et propagation des singularités pour les équations aux dérivées partielles non linéaires*, Ann. Sc. Ec. Norm. Sup. **14** (1981), 209–246.

[2] C. Fefferman and D.H. Phong, *On positivity of pseudo-differential operators*, Proc. Nat. Acad. Sci. **75** (1978), 4673–4674.

[3] _____, *The uncertainty principle and sharp Gårding inequalities*, Comm. Pure Appl. Math. **34** (1981), 285–331.

[4] L. Gårding, *Dirichlet's problem for linear elliptic partial differential equations*, Math. Scand. **1** (1953), 55–72.

[5] _____, *Solution directe du problème de Cauchy pour les équations hyperboliques*, Coll. Int. CNRS, Nancy 1956, pp. 71–90.

[6] L. Hörmander, *Pseudo-differential operators and non-elliptic boundary problems*, Ann. of Math. **83** (1966), 129–209.

[7] _____, *The Cauchy problem for differential equations with double characteristics*, J. Analyse Math. **32** (1977), 118–196.

[8] _____, *The Weyl calculus of pseudo-differential operators*, Comm. Pure Appl. Math. **32** (1979), 359–443.

[9] _____, *The Analysis of Linear Partial Differential Operators III*, Springer Verlag, 1985.

[10] _____, *Propagation of singularities and semiglobal existence theorems for (pseudo-)differential operators of principal type*, Ann. of Math. **108** (1978), 569–609.

[11] P.D. Lax and L. Nirenberg, *On stability for difference schemes: a sharp form of Gårding's inequality*, Comm. Pure Appl. Math. **19** (1966), 473–492.

[12] A. Melin, *Lower bounds for pseudo-differential operators*, Ark. Mat. **9** (1971), 117–140.

[13] Li-yeng Sung, *Positivity of a system of differential operators*, Preprint 1985.

SYMPLECTIC GEOMETRY AND DIFFERENTIAL EQUATIONS

LARS HÖRMANDER

0. Introduction. [17] From the beginning symplectic geometry has been very intimately connected with first order differential equations. However, it is only during the past 25 years or so that the great importance of symplectic geometry for the study of linear differential equations of high order has become clear. In these survey lectures I shall try to give a sketch of these developments as they look to me in retrospect. First I shall recall the background in classical mechanics, adding some remarks on the recent work on symplectic capacities. Then I shall recall the equally classical Hamilton-Jacobi theory in a form which is suitable for the third part, which is devoted to linear differential equations.

1. Classical mechanics. For a classical mechanical system with energy equal to $E(x, \xi)$ where $x \in \mathbf{R}^n$ are the coordinates of the system and $\xi \in \mathbf{R}^n$ the momenta, the Hamilton equations for the motion have the form

$$\frac{dx}{dt} = \frac{\partial E(x, \xi)}{\partial \xi}, \quad \frac{d\xi}{dt} = -\frac{\partial E(x, \xi)}{\partial x}.$$

(Physicists prefer the notation q instead of x and p instead of ξ.) Thus the mechanical system moves along the orbits of the Hamilton vector field

$$H_E = \frac{\partial E(x, \xi)}{\partial \xi} \frac{\partial}{\partial x} - \frac{\partial E(x, \xi)}{\partial x} \frac{\partial}{\partial \xi}.$$

The differential of E, on the other hand, is the one form (covector)

$$dE(x, \xi) = \langle d\xi, \partial E(x, \xi)/\partial \xi \rangle + \langle dx, \partial E(x, \xi)/\partial x \rangle,$$

which means that

(1.1) $$\langle dE, X \rangle = \sigma(X, H_E), \quad X \in \mathbf{R}^n \times \mathbf{R}^n,$$

where σ is the bilinear form

$$\sigma((x, \xi), (y, \eta)) = \langle \xi, y \rangle - \langle x, \eta \rangle, \quad (x, \xi) \in \mathbf{R}^n \times \mathbf{R}^n, \ (y, \eta) \in \mathbf{R}^n \times \mathbf{R}^n,$$

that is, $\sigma = \sum_1^n d\xi_j \wedge dx_j$. This two form is called the *standard symplectic form*. There is a dual bilinear form in the cotangent space called the *Poisson bracket*,

$$\{\varphi, \psi\} = H_\varphi \psi = -H_\psi \varphi = \langle \partial \varphi/\partial \xi, \partial \psi/\partial x \rangle - \langle \partial \varphi/\partial x, \partial \psi/\partial \xi \rangle.$$

---

[17]Outline of lectures in Augsburg, February 25 and 26, 1991.

© Springer International Publishing AG, part of Springer Nature 2018
L. Hörmander, *Unpublished Manuscripts*,
https://doi.org/10.1007/978-3-319-69850-2_20

The advantage of this way of looking at the Hamilton equations is that it makes the transformation laws transparent.

DEFINITION 1.1. A (C^∞) manifold M of dimension $2n$ is called a symplectic manifold if it is equipped with a *closed* two form σ which is non-degenerate at every point.

By a classical theorem of Darboux one can introduce local coordinates (x, ξ) at every point such that $\sigma = \sum_1^n d\xi_j \wedge dx_j$. One calls n the number of degrees of freedom.

If E is a C^2 function on M, then (1.1) (where X is now a tangent vector of M) defines a Hamiltonian vector field H_E, and the Hamilton equations consist in integrating this vector field. The transformation laws are now quite obvious. Suppose that $\widetilde{M}, \widetilde{\sigma}$ is another symplectic manifold with dim $\widetilde{M} = $ dim M, and that $\varphi : \widetilde{M} \to M$ is a *symplectic embedding*, that is, a diffeomorphism on an open subset of M, such that

$$(1.2) \qquad\qquad\qquad \varphi^*\sigma = \widetilde{\sigma}.$$

Let E be a C^2 function on M, and let $\widetilde{E} = \varphi^* E = E \circ \varphi$ be the pullback to $\widetilde{M}$. Then $\varphi_* H_{\widetilde{E}} = H_E$, so the orbits of the Hamilton field of $\widetilde{E}$ are mapped by φ on those of E, including the parameter. Thus the Hamilton equations are transformed by just transforming the corresponding Hamilton function, which is much simpler than transforming a vector field.

A major task of classical mechanics was to find symplectic maps which simplify a given Hamiltonian E. The simplest case is a Hamiltonian E with $dE \neq 0$ at (x_0, ξ_0). In a neighborhood it is then possible to transform E to η_1 by a symplectic map; the orbits of the Hamilton equations are then the parallels of the y_1 axis. A positive definite quadratic form can always be diagonalized by a linear symplectic map. If E vanishes of second order at the origin and has positive definite Hessian, it can therefore be reduced by a symplectic transformation to the form

$$E(y, \eta) = \tfrac{1}{2} \sum \lambda_j (y_j^2 + \eta_j^2) + O(|y|^3 + |\eta|^3),$$

where all $\lambda_j > 0$. By a theorem of Birkhoff, if $\sum \lambda_j k_j \neq 0$ for all integers k_j (not all equal to 0), then the remainder term can be made a function of $y_j^2 + \eta_j^2$, $j = 1, \ldots, n$, with an error vanishing of infinite order. The functions $y_j^2 + \eta_j^2$ are thus nearly invariants, that is, constant on the Hamilton flow. If for a given Hamiltonian E one can find invariants I_j, $j = 1, \ldots, n$, thus $\{E, I_j\} = 0$, which are *in involution*, that is, $\{I_j, I_k\} = 0$ for all j, k, and if T is a compact n dimensional manifold where $I_j = 0$ for $j = 1, \ldots, n$ and the differentials dI_j are linearly independent, then T is a torus and by a symplectic map one can reduce a neighborhood to $(\mathbf{R}^n / \mathbf{Z}^n) \times \{\eta \in \mathbf{R}^n, |\eta| < \delta\}$ so that $E = F(\eta)$ and $I_j = I_j(\eta)$. The orbits of the Hamilton flow are then defined by $\eta = $ constant and $dx/dt = F'(\eta)$, where $x \in \mathbf{R}^n / \mathbf{Z}^n$. This is the so called integrable case. The search for integrals in involution of the n body problem, in addition to those known from physics, was stopped one hundred years ago by a proof of non-existence. It is now known that few of the natural problems encountered in mechanics are integrable, and every discovery of a new one is a major event.

A case of the old transformation problem which has attracted much attention during the last few years is the problem of transforming sets (that is, characteristic functions) by symplectic maps. Before making some remarks on that, I wish to mention one of the methods for constructing symplectic maps. One can do so using Hamilton vector fields. Let q_t be Hamiltonians depending on a parameter t, equal to 0 outside a compact set for the sake of simplicity. Then the Hamilton equations (in local coordinates)

$$\frac{dx}{dt} = \frac{\partial q_t(x, \xi)}{\partial \xi}, \quad \frac{d\xi}{dt} = -\frac{\partial q_t(x, \xi)}{\partial x}, \quad x = y, \xi = \eta \quad \text{when } t = 0,$$

define at least for small t a diffeomorphism $\Phi_t : (y, \eta) \mapsto (x, \xi)$ in M, and Φ_t is symplectic, that is, $\Phi_t^* \sigma = \sigma$. It suffices to prove that the derivative of $\Phi_t^* \sigma$ with respect to t vanishes when $t = 0$. The derivative is for $t = 0$

$$d(-\partial q_0/\partial x) \wedge dx + d\xi \wedge d(\partial q_0/\partial \xi) = -(\partial^2 q_0/\partial x \partial x \, dx) \wedge dx - (\partial^2 q_0/\partial x \partial \xi \, d\xi) \wedge dx$$
$$+ d\xi \wedge \partial^2 q_0/\partial \xi \partial x \, dx + d\xi \wedge \partial^2 q_0/\partial \xi \partial \xi \, d\xi = 0$$

because one can interchange orders of differentiation.

A map satisfying (1.2) is automatically a local diffeomorphism, for taking the nth exterior power we obtain

$$(1.2)' \qquad\qquad\qquad \varphi^* \sigma^n = \widetilde{\sigma}^n,$$

and in canonical coordinates we have $\sigma^n/n! = d\xi_1 \wedge dx_1 \cdots \wedge d\xi_n \wedge dx_n$, which is the volume form. In $\mathbf{R}^{2n}$ with the standard symplectic structure, for example, this means that the Jacobian of φ is equal to 1. Hence a symplectic manifold has a natural volume element and orientation which are preserved by symplectic maps (Liouville's theorem). For $n = 1$ we almost have a converse statement: If $\Omega, \widetilde{\Omega}$ are open sets in $\mathbf{R}^2$ and $\varphi : \Omega \to \widetilde{\Omega}$ is a C^1 map such that $m(\varphi(B)) = m(B)$ for all small discs $\subset \Omega$, then φ is a local diffeomorphism and $|\det \varphi'| = 1$, so φ is symplectic or *antisymplectic*. For $n > 1$ the same argument still shows that volume preservation implies that φ is a local diffeomorphism with $|\det \varphi'| = 1$, but that is far less than being a symplectic map. (However, the volume preservation (Liouville's theorem) of symplectic maps alone has important consequences in ergodic theory.)

The simplest Hamilton equations to integrate come from a harmonic oscillator,

$$(1.3) \qquad\qquad\qquad E(x, \xi) = \tfrac{1}{2} \sum_1^n a_j x_j^2 + \tfrac{1}{2} \sum_1^n b_j \xi_j^2,$$

where $a_j > 0$, $b_j > 0$. The corresponding Hamilton equations are

$$\frac{dx_j}{dt} = b_j \xi_j, \qquad \frac{d\xi_j}{dt} = -a_j x_j,$$

which imply $d^2 x_j/dt^2 + a_j b_j x_j = 0$ and similarly for ξ_j. Thus (x_j, ξ_j) describes an ellipse $a_j x_j^2 + b_j \xi_j^2 = $ constant with frequency $\sqrt{a_j b_j}$, that is, period $2\pi/\sqrt{a_j b_j}$. The smallest period is 1 if $\max \sqrt{a_j b_j} = 2\pi$; if $a_j = b_j$ this means that $\max E(x, \xi)/(|x|^2 + |\xi|^2) = \pi$. The situation is quite similar for arbitrary non-negative quadratic forms E. There is always a linear symplectic map transforming it to the form (1.3). — It is also easy to integrate the Hamilton equations if $E(x, \xi) = f(|x|^2 + |\xi|^2)$ (with Euclidean norms). They are

$$\frac{dx_j}{dt} = 2\xi_j f'(|x|^2 + |\xi|^2), \qquad \frac{d\xi_j}{dt} = -2x_j f'(|x|^2 + |\xi|^2),$$

and imply that $|x|^2 + |\xi|^2 = R^2$ is constant on an orbit, hence $d^2 x_j/dt^2 = -4x_j f'(R^2)^2$, so the period is $\pi/|f'(R^2)|$ on a non-constant orbit.

Gromov [4] has proved a theorem which shows in a very striking way how symplectic maps are much more special than volume preserving maps; it can be considered as a very precise expression of the uncertainty principle in quantum mechanics:

THEOREM 1.1 (GROMOV). *It is not possible to embed the ball $B(r) = \{(x,\xi) \in \mathbf{R}^{2n}; |x|^2 + |\xi|^2 < r^2\}$ symplectically in the cylinder $Z(r') = \{(x,\xi) \in \mathbf{R}^{2n}, x_1^2 + \xi_1^2 < r'^2\}$ if $r > r'$.*

A very interesting proof of this is given by the following theorem of Hofer and Zehnder [6], which also contains the solution of a conjecture of A. Weinstein [12] on the existence of periodic orbits of Hamilton fields (see also [1, 2, 7, 8, 9, 10, 11]):

THEOREM 1.2 (HOFER-ZEHNDER). *Let E be a smooth function in $Z(1)$ with $E = 0$ in an open subset, $E = m > \pi$ outside a compact set, and $0 \le E \le m$ in $Z(1)$. Then the Hamilton field H_E of E has a non-constant orbit of period 1 in $Z(1)$.*

To prove that Theorem 1.2 implies Theorem 1.1 we may assume that $r' = 1$, $r > 1$. Take a function f with $0 \le f' < \pi$ which is equal to 0 near 0 and a constant $m > \pi$ near r^2. Then the non-constant orbits of the Hamiltonian $E(x,\xi) = f(|x|^2 + |\xi|^2)$ have period > 1. If there did exist a symplectic embedding φ of $B(r)$ in $Z(1)$, we could use it to transplant E to $Z(1)$, extending the definition as m outside the embedding of $B(r)$. Then the non-constant orbits have period > 1, and $E = 0$ near $\varphi(0)$, $E = m > \pi$ outside a compact subset of $Z(1)$, which contradicts Theorem 1.2.

2. Hamilton-Jacobi theory. In this section we shall discuss the integration of a non-linear partial differential equation

$$(2.1) \qquad\qquad E(x, \partial u/\partial x) = 0$$

where E is a given function in $\mathbf{R}^{2n}$ or in part of $\mathbf{R}^{2n}$. Write $\xi = \partial u/\partial x$, which is then also a function of x if u is a solution of (2.1). Differentiation of (2.1) with respect to x_j gives

$$\sum_{k=1}^n \partial E(x,\xi)/\partial\xi_k \partial\xi_j/\partial x_k = -\partial E(x,\xi)/\partial x_j,$$

where we have exchanged the order of differentiation. Hence the first set of equations

$$(2.2) \qquad\qquad \frac{dx_k}{dt} = \partial E(x,\xi)/\partial\xi_k, \quad \frac{d\xi_k}{dt} = -\partial E(x,\xi)/\partial x_k$$

implies the second set of equations. This means that if a solution of the Hamilton equations (2.2) starts at $(x_0, u'(x_0))$, then $\xi(t) = u'(x(t))$ along the entire curve, which allows one to solve the Cauchy problem for the equation (2.1), at least locally. Suppose we are given a function $u_0(x')$, $x' = (x_2, \ldots, x_n)$ and want to find a solution of (2.1) with $u = u_0$ when $x_1 = 0$. Then we only know the ξ' component of $\xi = u'(x)$ when $x_1 = 0$, but (2.1) is an equation for ξ_1 then. If ξ_1 is prescribed at $x' = 0$ say, and $\partial E(x,\xi)/\partial\xi_1 \ne 0$ there, then the implicit function theorem allows us to determine ξ_1 as a function of x' in a neighborhood of 0, and we can then take the integral curves of (2.2) starting at $(0, x', \xi(x'))$ for an arbitrary $x' \ne 0$. They form a manifold Λ of dimension n in $\mathbf{R}^{2n}$ in a neighborhood of 0, and the projection $\Lambda \ni (x,\xi) \mapsto x$ has full rank since the range contains both the plane $x_1 = 0$ and one transversal vector. Hence Λ is a manifold of the form $\xi = \varphi(x)$. If there is a local solution of (2.1) with the given initial data we must have $u'(x) = \varphi(x)$, and such a solution exists (locally) if and only if the form $\sum_1^n \varphi_j(x) dx_j$ is closed, that is, $\sum d\varphi_j \wedge dx_j = 0$, or equivalently, the standard symplectic form vanishes on Λ. Now this is true when $x_1 = 0$ for the tangent plane T of Λ at $\Lambda_0 = \{(0, x', \xi(x'))\}$ is spanned by H_p and the tangent plane T_0 of Λ_0. Since $\xi'(x') = \partial u_0(x')/\partial x'$, the plane T_0 is isotropic, that is, the symplectic form vanishes in $T_0 \times T_0$. We have $\sigma(T, H_p) = \langle dp, T \rangle = 0$, so it follows that σ vanishes in $T \times T$, and since the Hamilton flow is symplectic it is true everywhere. Since E is constant along the Hamilton flow, the equation (2.1) is fulfilled. (The integration constant in passing from u' to u is determined by the initial data.)

In the course of this proof we encountered an important concept, a manifold Λ in a symplectic manifold with dimension equal to the number of degrees of freedom and on which the restriction of the symplectic form is equal to 0. Such a manifold is called *Lagrangian*. The solution of the Cauchy problem really consisted of two steps: The Cauchy data gave an isotropic manifold Λ_0 of dimension $n - 1$ contained in $\{(x,\xi); E(x,\xi) = 0\}$ which had to be contained in Λ. The Hamilton field of E was symplectically orthogonal to Λ_0 since $E = 0$ on Λ_0, and as a consequence the flowout Λ of Λ_0 along the Hamilton field of E was Lagrangian. This statement contains no reference to the rank of the projection $\Lambda \ni (x,\xi) \mapsto x$, and it is therefore fully invariant under symplectic maps.

In all this we could have allowed x to belong to a manifold X of dimension n provided that $E(x,\xi)$ is defined in a suitable open subset of the cotangent bundle T^*X. The symplectic form is invariantly defined in T^*X; indeed, the one form $\langle \xi, dx \rangle$ is invariantly defined on T^*X, and the symplectic form $\sum d\xi_j \wedge dx_j$ is its differential.

3. Propagation of singularities. In geometrical optics the phase $\varphi(t,x)$ of a light wave is supposed to satisfy the eiconal equation in $\mathbf{R}^{1+3}$

$$c^{-2}(\partial\varphi/\partial t)^2 - |\partial\varphi/\partial x|^2 = 0. \tag{3.1}$$

Here c is the speed of light. The equation means that the level surfaces where $\varphi(t,\cdot) = \gamma$ move with the speed c in the normal direction as t varies. We know from Section 2 how such an equation can be integrated using the Hamilton equations. With the parameter denoted by s they are $\xi, \tau = $ constant, where $\tau^2 = c^2|\xi|^2$, and

$$dt/ds = 2\tau, \quad dx/ds = -2\xi. \tag{3.2}$$

The solutions are the light rays of geometrical optics while the level surfaces of φ are surfaces of constant phase. The equation (3.1) is motivated from the wave equation

$$\Box u = c^{-2}\partial^2 u/\partial t^2 - \Delta u = 0 \tag{3.3}$$

by considering highly oscillatory solutions $u(t,x) = a(t,x)e^{i\omega\varphi(t,x)}$ and ignoring all terms except the highest in ω as $\omega \to \infty$; (3.1) is the coefficient of ω^2. The next term gives a differential equation for the amplitude $a(t,x)$ along the light rays. In the end of the 1940's one started to improve this to higher order approximations by looking also at the following equations with a replaced by $a_0(t,x) + \omega^{-1}a_1(t,x) + \ldots$, which actually yields an asymptotic series. This is the procedure which has slowly been integrated into the mainstream of the theory of linear partial differential equations. It took a surprisingly long time after the initial step taken by Peter Lax [12] in 1957.

The level surfaces of a solution of (3.1) are called *characteristic surfaces* for the wave equation. Classically these were regarded as the only carriers of singularities for solutions of the wave equation (3.3). That they can carry singularities is seen most simply in the example of a linear function $\varphi(t,x) = t\tau + \langle x,\xi \rangle$ satisfying (3.1), that is, $c^{-2}\tau^2 - |\xi|^2 = 0$. Then $u(t,x) = f(t\tau + \langle x,\xi \rangle)$ satisfies (3.3) for any function (or distribution) on $\mathbf{R}$, and the singularities can then be placed on any closed set of parallel characteristic planes. In 1960 it was still the prevailing opinion that characteristic surfaces should be the *only* carriers of singularities. When Fritz John in 1960 had constructed a solution of (3.3) (for the case of two space variables) which was C^∞ inside the cylinder $\{(t,x); x_1^2 + x_2^2 < 1\}$ but only in

C^m for a given m at every boundary point, he commented (John [14, p. 574]): "What is remarkable is that this cylinder is not a characteristic surface for the differential equation. Apparently not *all* types of singularities propagate along characteristic surfaces." It would not be hard to find other similar quotations confirming that at the time singularities were always associated with characteristic surfaces, but it should suffice with documenting that a mathematician of the great power and insight of Fritz John was surprised to find a result in conflict with this tradition. However, geometrical optics says that light propagates along *light rays*, and about 1960 it finally dawned on mathematicians that this is true for the singularities of solutions of the wave equation (3.3). (See e. g. Zerner [14].) Examples are in fact very easy to find. The difference U between the forward and backward fundamental solutions of the wave equation is a solution of the homogeneous equation with a very simple kind of transversal singularity only on the wave cone. This is of course a characteristic surface, but if we convolve U with a smooth density on a light ray, the singularity disappears where the light ray is transversal to the wave cone, that is, everywhere except on the parallel generator of the light cone. The construction is easy to generalize to large classes of constant coefficient operators at least.

Grušin [15] proved shortly afterwards that if u satisfies (3.3) and on every line through the origin with the speed of light there is some point where $u \in C^\infty$, then $u \in C^\infty$ at the origin. In fact, he proved the analogue for general differential equations $P(D)u = f \in C^\infty$ with constant coefficients, where $D = -i\partial/\partial x$, and P is a polynomial with real principal part P_m such that $P_m(\xi) = 0$ implies $P'_m(\xi) \neq 0$; the lines with the speed of light are then replaced by the lines with such *bicharacteristic* direction. They are of course solutions of the Hamilton equations for the Hamiltonian $P_m(\xi)$. An equivalent but more suggestive formulation is that if u is not in C^∞ at x , then there is a whole bicharacteristic line through x such that u is singular at every point on it. Thus singularities propagate along bicharacteristic lines. However, the proof attacks the first statement by constructing for the given u a fundamental solution which is only singular on rays such that there is a point of smoothness for u on the opposite ray. This complication of the proof became quite important when I worked out in 1968-69 a *local theory of Fourier integral operators* which allowed the same result to be proved for operators with variable coefficients, still of *real principal type*, that is, real principal part with only simple zeros. The bicharacteristic lines are then replaced by projections in the x variables of the orbits of the Hamilton equations for the principal symbol $P_m(x, \xi)$ on which $P_m(x, \xi) = 0$. (Note that by the Hamilton-Jacobi theory every characteristic surface is generated by bicharacteristic curves.) At first this was just a local theory designed to permit a local construction of fundamental solutions. However, knowing that the singularities must propagate along a bicharacteristic curve locally, one can only conclude that singularities must continue along a curve which is composed of bicharacteristic arcs, but it is conceivable that the hidden variable ξ makes occasional jumps. To eliminate this possibility I decided to build up a *global theory of Fourier integral operators*. Just about the time when it was finished in 1970 I realised that it is possible to avoid the question entirely if one does not consider ξ as hidden variables but refines the question about the position of singularities of u to a question about the frequencies of the oscillations of u which cause them. This can be defined in the following way:

A distribution u in $\mathbf{R}^n$ belongs to C^∞ at a point x_0 if for some $\chi \in C_0^\infty$ with $\chi(x_0) \neq 0$ the product $\chi u \in C_0^\infty$, that is, the Fourier transform $\widehat{\chi u}(\xi)$ is rapidly decreasing as $\xi \to \infty$. It is natural to say that u is C^∞ in the direction $\xi_0 \in \mathbf{R}^n \setminus 0$ at x_0 if this is true in a conic neighborhood of ξ_0. If one interprets (x_0, ξ_0) as a cotangent vector, that is, considers the one form $\langle \xi_0, dx \rangle$ at x_0, this turns out to be an invariant notion so it makes sense for distributions on a manifold X. The complement in $T^*X \setminus 0$ of the set of such (x_0, ξ_0) is called the *wave front set* of u, and denoted $WF(u)$. Its projection in X is sing supp u, the complement of the set where $u \in C^\infty$. There is an analogous notion WF_A with C^∞ replaced by real analytic functions introduced even a bit earlier by Sato [26], and which is defined also for hyperfunctions. (See also [27].) (Notations such as SS or sing spec are also used.) Just as one can introduce germs of functions or singularities at a point in X one can introduce such germs at $(x_0, \xi_0) \in T^*X \setminus 0$, called microdistributions, as elements in the quotient space

$$\mathcal{D}'(X)/\{u \in \mathcal{D}'(X); (x_0, \xi_0) \notin WF(u)\}.$$

Now the propagation theorem for singularities takes the simple form: If $P(x, D)u = f$ and the principal part of P is real, then $(x_0, \xi_0) \in WF(u)$ implies that the bicharacteristic strip through (x_0, ξ_0), that is, the solution of the Hamilton equations with Hamiltonian $P_m(x, \xi)$ passing through (x_0, ξ_0), must remain in $WF(u)$ until it encounters $WF(f)$. This is a result which it suffices to prove locally; the global statement follows at once since there is no longer any hidden variable which makes it uncertain in which direction the singularity moves. The notion of wave front set implements what everyone has seen in elementary physics texts explaining Huyghens principle on the propagation of light: One always draws a point and an oriented surface through it, marking the position of the wave front at a certain time and the direction in which it moves, and then one draws the surface translated in the normal direction. A point with an oriented hyperplane in the tangent space is just the same as a ray in the cotangent bundle.

The propagation theorem just mentioned is very easy to prove in a variety of ways; I shall give much more sophisticated results later. However, this is a good place to mention the main techniques. First a word about *pseudo-differential* operators. If $P(x, D)$ is a differential operator in $\mathbf{R}^n$, then we can express $P(x, D)u$ when $u \in C_0^\infty$ by using the Fourier inversion formula to write $D^\alpha u$ in terms of $\hat{u}$, which gives

$$(3.4) \qquad P(x, D)u(x) = (2\pi)^{-n} \int e^{i\langle x, \xi \rangle} P(x, \xi) \hat{u}(\xi)\, d\xi, \quad u \in C_0^\infty(\mathbf{R}^n).$$

A pseudo-differential operator is of the same form but $P(x, \xi)$ need not be a polynomial in ξ but is a C^∞ function such that, in the simplest case, for all multi-indices α and β,

$$|D_x^\beta D_\xi^\alpha P(x, \xi)| \leq C_{\alpha\beta}(1 + |\xi|)^{m-|\alpha|},$$

as one would expect if P at infinity is a homogeneous function of degree m in ξ depending smoothly on x. Here m may be any real number. These operators merit to be called pseudo-differential for they act essentially as differential operators on highly oscillatory functions:

$$e^{-i\omega\psi(x)} P(x, D)(e^{i\omega\psi(x)} u(x))$$
$$= P(x, \omega\psi'(x))u(x) + \text{ terms of order } \omega^{m-1} \text{ or lower} \qquad \text{as } \omega \to \infty.$$

We say that P has a homogeneous principal part P_m if $P - P_m$ is of order $< m$ (for $|\xi| > 1$, say), and P is non-characteristic at (x, ξ) if $P_m(x, \xi) \neq 0$. The complement $\text{Char}(P(x, D))$ is the set of zeros of P_m. The definition of the wave front set can be restated as follows:

$$WF(u) = \bigcap_{a(x,D)u \in C^\infty} \text{Char}(a(x, D)).$$

This is how I defined it originally, in analogy to the fact that

$$\text{sing supp}\, u = \bigcap_{\varphi u \in C^\infty} \text{supp}\, \varphi; \quad \text{here } \varphi \in C^\infty.$$

It follows that $WF(Pu) \subset WF(u)$, so P acts on microdistributions.

The *energy method* in its simplest form consists in estimating the solution of an ordinary differential equation

$$du/dt + au = f$$

where $a \geq 0$ by integrating from 0 to $t > 0$ after multiplication by $2u$, which gives

$$u(t)^2 \leq u(0)^2 + 2 \int_0^t f(s)u(s)\, ds \leq u(0)^2 + 2 \sup_{0<s<t} |u(s)| \int_0^t |f(s)|\, ds$$

$$\leq u(0)^2 + \tfrac{1}{2} \cdot \sup_{0<s<t} |u(s)|^2 + 2(\int_0^t |f(s)|\, ds)^2,$$

or after replacing the left-hand side by the supremum over $(0,t)$,

$$\sup_{0<s<t} |u(s)|^2 \leq 2u(0)^2 + 4(\int_0^t |f(s)|\, ds)^2.$$

An analogue for estimating solutions of the wave equation is classical, used by Kurt Friedrichs and Hans Lewy already in the 1920's. It consists of multiplying (3.3) by $2\partial u/\partial t$ and integrating by parts. By introducing pseudo-differential operators in such a way that one estimates in a narrowing neighborhood of a bicharacteristic starting from a point of smoothness, one can obtain an analogue of the condition $a \geq 0$ above and prove the propagation theorem also for many operators with complex coefficients in the principal part. Naturally one proves such results for pseudo-differential operators at the same time with no additional effort.

One can also use Fourier integral operators. It would take too long to explain the technical details here (see [18, Chap. 25]), but the simplest ones are defined by putting in (3.4) a more general exponential factor $e^{i\varphi(x,\xi)}$, homogeneous of degree 1 with respect to ξ; it is assumed that $\partial^2 \varphi(x,\xi)/\partial x \partial \xi$ is non-singular. This means that φ is locally the *generating function* of a homogeneous symplectic map

$$\chi : (\partial\varphi(x,\xi)/\partial\xi, \xi) \mapsto (x, \partial\varphi(x,\xi)/\partial x).$$

Such operators affect the wave front set so that $WF(Pu) \subset \chi WF(u)$; in the case of a pseudo-differential operator the symplectic map is the identity so the wave front set is preserved (the *pseudo-local property*). Replacing an operator P of real principal type with APB where A and B are Fourier integral operators corresponding to inverse symplectic maps χ^{-1}, χ, and $P_m \circ \chi(x,\xi) = |\xi|^{m-1}\xi_1$ it is in fact possible to reduce the proof of the propagation theorem to operators of the form $D_1 u = f$ where it is quite trivial. The proof starts from the classical fact, mentioned above, that there P_m can locally be transformed to ξ_1 by a symplectic map. (The homogeneity conditions must be taken into account, and one must require that the Hamilton field does not have the direction of the radial vector field $\xi \partial/\partial\xi$.) This is a first example of how the local theory of Fourier integral operators can be used to simplify the study of a differential operator: One first uses more or less classical results from analytical mechanics to find symplectic transformations χ

which simplify the principal symbol. Then one chooses local Fourier integral operators A and B corresponding to χ^{-1} and χ with amplitudes chosen by a successive approximation argument to make the lower order symbols of APB also as simple as possible.

Solutions which are singular precisely on a given bicharacteristic are also obtained at once from such a transformation provided that one makes it global along a chosen bicharacteristic.

4. Solvability of differential equations. In the early fifties it was shown, primarily by the work of Leon Ehrenpreis and Bernard Malgrange that differential equations with constant coefficients always have solutions, at least in all of $\mathbf{R}^n$. My thesis [16] was mainly devoted to operators with constant coefficients, but in the last chapter a version of the energy integral method was used to prove that for an operator $P(x, D)$ of order m of real principal type and smooth coefficients at a point x_0, there is a neighborhood Ω of x_0 such that the equation $P(x, D)u = f$ has a distribution solution $u \in L^2(\Omega)$ for every $f \in L^2(\Omega)$ (or more generally $H_{(1-m)}(\Omega)$). In the introduction I remarked that the proof could be extended to the case where $[P(x, D), \bar{P}(x, D)]$ is of order $2m-2$. Without knowing it, I had in fact encountered a symplectic condition for the first time. Indeed, a basic principle in quantum mechanics states that where the Poisson bracket of two Hamiltonians occurs in classical mechanics one should expect a commutator of two operators in quantum mechanics. This can be understood as a simple consequence of Leibniz' formula for computing the composition of two operators: If $P(x, D)$ and $Q(x, D)$ are two (pseudo-)differential operators, of order m and μ with principal symbols $p(x, \xi)$ and $q(x, \xi)$ homogeneous in ξ of order m and μ respectively, then the commutator $[P(x, D), Q(x, D)] = P(x, D)Q(x, D) - Q(x, D)P(x, D)$ is of order $m + \mu - 1$ with principal symbol

$$-i \sum_1^n (\partial p(x, \xi)/\partial \xi_j \partial q(x, \xi)/\partial x_j - \partial p(x, \xi)/\partial x_j \partial q(x, \xi)/\partial \xi_j) = -i\{p(x, \xi), q(x, \xi)\}$$

where $\{\cdot, \cdot\}$ is the Poisson bracket, which is dual to the symplectic form. The sufficient condition for solvability mentioned in my thesis can therefore be written

$$\{p(x, \xi), \bar{p}(x, \xi)\} = -2i\{\operatorname{Re} p(x, \xi), \operatorname{Im} p(x, \xi)\} = 0,$$

which in the terminology of analytical mechanics means that $\operatorname{Re} p$ and $\operatorname{Im} p$ are in *involution*. At the time I took it for granted that it was only a flaw in the method which forced me to make that assumption, but this idea had to be revised soon, for H. Lewy [20] showed that the equation

(4.1) $$(D_1 + iD_2 + i(x_1 + ix_2)D_3)u = f$$

has no classical C^1 solution in any open set for most choices of $f \in C^\infty$. (This is not really a contrived example, for it is the tangential Cauchy-Riemann operator on the boundary of the pseudo-convex domain $\{(z_1, z_2) \in \mathbf{C}^2; |z_1|^2 + 2\operatorname{Im} z_2 < 0\}$.) As soon as it became clear that there is no distribution solution either, this showed that the troublesome condition in my thesis was in no way superfluous, and I proved shortly afterwards that the phenomenon observed by Hans Lewy always occurs for the equation $P(x, D)u = f$ with principal symbol p in

$$\{x; \{p(x, \xi), \bar{p}(x, \xi)\} \neq 0 \text{ for some real } \xi \text{ with } p(x, \xi) = 0\}.$$

In this result there are also "hidden variables" ξ, so it is natural to try to put it in another form involving the wave front set. Quite elementary functional analysis proves that if X is an open set in $\mathbf{R}^n$ (or a manifold) and K is a compact subset, then the following conditions are equivalent:

(1) P is *solvable* at K in the sense that for every f in a subspace of $C^\infty(X)$ of finite codimension there is a distribution u in X such that $Pu = f$ in a neighborhood of K.

(2) There is an integer N such that for every f with $D^\alpha f \in L^2(X)$ when $|\alpha| \leq N$ there is a distribution u in X with $Pu - f \in C^\infty$ in a neighborhood of K.

In analogy with the second condition we can say that P is solvable at a compact subset K of $T^*X \setminus 0$ if one can find u such that $K \cap WF(Pu - f) = \emptyset$. This condition becomes stricter when K increases. The proof of non-solvability in [16] can actually be modified to show that the (pseudo-)differential operator P is not solvable at a characteristic point (x, ξ) where $\{\operatorname{Re} p(x, \xi), \operatorname{Im} p(x, \xi)\} > 0$. However, it took twenty more years before such a formulation was made (see Hörmander [18, Sect. 26.4]) so that one could take full advantage of the microlocal analysis machinery, including Fourier integral operators.

There was also a positive result in [16]: the operator $P(x, D)$ is solvable if

$$\{p(x, \xi), \bar{p}(x, \xi)\} = i \operatorname{Re} q(x, \xi) p(x, \xi)$$

for some q which is a polynomial of degree $m - 1$ in ξ. (This was before the advent of pseudo-differential operators; when they came on the scene the same arguments proved solvability when $\{\operatorname{Re} p(x, \xi), \operatorname{Im} p(x, \xi)\} \leq \operatorname{Re} p(x, \xi) q(x, \xi)$ for some smooth q.)

The next step taken by Nirenberg and Treves [21] concerned first order differential operators. Writing the operator in the form $D_t + i\langle a(t, x), D_x \rangle + c$ they showed that solvability requires that the direction of a does not depend on t. They returned to the problem for operators of higher order in [22]. Local Fourier integral operators were then available and they were used to reduce the principal symbol to the form $\tau + iq(t, x, \xi)$, with the variables denoted (t, x). It was proved that solvability implies that q never changes sign from $-$ to $+$ for fixed x, ξ at a zero of finite order with respect to t. To require that no such sign change occurs is called condition (Ψ); in the case of differential operators the symmetry between ξ and $-\xi$ implies that there can be no change of sign at all at a zero of finite order. This is called condition (P). A full proof of the necessity of condition (Ψ) was outlined by Moyer in 1978 (see Hörmander [17]). Nirenberg and Treves also showed that condition (P) implies local solvability if the coefficients in the principal part are real analytic, a restriction removed by Beals and Fefferman [23].

For some of these results one can give very precise and instructive proofs by reduction to normal forms. The first step is to simplify the principal symbol by multiplication with a non-vanishing homogeneous factor and composition with a homogeneous symplectic transformation. The following cases are accessible in that way, and contain the necessary conditions of Nirenberg and Treves as well as some of the early sufficient conditions. We denote by p the principal symbol, assumed smooth and homogeneous, and by (x_0, ξ_0) a point where $p(x_0, \xi_0) = 0$. We write $\varepsilon_n = (0, \ldots, 0, 1)$.

(1) If p is real valued and the Hamilton vector H_p at (x_0, ξ_0) does not have the radial direction $(0, \xi_0)$, one can transform P to ordinary differentiation D_1 with parameters; (x_0, ξ_0) is mapped to $(0, \varepsilon_n)$.

(2) If $\operatorname{Re} H_p$, $\operatorname{Im} H_p$ and the radial direction are linearly independent and $\{p, \bar{p}\} = 0$ in a neighborhood of (x_0, ξ_0) when $p = 0$, which is a consequence of condition (P), then one can transform P to the Cauchy-Riemann operator $D_1 + iD_2$ with parameters; (x_0, ξ_0) is mapped to $(0, \varepsilon_n)$.

(3) If $\{\operatorname{Re} p, \operatorname{Im} p\} \gtrless 0$ at (x_0, ξ_0) then one can transform P to $D_1 + ix_1 D_n$ at $(0, \pm\varepsilon_n)$.

(4) More generally, if $p = p_1 + i p_2$, $H_{p_1} \neq 0$ at (x_0, ξ_0), and p_2 has a zero of fixed order $k > 1$ nearby on each integral curve of H_{p_1} in $p_1^{-1}(0)$, then P can be transformed to $D_1 + i x_1^k D_n$ at $(0, \varepsilon_n)$ unless k is odd and $\operatorname{Im} p$ changes sign from $+$ to $-$ along the integral curves of H_{p_1}; in that case (x_0, ξ_0) must be mapped to $(0, -\varepsilon_n)$.

My original necessary conditions for solvability as well as the improvement due to Nirenberg and Treves [22] follow by an explicit study of the last two cases. (See Duistermaat and Sjöstrand [25] and also Hörmander [18, Chapter XXVI] for proofs and further references to the literature.) The first case covers entirely the original existence results which are thus reduced to the existence of primitive functions — modulo a considerable amount of sophisticated techniques.

The second case, called the *involutive case*, suggests a connection with the theory of analytic and harmonic functions which was developed by Duistermaat and Hörmander [24] to a theorem on propagation of singularities in that situation. First of all, the definition of the involutive case means that $\{\operatorname{Re} p, \operatorname{Im} p\} = q_1 \operatorname{Re} p + q_2 \operatorname{Im} p$ locally for some smooth q_1, q_2, and this implies that $[H_{\operatorname{Re} p}, H_{\operatorname{Im} p}] = q_1 H_{\operatorname{Re} p} + q_2 H_{\operatorname{Im} p}$ in the manifold of codimension two defined locally by $p = 0$. By the Frobenius theorem it is therefore foliated by two dimensional manifolds such that the tangent plane is spanned by $H_{\operatorname{Re} p}$ and $H_{\operatorname{Im} p}$, which means that the complex vector field H_p gives a structure of Riemann surface to the leaves of the foliation, which we shall also call bicharacteristics. Hence it makes sense to talk about analytic, harmonic or superharmonic functions there.

So far we have stated the results on propagation of singularities for operators of real principal type in a fairly weak form. To be more precise we observe that one can also define that $u \in H_{(s)}^{\text{loc}}$ (that is, has $s \in \mathbf{R}$ L^2 derivatives) at $(x_0, \xi_0) \in T^* X$ if $a(x, D) u \in L^2$ for some pseudo-differential operator of order s which is non-characteristic at (x_0, ξ_0). If $P(x, D)$ is of real principal type, and $P(x, D) u = f$, then $u \in H_{(s)}^{\text{loc}}$ at a characteristic point (x_0, ξ_0) implies that this remains true on the bicharacteristic through (x_0, ξ_0) until it reaches a point where $f \notin H_{(s+1-m)}^{\text{loc}}$. The analogue in a two dimensional bicharacteristic leaf is that if s is a harmonic function near a compact subset K of a leaf and $f \in H_{(s+m-1)}^{\text{loc}}$ in K, then $u \in H_{(s-\varepsilon)}^{\text{loc}}$ in K for every $\varepsilon > 0$ if this is true on the boundary. If we define

$$s_u^*(x, \xi) = \sup\{s; u \in H_{(s)}^{\text{loc}} \text{ at } (x, \xi)\},$$

then s_u^* is a superharmonic function. Combined with elementary properties of superharmonic functions this proves that if the leaf is connected and does not intersect $WF(f)$, then it is either contained in $WF(u)$ or disjoint with $WF(u)$.

For a general operator satisfying condition (P) the geometry of the characteristic set is not covered by the cases listed above. However, it turns out that condition (P) implies that the flowout from the involutive set in case (2) above along all Hamilton fields $H_{\operatorname{Re} qp}$ remains an involutive manifold of codimension two, so it is foliated by two dimensional leaves. In these we have a degenerate analytic structure; the Cauchy-Riemann operator becomes a directional derivative outside the non-degenerate subset (2). Existence theorems for these degenerate Cauchy-Riemann equations were proved in [19], and it was proved there that with harmonic functions replaced by the real parts of the solutions, the result above remains valid for the extended leaves.

What remains under condition (P) is to study propagation of singularities in the part of the characteristic set which cannot be connected to the non-degenerate involutive set. This contains the set where case (1) occurs and also the set where the imaginary part of p has a constant sign. It is not possible to go through all the results here, but roughly speaking the regularity function of a solution is on a one dimensional bicharacteristic (where the tangent is proportional to H_p) either concave, semi-concave or quasi-concave. A function on an affine line is concave if it satisfies the minimum principle with respect to all linear functions, on an oriented affine line it is called semi-concave if this is true for all decreasing linear functions, and a quasi-concave function just satisfies the minimum principle with respect to constants. In the cases where the geometry allows one to define invariantly an affine structure (and an orientation) the regularity of solutions is a concave function (semi-concave), and otherwise it is just quasi-concave. Thus there is a perfect fit between the geometry and the analysis. All these results can be found in detail in [18, Chap. 26]. Enough was proved already in [19] to make it possible to conclude using elementary functional analysis that under condition (P) the inhomogeneous equation $Pu = f$ has, say C^∞ solutions on any compact subset of X provided that $f \in C^\infty$ satisfies a finite number of compatibility conditions and a geometric convexity condition is fulfilled which states roughly that no bicharacteristics are trapped by compact subsets of X.

In case of condition (Ψ) the situation is much less satisfactory. Only in the two dimensional case has the sufficiency been completely proved, although the literature contains quite a few incorrect "proofs" of the general sufficiency. The sufficiency is known microlocally in many cases though, and no counterexample has been found. Among the cases which are understood are those where the adjoint is subelliptic; there one also has an interesting symplectic classification of the possible Taylor expansions.

Unfortunately lack of time also makes it impossible to discuss the interesting symplectic geometry which occurs in the study of mixed boundary problems; I must content myself with a reference to Chapter 24 in [18], where all the topics discussed above are studied in much greater detail. Much more complete references and historial remarks can be found in Chapters 18, 21 and 24–27 there.

References

[1] I Ekeland and H. Hofer, *Symplectic topology and Hamiltonian dynamics*,   Math. Z. **200** (1989), 355–378.

[2] _____ , *Symplectic topology and Hamiltonian dynamics II*, Math. Z..

[3] M. Gromov, *Partial Differential Relations*, Springer-Verlag, 1986.

[4] _____ , *Pseudoholomorphic curves in symplectic manifolds*, Inv. Math. **82** (1985), 307–347.

[5] _____ , *Soft and hard symplectic geometry*, Proc. Intern. Congr. of Math. 1986, vol. 1, Amer. Math. Soc., 1987, pp. 81–98.

[6] H. Hofer and E. Zehnder, *A new capacity for symplectic manifolds*, Analysis, et cetera, Academic Press, Boston, San Diego, New York, Berkeley, London, Sydney, Tokyo, Toronto, 1990, pp. 405 –427.

[7] _____ , *Periodic solutions on hypersurfaces and a result by C. Viterbo*, Inv. Math. **90** (1987), 1–9.

[8] P. H. Rabinowitz, *Periodic solutions of a Hamiltonian system on a prescribed energy surface*, J. of diff. eq. **33** (1979), 336–352.

[9] V. Benci and P. H. Rabinowitz, *Critical point theorems for indefinite functionals*, Inv. math. **52** (1979), 241–274.

[10] C. Viterbo, *A proof of Weinstein's conjecture in* $\mathbf{R}^{2n}$, Ann. Inst. H. Poincaré, Anal. nonlinéaire **4** (1987), 337–356.

[11] A. Weinstein, *Normal modes for nonlinear Hamiltonian systems*, Inv. Math. **20** (1973), 47–57.

[12] _____ , J. of diff. eq. **33** (1979), 353–358.

[12] P.D. Lax, *Asymptotic solutions of oscillatory initial value problems*, Duke Math. J. **24** (1957), 627–646.

[13] F. John, *Continuous dependence on data for solutions of partial differential equations with a pre-scribed bound*, Comm. Pure Appl. Math. **13**, 551–585.

[14] M. Zerner, *Solutions singulières d'équations aux dérivées partielles*, Bull. Soc. Math. France **91** (1963), 203–226.

[15] V. Grušin, *The extension of smoothness of solutions of differential equations of principal type*, Dokl. Akad. Nauk SSSR **148** (1963), 1241–1248, Translation in Soviet Math. Doklady 4(1963), 248–252..

[16] L. Hörmander, *On the theory of general partial differential operators*, Acta Math. **94** (1955), 161–248.

[17] _____, *Pseudo-differential operators of principal type*, Nato Adv. Study Inst. on Sing. in Bound. Value Prolems, Reidel Publ. Co., Dordrecht, 1981, pp. 69–96.

[18] _____, *The analysis of linear partial differential operators I–IV*, Springer Verlag, 1983–1985.

[19] _____, *Propagation of singularities and semiglobal existence theorems for (pseudo-)differential operators of principal type*, Ann. of Math. **108** (1978), 569–609.

[20] H. Lewy, *An example of a smooth linear partial differential equation without solution*, Ann. of Math. **66** (1957), 155–158.

[21] L. Nirenberg and F. Treves, *Solvability of a first order linear partial differential equation*, Comm. Pure Appl. Math. **16** (1963), 331–351.

[22] _____, *On local solvability of linear partial differential equations. I. Necessary conditions. II. Sufficient conditions. Correction*, Comm. Pure Appl. Math. **23, 24** (1970,1971), 459–509, 279–288.

[23] R. Beals and C. Fefferman, *On local solvability of linear partial differential equations*, Ann. of Math. **97** (1973), 482–498.

[24] J.J. Duistermaat and L. Hörmander, *Fourier integral operators II*, Acta Math. **128** (1972), 183–269.

[25] J.J. Duistermaat and J. Sjöstrand, *A global construction for pseudo-differential operators with non-involutive characteristics*, Invent. Math. **20** (1973), 209–225.

[26] M. Sato, *Hyperfunctions and partial differential equations*, Proc. Int. Conf. on Funct. Anal. and Rel. Topics, Tokyo Univesity Press, Tokyo, 1969, pp. 91–94.

[27] M. Kashiwara, T. Kawai and T. Kimura, *Foundations of algebraic analysis*, Princeton University Press, Princeton, N.J., 1986.

GENERAL MEHLER FORMULAS AND THE WEYL CALCULUS

LARS HÖRMANDER

When Iz Singer called to invite me to give a lecture at this symposium I was very doubtful since I did not see that anything I had been thinking of recently was related to the work of Norbert Wiener. However, I quickly realised when browsing through the collected works that I had underestimated the wide scope of his work, for there is a paper [Wie] which is connected with a recent paper of mine [H].

Wiener states that his purpose is "to extend certain results of Hermann Weyl leading to Fourier developments of fractional order", and he quotes [Wey], a forerunner of the book with almost the same title. This paper is also the basic background for what else I want to discuss so let me start by reviewing the relevant parts.

For the self-adjoint operators p and q in $L^2(\mathbf{R})$ defined by $D = -i d/dx$ and multiplication by x which satisfy the commutation relation $[p, q] = -i$ in a formal sense Weyl wrote down the commutation relations in the rigorous integral form

$$(1) \qquad \exp(i\sigma p)\exp(i\tau q)\exp(-i\sigma p)\exp(-i\tau q) = e^{i\tau\sigma}, \quad \tau, \sigma \in \mathbf{R},$$

and defined

$$(2) \quad \exp(i(\sigma p + \tau q)) = \exp(-i\tau\sigma/2)\exp(i\sigma p)\exp(i\tau q) = \exp(i\tau\sigma/2)\exp(i\tau q)\exp(i\sigma p),$$

which agrees with (1). He then went on to define a general function $f(p, q)$ of p and q by

$$(42) \qquad f(p, q) = \iint \exp(i(\sigma p + \tau q))\xi(\sigma, \tau)\,d\sigma\,d\tau$$

where ξ is the Fourier transform of f normalized so that this is true for real variables p, q. He observes (p. 117, Collected works): "Die Einschränkungen, denen $f(p, q)$ unterworfen sein muß, damit sie eine Entwicklung des Typus (42) gestattet, könnten noch Bedenken erregen. Nun wissen wir aber, daß es eigentlich gilt, $\exp(ikf(p, q))$ so zu entwickeln (k irgend eine reelle Konstante), und in dieser Fassung läßt sich die Aufgabe nach neueren Untersuchungen von N. Wiener, Bochner und Hardy in zwingender Weise eindeutig erledigen." This must have been a very optimistic view of Fourier analysis in 1927 but not today, and it is the basis of what is now called the Weyl calculus, which I shall describe briefly.

With the notation a instead of f and $\xi = \hat{a}/(2\pi)^2$ in (42), where $\hat{a}$ is the Fourier transform of a, we obtain formally using (2) when $f(p, q)$ operates on say $u \in C_0^\infty$

$$a(x, D)u = (2\pi)^{-2} \iint \hat{a}(\tau, \sigma)e^{i\tau\sigma/2}u(x + \sigma)e^{i\tau x}\,d\tau d\sigma$$

$$= \frac{1}{2\pi} \iint a(x + \sigma/2, \xi)e^{-i\sigma\xi}u(x + \sigma)\,d\sigma d\xi,$$

© Springer International Publishing AG, part of Springer Nature 2018

L. Hörmander, *Unpublished Manuscripts*,

https://doi.org/10.1007/978-3-319-69850-2_21

by Fourier's inversion formula. A change of integration variables gives

$$(3) \qquad a(x,D)u = \frac{1}{2\pi} \iint a(\tfrac{1}{2}(x+y),\xi)e^{i(x-y)\xi}u(y)\,dyd\xi,$$

which means that the relation between the "symbol" a and the kernel $\mathcal{A}$ of $a(x,D)$ is given by

$$(4) \qquad \mathcal{A}(x,y) = \frac{1}{2\pi} \int a(\tfrac{1}{2}(x+y),\xi)e^{i(x-y)\xi}\,d\xi.$$

This formula makes perfectly good sense for any $a \in \mathcal{S}'(\mathbf{R}^2)$, and by Fourier's inversion formula it defines an isomorphism of $\mathcal{S}'(\mathbf{R}^2)$ with inverse

$$(5) \qquad a(x,\xi) = \int_{\mathbf{R}} \mathcal{A}(x+\tfrac{1}{2}t, x-\tfrac{1}{2}t)e^{-it\xi}\,dt.$$

Thus $a(x,D)$ is well defined as a continuous operator from $\mathcal{S}(\mathbf{R})$ to $\mathcal{S}'(\mathbf{R})$ then. From now on I shall use the notation $a^w(x,D)$ for the operator defined by (3), (4) to distinguish it from the more customary definitions of pseudo-differential operators. The definitions are the same in $\mathbf{R}^n$ as in $\mathbf{R}$ apart from the fact that 2π is replaced by $(2\pi)^n$ and $x\xi$ should be read as the scalar product $\langle x,\xi\rangle = \sum_1^n x_j\xi_j$.

Let $\sigma = \sum_1^n d\xi_j \wedge dx_j$ be the standard symplectic form in $\mathbf{R}^{2n} = T^*(\mathbf{R}^n)$. If χ is a linear (or affine) symplectic map $\mathbf{R}^{2n} \to \mathbf{R}^{2n}$, that is, $\chi^*\sigma = \sigma$, then it was observed 30 years ago by Irving Segal [S] that there is a corresponding unitary operator U_χ in $L^2(\mathbf{R}^n)$, uniquely determined up to a factor of absolute value 1, such that

$$(6) \qquad U_\chi^{-1}a^w(x,D)U_\chi = (a \circ \chi)(x,D).$$

Thus the map $\chi \mapsto U_\chi$ is a projective representation of the symplectic group; I shall return later to the more precise definition due to Shale and Weil of U_χ as a representation of the metaplectic covering group. At the moment I content myself with pointing out that by the definition of the Weyl operators it suffices to establish (6) for linear functions a. Hence for the symplectic map $\chi(x,\xi) = (\xi,-x)$ we can take U as the Fourier transformation, and for the symplectic map $\chi(x,\xi) = (x,\xi + Q'(x))$ where Q is a real quadratic form we can take for U the multiplication by e^{iQ}. These cases generate the linear symplectic group and suffice to establish the existence of a unitary U such that (6) is valid. As a final property of the Weyl calculus let us note that

$$a^w(x,D)b^w(x,D) = c^w(x,D) \quad \text{is equivalent to}$$
$$(7)$$
$$c(x,\xi) = e^{\frac{1}{2}i\sigma(D_x,D_\xi;D_y,D_\eta)}a(x,\xi)b(y,\eta)_{(x,\xi)=(y,\eta)},$$

under suitable conditions on the symbols a and b. I shall content myself with two cases: If a is linear then $c = ab + \{a,b\}/2i$ where $\{a,b\}$ is the Poisson bracket. This underlines the symplectic invariance of the Weyl calculus. If a is quadratic there is another more complicated term

$$-\tfrac{1}{8}\Big(\sum_{j,k=1}^n \partial^2 a/\partial\xi_j\partial\xi_k\partial^2/\partial x_j\partial x_k + \sum_{j,k=1}^n \partial^2 a/\partial x_j\partial x_k\partial^2/\partial\xi_j\partial\xi_k$$

$$-2\sum_{j,k=1}^n \partial^2 a/\partial x_j\partial\xi_k\partial^2/\partial\xi_j\partial x_k\Big)b,$$

which is a second order differential operator with constant coefficients acting on b. More generally, if a or b is a polynomial then c is always obtained by a Taylor expansion of the exponential with finitely many terms.

Let us now return to Wiener's paper. He sets out to find an operator $\mathcal{K}_\theta$ which multiplies the nth Hermite function $H_n(x)e^{-x^2/2}$ by ω^n where $\omega = e^{i\theta}$, $\theta \in \mathbf{R}$. Since the Hermite functions are an orthogonal basis in $L^2(\mathbf{R})$ this will be a unitary operator, and if we note that they are eigenfunctions of the harmonic oscillator $D^2 + x^2$ with eigenvalue $2n + 1$, the operator will be $\exp(\frac{1}{2}i\theta(D^2 + x^2 - 1))$. Wiener observes that if K_θ is the kernel of $\mathcal{K}_\theta$ then multiplication of the desired equation

$$\omega^n H_n(x)e^{-\frac{1}{2}x^2} = \int_{\mathbf{R}} K_\theta(x,y)H_n(y)e^{-\frac{1}{2}y^2}\,dy$$

by $(it)^n/n!$ and summation gives, using the standard generating function of the Hermite functions, the integral equation

$$\int_{\mathbf{R}} (K_\theta(x,y)e^{\frac{1}{2}(x^2-y^2)})e^{2ity}\,dy = e^{-t^2(1-\omega^2)+2it\omega x}, \quad t \in \mathbf{R},$$

which can be solved by inverting the Fourier transformation when $\operatorname{Re}(1 - \omega^2) > 0$, that is, when θ is not a multiple of π. In particular, the operator becomes the Fourier transformation when $\theta = -\pi/2$, which was of course a known formula for the Fourier transforms of Hermite functions which was the point of departure for defining "Fourier developments of fractional order". Wiener then uses the explicit form of K_θ to prove that

(8) $$\mathcal{K}_\theta x = (x \cos\theta + D \sin\theta)\mathcal{K}_\theta$$

and verifies in the same way that

$$\mathcal{K}_\theta e^{i\mu x} = e^{i\mu(x\cos\theta + D\sin\theta)}\mathcal{K}_\theta$$

with the definition of Weyl for the exponential in the right-hand side. The calculations in the paper could also have given

(9) $$\mathcal{K}_\theta D = (-x\sin\theta + D\cos\theta)\mathcal{K}_\theta,$$

so $\mathcal{K}_\theta$ is a unitary operator corresponding to rotation by $-\theta$ in $\mathbf{R}^2$, which is a symplectic linear map.

Wiener's formulas for K_θ are of course limiting cases of the Mehler formula [M] from 1866, which is now again very famous after the work of Getzler on the index theorem and work by many people on L^p estimates for operators with Gaussian kernels (see Lieb [L]). It is not mentioned by Wiener so it seems not to have been equally well known in the 1920's. Mehler's formula states that for $t > 0$ the kernel of $\exp(-t(D^2 + x^2))$ is

(10) $$(x,y) \mapsto \exp(-\tfrac{1}{2}((x^2 + y^2)\cosh(2t) - 2xy)/\sinh(2t))/\sqrt{2\pi\sinh(2t)}.$$

The formula is of course valid by analytic continuation when $\operatorname{Re} t > 0$ and has boundary values when $\operatorname{Re} t = 0$ except when t is a multiple of $\pi i/2$ which yields Wiener's formula. Mehler's proof starts from the expressions

$$H_n(x) = e^{x^2}(-d/dx)^n e^{-x^2} = (-2i)^n \frac{1}{\sqrt{\pi}} \int_{\mathbf{R}} e^{-(t-xi)^2} t^n\,dt;$$

the first equality is a definition and the second equality follows from the Fourier representation $e^{-x^2} = \int e^{-t^2 + 2ixt} \, dt/\sqrt{\pi}$. Hence Taylor's formula gives if $0 < \varrho < 1$

$$
\sum_0^\infty H_n(x) H_n(y) \varrho^n / n! 2^n = \frac{e^{x^2}}{\sqrt{\pi}} \int_{\mathbf{R}} \left(\sum_0^\infty (d/dx)^n e^{-x^2} (it\varrho)^n / n! \right) e^{-(t-yi)^2} \, dt
$$

$$
= \frac{e^{x^2}}{\sqrt{\pi}} \int e^{-(x+it\varrho)^2 - (t-yi)^2} \, dt = (1 - \varrho^2)^{-\frac{1}{2}} \exp \left((2\varrho xy - \varrho^2(x^2 + y^2))/(1 - \varrho^2) \right)
$$

which reduces to (10) when $\varrho = e^{-2t}$ if we multiply by $e^{-t - \frac{1}{2} x^2 - \frac{1}{2} y^2} / \sqrt{\pi}$. Actually Mehler arrived at these formulas by solving the Dirichlet problem in the unit ball in $\mathbf{R}^n$ when $n \to \infty$ when the given boundary data were of the form $F(x_1 \sqrt{n/2}, \ldots, x_\nu \sqrt{n/2})$ for a fixed ν. In his paper ϱ appears as the radial variable in the ball. Mehler's problem was in fact the opposite of the preceding discussion; he knew the Mehler formula from the solution of the Dirichlet problem and wanted to expand it in a power series in ϱ.

The starting point of my interest in these matters was a paper by Derezinski [D] where he observed that since the symbol $x^2 + \xi^2$ of $x^2 + D^2$ is rotation invariant it follows from the symplectic invariance of the Weyl calculus that the Weyl symbol of $\exp(-t(x^2 + D^2))$ must be rotation invariant. It is

(11) $$ (x, \xi) \mapsto \exp(-(x^2 + \xi^2) \tanh t)/ \cosh t, $$

which obviously looks much simpler than (10). The equation (11) follows from (10) and (5) by evaluation of a Gaussian integral; however, we shall see that (11) is more natural and general if properly interpreted, and (10) follows from (11) and (4). Derezinski went on to determine the general quadratic forms Q such that the Weyl symbol of a function of Q^w is always a function of Q. This happens rarely, and I shall return to his result later.

The remark of Derezinski spurred me to look at the Weyl symbol of $\exp(Q^w)$ for a general quadratic form such that the exponential can be defined. Let us first look at this from a formal point of view, and denote by e^{q_t} the putative Weyl symbol of $\exp(tQ^w)$. Differentiation with respect to t should give

$$
(\partial q_t / \partial t \exp q_t)^w = Q^w \exp(tQ^w) = Q^w (\exp q_t)^w,
$$

which by the calculus formula (7) means that

$$
\partial q_t / \partial t \exp q_t = e^{\frac{i}{2} \sigma(D_x, D_\xi; D_y, D_\eta)} Q(x, \xi) \exp q_t(y, \eta)|_{(x, \xi) = (y, \eta)}
$$

$$
= Q \exp q_t + \frac{1}{2i} \{Q, \exp q_t\} - \frac{1}{8} \Bigg(\sum_{j,k=1}^n \partial^2 Q/\partial \xi_j \partial \xi_k \partial^2 /\partial x_j \partial x_k
$$

$$
+ \sum_{j,k=1}^n \partial^2 Q/\partial x_j \partial x_k \partial^2 /\partial \xi_j \partial \xi_k - 2 \sum_{j,k=1}^n \partial^2 Q/\partial x_j \partial \xi_k \partial^2 /\partial \xi_j \partial x_k \Bigg) \exp q_t.
$$

From (11) we are led to try to find a solution of the form $q_t = g_t + h(t)$ where g_t is a quadratic form in (x, ξ). Apart from the exponential we then have only terms of degree 2 and of degree 0 in the two sides and separation gives the system of equations

$$\partial g_t/\partial t = Q + \tfrac{1}{2i}\{Q, g_t\} - \tfrac{1}{8}\Big(\sum_{j,k=1}^{n} \partial^2 Q/\partial\xi_j\partial\xi_k\partial g_t/\partial x_j\partial g_t/\partial x_k$$

$$+ \sum_{j,k=1}^{n} \partial^2 Q/\partial x_j\partial x_k\partial g_t/\partial\xi_j\partial g_t/\partial\xi_k - 2\sum_{j,k=1}^{n} \partial^2 Q/\partial x_j\partial\xi_k\partial g_t/\partial\xi_j\partial g_t/\partial x_k\Big),$$

(12)

$$h'(t) = -\tfrac{1}{8}\Big(\sum_{j,k=1}^{n} \partial^2 Q/\partial\xi_j\partial\xi_k\partial^2 g_t/\partial x_j\partial x_k$$

$$+ \sum_{j,k=1}^{n} \partial^2 Q/\partial x_j\partial x_k\partial^2 g_t/\partial\xi_j\partial\xi_k - 2\sum_{j,k=1}^{n} \partial^2 Q/\partial x_j\partial\xi_k\partial^2 g_t/\partial\xi_j\partial x_k\Big).$$

This can be regarded as a system of non-linear ordinary differential equations for h and the $n(2n + 1)$ coefficients of g_t, so for small t there is a unique solution with $q_0 = 0$, that is, $g_0 = 0$ and $h(0) = 0$. We want to examine if there is a global solution. In doing so we can clearly work over the complex numbers since the preceding equations are valid over $\mathbf{C}$, and we look for g_t and $h(t)$ as analytic functions of t.

First consider the non-degenerate case with one degree of freedom. It is then enough to study the case $Q(x,\xi) = 2\lambda x\xi$ for some $\lambda \in \mathbf{C} \setminus \{0\}$. The equation for g_t is

$$\partial g_t/\partial t = 2\lambda(x\xi + \tfrac{1}{2i}\{x\xi, g_t\} + \tfrac{1}{4}\partial g_t/\partial\xi\partial g_t/\partial x),$$

which is solved by $g_t = \gamma(t)x\xi$ where

$$\gamma'(t) = 2\lambda(1 + \tfrac{1}{4}\gamma(t)^2), \quad \gamma(0) = 0; \quad \text{thus } \gamma(t) = 2\tan(\lambda t).$$

The equation $h'(t) = \lambda\gamma(t)/2 = \lambda\tan(\lambda t)$ has the solution $h(t) = -\log\cos(\lambda t)$ so

$$\exp q_t(x, \xi) = \frac{e^{2x\xi\tan(\lambda t)}}{\cos(\lambda t)},$$

which is analytic except at the zeros of $\cos(\lambda t)$.

Suppose now that we have n degrees of freedom and that

(13)
$$Q(x,\xi) = \sum_{1}^{n} 2\lambda_j x_j\xi_j.$$

A generic form is symplectically equivalent to such a form. Then we have

$$g_t(x,\xi) = 2\sum_{1}^{n} x_j\xi_j \tan(\lambda_j t), \quad h'(t) = \sum_{1}^{n} \lambda_j \tan(\lambda_j t), \quad \text{thus } \exp(-h(t)) = \prod_{1}^{n} \cos(\lambda_j t).$$

For every quadratic form Q in a symplectic vector space S, with symplectic form σ, the Hamilton vector field of $\tfrac{1}{2}Q$ defines a linear Hamilton map F by

$$\sigma(Y, FX) = Q(Y, X), \quad X, Y \in S,$$

where the right-hand side contains the polarized form of Q. The symmetry of Q means that F is skew symmetric with respect to σ. For the special case (13) the Hamilton map is

$$F = \begin{pmatrix} \Lambda & 0 \\ 0 & -\Lambda \end{pmatrix}, \quad \Lambda = \operatorname{diag}(\lambda_1, \ldots, \lambda_n).$$

If $X = (x, \xi)$ then $\sigma(X, \tan(tF)X) = 2\langle \xi, \tan(t\Lambda)x \rangle = g_t(X)$, so we obtain

(14) $$g_t(X) = \sigma(X, \tan(tF)X), \quad \exp(-h(t)) = \prod_1^n \cos(\lambda_j t),$$

except at the zeros of the product.

For an arbitrary quadratic form Q in $T^*\mathbf{C}^n$, with Hamilton map F, it is clear that $\sin F$ and $\cos F$ are entire analytic functions of (the coefficients of) F, and $\det(\cos F) = \prod_1^{2n} \cos \lambda_j$ where λ_j are all the $2n$ eigenvalues of F with multiple ones repeated. Now the secular equation $\det(F - \lambda I) = 0$ is even in λ, for in the generic special case above it is equal to the product $\prod(\lambda^2 - \lambda_i^2)$, taken over one half of the zeros. Let $\mu_1, \ldots, \mu_n$ be the zeros of $\det(F - \lambda I)$ as a polynomial in $\mu = \lambda^2$. Then $\det(\cos F) = \prod_1^n (\cos \sqrt{\mu_j})^2$, and the square root $\sqrt{\det(\cos F)} = \prod_1^n \cos \sqrt{\mu_j}$ is an analytic function of $\mu_1, \ldots, \mu_n$ which is symmetric under permutations, hence an entire function of the elementary symmetric functions which are polynomials in F. Thus $\sqrt{\det(\cos F)}$ is an entire analytic function of F. Outside the zeros it is clear that $\tan F = \sin F (\cos F)^{-1}$ is analytic.

THEOREM 1. *For every quadratic form Q in $T^*\mathbf{C}^n$ the system of differential equations (12) for a quadratic form $g_t(x, \xi)$ and a scalar function $h(t)$ has a unique solution with vanishing initial values such that g_t and $\exp h(t)$ are meromorphic functions,*

(15) $$\exp(q_t(X)) = \exp(g_t(X) + h(t)) = \exp(\sigma(X, \tan(tF)X)) / \sqrt{\det(\cos tF)}$$

where F is the Hamilton map of Q. It is analytic in t and F except at the zeros of the denominator, that is, where $t\lambda_j = \frac{1}{2}\pi + k\pi$ for some eigenvalue λ_j of F and some $k \in \mathbf{Z}$.

Mehler's formula in the version (11) is of course an obvious special case, and it is clear that q_t is a function of Q if and only if $F^3 = cF$ for some constant c, which is the result of Derezinski. Then $\tan(tF) = F \tan(t\sqrt{c})/\sqrt{c}$ ($= tF$ if $c = 0$). However, so far the arguments have been formal, and we must give them a rigorous interpretation. For any quadratic form Q in $T^*(\mathbf{R}^n)$ the closure in $L^2(\mathbf{R}^n)$ of $Q^w(x, D)$ defined on $\mathcal{S}(\mathbf{R}^n)$ or just on $C_0^\infty(\mathbf{R}^n)$ is defined for all $u \in L^2(\mathbf{R}^n)$ such that $Q^w(x, D)u \in L^2(\mathbf{R}^n)$ when defined in the sense of distribution theory. The adjoint of the operator is defined by $\overline{Q}$. We shall denote the operator defined by $Q^w(x, D)$ in $L^2(\mathbf{R}^n)$ also by Q^w. If $\operatorname{Re} Q \le 0$ then

$$\operatorname{Re}(Q^w(x, D)u, u) = ((\operatorname{Re} Q)^w(x, D)u, u) \le 0$$

for u in the L^2 domain of Q^w. This follows from the symplectic invariance of the Weyl calculus, for $\operatorname{Re} Q$ is symplectically equivalent to a sum of (possibly degenerate) harmonic oscillators. Thus the operator $Q^w(x, D)$ in L^2 and its adjoint are dissipative, so $Q^w(x, D)$ generates a contraction semigroup, which will be denoted by $\exp(tQ^w)$, $t \ge 0$.

THEOREM 2. *If $\operatorname{Re} Q$ is negative definite, then the Weyl symbol A_t of $\exp(tQ^w)$ is for $t > 0$ a function in $\mathcal{S}(T^*\mathbf{R}^n)$ given by (15) where the quadratic form g_t and $\exp(h(t))$ are always finite, and g_t has negative definite real part. When $\operatorname{Re} Q$ is just negative semidefinite, the map $Q \mapsto \exp(Q^w)u$ is a continuous function (resp. C^∞ function) of Q with* *values in* $L^2(\mathbf{R}^n)$ *(resp.* $\mathcal{S}(\mathbf{R}^n)$*) if* $u \in L^2(\mathbf{R}^n)$ *(resp.* $u \in \mathcal{S}(\mathbf{R}^n)$*), and the Weyl symbol of* $\exp(Q^w)$ *is a continuous function of* Q *with values in* $\mathcal{S}'(\mathbf{R}^{2n})$ *given by* (15) *when* $\det(\cos F) \neq 0$.

The verification is easy. The question is now what happens when (15) does not make sense any longer. Let us first look again at the harmonic oscillator, that is, return to the Mehler and Wiener formulas. The Weyl symbol of $\exp(-t(x^2 + D^2))$ is given by (11) when $\operatorname{Re} t \geq 0$ and $\cosh t \neq 0$. In particular, if $t = is$ with $s \in \mathbf{R}$ and $\cos s \neq 0$, it is given by

$$(16) \qquad (x, \xi) \mapsto \exp(-i(x^2 + \xi^2)\tan s)/\cos s.$$

If $s = \pi/2 + k\pi$ for some integer k, then the Weyl symbol is the limit in $\mathcal{S}'(\mathbf{R}^2)$ as $\varepsilon \to +0$ of the symbol for $t = is + \varepsilon$,

$$(x, \xi) \mapsto \exp(-(x^2 + \xi^2)/\tanh \varepsilon)i(-1)^{k+1}/\sinh \varepsilon,$$

that is, $i(-1)^{k+1}\pi\delta_0$. The corresponding Weyl operator has the kernel $i(-1)^{k+1}\delta_0(x + y)$, which defines a reflection operator. When $\sin s = 0$ then the Weyl symbol (16) is ± 1, and the corresponding kernel is $\pm\delta_0(x - y)$, so we have $\pm$ the identity operator. When $2\sin s \cos s = \sin(2s) \neq 0$ the kernel of $e^{-is(x^2+D^2)}$ is easily obtained from (10); it is

$$(17) \qquad (x, y) \mapsto \exp\left(\tfrac{i}{2}((x^2 + y^2)\cos(2s) - 2xy)/\sin(2s)\right)/\sqrt{2\pi i \sin(2s)}.$$

Since $\sqrt{2\pi \sinh(2t)} = \cosh t\sqrt{4\pi \tanh t}$ and $\operatorname{Re}\tanh t > 0$ when $\operatorname{Re} t > 0$, the square root in (17) should be taken in the right (left) half plane when $\cos s > 0$ ($\cos s < 0$). When $\cos(2s) = 0$, hence $\sin(2s) = \pm 1$, the exponential reduces to $\exp(\mp ixy)$, so $e^{-is(x^2+D^2)}$ becomes the (inverse) Fourier transformation apart from a factor $\pm e^{\pm \pi i/4}$. In particular, when $s = \pi/4$ then the operator is $e^{-\pi i/4}$ times the Fourier transformation, and from the group property and Fourier's inversion formula we obtain the values at $s = \nu\pi/4$ for any integer ν which we have already given.

We turn now to the case of a general quadratic form Q in $T^*\mathbf{R}^n$ with $\operatorname{Re} Q \leq 0$. First we shall rewrite the result already proved when $\det \cos F \neq 0$. Then $\sec F = (\cos F)^{-1}$ is defined, and

$$(18) \qquad \sigma_F(X, Y) = \sigma(X, (\sec F)Y)$$

is also a symplectic form in $T^*(\mathbf{C}^n)$. The corresponding measure vol_{σ_F}, defined by the nth power of σ_F and the standard orientation, is equal to $\mathrm{vol}_\sigma/\sqrt{\det \cos F}$ where vol_σ is defined by the standard symplectic form. That σ_F is skew symmetric is obvious since $\cos F$ is even in F, and it is non-degenerate since $\sec F$ is bijective. When $F = \begin{pmatrix} \Lambda & 0 \\ 0 & -\Lambda \end{pmatrix}$ with $\Lambda = \operatorname{diag}(\lambda_1, \ldots, \lambda_n)$ then

$$\sigma_F = \sum_1^n d\xi_j \wedge dx_j/\cos \lambda_j$$

which proves that $\mathrm{vol}_{\sigma_F} = \mathrm{vol}_\sigma/\prod_1^n \cos \lambda_j$ as claimed. Since this situation is symplectically equivalent to the generic case, the statement follows. The result of Theorem 2 is now that

$$(19) \qquad A_1(X)\,\mathrm{vol}_\sigma = \exp(\sigma(X, (\tan F)X))\,\mathrm{vol}_{\sigma_F}.$$

Our next goal is to give a sense to the two factors in the right-hand side when $\det \cos F = 0$. As suggested by the example of the harmonic oscillator vol_{σ_F} will be replaced by a translation invariant measure on the range of $\cos F$, and the quadratic form in the exponential will only be defined there. We need some preliminaries:

LEMMA 3. *If $Q = Q_1 + iQ_2$ where $Q_1 \leq 0$, then $\mathrm{Ker}(F - \lambda)$ is the complex conjugate of $\mathrm{Ker}(F + \lambda)$ for every $\lambda \in \mathbf{R}$, and if $F = F_1 + iF_2$ then $F_1 \mathrm{Ker}(F \pm \lambda) = 0$. Thus $\mathrm{Ker}(F - \lambda) \oplus \mathrm{Ker}(F + \lambda)$, $0 \neq \lambda \in \mathbf{R}$ is the complexification of its intersection with $T^*\mathbf{R}^n$, and so is $\mathrm{Ker}\, F$.*

PROOF. Assume that $X \in T^*\mathbf{C}^n$ and that $(F - \lambda)X = 0$. This means that

$$Q(Y, X) = \sigma(Y, FX) = \lambda\sigma(Y, X)$$

for every Y, thus

$$Q(\overline{X}, X) = \lambda\sigma(\overline{X}, X) = 2\lambda i\sigma(\mathrm{Re}\, X, \mathrm{Im}\, X),$$

so $Q_1(\overline{X}, X) = 0$. Since Q_1 is semidefinite it follows that $Q_1(Y, X) = 0$ for arbitrary Y, that is, $F_1 X = 0$. Thus $(iF_2 - \lambda)X = 0$ so $(-iF_2 - \lambda)\overline{X} = 0$, and $(F + \lambda)\overline{X} = 0$. The proof is complete.

A simple example shows that the generalized eigenspaces need not be complex conjugates, but only the genuine eigenspaces are important in the following:

PROPOSITION 4. *If $\mathrm{Re}\, Q \leq 0$ then the kernel (resp. range) of $\cos F$ is the complexification of its intersection $K(F)$ (resp. $W(F)$) with $T^*\mathbf{R}^n$. The restriction of $(\sin F)/i$ to $K(F)$ is a bijection with square equal to minus the identity, so it defines a complex vector space structure in $K(F)$ and the determinant is equal to 1. The corresponding orientation induces an orientation in $W(F)$.*

We can define a symplectic form in the complex vector space $W(F)_\mathbf{C}$ by

$$(18)' \qquad \sigma_F((\cos F)X, (\cos F)Y) = \sigma((\cos F)X, Y), \quad X, Y \in T^*\mathbf{C}^n.$$

This agrees with (18) when $\det \cos F \neq 0$. The definition $(18)'$ is unique, for $\cos F$ is even in F so the right-hand side is equal to $\sigma(X, (\cos F)Y)$, hence equal to 0 if $(\cos F)X = 0$ or $(\cos F)Y = 0$. It is obvious that σ_F is skew symmetric. If the right-hand side of $(18)'$ vanishes for every Y then $(\cos F)X = 0$ which proves that the form is non-degenerate. Thus we have a symplectic form defined in the range $W(F)_\mathbf{C}$ of $\cos F$, and the corresponding volume form defines a translation invariant measure $\mathrm{vol}\,\sigma_F \neq 0$ in $W(F)$ when combined with the orientation defined in Proposition 4.

We define the quadratic form similarly. The quadratic form in (19) corresponds to the symmetric bilinear form

$$E_F(X, Y) = \sigma(X, (\tan F)Y),$$

and we extend the definition to the case where $\det \cos F = 0$ by

$$(20) \qquad E_F((\cos F)X, (\cos F)Y) = \sigma((\cos F)X, (\sin F)Y), \quad X, Y \in T^*\mathbf{C}^n.$$

Since $\cos F$ is symmetric and $\sin F$ is skew symmetric with respect to σ the right-hand side is equal to $\sigma((\cos F)Y, (\sin F)X)$, which proves that the form is uniquely defined and symmetric. We can now state a general Mehler formula:

THEOREM 5. *If $\mathrm{Re}\, Q \leq 0$ then $(18)'$ defines a symplectic form in the range $W(F)_\mathbf{C}$ of the Hamilton map F of Q, and (20) defines a symmetric bilinear form with*

$$\mathrm{Re}\, E_F(X, X) \leq 0 \quad \text{when } X \in W(F).$$

The product of the Weyl symbol of $\exp(Q^w)$ and vol_σ is equal to $(\pi i)^\nu \exp(E_F)\, \mathrm{vol}_{\sigma_F}$ where 2ν is the dimension of $K(F)$. Here vol_σ is the positive measure defined by the symplectic *form, and* vol_{σ_F} *is the measure in* $W(F)$ *defined by the form* σ_F *and the orientation of* $W(F)$ *in Proposition 4.*

The proof is too lengthy to give here, but it does not differ in principle from the case of the harmonic oscillator. It is not hard to extend to quadratic polynomials with lower order terms. This covers some known formulas such as that of Avron and Herbst [AH] for the exponential of $D^2 + x$, the Stark operator. Instead of discussing this I would like to spend the remaining moments on the connections with the infinitesimal form of the calculus of Fourier integral operators.

The symbol obtained in Theorem 5 is a Gaussian in a generalized sense also if $\det \cos F = 0$: we shall say that a distribution $u \in \mathcal{D}'(\mathbf{R}^n) \setminus \{0\}$ is a *Gaussian* if every distribution annihilated by

$$\mathcal{L}_u = \{L; L(x,D)u = 0\}, \quad \text{where } L(x,D) = \sum_1^n a_j D_j + \sum_1^n b_j x_j$$

is a multiple of u. Since a commutator $[L_1, L_2]$ with $L_j \in \mathcal{L}_u$ is a constant it must be equal to 0, so

$$\lambda_u = \{(x,\xi) \in T^* \mathbf{C}^n; L(x,\xi) = 0 \ \forall \ L \in \mathcal{L}_u\}$$

is an involutive subspace. It is in fact a complex *Lagrangian* plane, that is, of dimension n, and $V = \{\xi \in \mathbf{C}^n; (0,\xi) \in \lambda_u\}$ is invariant under conjugation, hence generated by $V \cap \mathbf{R}^n$. Conversely, for every such Lagrangian plane λ a distribution u such that $L(x,D)u = 0$ for every L vanishing on λ is a Gaussian, and $u = ce^q d$ where d is a δ function on a linear subspace and q is a quadratic form, both determined by λ, and $c \in \mathbf{C}$. This condition on λ is not invariant under the linear symplectic group. If $\lambda \subset T^* \mathbf{C}^n$ is a complex Lagrangian plane then for every real Lagrangian plane $\mu \subset T^* \mathbf{R}^n$ with complexification $\mu_{\mathbf{C}}$ the intersection $\lambda \cap \mu_{\mathbf{C}}$ is invariant under complex conjugation if and only if $\lambda \ni X \mapsto i\sigma(\overline{X}, X)$ is semi-definite. Note that

$$i\sigma(\overline{Y}, X), \quad X, Y \in T^* \mathbf{C}^n,$$

is Hermitian symmetric.

DEFINITION. If S is a real symplectic vector space with complexification $S_{\mathbf{C}}$, then a Lagrangian $\lambda \subset S_{\mathbf{C}}$ is said to be positive if $i\sigma(\overline{X}, X) \geq 0$, $X \in \lambda$, and strictly positive if $i\sigma(\overline{X}, X) > 0$ when $0 \neq X \in \lambda$.

If Λ^+ is the set of strictly positive Lagrangian planes then the closure is the set of positive Lagrangian planes. A Gaussian is temperate if and only if the corresponding Lagrangian plane is positive, and the temperate Gaussians with the topology induced by $\mathcal{S}'$ form a line bundle over $\overline{\Lambda_+}$ with the zero section removed.

If $Q(x,\theta)$ is a complex valued quadratic form in $\mathbf{R}^n \oplus \mathbf{R}^N$ such that $\operatorname{Im} Q \geq 0$ and the linear forms $\partial Q / \partial \theta_j$, $j = 1, \ldots, N$, are linearly independent over $\mathbf{C}$, then

$$u = \int e^{iQ(x,\theta)} \, d\theta$$

can be interpreted as a Gaussian belonging to the positive Lagrangian

$$\lambda = \{(x, \partial Q(x,\theta)/\partial x); \partial Q(x,\theta)/\partial \theta = 0\},$$

and every positive Lagrangian can be defined in this way.

Now we specialize to positive Lagrangians $\lambda \subset T^*\mathbf{C}^n \oplus T^*\mathbf{C}^n \cong T^*\mathbf{C}^{2n}$ with bijective projection in each of the two factors. Then the twisted Lagrangian

$$C = \lambda' = \{(X, Y'); (X, Y) \in \lambda\}, \quad (y, \eta)' = (y, -\eta),$$

is the graph of a symplectic linear bijection T in $T^*\mathbf{C}^n$ (from the second to the first factor) which is positive in the sense that

$$i(\sigma(\overline{TY}, TY) - \sigma(\overline{Y}, Y)) \geq 0, \quad Y \in T^*\mathbf{C}^n.$$

With T we associate the Gaussian distribution

$$(21) \qquad K_T = (2\pi)^{-(n+N)/2} \sqrt{\det \begin{pmatrix} Q''_{\theta\theta}/i & Q''_{\theta y} \\ Q''_{x\theta} & iQ''_{xy} \end{pmatrix}} \int e^{iQ(x,y,\theta)}\, d\theta \in \mathcal{S}'(\mathbf{R}^n \times \mathbf{R}^n),$$

where Q is a quadratic form defining $\lambda = C'$. We do not prescribe the sign of the square root so K_T is only determined up to the sign, but apart from that K_T is uniquely determined, and it defines a continuous map $\mathcal{K}_T : \mathcal{S}(\mathbf{R}^n) \to \mathcal{S}(\mathbf{R}^n)$ which extends continuously to a map $\mathcal{S}'(\mathbf{R}^n) \to \mathcal{S}'(\mathbf{R}^n)$. If T_1 and T_2 are two positive symplectic bijections in $T^*\mathbf{C}^n$ then $T_1 T_2$ is a positive symplectic bijection and $\mathcal{K}_{T_1 T_2} = \pm \mathcal{K}_{T_1} \mathcal{K}_{T_2}$.

PROPOSITION 6. *If Q is a quadratic form in $T^*\mathbf{R}^n$ with $\operatorname{Im} Q \geq 0$ and the Hamilton map F of Q does not have the eigenvalues ± 1, then*

$$(22) \qquad \sqrt{\det(I - F)}(e^{iQ})^w = \mathcal{K}_{(I-F)/(I+F)},$$

where $\mathcal{K}_T$ is the operator with kernel K_T defined in (21).

We shall apply this result to e^{iQ^w} where Q is a quadratic form with Hamilton map F and $\operatorname{Im} Q \geq 0$. When $\det \cosh F \neq 0$ we know from Theorem 2 that the Weyl symbol is equal to $\exp(i\widetilde{Q})/\sqrt{\det \cosh F}$ where $\widetilde{Q}(X) = \sigma(X, (\tanh F)X)$. Thus the Hamilton map of $\widetilde{Q}$ is $\tanh F$, which according to Proposition 6 is associated with the linear symplectic map

$$(I - \tanh F)(I + \tanh F)^{-1} = e^{-2F}.$$

We have $\det(I - \tanh F) = \det(\cosh F - \sinh F)/\det \cosh F = 1/\det \cosh F$, for the determinant of e^{-F} is equal to 1 since it is a symplectic map. Now it follows from Proposition 6 that the Weyl operator defined by $\exp(i\widetilde{Q})/\sqrt{\det \cosh F}$ is equal to $\mathcal{K}_{\exp(-2F)}$, which means that e^{iQ^w} is equal to $\mathcal{K}_{\exp(-2F)}$. We can therefore give another interpretation of the Mehler formulas:

THEOREM 7. *If Q is a quadratic form in $T^*\mathbf{R}^n$ with $\operatorname{Im} Q \geq 0$, then $\exp(iQ^w) = \mathcal{K}_{\exp(-2F)}$ where $\mathcal{K}_T$ is the operator with kernel K_T defined by* (21) *when T belongs to the semigroup $\mathcal{C}_+$ of positive symplectic linear maps in $T^*\mathbf{C}^n$. The semigroup generated by the contraction operators $\exp(iQ^w)$ consists of all operators $\mathcal{K}_T$ with $T \in \mathcal{C}_+$; it is a double cover of $\mathcal{C}_+$. The invertible elements in the group are those with $T \in \mathcal{C}_0 \subset \mathcal{C}_+$, where $\mathcal{C}_0$ is the real symplectic group. They form a double cover of $\mathcal{C}_0$, generated by $\exp(iQ^w)$ when Q is a real quadratic form.*

The group of operators $\{\mathcal{K}_T\}$ with $T \in \mathcal{C}_0$ is isomorphic to the metaplectic group. If Q is a real quadratic form with Hamilton map F, then

$$(6)' \qquad e^{-i\theta Q^w(x,D)} a^w(x, D) e^{i\theta Q^w(x,D)} = (a \circ e^{-2\theta F})^w(x, D), \quad a \in \mathcal{S}'(\mathbf{R}^{2n}).$$

When $Q(x,\xi) = x^2 + \xi^2$ then $F(x,\xi) = (\xi, -x)$ and $e^{-2\theta F}$ is rotation by the angle 2θ, which gives the formulas (8), (9) from Wiener's paper. It seems natural to call the semigroup of operators $\{\mathcal{K}_T\}$ with $T \in \mathcal{C}_+$ the *metaplectic semigroup*. If $U = \mathcal{K}_T$ with $T \in \mathcal{C}_0$ and if $\operatorname{Im} Q \geq 0$, then the metaplectic invariance of the Weyl calculus (see (6)) implies that

$$U^{-1}(e^{iQ})^w U = (e^{i\widetilde{Q}})^w,$$

where $\widetilde{Q}(X) = Q(TX)$ has the Hamilton map $\widetilde{F} = T^{-1}FT$, if F is the Hamilton map of Q. This is also a consequence of Proposition 6, but the full metaplectic invariance does not follow from Proposition 6 since all Gaussian symbols are even.

Examples show that the exponential map $Q \mapsto \exp(iQ^w)$, defined when Q is a quadratic form in $\mathbf{R}^n$ with $\operatorname{Im} Q \geq 0$, is not a surjection of a neighborhood of 0 on a neighborhood of the identity in $\{\mathcal{K}_T; T \in \mathcal{C}_+\}$ which is an essential difference from the real case.

Appendix: Mehler's Dirichlet problem. The solution of the Dirichlet problem in the unit ball in $\mathbf{R}^n$ with boundary data f is given by

$$u(x) = \frac{1 - |x|^2}{c_n} \int_{S^{n-1}} f(y)|x - y|^{-n} \, dS(y), \quad |x| < 1,$$

where dS is the surface measure and c_n the total mass. With polar coordinates we obtain

$$u(\varrho x) = \frac{1 - \varrho^2}{c_n} \int_{S^{n-1}} f(y)(1 + \varrho^2 - 2\varrho(x,y))^{-\frac{n}{2}} \, dS(y), \quad |x| = 1, \ 0 < \varrho < 1.$$

Mehler considers the case where f only depends on the first ν coordinates but $n \to \infty$. We write $x = (x', x'')$ and $y = (y', y'')$ where x', y' denote the first ν coordinates and x'', y'' the last $\mu = n - \nu$ coordinates.

If $g \in C_0(\mathbf{R}^{\nu+\mu})$ then the integral over the unit sphere is the limit as $\varepsilon \to 0$ of

$$\frac{1}{\varepsilon} \iint_{1-2\varepsilon < |x'|^2 + |x''|^2 < 1} g(x', x'') \, dx' \, dx''$$

$$= \int (1 - |x'|^2)^{\frac{\mu}{2}} \, dx' \frac{1}{\varepsilon} \int_{1-2\varepsilon/(1-|x'|^2) < |x''|^2 < 1} g(x', \sqrt{1 - |x'|^2}x'') dx'',$$

which proves that

$$\int_{S^{\nu+\mu-1}} g(x', x'') dS = \int (1 - |x'|^2)^{\frac{\mu}{2}-1} \int_{S^{\mu-1}} g(x', \sqrt{1 - |x'|^2}x'') dS(x'').$$

Hence the solution considered by Mehler is

$$u(\varrho x) = \frac{1 - \varrho^2}{c_{\nu+\mu}} \int_{|y'|<1} f(y')(1 - |y'|^2)^{\frac{\mu}{2}-1} dy' \cdot$$

$$\cdot \int_{S^{\mu-1}} (1 + \varrho^2 - 2\varrho((x',y') + \sqrt{1 - |y'|^2}(x'', y'')))^{-\frac{n}{2}} dS(y'').$$

With t denoting the component of y'' in the direction x'' on $S^{\mu-1}$ we write the surface measure as $c_{\mu-1}(1 - t^2)^{\frac{\mu-3}{2}} dt$ and obtain

$$u(\varrho x) = (1 - \varrho^2)\frac{c_{\mu-1}}{c_{\nu+\mu}} \int_{|y'|<1} f(y')(1 - |y'|^2)^{\frac{\mu}{2}-1} dy' \int_{-1}^{1} (1 - t^2)^{\frac{\mu-3}{2}} \cdot$$

$$\cdot (1 + \varrho^2 - 2\varrho((x',y') + t\sqrt{1 - |x'|^2}\sqrt{1 - |y'|^2}))^{-\frac{n}{2}} \, dt.$$

Here $c_{\mu-1}/c_{\nu+\mu} \sim (\mu/2\pi)^{\frac{\nu+1}{2}}$.

Now Mehler changes scales for f, replacing $f(y')$ by $f(y'\sqrt{n/2})$ where $f \in C(\mathbf{R}^\nu) \cap L^\infty(\mathbf{R}^\nu)$, say, and correspondingly x' is replaced by $x'\sqrt{2/n}$. The resulting function is

$$U_n(x',\varrho) = (1-\varrho^2)\frac{c_{\mu-1}}{c_{\nu+\mu}}(2/n)^{\frac{\nu}{2}}\int f(x')(1-2|x'|^2/n)^{\frac{\mu}{2}-1}\int(1-t^2)^{-\frac{\nu+3}{2}}A^{-\frac{n}{2}}\,dt$$

where

$$A = \left(1+\varrho^2-2\varrho((2/n)(x',y')+t\sqrt{1-2|x'|^2/n}\sqrt{1-2|y'|^2/n})\right)/(1-t^2)$$
$$= 1 + \left((\varrho-t)^2 - \varrho(2/n)(2(x',y')-(|x'|^2+|y'|^2)t+O(1/n))\right)/(1-t^2)$$

If we set $t = \varrho+\theta\sqrt{2/n}$ then $A^{-\frac{n}{2}}$ will converge to the exponential of $(-\theta^2+2\varrho(x',y') - \varrho^2(|x'|^2+|y'|^2))/(1-\varrho^2)$, so the inner integral will be asymptotic to $\sqrt{2/n}\sqrt{\pi(1-\varrho^2)}$ times $(1-\varrho^2)^{-\frac{\nu+3}{2}}$ times the exponential of the terms not involving θ. Summing up, we find that

$$\lim_{n\to\infty}U_n(x',\varrho) = (1-\varrho^2)^{-\frac{\nu}{2}}\pi^{-\frac{\nu}{2}}\int f(y')e^{(2\varrho(x',y')-\varrho^2(|x'|^2+|y'|^2))/(1-\varrho^2)}e^{-|y'|^2}\,dy'$$
$$= (1-\varrho^2)^{-\frac{\nu}{2}}\pi^{-\frac{\nu}{2}}\int f(y')e^{-|\varrho x'-y'|^2/(1-\varrho^2)}\,dy'$$

which is Mehler's result. Thus the Poisson integral becomes convolution with the Gaussian measure

$$\prod((1-\varrho^2)^{-\frac{1}{2}}\pi^{-\frac{1}{2}}e^{-y_j^2/(1-\varrho^2)}\,dy_j),$$

with variance $(1-\varrho^2)/2$, followed by a dilation. This is already a hint of infinite dimensional integration. The Hermite functions appear when the power series expansion with respect to ϱ is formed.

References

[AH] J. E. Avron and I. Herbst, *Spectral and scattering theory of Schrödinger operators related to the Stark effect*, Comm. Math. Phys. **52** (1977), 239–254.

[D] Jan Derezinski, *Some remarks on Weyl pseudo-differential operators*, Journées "Equations aux dérivées partielles" Saint-Jean-de Monts 1993, Exp. XII.

[H] L. Hörmander, *Symplectic classification of quadratic forms, and general Mehler formulas*, Math. Z. (to appear).

[L] E. H. Lieb, *Gaussian kernels have only Gaussian maximizers*, Inv. Math. **102** (1990), 179–208.

[M] E. Mehler, *Reihenentwicklungen nach Laplaceschen Funktionen höherer Ordnung*, J. Reine Angew. Math. **66** (1866), 161–176.

[S] I. Segal, *Transforms for operators and symplectic automorphisms over a locally compact abelian group*, Math. Scand. **13** (1963), 31–43.

[Wey] H. Weyl, *Quantenmechanik und Gruppentheorie*, Zeitschrift für Physik **46** (1927), 1–47, Collected works, Vol. III, pp. 90–135.

[Wie] N. Wiener, *Hermitian polynomials and Fourier analysis*, J. Math. and Phys. **8** (1929), 70–73, Collected works Vol. II, pp. 914–918.

GUIDE TO THE MATHEMATICAL MODELS AT THE DEPARTMENT OF MATHEMATICS IN LUND

LARS HÖRMANDER

1. Introduction. The mathematics department in Lund has a rather large collection of mathematical models dating from the end of the 19th century. For many years they were carelessly stored in crates in the air raid shelters in the basement and many show signs of this treatment.

The great activity in construction of models of surfaces in Germany seems to have been started by Kummer and by Plücker in the 1860's. Of Plücker's 27 models intended to illustrate the theory of quadratic line complexes there are 24 in Lund. According to his student Felix Klein [Kll II, p. 7], whose first mathematical work was in this area, Plücker was stimulated by Faraday who had used models to make mathematical formulas understandable to himself as a nonprofessional. Klein also commented that the Plücker series was not complete and of uneven quality, which made him produce in 1871 three models representing the main cases, two of which are still in Lund. He also made a model of the related Kummer surface for which Kummer himself had made a wire model earlier. A large number of models illustrating quadratic surfaces, the classification of cubic surfaces, analytic function theory and differential geometry were made in Germany during the following two decades. A survey of the situation in the early 1890's can be found in [Dy].

The models are obviously not important for current research; algebraic geometry, for example, is now a much more abstract and esoteric discipline than 100 years ago. However, in addition to their historical interest, they can still provide valuable intuitive understanding of basic facts on various topics such as differential geometry. In the last few years there has also been an international revival of interest in such models in spite of the development of computer graphics, or perhaps because of it. An outstanding example of this is the books by G. Fischer published by Vieweg Verlag in 1986. One volume [FP] contains beautiful photographs taken at various collections in Germany and France, the other [FC] consists of excellent mathematical comments with references written by a number of experts. These books have been very useful in the preparation of this introduction to the local collection and should be consulted for more information.

Another important source has been the volume "Ytmodellerna" in the section "Allmänt" of the library, quoted as [YTM] below. It contains catalogs from the original German producers of most of the models and comments supplied with them. The contents are as follows:

R. Diesel: Die Krümmungslinien auf den Mittelpunktsflächen zweiter Ordnung, 7 pp. (1878).

R. Diesel: Die Krümmungslinien auf dem Paraboloid, 6 pp. (1878).

E. Lange: Die vier Arten der Raumcurven dritter Ordnung, 4 pp. (1880).

W. Dyck, Die Centrafläche des einschaligen Hyperboloids, 6 pp.

© Springer International Publishing AG, part of Springer Nature 2018
L. Hörmander, *Unpublished Manuscripts*,
https://doi.org/10.1007/978-3-319-69850-2_22

L. Schleiermacher, Die Brennfläche eines Strahlensystems welche mit der Fläche der Krümmungscentra des elliptischen Paraboloids in collinearer Verwandtschaft steht, 7pp.

K. Rohn, Die geodätischen Linien auf dem Rotationsellipsoid, 4pp.

K. Rohn, Die geodätischen Linien durch die Nabelpunkte auf dem dreiaxigen Ellipsoid, 4pp.

C. Rodenberg, Modelle von Flächen dritter Ordnung, 33 pp.

J. Bacharach, Fläche 3. Ordnung mit 4 reellen, konischen Knotenpunkten nebst Haupttangentencurven, 2pp.

K. Rohn, Drei Modelle der Kummer'schen Fläche, 3 pp. (1877)

W. Dyck, Modelle zur Functionentheorie, 13+5 pp. (1886)

A. v. Braunmühl, Die geodätischen Linien der Rotationsflächen constanter mittlerer Krümmung, 5 pp. (1877)

Th. Kuen und Chr. Wolff, Darstellung der elliptischen Funktion $\varphi = am(u, k)$ durch eine Fläche, 3 pp. (1880)

J. Bacharach, Die Rotationsfläche der Tractrix mit geodätischen und Haupttangenten-Curven, 4 pp.

J. Bacharach, Rotationsfläche von constantem negativem Krümmungsmass. Kegel-Typus, 4 pp. (1877)

W. Dyck, Rotationsfläche von constantem negativen Krümmungsmass. Hyperboloid-Typus, 10 pp. (1877)

P. Vogel, Schraubenfläche von constantem negativen Krümmungsmass, 2 pp. (1880)

Th. Kuen, Ueber die auf die Kugel abwickelbaren Schrauben- und Umdrehungs-Flächen, 6 pp. (1880)

Modelle der Plücker'schen Flächen, 2 pp. (List of the models in the series)

Vier Modelle zur Theorie der Linien-Complexe zweiten Grades, 3 pp. (Catalog from Joh. Eigel Sohn, Mechanische Werkstätte, Cöln a. Rh.)

A few short papers on developable surfaces, 20+6 pp.

C. F. E. Björling, Om algebraiska rymdkurvors singulariteter och polardeveloppabelns karakterer, 26 pp. (1881). Särtryck ur Öfversigt af Kongl. Vetenskaps-Akademiens Förhandlingar 1881, No. 4

Chr. Wiener, Die Abhängigkeit der Rückkehrelemente der Projektionen einer unebenen Kurve von denen der Kurve selbst, 3pp. (1883)

Drahtgestelle zur Darstellung von Minimalflächen, 1p.

Th. Kuen, Fläche von constantem negativen Krümmungsmaass nach L. Bianchi, 8pp. (1881)

L. Brill, Catalog mathematischer Modelle, Vierte Auflage, 6+62 pp. (1888) (referred to as [Brill] below).

With the exception of the reprint by Björling, who was professor in Lund, and the short papers preceding it (also by him?), all these papers are relevant for the collection and give interesting information on the design of the models. Part I of the catalog at the end lists the models in the order of publication while part II lists them according to subject in the following groups:

I. Flächen zweiter Ordnung

II. Algebraische Flächen höherer Ordnung

III. Modelle zur Functionentheorie, transcendente Flächen

IV. Raumcurven

V. Krümmung der Flächen

VI. Modelle zur darstellenden Geometrie, Raumlehre, Physik und Mechanik

The collection in Lund contains nothing from section VI, but it is otherwise fairly extensive.

In addition there are some models made in Paris and a rather large series of models of singularities, with unknown origin possibly local at least in part. For most of the latter there are labels which describe their purpose. However, for the models from Paris there is no catalogue preserved and they are only labelled "Collection Muret".

In most models the boundaries contain parts which do not belong to the surface being modelled. In some cases this has been indicated by painting these parts in black or brown, but this is in no way systematic.

2. Quadratic surfaces. A quadratic surface in $\mathbf{R}^3$, with coordinates $x = (x_1, x_2, x_3)$, is defined by an equation $P(x) = 0$ where P is a real polynomial of degree 2, that is,

$$(2.1) \qquad P(x) = \sum_{j,k=1}^{3} a_{jk} x_j x_k + \sum_{j=1}^{3} a_{0j} x_j + a_{00},$$

where all a_{jk} are real constants. The same surface is of course defined by $cP(x)$ if c is a constant $\neq 0$. We shall now discuss the classification of the surfaces from several different points of view.

2.1. The Euclidean point of view. When we consider $\mathbf{R}^3$ as a Euclidean space, with the Euclidean metric $|x| = (\sum_1^3 x_j^2)^{\frac{1}{2}}$, then two surfaces are regarded as equal if one can be transformed to the other by an orthogonal transformation and a translation. By an orthogonal transformation we can reduce the terms of second order to diagonal form, thus with new coefficients

$$(2.2) \qquad P(x) = \sum_{j=1}^{3} (a_{jj} x_j^2 + a_{0j} x_j) + a_{00}.$$

When $a_{jj} \neq 0$ we have $a_{jj} x_j^2 + a_{0j} x_j = a_{jj}(x_j + \frac{1}{2} a_{0j}/a_{jj})^2 - \frac{1}{4} a_{0j}^2/a_{jj}$. By a translation we can therefore achieve that $a_{0j} = 0$ when $a_{jj} \neq 0$. If $a_{jj} = 0$ for more than one j we can make all but one of the corresponding a_{0j} vanish by an orthogonal transformation, and we can make a_{00} equal to 0 unless these a_{0j} vanish. The geometrical meaning of the surfaces is now clarified by distinguishing cases depending on the signs of the a_{jj}. This gives the following list of quadratic surfaces where we use the classical notation (x, y, z) instead of (x_1, x_2, x_3) for the coordinates:

$$\text{ellipsoid:} \quad \frac{x^2}{a^2} + \frac{y^2}{b^2} + \frac{z^2}{c^2} = 1, \quad a \geq b \geq c > 0,$$

$$\text{hyperboloid with one sheet:} \quad \frac{x^2}{a^2} + \frac{y^2}{b^2} - \frac{z^2}{c^2} = 1, \quad a \geq b > 0,\ c > 0,$$

$$\text{hyperboloid with two sheets:} \quad \frac{x^2}{a^2} - \frac{y^2}{b^2} - \frac{z^2}{c^2} = 1, \quad a > 0,\ b \geq c > 0,$$

$$\text{quadratic cone:} \quad \frac{x^2}{a^2} + \frac{y^2}{b^2} - z^2 = 0, \quad a \geq b > 0,$$

$$\text{elliptic paraboloid:} \quad \frac{x^2}{a^2} + \frac{y^2}{b^2} = 2z, \quad a \geq b > 0,$$

$$(2.3) \qquad \text{hyperbolic paraboloid:} \quad \frac{x^2}{a^2} - \frac{y^2}{b^2} = 2z, \quad a \geq b > 0,$$

$$\text{elliptic cylinder:} \quad \frac{x^2}{a^2} + \frac{y^2}{b^2} = 1, \quad a \geq b > 0,$$

$$\text{hyperbolic cylinder:} \quad \frac{x^2}{a^2} - \frac{y^2}{b^2} = 1, \quad a > 0,\ b > 0,$$

$$\text{parabolic cylinder:} \quad \frac{x^2}{a^2} = 2z, \quad a > 0,$$

$$\text{two intersecting planes:} \quad \frac{x^2}{a^2} - z^2 = 0, \quad a > 0,$$

$$\text{two parallel planes:} \quad \frac{x^2}{a^2} = 1, \quad a > 0,$$

$$\text{a double plane:} \quad x^2 = 0.$$

No two of the surfaces above are equivalent under Euclidean motions.

2.2. The affine point of view. The list (2.3) is very long although we have left out the cases which would only give the empty set, a point or a line. If we adopt an affine point of view and consider two surfaces as equal if they can be brought to coincide with a linear transformation and a translation, then different choices of the constants a, b, c give equivalent surfaces, so we could put $a = b = c = 1$ in (2.3). Apart from that the list is not shortened, but a direct proof would be easier in the affine case, for instead of relying on the reduction of the leading quadratic form to diagonal form by an orthogonal transformation, it suffices to use completion of squares to attain the result.

2.3. The projective point of view. The real projective n dimensional space $P_{\mathbf{R}}^n$ is defined as the set of all straight lines through the origin in $\mathbf{R}^{1+n}$. Every $x = (x_0, \ldots, x_n) \in \mathbf{R}^{1+n} \setminus \{0\}$ defines a straight line through the origin, $L_x = \{\lambda x, \lambda \in \mathbf{R}\}$. We have $L_x = L_y$ if and only if $y = \lambda x$ for some $\lambda \in \mathbf{R} \setminus \{0\}$, so $P_{\mathbf{R}}^n$ is the set of equivalence classes of points in $\mathbf{R}^{1+n} \setminus 0$ when proportional elements are regarded as equivalent. If $x_0 \neq 0$ and $L_x = L_y$, then $y_0 \neq 0$ and

$$(x_1/x_0, \ldots, x_n/x_0) = (y_1/y_0, \ldots, y_n/y_0).$$

Thus L_x can be identified with a point in $\mathbf{R}^n$. The remaining equivalence classes lie in the subspace of $\mathbf{R}^{1+n}$ where $x_0 = 0$, so they form a space $P_{\mathbf{R}}^{n-1}$, which can be thought of as a plane at infinity in $\mathbf{R}^n$: the point $(x_1, \ldots, x_n)/x_0$ tends to infinity in $\mathbf{R}^n$ in the direction $(x_1, \ldots, x_n) \neq 0$ if $x_0 \to 0$. We shall use the notation $(x_0 : x_1 : \cdots : x_n)$ in general for the point in $P_{\mathbf{R}}^n$ defined by $(x_0, x_1, \ldots, x_n) \in \mathbf{R}^{1+n} \setminus \{0\}$. One calls $(x_0, x_1, \ldots, x_n)$ *homogeneous coordinates.*

An affine hyperplane in $\mathbf{R}^n$ with equation $\sum_1^n a_j x_j + a_0 = 0$ extends to a hyperplane through the origin in $P_{\mathbf{R}}^n$ with equation $\sum_0^n a_j x_j = 0$ and conversely every such hyperplane defines a hyperplane in $\mathbf{R}^n$ unless a_0 is the only nonzero coefficient; then it defines the hyperplane at infinity. More generally, affine subspaces of $\mathbf{R}^n$ of dimension ν define subspaces of dimension $\nu + 1$ through the origin in $\mathbf{R}^{1+n}$. Conversely, such subspaces define affine subspaces of $\mathbf{R}^n$ unless they are contained in the plane at infinity.

A linear transformation in $\mathbf{R}^{1+n}$ induces a transformation in $\mathbf{R}^n$ of the form

$$x = (x_1, \ldots, x_n) \mapsto (L_1(x)/L_0(x), \ldots, L_n(x)/L_0(x))$$

where L_j are linearly independent affine linear functions. When L_0 is a constant, these are just the affine transformations used above, but otherwise they are not defined everywhere with values in $\mathbf{R}^n$, for $\{x \in \mathbf{R}^n, L_0(x) = 0\}$ is mapped to the plane at infinity.

A quadratic surface in $\mathbf{R}^3$ defined by $P(x_1, x_2, x_3) = 0$ is similarly extended to a quadratic cone in $\mathbf{R}^4$ with equation $x_0^2 P(x_1/x_0, x_2/x_0, x_3/x_0) = 0$. Thus equation (2.1) is changed to

$$(2.1)' \qquad\qquad \sum_{j,k=0}^{3} a_{jk} x_j x_k = 0.$$

By completion of squares it follows that the quadratic form can be brought to diagonal form $\sum_0^4 d_j x_j^2 = 0$ where $d_j = \pm 1$ or 0 by a linear transformation of the homogeneous coordinates. Changing the sign does not affect the surface, so by a projective transformation, that is, a linear transformation in $\mathbf{R}^4$ the equation can be reduced to one in the following list

$$\text{(2.3a)} \qquad\qquad x_1^2 + x_2^2 + x_3^2 - x_0^2 = 0,$$

$$\text{(2.3b)} \qquad\qquad x_1^2 + x_2^2 - x_3^2 - x_0^2 = 0,$$

$$\text{(2.3c)} \qquad\qquad x_1^2 + x_2^2 - x_0^2 = 0,$$

$$\text{(2.3d)} \qquad\qquad x_1^2 - x_0^2 = 0,$$

$$\text{(2.3e)} \qquad\qquad x_1^2 = 0,$$

where we have omitted cases defining a set contained in a line in $P_{\mathbf{R}}^3$. Here

(2.3a) covers ellipsoids, elliptic paraboloids and hyperboloids with two sheets,
(2.3b) covers hyperboloids with one sheet and hyperbolic paraboloids,
(2.3c) covers quadratic cones as well as elliptic, parabolic and hyperbolic cylinders,
(2.3d) covers parallel and intersecting planes,
(2.3e) is a double plane.

Thus 12 cases have been reduced to 5.

The equation (2.3b) can be written

$$(x_1 - x_3)(x_1 + x_3) = (x_0 - x_2)(x_0 + x_2),$$

so it is fulfilled in the two dimensional planes

$$x_1 - x_3 = \mu(x_0 - x_2), \quad x_1 + x_3 = \mu^{-1}(x_0 + x_2), \quad \text{and}$$
$$x_1 - x_3 = \nu(x_0 + x_2), \quad x_1 + x_3 = \nu^{-1}(x_0 - x_2),$$

where μ and ν are constants $\neq 0$. Two planes in the same family intersect only at the origin whereas two planes in different families intersect in the line $(\mu + \nu : \mu\nu + 1 : \mu - \nu : 1 - \mu\nu)$. This means that the hyperboloid of one sheet in (2.3) is generated by the lines

$$\frac{x}{a} - \frac{z}{c} = \mu\left(1 - \frac{y}{b}\right), \quad \frac{x}{a} + \frac{z}{c} = \mu^{-1}\left(1 + \frac{y}{b}\right), \quad \text{and}$$
$$\frac{x}{a} - \frac{z}{c} = \nu\left(1 + \frac{y}{b}\right), \quad \frac{x}{a} + \frac{z}{c} = \nu^{-1}\left(1 - \frac{y}{b}\right),$$

with lines in the same family skew but meeting those in the other family. Starting from three mutually skew lines the surface is generated by the lines intersecting all three unless their points at infinity are on a straight line. For the hyperbolic paraboloid in (2.3) the corresponding generators are

$$\frac{x}{a} + \frac{y}{b} = 2\mu, \quad \frac{x}{a} - \frac{y}{b} = \frac{z}{\mu}, \quad \text{and}$$
$$\frac{x}{a} - \frac{y}{b} = 2\nu, \quad \frac{x}{a} + \frac{y}{b} = \frac{z}{\nu}$$

The points at infinity are on the lines $x/a + y/b = 0$ and $x/a - y/b = 0$ respectively.

From an algebraic point of view the preceding discussion can equally well be carried out when the coefficients are complex. The equivalence classes of the corresponding surfaces in the complex projective space $P_{\mathbf{C}}^3$ are then determined by the rank alone, so we can drop (2.3a) or (2.3b) from the list above and change all minus signs to plus signs if we prefer that.

2.4. The cubic space curves. If $p_0(t), \ldots, p_n(t)$ is a basis for the homogeneous polynomials in $t = (t_1, t_2)$ of degree n then

$$\mathbf{R}^2 \setminus \{0\} \ni t \mapsto p(t) = (p_0(t) : p_1(t) : \cdots : p_n(t)) \in P_{\mathbf{R}}^n$$

is a curve not contained in any projective subspace. Since $t_1^{n-j} t_2^j = \sum_{k=0}^n a_{jk} p_k(t)$ with an invertible matrix (a_{jk}), all such curves are projectively equivalent. However, we shall now discuss the curves from an affine point of view when $n = 2$ or $n = 3$.

When $n = 2$ the curve $t \mapsto (t_1^2 : t_1 t_2 : t_2^2) = (x_0 : x_1 : x_2)$ has the equation $x_1^2 = x_0 x_2$ so it is a non-singular conic. Thus $t \mapsto (p_0(t) : p_1(t) : p_2(t))$ is an ellipse if $p_0(t) \neq 0$ for real $t \neq 0$, a hyperbola if $p_0(t)$ vanishes on two different real lines, and a parabola if $p_0(t)$ vanishes of second order on a real line.

Now let $n = 3$. The curve $\Gamma : t \mapsto p(t) = (p_0(t) : p_1(t) : p_2(t) : p_3(t)) \in P_{\mathbf{R}}^3$ is then called a *cubic space curve*. From an affine point of view the curve is determined by $p_0(t)$, and it does not change if one makes a linear transformation of the parameters t. To study the curve Γ in $\mathbf{R}^3$ we shall examine the projection in the direction of the x_3 axis on the $x_1 x_2$ plane, assuming that the point $(0 : 0 : 0 : 1)$ at infinity on the x_3 axis is on Γ. We may assume that it is attained when $t = (0, 1)$, so then we have

$$p_j(t) = t_1 q_j(t), \quad j = 0, 1, 2; \quad p_3(t) = t_1 q_3(t) + t_2^3$$

where q_j, $j = 0, \ldots, 3$, are quadratic forms and q_0, q_1, q_2 are linearly independent. The projection is

$$\{(p_1(t)/p_0(t), p_2(t)/p_0(t)); p_0(t) \neq 0\} = \{(q_1(t)/q_0(t), q_2(t)/q_0(t)); q_0(t) \neq 0, t_1 \neq 0\},$$

so it follows from the two dimensional case that it is an ellipse if $q_0(t) \neq 0$ for $t \neq 0$, a hyperbola if $q_0(t)$ vanishes on two different lines, and a parabola if $q_0(t)$ is the square of a linear form. Thus the following four cases can occur:

a) If p_0 has a simple real zero τ and no real zeros $\notin \mathbf{R}\tau$, then Γ lies on an elliptic cylinder with axis in the direction $p(\tau)$. Then Γ is called a *cubic ellipse*.

b) If p_0 has three real zeros τ^1, τ^2, τ^3 with different directions then Γ lies on a hyperbolic cylinder with axis direction $p(\tau^j)$ for $j = 1, 2, 3$. Then Γ is called a *cubic hyperbola*.

c) If p_0 has a simple zero τ^1 and a double zero τ^2, then Γ lies on a hyperbolic cylinder with axis direction $p(\tau^2)$ and also on a parabolic cylinder with axis direction $p(\tau^1)$. Then Γ is called a *cubic hyperbolic parabola*.

d) If p_0 has a triple zero $\tau \neq 0$, then Γ lies on a parabolic cylinder with axis $p(\tau)$ and is called a *cubic parabola*.

In view of the projective equivalence of all cubic space curves, such a curve Γ lies on a quadratic cone with vertex at any point of it. We shall now examine the projection on a plane from a point outside the curve. Clearly it has to be a cubic curve since it intersects any line in the plane at three points at most. However, it is not an arbitrary cubic curve.

First we shall prove that through every point outside Γ there is exactly one line which intersects Γ in two points, which may coincide so that the line is a tangent. In the proof it is convenient to take Γ in the standard form

$$\mathbf{R}^2 \setminus \{0\} \ni t = (t_1, t_2) \mapsto (x_0 : x_1 : x_2 : x_3), \quad x_j = t_1^{3-j} t_2^j.$$

The curve is also defined by

(2.4) $$x_1^2 = x_0 x_2, \quad x_2^2 = x_1 x_3, \quad x_0 x_3 = x_1 x_2;$$

the third equation is a consequence of the first two unless $x_1 x_2 = 0$. Assume now that $a = (a_0 : a_1 : a_2 : a_3) \notin \Gamma$. Points on the line from a to a general point on Γ have the homogeneous coordinates $\tau_1^{3-j} \tau_2^j + s a_j$ where $s \in \mathbf{R}$, and at the intersections with Γ we have by (2.4)

$$s\big((a_0 a_2 - a_1^2)s + \tau_1^3 a_2 + \tau_1 \tau_2^2 a_0 - 2\tau_1^2 \tau_2 a_1\big) = 0,$$
$$s\big((a_1 a_3 - a_2^2)s + \tau_1^2 \tau_2 a_3 + \tau_2^3 a_1 - 2\tau_1 \tau_2^2 a_2\big) = 0$$
$$s\big((a_0 a_3 - a_1 a_2)s + \tau_1^3 a_3 + \tau_2^3 a_0 - \tau_1^2 \tau_2 a_2 - \tau_1 \tau_2^2 a_1\big) = 0.$$

A second intersection requires that the parentheses are proportional, which gives

$$(\tau_1^3 a_2 + \tau_1 \tau_2^2 a_0 - 2\tau_1^2 \tau_2 a_1)(a_1 a_3 - a_2^2) - (\tau_2^3 a_1 + \tau_1^2 \tau_2 a_3 - 2\tau_1 \tau_2^2 a_2)(a_0 a_2 - a_1^2)$$
$$= (\tau_1 a_2 - \tau_2 a_1)(\tau_2^2(a_0 a_2 - a_1^2) + \tau_1 \tau_2(a_2 a_1 - a_0 a_3) + \tau_1^2(a_1 a_3 - a_2^2))$$

when we eliminate between the first two equations. For the other eliminations the first order factor is replaced by $a_1 \tau_1 - a_0 \tau_2$ and $a_2 \tau_2 - a_3 \tau_1$. They cannot all vanish since a defines a point outside Γ so we must have

$$\tau_2^2(a_0 a_2 - a_1^2) + \tau_1 \tau_2(a_2 a_1 - a_0 a_3) + \tau_1^2(a_1 a_3 - a_2^2) = 0$$

which gives the two roots defining the chord through the given point, the tangent in case the roots are equal, that is,

$$a_0^2 a_3^2 - 3a_1^2 a_2^2 - 6a_0 a_1 a_2 a_3 + 4a_0 a_2^3 + 4a_3 a_1^3 = 0.$$

Now consider a general cubic space curve

$$\Gamma : t \mapsto (p_0(t) : p_1(t) : p_2(t) : p_3(t))$$

such that the point at infinity e_3 on the x_3 axis is not on Γ. Thus $(p_0(t), p_1(t), p_2(t)) \neq 0$ when $0 \neq t \in \mathbf{R}^2$, so this defines a curve $\widetilde{\Gamma}$ in $P_{\mathbf{R}}^2$.

a) If there is a chord of Γ passing through e_3 then $\widetilde{\Gamma}$ will have a self intersection. We can choose the parametrization so that it occurs for $t = (1, 0)$ and $t = (0, 1)$ and choose the affine coordinates so that the intersection is at the origin. Then $p_1(t)$ and $p_2(t)$ are divisible by t_1 and by t_2, and since they are linearly independent we achieve by another affine change of coordinates that

$$p_1(t) = t_1^2 t_2, \quad p_2(t) = t_1 t_2^2, \quad p_0(1, 0) \neq 0, \ p_0(0, 1) \neq 0.$$

With $x_j = p_j(t)$ we have $t_1^3 = x_1^2 / x_2$ and $t_2^3 = x_2^2 / x_1$, hence

$$x_0 = \sum_0^3 c_j t_1^{3-j} t_2^j = c_0 x_1^2 / x_2 + c_1 x_1 + c_2 x_2 + c_3 x_2^2 / x_1, \quad c_0 c_3 \neq 0.$$

Taking $x_0 - c_1 x_1 - c_2 x_2$ and suitable multiples of x_1 and x_2 as new coordinates we obtain the projectively equivalent equation

$$x_0 x_1 x_2 = x_1^3 + x_2^3,$$

that is, $x_1 x_2 = x_1^3 + x_2^3$ in inhomogeneous coordinates. This is called a *nodal curve*.

b) If there is a tangent to Γ through e_3 we may assume that it is at the point with $t_1 = 0$ and that $p_1 = p_2 = 0$ there. Then we have after an affine change of coordinates

$$p_1(t) = t_1^2 t_2, \quad p_2(t) = t_1^3, \quad p_0(0,1) \neq 0,$$

for p_1 and p_2 must vanish of second order at $(0,1)$. Set $x_j = p_j(t)$. Then

$$x_0 = \sum_0^3 c_j t_1^{3-j} t_2^j = c_0 x_2 + c_1 x_1 + c_2 x_1^2/x_2 + c_3 x_1^3/x_2^2, \quad c_3 \neq 0,$$

which can be written

$$(x_0 - c_0 x_2 - c_1 x_1) x_2^2 = c_2 x_1^2 x_2 + c_3 x_1^3 = c_3 (x_1 + c_2 x_2/3c_3)^3 - x_2^2 (x_1 c_2^2/3c_3 + x_2 c_2^3/27c_3^2).$$

The curve is therefore projectively equivalent to the *cusp*

$$x_0 x_2^2 = x_1^3.$$

The nonsingular cubic curves, which can be reduced to the form

(2.5) $$x_0 x_2^2 = 4x_1^3 - g_2 x_1 - g_3$$

and depend effectively on one parameter g_3^2/g_2^3, cannot occur as projection precisely because the chords and tangents of a cubic space curve cover the whole complement.

The models in the collection

Q1. Ellipsoid of revolution with three geodesic lines in colors [Brill #102, p. 53]?

Q2. General ellipsoid with some plane (elliptic) sections through the center marked with threads.

Q3. General ellipsoid with lines of curvature marked [Brill #95, p.52].

Q4. General ellipsoid with lines of curvature marked [Brill #96, p. 52].

Q5. General plain ellipsoid [Collection Muret].

Q6. Elliptic paraboloid with lines of curvature marked [Brill #100, p. 53]?

Q7. One half of a quadratic cone with generators and lines of curvature (= geodesic circles centered at the vertex) [Brill #98, p. 53].

Q8. Hyperbolic paraboloid with horizontal plane intersections [Brill #24, p. 39].

Q9. Hyperbolic paraboloid with generating lines [Brill #25, p. 39].

Q10. Another model of a hyperbolic paraboloid with generating lines.

Q11. Hyperboloid with one sheet with lines of curvature marked [Brill 97, p. 53].

Q12. Wire model of a hyperboloid with one sheet.

Q13. Elliptic cylinder with a cubic ellipse marked [Brill #90a, P. 51].

Q14. Hyperbolic cylinder with a cubic hyperbola marked [Brill #90b, p. 51]. Note the other two directions at infinity.

Q15. Parabolic cylinder with a cubic hyperbolic parabola marked [Brill #90c, p.51]. The other point at infinity is the axis of the parabola and one sees the hyperbolic cylinder containing the curve by looking at the model from above.

3. Local classification of surfaces. A hypersurface in $\mathbf{R}^n$ passing through the origin is defined by an equation $f(x) = 0$, $x = (x_1, \ldots, x_n)$, where $f(0) = 0$. (In a moment n will become equal to 3.) Here f is assumed real valued and smooth (or sometimes even analytic) near 0.

3.1. The Euclidean point of view. From a Euclidean point of view two surfaces are equal if there is an orthogonal transformation of the coordinates mapping one to the other. Assume that $f'(0) \neq 0$, that is, that $\partial f(0)/\partial x_j \neq 0$ for some j. Then we can make an orthogonal transformation such that $-\langle x, f'(0)\rangle/|f'(0)|$ becomes the last coordinate. In the new coordinates the equation $f(x) = 0$ can be written in the form

$$g(x) - x_n = 0$$

where $g'(0) = 0$. By the implicit function theorem the equation $g(x) - x_n = 0$ can be solved for x_n in a neighborhood of the origin as

$$x_n = \varphi(x'), \quad x' = (x_1, \ldots, x_{n-1}),$$

where $\varphi(x') = O(|x'|^2)$ as $x' \to 0$; we have $\varphi(x') = g(x', 0) + O(|x'|^3)$. By an orthogonal transformation of the x' variables we can diagonalize the quadratic form which starts the Taylor expansion of φ, and then we have

(3.1) $$\varphi(x') = \tfrac{1}{2} \sum_{1}^{n-1} \kappa_j x_j^2 + O(|x'|^3), \quad x' \to 0.$$

The numbers κ_j are the *principal curvatures* at 0 of the surface, and the unit vectors along the x_j axes are called *principal directions of curvature* at 0, for $j = 1, \ldots, n-1$. Note that the unit normal of the surface is $(\varphi'(x'), -1)/\sqrt{1 + |\varphi'(x')|^2}$. At 0 the derivative of the normal in the tangential direction $(t', 0)$ has the components

$$\sum_{k=1}^{n-1} \partial^2 \varphi/\partial x_j \partial x_k t_k = \kappa_j t_j, \quad j = 1, \ldots n-1, \quad 0, \text{ for } j = n,$$

which is in the direction $(t', 0)$ precisely when $(t', 0)$ is a direction of principal curvature. (If the normal is not chosen of unit length this remains true for the tangential component of the derivative of the normal.)

The principal curvatures are uniquely defined apart from a permutation and so are the principal directions of curvature if all the principal curvatures are different. (The surface has been oriented with normal in the direction $-f'(0)$. Changing the sign changes the signs of all principal curvatures.) The symmetric functions are always unambiguously defined, such as

the total curvature: $$\prod_{j=1}^{n-1} \kappa_j = \det \left(\partial^2 \varphi(0)/\partial x_j \partial x_k \right)_{j,k=1}^{n-1},$$

the mean curvature: $$\frac{1}{n-1} \sum_{j=1}^{n-1} \kappa_j = \frac{1}{n-1} \operatorname{Tr} \varphi''(0) = \frac{1}{n-1} \sum_{j=1}^{n-1} \partial^2 \varphi(0)/\partial x_j^2.$$

The total curvature is independent of the orientation when n is odd. In these formulas the quadratic part of φ need not be diagonalized, and we may even replace φ by g.

When $n = 3$ the surface thus behaves infinitesimally as an elliptic or hyperbolic paraboloid, a parabolic cylinder or a plane defined by

$$2x_3 = \sum_1^2 \kappa_j x_j^2.$$

The intersection with a plane $x_1 = r\cos\theta$, $x_2 = r\sin\theta$, through the x_3 axis has the equation $2x_3 = r^2(\kappa_1 \cos^2\theta + \kappa_2 \sin^2\theta)$, the curvature at 0 is $\kappa_1 \cos^2\theta + \kappa_2 \sin^2\theta$, and the circle of curvature there is given by $2x_3 = (r^2 + x_3^2)(\kappa_1 \cos^2\theta + \kappa_2 \sin^2\theta)$. Hence the circles of curvature generate the surface

$$2x_3(x_1^2 + x_2^2) = (x_1^2 + x_2^2 + x_3^2)(\kappa_1 x_1^2 + \kappa_2 x_2^2),$$

except the part of the x_3 axis where $x_3 \notin 2/[\kappa_1, \kappa_2]$.

We shall return to a discussion of curvatures in the section on differential geometry, but already now we shall explain how the lines of curvature in the models of quadratic surfaces were obtained.

Consider in $\mathbf{R}^3$ a surface of the form $\sum_1^3 x_j^2/A_j = 1$ with $A_1 < A_2 < A_3$. It is embedded in the *confocal* family of surfaces $\sum_1^3 x_j^2/(A_j - \lambda) = 1$, which are ellipsoids if $\lambda < A_1$, hyperboloids with one sheet if $A_1 < \lambda < A_2$ and hyperboloids with two sheets if $A_2 < \lambda < A_3$. For arbitrary x with $x_1 x_2 x_3 \neq 0$ the cubic equation $\sum x_j^2/(A_j - \lambda) = 1$ has one root $\lambda_1(x) \in (-\infty, A_1)$, one root $\lambda_2(x) \in (A_1, A_2)$ and one root $\lambda_3(x) \in (A_2, A_3)$. The differential of λ_ν is given by

$$\sum_1^3 2x_j\, dx_j/(A_j - \lambda_\nu(x)) + d\lambda_\nu(x) \sum_1^3 x_j^2/(A_j - \lambda_\nu(x))^2 = 0$$

so $\lambda_\nu'(x)$ has the direction $(x_1/(A_1 - \lambda_\nu(x)), x_2/(A_2 - \lambda_\nu(x)), x_3/(A_3 - \lambda_\nu(x)))$. Since

$$0 = \sum_1^3 \frac{x_j^2}{A_j - \lambda_\nu(x)} - \sum_1^3 \frac{x_j^2}{A_j - \lambda_\mu(x)} = (\lambda_\nu(x) - \lambda_\mu(x)) \sum_1^3 \frac{x_j^2}{(A_j - \lambda_\nu(x))(A_j - \lambda_\mu(x))}$$

it follows that $\lambda_1'(x), \lambda_2'(x), \lambda_3'(x)$ are mutually orthogonal and not 0. For the local inverse $\lambda \mapsto f(\lambda)$ of $x \mapsto \lambda(x)$ it follows that $\partial f/\partial\lambda_1, \partial f/\partial\lambda_2, \partial f/\partial\lambda_3$ are proportional to λ_1', λ_2', λ_3', hence mutually orthogonal. Differentiation of an equation $(\partial f/\partial\lambda_j, \partial f/\partial\lambda_k) = 0$, $j \neq k$, gives $(\partial^2 f/\partial\lambda_j\partial\lambda_l, \partial f/\partial\lambda_k) + (\partial f/\partial\lambda_j, \partial^2 f/\partial\lambda_k\partial\lambda_l) = 0$, hence

$$(\partial^2 f/\partial\lambda_1\partial\lambda_2, \partial f/\partial\lambda_3) = -(\partial f/\partial\lambda_1, \partial^2 f/\partial\lambda_2\partial\lambda_3)$$
$$= (\partial^2 f/\partial\lambda_3\partial\lambda_1, \partial f/\partial\lambda_2) = -(\partial f/\partial\lambda_3, \partial^2 f/\partial\lambda_1\partial\lambda_2)$$

so $\partial^2 f/\partial\lambda_1\partial\lambda_2$ is orthogonal to $\partial f/\partial\lambda_3$, hence a linear combination of $\partial f/\partial\lambda_1$ and $\partial f/\partial\lambda_2$. This means that on a surface where λ_1 is constant (thus $\partial f/\partial\lambda_1$ is a normal) the lines where λ_3 is constant (thus $\partial f/\partial\lambda_2$ are tangents) are lines of curvature. (Note that this argument applies to any "triply orthogonal" system and its analogue in higher dimensions.) In particular, the lines of curvature on a quadratic surface with center are the intersections with the confocal quadratic surfaces. Their projections on the coordinate planes are ellipses or hyperbolas.

An elliptic or hyperbolic paraboloid $\sum_1^2 x_j^2/A_j = 2Cx_3$ is the limit of the surfaces

$$\sum_1^2 x_j^2/A_j + (x_3 - CA_3)^2/A_3 = C^2 A_3$$

with center $(0, 0, CA_3)$ as $A_3 \to \infty$. The corresponding family of confocal surfaces

$$\sum_1^2 x_j^2/(A_j - \lambda) + (x_3 - CA_3)^2/(A_3 - \lambda) = C^2 A_3$$

converges to the family $\sum_1^2 x_j^2/(A_j - \lambda) = 2Cx_3 - C^2\lambda$ as $A_3 \to \infty$. Hence these surfaces intersect along lines of curvature. The projections on the $x_1 x_2$ plane are ellipses or hyperbolas but the projections on the $x_1 x_3$ or $x_2 x_3$ planes are parabolas.

3.2. The affine point of view. If we allow not only orthogonal transformations but arbitrary linear transformations, then the numbers κ_j can be reduced to ± 1 or 0. When $n = 3$ one calls the origin a *saddle point* if the two curvatures have opposite signs. If they have the same sign the point is *convex* or *concave* depending on the orientation of the normal.

3.3. The differentiable point of view. We shall now discuss equivalence of surfaces when one allows arbitrary smooth changes of coordinates preserving the origin. If $f'(0) \neq 0$, that is, $\partial f(0)/\partial x_j \neq 0$ for some j, then one can introduce new local coordinates $\xi_k = x_k$ when $k \neq j$, $\xi_j = f(x)$ which makes the surface defined by $\xi_j = 0$. Such non-singular surfaces can therefore be transformed to a plane.

If $f(0) = f'(0) = 0$, then Taylor's formula gives $f(x) = A(x) + O(|x|^3)$ where A is a quadratic form. If A is non-degenerate then the *Morse lemma* states that one can choose new local coordinates ξ such that $f(x) = \sum_1^n \varepsilon_j \xi_j^2$ where $\varepsilon_j = \pm 1$. The proof is simple: By a linear change of x variables we can make $A(x) = \sum_1^n \varepsilon_j x_j^2$ with $\varepsilon_j = \pm 1$. Using Taylor's formula we can then write $f(x) = \sum_{j,k=1}^n f_{jk}(x) x_j x_k$, where $f_{jk} = f_{kj}$ is smooth and $f_{jj}(0) = \varepsilon_j$, $f_{jk}(0) = 0$ when $j \neq k$. One can then argue by completion of squares as in the case where f is exactly equal to a quadratic form. Thus $f(x) = 0$ is equivalent to $\sum_1^n \varepsilon_j \xi_j^2 = 0$, which is a quadratic cone (possibly reduced to the origin). If A is degenerate, of rank $k < n$, one can achieve by a linear transformation of variables that A is non-degenerate in $x' = (x_1, \ldots, x_k)$ when $x'' = (x_{k+1}, \ldots, x_n) = 0$. By the implicit function theorem the equations $\partial f(x)/\partial x_j = 0$, $j = 1, \ldots, k$, define x' as a function $\psi(x'')$, and the Morse lemma (with parameters now) shows that $f(x' + \psi(x''), x'') = f(\psi(x''), x'') + \sum_1^k \varepsilon_j \xi_j^2$ where $\xi_1, \ldots, \xi_k$ are new local coordinates replacing $x_1, \ldots, x_k$.

Let us now specialize to the case $n = 3$ and denote the coordinates by (x, y, z). Then A has rank 0, 1, 2 or 3. If the rank is 3 then the surface defined by $f = 0$ has an isolated point or a conic point at 0, with the classical notation C_2 and modern notation A_1. Assume now that the rank is 2. We have then seen that there are new local coordinates (ξ, η, ζ) such that the surface is defined by $\xi^2 \pm \eta^2 + g(\zeta) = 0$ where $g(\zeta) = O(\zeta^3)$. If $g(\zeta)$ vanishes of precisely order m at the origin, then $(\pm g(\zeta))^{1/m}$ is a smooth function (by Taylor's formula), and replacing ζ by this new function we find that the surface is defined by

$$\xi^2 \pm \eta^2 \pm \zeta^m = 0$$

where $m \geq 3$. In the classical literature this singularity is called a *biplanar double point* of type B_m; the current notation is A_{m-1}. The term *biplanar* refers to the fact that the equation $\xi^2 \pm \eta^2 = 0$ (or $A(x, y, z) = 0$ in the original coordinates) defines two distinct planes, $\xi = \pm\eta$ for the lower sign; for the upper sign they define the complex planes $\xi = \pm i\eta$. If f is analytic, that is, can be expanded in a convergent power series at the origin, then g is analytic so if g vanishes of infinite order at 0 then g is identically 0. Thus the origin is not an isolated singularity then.

If A has rank 1 there are new local coordinates (ξ, η, ζ) such that the surface is defined by

$$\xi^2 + g(\eta, \zeta) = 0, \quad g(\eta, \zeta) = g_3(\eta, \zeta) + O(|\eta| + |\zeta|)^4,$$

where g_3 is homogeneous of degree 3. Assume that g is not identically 0. After a linear transformation of the variables η, ζ we have one of the following cases:

$$g_3(\eta, \zeta) = \begin{cases} \zeta(\eta^2 \pm \zeta^2) \\ \zeta\eta^2 \\ \eta^3. \end{cases}$$

Since a real cubic equation has either three distinct real roots, one real root and two complex conjugate ones, one simple and one double real root, or a triple real root, this is obvious except when all roots are simple. With one placed where $\zeta = 0$ we have then

$$g_3(\eta, \zeta) = \zeta(a\eta^2 + 2b\eta\zeta + c\zeta^2);$$

$a \neq 0$ and $b^2 - ac \neq 0$ since the roots are simple. Taking $\eta + b\zeta/a$ and ζ as new variables we may assume that $b = 0$, and since $ac \neq 0$ a suitable dilation of the variables η and ζ then reduce to the form $\zeta(\eta^2 \pm \zeta^2)$.

a) Let $g_3(\eta, \zeta) = \zeta(\eta^2 \pm \zeta^2) = G(\eta, \zeta)$ and set $g(\eta, \zeta) = g_3(\eta, \zeta) + g_4(\eta, \zeta) + O(|\eta| + |\zeta|)^5$ where g_4 is homogeneous of degree 4. If $Y(\eta, \zeta)$ and $Z(\eta, \zeta)$ are smooth and vanish of second order at 0, then $(\eta, \zeta) \mapsto (\eta - Y(\eta, \zeta), \zeta - Z(\eta, \zeta))$ is a diffeomorphism near the origin, so changing local coordinates we can replace $g(\eta, \zeta)$ by

$$g(\eta - Y(\eta, \zeta), \zeta - Z(\eta, \zeta))$$
$$= g_3(\eta, \zeta) - Y(\eta, \zeta)\partial G(\eta, \zeta)/\partial\eta - Z(\eta, \zeta)\partial G(\eta, \zeta)/\partial\zeta + g_4(\eta, \zeta) + O(|\eta| + |\zeta|)^5.$$

Now

$$\partial G(\eta, \zeta)/\partial\eta = 2\zeta\eta, \quad \partial G(\eta, \zeta)/\partial\zeta = \eta^2 \pm 3\zeta^2, \quad \text{hence}$$
$$\eta^4 = \eta^2 \partial G(\eta, \zeta)/\partial\zeta \mp \tfrac{3}{2}\zeta\eta\partial G(\eta, \zeta)/\partial\eta, \quad \zeta^4 = \pm\tfrac{1}{3}\zeta^2 \partial G(\eta, \zeta)/\partial\zeta \mp \tfrac{1}{6}\zeta\eta\partial G(\eta, \zeta)/\partial\eta,$$

and $\eta^\mu\zeta^\nu = \tfrac{1}{2}\eta^{\mu-1}\zeta^{\nu-1}\partial G(\eta, \zeta)/\partial\eta$ if $\mu > 0$, $\nu > 0$. We can therefore choose $Y(\eta, \zeta)$ and $Z(\eta, \zeta)$ vanishing of second order at the origin so that $g(\eta + Y(\eta, \zeta), \zeta + Z(\eta, \zeta)) = g_3(\eta, \zeta) + O(|\eta| + |\zeta|)^5$. The argument can then be repeated so that terms of order 5 are eliminated, and by repeating the argument we obtain for any N new local coordinates in which $g(\eta, \zeta) = g_3(\eta, \zeta) + O(|\eta| + |\zeta|)^N$. General results (see Tougeron [T], Mather [M]) prove that the remainder term can be completely removed. Thus there are new local coordinates such that

(3.2) $$f = \xi^2 + \zeta(\eta^2 \pm \zeta^2).$$

In the classical literature the singularity (3.2) is called a *uniplanar double point* and it is denoted by U_6; in current terminology it is denoted D_4 (see Arnol'd [A]).

b) Now let $g_3(\eta, \zeta) = \zeta \eta^2$. Then

$$\partial g_3(\eta, \zeta)/\partial \eta = 2\zeta\eta, \quad \partial g_3(\eta, \zeta)/\partial \zeta = \eta^2.$$

A polynomial g_4 homogeneous of degree 4 can be written in the form $Y(\eta, \zeta)\partial g_3(\eta, \zeta)/\partial \eta + Z(\eta, \zeta)\partial g_3(\eta, \zeta)/\partial \eta$ if and only if $g_4(0, \zeta) \equiv 0$, that is, there is no ζ^4 term. In that case we can eliminate the terms of degree 4 as in part a) and continue to eliminate the terms of degree 5, ... until for some $k \geq 4$ we have with new coordinates

$$g(\eta, \zeta) = g_3(\eta, \zeta) + c\zeta^k + O(|\zeta| + |\eta|)^{k+1}$$

where $c \neq 0$. Then we set $G(\eta, \zeta) = g_3(\eta, \zeta) + c\zeta^k$ and note that

$$\partial G(\eta, \zeta)/\partial \eta = 2\zeta\eta, \quad \partial G(\eta, \zeta)/\partial \zeta = \eta^2 + ck\zeta^{k-1}.$$

If $\nu > k \geq 4$ then

$$\eta^\nu = \eta^{\nu-2}\partial G(\eta, \zeta)/\partial \zeta - \tfrac{1}{2}ck\eta^{\nu-3}\zeta^{k-2}\partial G(\eta, \zeta)/\partial \eta,$$

$$ck\zeta^\nu = \zeta^{\nu+1-k}\partial G(\eta, \zeta)/\partial \zeta - \tfrac{1}{2}\eta\zeta^{\nu-k}\partial G(\eta, \zeta)/\partial \eta,$$

so every homogeneous polynomial of degree $> k$ can be written in the form

$$Y(\eta, \zeta)\partial G(\eta, \zeta)/\partial \eta + Z(\eta, \zeta)\partial G(\eta, \zeta)/\partial \zeta$$

with $Y(\eta, \zeta)$ and $Z(\eta, \zeta)$ vanishing of second order at the origin. As in part a) we can therefore successively eliminate terms in the Taylor expansion of order $k + 1$, $k + 2$, ... After a change of scales for ζ and η it follows that for suitable local coordinates

(3.3) $$f = \xi^2 + \zeta(\eta^2 \pm \zeta^{k-1}).$$

When $k = 4$ this is classically known as the uniplanar double point U_7; in modern notation it is D_{k+1} for any $k \geq 3$ (with (3.2) as a special case).

c) Now let $g_3(\eta, \zeta) = \eta^3$. Then

$$\partial g_3(\eta, \zeta)/\partial \eta = 3\eta^2, \quad \partial g_3(\eta, \zeta)/\partial \zeta = 0.$$

Set $g_4(\eta, \zeta) = \sum_0^4 a_j \eta^j \zeta^{4-j}$.

(i) If $a_0 \neq 0$ then $g_4(\eta, \zeta - a_1\eta/4a_0) = a_0\zeta^4 + Y(\eta, \zeta)\partial g_3(\eta, \zeta)/\partial \eta$ where Y is a quadratic form. Hence

$$g(\eta - Y(\eta, \zeta), \zeta - a_1\eta/4a_0) = G(\eta, \zeta) + O(|\eta| + |\zeta|)^5, \quad G(\eta, \zeta) = \eta^3 + a_0\zeta^4.$$

Since $\partial G(\eta, \zeta)/\partial \eta = 3\eta^2$ and $\partial G(\eta, \zeta)/\partial \zeta = 4a_0\zeta^3$ it is clear that every monomial $\eta^\mu \zeta^\nu$ with $\mu + \nu \geq 5$ can be written $Y(\eta, \zeta)\partial G(\eta, \zeta)/\partial \eta + Z(\eta, \zeta)\partial G(\eta, \zeta)/\partial \zeta$ with $Y(\eta, \zeta)$ and $Z(\eta, \zeta)$ vanishing of second order at the origin. As before we can then successively remove the terms of order 5, 6, ... For suitable local coordinates we therefore have

$$f = \xi^2 + \eta^3 \pm \zeta^4.$$

This is classically known as the uniplanar double point U_8; the modern notation is E_6.

(ii) If $a_0 = 0$ but $a_1 \neq 0$ then $g_4(\eta, \zeta) = a_1 \zeta^3 \eta + Y(\eta, \zeta) \partial g_3(\eta, \zeta)/\partial \eta$, and we obtain

$$g(\eta - Y(\eta, \zeta), \zeta) = G(\eta, \zeta) + O(|\eta| + |\zeta|)^5, \quad G(\eta, \zeta) = \eta^3 + a_1 \zeta^3 \eta.$$

Introducing $a_1^{\frac{1}{3}} \zeta$ as new variable instead of ζ we may assume that $a_1 = 1$. Then

$$\partial G(\eta, \zeta)/\partial \eta = 3\eta^2 + \zeta^3, \quad \partial G(\eta, \zeta)/\partial \zeta = 3\zeta^2 \eta,$$

and we obtain

$$\eta^5 = \tfrac{1}{3}\eta^3 \partial G(\eta, \zeta)/\partial \eta - \tfrac{1}{9}\eta^2 \zeta \partial G(\eta, \zeta)/\partial \zeta, \quad \eta^4 \zeta = \tfrac{1}{3}\eta^2 \zeta \partial G(\eta, \zeta)/\partial \eta - \tfrac{1}{9}\eta \zeta^2 \partial G(\eta, \zeta)/\partial \zeta,$$
$$\zeta^5 = \zeta^2 \partial G(\eta, \zeta)/\partial \eta - \eta \partial G(\eta, \zeta)/\partial \zeta, \quad \eta^\mu \zeta^\nu = \tfrac{1}{3}\eta^{\mu-1} \zeta^{\nu-2} \partial G(\eta, \zeta)/\partial \zeta, \quad \text{if } \mu \geq 1, \ \nu \geq 2.$$

When $\mu + \nu \geq 5$ all the factors in front of the derivatives of G vanish of second order at the origin except the factor $-\eta$ in the representation of ζ^5. However, if Y vanishes of second order at the origin, then the Jacobian $D(\eta - Y(\eta, \zeta), \zeta - Z(\eta, \zeta))/D(\eta, \zeta) = 1 - \partial Z(\eta, \zeta)/\partial \zeta$ at $(0,0)$, so we get a diffeomorphism if only $\partial Z(0,0)/\partial \zeta = 0$. Thus the terms of degree 5, 6, ... can be eliminated as before, and with new local coordinates we obtain

(3.4) $$f = \xi^2 + \eta^3 + \zeta^3 \eta.$$

This singularity is denoted by E_7.

(iii) If $a_0 = a_1 = 0$ we can reduce g_4 to 0 and look at terms in the next homogeneous polynomial in the Taylor expansion, $g_5(\eta, \zeta) = \sum_0^5 b_j \eta^j \zeta^{5-j}$. If $b_0 \neq 0$ we can as before reduce to $g_5 = \zeta^5$, and it follows that with suitable local coordinates

(3.5) $$f = \xi^2 + \eta^3 + \zeta^5.$$

This singularity is denoted by E_8, and it is the last of the *simple singularities*.

The preceding discussion has been taken from Arnol'd [A] where much more information on these matters can be found.

If A has rank 0, then $f(x, y, z) = f_3(x, y, z) + O(|x| + |y| + |z|)^4$ where f_3 is a homogeneous polynomial. If we regard (x, y, z) as homogeneous coordinates in $P_{\mathbf{R}}^2$ then the equation $f_3(x, y, z) = 0$ is a cubic curve in $P_{\mathbf{R}}^2$, so the surface is approximated by a cone over a cubic curve.

The models in the collection

There is a series of surfaces made in wood, painted in black and white, which illustrate the preceding facts. The models are supposed to be seen with the labels in front.

S1. A hyperbolic paraboloid with two generators marked; the equation is $az = y^2 - x^2$.
S2. A parabolic point, with equation $az = x^4 - b^2 y^2$.
S3. Another parabolic point, with equation $z = ax^2 - by^3$.

The origin is a regular point on these surfaces, so the models illustrate some of the features which are ignored when one allows a change of coordinates making them plane.

S4. A conic double point (C_2) with equation $y^2 - x^2 = z^2(z + a)$. When $a \to 0$ it degenerates to a biplanar double point of type B_3.
S5. A biplanar double point of type B_3, with equation $x^2 - y^2 + az(x^2 + y^2) + bz^3 = 0$ where $ab < 0$. The intersections with a plane $x \pm y = 0$ consist of three different lines.
S6. The equivalent surface $x^2 - y^2 = az^3$ with biplanar double point of type B_3.

S7. Another equivalent surface $(x + az^2)^2 = y^2 + bz^3$.

S8. A biplanar double point of type B_4, with equation $x^2 - y^2 = axz^2$. (The equation can be written $(x - az^2/2)^2 - y^2 = a^2z^4/4$.

S9. The equivalent surface $x^2 - y^2 = az^4$.

S10. A biplanar double point of type B_5, with equation

$$x^2 - y^2 + a(x - y)z^2 + bxyz = 0.$$

S11. A biplanar double point of type B_6, with equation $xy + axz^2 + by^3 = 0$; it is equivalent to $\xi\eta = -ba^3z^6$.

S12. An equivalent surface $(x - az^2)^2 = y^2 + bz^6$

S13. A uniplanar double point of type U_6 with equation $ay^2 = z(x^2 - z^2)$. The intersection with the plane $y = 0$ consists of three separate lines.

S14. A uniplanar double point of type U_7, with equation $ay^2 = -2x^2z + yz^2$. The intersection with the plane $y = 0$ has a double line.

S15. A uniplanar double point of type U_8, with equation $ay^2 = bz^3 + x^2y$. The equation can be written $a(y - x^2/2a)^2 = bz^3 + x^4/4a$.

S16. A degenerate form of the singularity U_8, with equation $z^3 = y^6 - ax^2$, $a > 0$, called the uncomfortable saddle in the old labels. It is not a simple singularity.

S17. The surface $xyz = a^3$ which has no singularity when $a \neq 0$.

S18. A triple point with equation $x^2(ay + x^2) = (y^2 - bz)^2$, called a pinch point. It is not an isolated singularity, for every point with $x = 0$ and $y^2 = bz$ is a double point. The origin is called a pinch point because the rank of the Hessian is lower there. With $ay + x^2$ and $bz - y^2$ as new variables instead of y and z it is a cubic ruled surface which will be discussed in subsection 4.2.

S19. A triple point with equation

$$(x^2 + y^2 + z^2)(x^2/a^2 + y^2/b^2) = 2z(x^2 + y^2)$$

generated by the circles of curvature at 0 for the elliptic paraboloid $2z = x^2/a^2 + y^2/b^2$, $a \geq b$ (or any other surface with principal lines of curvature along the x and y axes at the origin and positive principal curvatures $1/a^2$ and $1/b^2$). The z axis is a double line which is isolated except when $z \in [2b^2, 2a^2]$. When $2b^2 < z < 2a^2$ the surface is then approximated by the lines $x\sqrt{2 - z/a^2} = \pm y\sqrt{z/b^2 - 2}$ and the surface consists of two smooth intersecting sheets. It can be considered as a simple topological model of the nonorientable projective plane P_R^2. This is easily seen if we note that topologically the surface is the same as if we replace the circles by half circles, with equal radii, and identify opposite points on the boundary of the resulting half sphere. This is equivalent to identifying opposite points on the sphere, that is, defining P_R^2 as lines through the origin in $\mathbf{R}^3$.

S20. A triple point, with equation

$$(x^2 + y^2 + z^2)(x^2/a^2 - y^2/b^2) = 2z(x^2 + y^2)$$

generated by the circles of curvature at 0 for the hyperbolic paraboloid $2z = x^2/a^2 - y^2/b^2$, thus the hyperbolic analogue of model S19. The z axis is a double line which is isolated when $-2b^2 < z < 2a^2$.

S21. A triple point with equation

$$(x^2 + y^2 + z^2)x^2/a^2 = 2z(x^2 + y^2)$$

generated by the circles of curvature at 0 for the parabolic cylinder $2z = x^2/a^2$, a degenerate case of the preceding two models. The z axis is a double line which is isolated when $0 \neq z < 2a^2$.

4. Cubic surfaces. A cubic equation written in homogeneous coordinates has $\binom{6}{3} = 20$ coefficients and a 4×4 matrix has only 16 components, so it is clear that even from the projective point of view cubic surfaces will depend on at least four parameters. The purpose of the models is to display the main patterns of singularities. Before discussing them it is useful to review the discussion of local isolated singularities in Section 3 for the special case of cubic surfaces.

4.1. Isolated singularities of cubic surfaces. Assume at first that the origin in $\mathbf{R}^3$ is a singular point such that the defining polynomial is of the form

$$F(x, y, z) = x^2 \pm y^2 + q(x, y, z)$$

in affine coordinates, where q is a homogeneous cubic polynomial. We assume that F is irreducible, for the surface is otherwise the union of a plane and a quadratic surface. Recall that the critical points with respect to x, y are defined by

(4.1) $$2x + \partial q(x, y, z)/\partial x = 0, \quad \pm 2y + \partial q(x, y, z)/\partial y = 0,$$

and that the biplanar singularity at 0 is determined by the order of vanishing at the origin of the value $g(z)$ of $F(x, y, z)$ at the critical point $(x(z), y(z), z)$. We shall see that it is at most equal to 6 unless the z axis is a double line. Since $x(z) = O(z^2)$ and $y(z) = O(z^2)$ we have $g(z) = z^3 q(0, 0, 1) + O(z^4)$, so the singularity is of type B_3 if $q(0, 0, 1) \neq 0$. If $q(0, 0, 1) = 0$ we can write

$$q(x, y, z) = L(x, y)z^2 + Q(x, y)z + R(x, y)$$

where L is linear, Q is quadratic and R is cubic. If $L \equiv 0$ then the z axis consists of double points, so assume that $L \not\equiv 0$ and write $L(x, y) = 2ax + 2by$, thus

$$F(x, y, z) = (x + az^2)^2 \pm (y \pm bz^2)^2 - (a^2 \pm b^2)z^4 + Q(x, y)z + R(x, y).$$

Since $x(z) = O(z^2)$ and $y(z) = O(z^2)$ it follows from (4.1) that

$$x + az^2 = O(z^3), \quad \pm y + bz^2 = O(z^3),$$

which gives $g(z) = -z^4(a^2 \pm b^2) + O(z^5)$. If $a^2 \pm b^2 \neq 0$ we conclude that the singularity is of type B_4. If $a^2 \pm b^2 = 0$ we must have the minus sign since we have assumed that $a^2 + b^2 \neq 0$, and we obtain

$$g(z) = Q(-a, b)z^5 + O(z^6).$$

The singularity is therefore of type B_5 unless $Q(-a, b) = 0$. Then Q has a factor $bx + ay$. If we write $Q(x, y) = 2(bx + ay)M(x, y)$ where M is a linear form, then

$$F(x, y, z) = (x + az^2 + bM(x, y)z)^2 - (y - bz^2 - aM(x, y)z)^2 + R(x, y).$$

For the critical point we obtain

$$x(z) + az^2 + bM(x(z), y(z))z = O(z^4), \quad y(z) - bz^2 - aM(x(z), y(z))z = O(z^4),$$

so $g(z) = R(-a, b)z^6 + O(z^7)$. The singularity is therefore of type B_6 unless $R(-a, b) = 0$. But F is then divisible by $bx + ay$ which is against the assumption. Thus the singularity is of type B_j for some $j = 3, \ldots, 6$ unless there is a double line through the origin.

Assume now that the origin is a singular point such that $F(x,y,z) = x^2 + q(x,y,z)$ where q is a homogeneous cubic polynomial. With homogeneous coordinates the corresponding polynomial can be written

$$wx^2 + q(x,y,z) = x^2(w + L(x,y,z)) + 2xQ(y,z) + R(y,z)$$

where L is linear, Q quadratic and R cubic. After a projective transformation introducing $w + L(x,y,z), x, y, z$ as new homogeneous coordinates the polynomial assumes the form

$$F(x,y,z) = x^2 + 2xQ(y,z) + R(y,z) = (x + Q(y,z))^2 + R(y,z) - Q(y,z)^2$$

in affine coordinates. The critical point is located on the quadratic surface where $x = -Q(y,z)$, and the critical value is $R(y,z) - Q(y,z)^2$. If $R(y,z)$ has precisely three different zeros we know from Section 3 that the singularity is of type U_6, in the classical notation. If $R(y,z) = zy^2$ after a linear change of yz-variables then the singularity is of type U_7 unless $Q(0,z) \equiv 0$, but then every point on the z axis would be a double point. If $R(y,z) = y^3$ we have a singularity of type U_8 unless $Q(0,z) \equiv 0$ which would again mean that the z axis is a double line. Finally, if $R \equiv 0$ then F is reducible which is against the hypothesis. Only the singularities U_6, U_7 and U_8 can therefore occur unless there is a double line through the origin.

The last possibility is that $F(x,y,z)$ is a homogeneous cubic polynomial in (x,y,z). Then the origin is not an isolated singularity in the complex domain unless the curve defined by F in $P_\mathbb{C}^2$ is free from singularities. As already mentioned it can then be given the standard form (2.5),

(4.2) $$4x^3 - g_2 xz^2 - g_3 z^3 - zy^2 = 0$$

called an *elliptic cone*.

Summing up, a singular point which is not on a double line is a double point of type C_2, B_j with $3 \le j \le 6$, U_6, U_7 or U_8, or the vertex of an elliptic cone.

If a line contains two singular points then F vanishes identically on it since it has four zeros. If there are four singular points in a hyperplane it follows that F vanishes in the whole plane. When F is irreducible, as we have assumed, then it follows that there are at most four isolated singularities in all. A list of the possible combinations is given in [FC, p. 13].

4.2. Cubic ruled surfaces. Suppose now that the z axis is a double line. With homogeneous coordinates such that $w = 0$ is the equation of the plane at infinity, the equation can then be written

$$f(x,y) + g(x,y)z + h(x,y)w = 0$$

where f is a cubic form and g, h are quadratic forms. By a linear transformation of z, w, corresponding to a projective transformation, we can replace g, h by two arbitrary linearly independent linear combinations of them.

a) If g and h are linearly dependent we can thus make $h = 0$ so the surface is a cone with vertex at the origin and the z axis as a double line. A linear transformation in x, y, z then reduces the equation to one of the following forms (cf. subsection 2.4)

$$zxy = x^3 + y^3 \quad \text{or} \quad zy^2 = x^3,$$

representing a cone over a nodal curve or a cusp.

b) If g and h are linearly independent the discriminant of the form $\lambda g + \mu h$ is a quadratic form $\not\equiv 0$ in (λ, μ). If it has two different real (complex) roots $\lambda : \mu$ then after a linear change of the coordinates x, y the forms $x^2 \pm y^2$ and $2xy$ are linear combinations. Thus the equation is projectively equivalent to

$$f(x, y) + (x^2 \pm y^2)z + xyw = 0.$$

Now $x^3 = x(x^2 \pm y^2) \mp y(xy)$ and $y^3 = \pm y(x^2 \pm y^2) \mp x(xy)$ so we can eliminate f by another projective transformation changing z and w. Then we end up with the equation

(4.3)
$$(x^2 \pm y^2)z + xy = 0$$

in affine coordinates.

c) If the discriminant has a double root $\lambda = 1, \mu = 0$ then $\pm g$ can be reduced to x^2 by a change of x, y variables while $h = xy$. We can then remove from f all terms except the y^3 term, and obtain in affine coordinates the equation

(4.4)
$$y^3 + zx^2 + xy = 0.$$

(Recall that we assume irreducibility.) Hence there are only three possibilities from the projective point of view.

To determine the geometrical meaning of the surfaces we observe that a plane through the double line must obviously intersect the surface in the double line counted twice and another line. In fact, if we set $y = tx$ we obtain in case (4.3) $z = -t/(1 \pm t^2)$ so we have a line connecting the point $(0, 0, -t/(1 \pm t^2))$ on the double line with the point at infinity in the direction $(1, t, 0)$, which lies in the line L at infinity defined by $z = 0$, called the *director line*. The line intersects a plane $x = -z$ at a point in the conic where $x^2 \pm y^2 - y = 0$, a circle or hyperbola depending on the sign, called a *director conic*, with x, y coordinates $(t/(1 \pm t^2), t^2/(1 \pm t^2))$. Note that L does not intersect the conic. For the plus sign in (4.3) the planes $z = \pm\frac{1}{2}$ through L which are tangent to the conic at $(\mp\frac{1}{2}, \frac{1}{2}, \pm\frac{1}{2})$ intersect the double line at the points where it starts to be isolated in the real part of the surface. In case (4.4) we obtain the line $y = tx$, $z = -t - t^3x$ through $(0, 0, t)$ on the double line and intersecting the ellipse $z = y$, $y^2 + x^2 + x = 0$ at a point with xy coordinates $(-1/(1 + t^2), -t/(1 + t^2))$.

One can think of a non-degenerate conic as $P_{\mathbf{R}}^1$, for it can be parametrized by the projection on a line from any point on it. Letting L be the double line in case (4.4) we then have in both cases a projective isomorphism between L and the conic section, and the surface consists of all the lines joining corresponding points. The case (4.4) is called *Cayley's ruled surface*.

We now have a minimum background for describing the models in the *Rodenberg series*, consisting originally of 26 models. The collection in Lund contains 21 of them.

The Rodenberg series

R1. Clebsch's diagonal surface is an example of a cubic surface with no singularities containing 27 real lines. See [FC] for a discussion of the interesting combinatorial properties of these lines and their intersections. (Another such surface is in the library on the fifth floor.) This is the generic case, the following ones are degenerations.

R2. Cubic with 4 real conic points.

R3, R4, R5, R6. These surfaces are projectively equivalent to R2.

R7. Cubic with three real conic points.

R8. The same cubic seen from the other side.

R9. Cubic with three real double points of type B_3, with real tangent planes at them.

R10. Cubic with one real double point of type B_3, with real tangent planes there.

R11. Cubic with one real double point of type B_3, but with complex tangent planes there.

R12. Cubic with two real conic double points and one real double point of type B_4, with real tangent planes.

R13. Cubic with two nonreal conic double points and one real double point of type B_4, with real tangent planes.

R14. Cubic with one real conic point and one real double point of type B_5, with real tangent planes. Missing in the local collection.

R15. Cubic with one real conic point and one double point of type B_6, with real tangent planes.

R16. Cubic with a real double point of type U_6; the tangent plane there intersects the surface in three real lines.

R17. Another cubic with a real double point of type U_6; the tangent plane intersects the surface in only one real line.

R18. Cubic with a real double point of type U_7.

R19. Another such cubic; the intersection with the tangent plane at the double point is different from the previous one.

R20. Ruled surface projectively equivalent to (4.3) with the minus sign. The green line is the director line, a director circle is marked as well as a number of generators.

R21. Ruled surface projectively equivalent to (4.3) with the plus sign. The green line is the director line, a director circle is also marked. Note the lines to the end of the part of the double line which is not isolated on the real surface.

R22. This is Cayley's ruled surface, projectively equivalent to (4.4). The director line is the double line, and a director circle is marked.

5. Surfaces of degree four. There are many more types of surfaces of degree four than of degree three, for the equation contains $\binom{7}{3} = 35$ coefficients instead of just 20 in the cubic case. There is no ambition to model the general shape of the surfaces but only a few types of special interest. Some have already been mentioned in the list of models of local singularities and in the discussion of the tangent surface of the cubic space curve.

5.1. Dupin's cyclides. The simplest cyclide is a *torus*, obtained by rotating a circle around an axis in the same plane. If the axis is the x_3 axis and the circle is in the x_1x_3 plane with center at $(a,0)$ and radius R, then the circle is defined by $x_1^2 + x_3^2 + a^2 - R^2 = 2ax_1$, so the equation of the torus is

$$(5.1) \qquad (x_1^2 + x_2^2 + x_3^2 + a^2 - R^2)^2 = 4a^2(x_1^2 + x_2^2).$$

If $|a| > R$ this is a standard torus, also called a *ring torus* to distinguish it from the *horn torus* obtained when $|a| = R$ and the *spindle torus* obtained when $|a| < R$. In these cases the circle is tangent to or intersects the axis of rotation which gives singularities there. The general Dupin cyclide is obtained from a torus by inversion in a point ξ, that is, a map $x \mapsto y = (x-\xi)/|x-\xi|^2$, thus $x = \xi + y/|y|^2$. (When using the inversion it is convenient to extend $\mathbf{R}^3$ by a point at infinity corresponding to the point ξ.) An inversion maps spheres and planes to spheres or planes, for an equation of the form $a|x|^2 + L(x) + b = 0$ where L is a linear form and a, b are constants becomes $a + L(y) + b|y|^2 = 0$ if $x = y/|y|^2$. Now an orthogonal transformation and a translation will obviously change (5.1) to an equation of the form

$$(5.2) \qquad\qquad |x|^4 + |x|^2 L(x) + Q(x) = 0,$$

where L is a first and Q is a second order polynomial. If we set $x = y/|y|^2$ this gives

$$1 + |y|^2 L(y/|y|^2) + |y|^4 Q(y/|y|^2) = a|y|^4 + |y|^2 \widetilde{L}(y) + \widetilde{Q}(y) = 0,$$

where $a = Q(0)$ vanishes precisely when the inversion is made with respect to a point on the torus. Thus the equation of every Dupin cyclide is of the form

$$(5.2)' \qquad\qquad a|x|^4 + |x|^2 L(x) + Q(x) = 0$$

with a constant a and L of first order, Q of second order.

The lines of curvature on a torus are the circles which are rotated and, orthogonally to them, the circles described by a fixed point on the circle during the rotation. There is a sphere tangent to the torus along any one of these circles. Since spheres are mapped to spheres (or planes) by an inversion it follows that for a Dupin cyclide there is at every point two tangent spheres which remain tangent along a full circle. (Spheres and circles may become planes and lines.) These are lines of curvature on the Dupin cyclide. Conversely every surface such that the lines of curvature are circles or straight lines is a Dupin cyclide.

All surfaces with an equation of the form (5.2)' are called cyclides but most of them are not Dupin cyclides. In fact, if $a \neq 0$ one can by a translation reduce to the case where $a = 1$ and $L = 0$, and by an orthogonal transformation one can put the quadratic part of Q in the diagonal form $A_1 x_1^2 + A_2 x_2^2 + A_3 x_3^2$. If $A_1 > A_2 > A_3$ the coefficients are then uniquely determined so there are 7 parameters altogether whereas a Dupin cyclide only depends on 4 parameters. Much more information on cyclides can be found in [FC] and the references given there.

The Dupin cyclides have an interesting role in the theory of the wave equation. By a theorem of Friedlander [F] the wave equation $\partial^2 u/\partial t^2 - \Delta u = 0$, where $\Delta = \sum_1^3 \partial^2/\partial x_j^2$, is satisfied by $u(t, x) = v(x) F(t - \varphi(x))$ for all F if and only if the level sets of φ are Dupin cyclides, $|\partial \varphi(x)/\partial x| = 1$, and the amplitude v is chosen to fit the function φ. Thus φ is essentially the distance function to a Dupin cyclide. Riesz [R] observed that the corresponding characteristic surface $\{(\varphi(x), x)\} \subset \mathbf{R}^4$ is of a very special kind: By a translation and a Lorentz transformation (that is, a linear transformation preserving the quadratic form $t^2 - |x|^2$) it can be brought to the ruled surface generated by the lines intersecting the Lorentz circles

$$\{(t, x_1, 0, 0); t^2 - x_1^2 = a^2\} \quad \text{and} \quad \{(0, 0, x_2, x_3), x_2^2 + x_3^2 = a^2\}$$

in two Lorentz orthogonal planes. A point (t, x) is on this surface if and only if

$$t^2 - x_1^2 = \lambda^2 a^2, \quad x_2^2 + x_3^2 = (1 - \lambda)^2 a^2$$

for some λ, which implies $t^2 - |x|^2 + a^2 = 2\lambda a^2$, hence $(t^2 - |x|^2 + a^2)^2 = 4a^2(t^2 - x_1^2)$ or

$$(t^2 - |x|^2)^2 - 2a^2(t^2 - x_1^2) - 2a^2(x_2^2 + x_3^2) + a^4 = 0.$$

Thus the general Dupin cyclide is obtained by a Lorentz transformation of this surface followed by intersection with a plane where t is constant.

Models in the collection

H1. A ring cyclide with the two families of lines of curvature marked.
H2. Half a ring cyclide with the two families of lines of curvature marked.
H3. A ring cyclide with lines of curvature and some geodesics marked.
H4. A quarter of a ring cyclide.

5.2. Kummer surfaces. After a paper on the refraction of light in the atmosphere of the earth Kummer published in 1859 a systematic study of two parameter families of linear rays in $\mathbf{R}^3$. Infinitesimally such a system of rays near the x_3 axis can generically be described as the lines with direction $(-t, 1)$ passing through $(At, 0) \in \mathbf{R}^3$, where $t \in \mathbf{R}^2$ and A is a linear transformation in $\mathbf{R}^2$. By a translation in the x_3 direction and an orthogonal transformation of t we may assume that the symmetric part of A has the form $\begin{pmatrix} \delta & 0 \\ 0 & -\delta \end{pmatrix}$, hence

$$A = \begin{pmatrix} \delta & \gamma \\ -\gamma & -\delta \end{pmatrix}$$

for some γ. The distance of the ray $\mathbf{R} \ni s \mapsto (At-st, s) \in \mathbf{R}^3$ to the x_3 axis is $\min_s |At-st|$, which is attained when $(At - st, t) = 0$, that is,

$$s = (At, t)/|t|^2 = (\delta t_1^2 - \delta t_2^2)/(t_1^2 + t_2^2) \in [-\delta, \delta].$$

Kummer calls the points $(0, 0, \pm\delta)$ limit points ("Grenzpunkte") and proceeds to examining focal points. The range of $t \mapsto At - \lambda t$ for a fixed $\lambda = x_3$ is a line if $0 = \det(A - \lambda \operatorname{Id}) = \lambda^2 - \delta^2 + \gamma^2$, which means that $\lambda = \pm\sqrt{\delta^2 - \gamma^2}$. When $|\gamma| < \delta$, then the *focal points* $(0, 0, \pm\lambda)$ are real; they lie between the limit points and have the same mean value. When $x_3 = \pm\lambda$ all the rays pass through the focal lines $\gamma x_1 = (\delta \mp \lambda)x_2$. Since $\gamma^2 = (\delta - \lambda)(\delta + \lambda)$ these lines are symmetric with respect to the bisectors of the x_1 and x_2 axes, and the angle α between them satisfies $\sin \alpha = \lambda/\delta$ so it is a right angle only when $\gamma = 0$, which is the case if the rays originate from the normals of a surface. A beam bounded when $x_3 = 0$ by an ellipse $\{x', Q(x') = 1\}$ where $x' = (x_1, x_2)$ and Q is a positive definite quadratic form will be defined by

$$Q((\operatorname{Id} - A^{-1}x_3)^{-1}x') = 1,$$

for $At - x_3 t = (\operatorname{Id} - A^{-1}x_3)At$. After multiplication by $\det(\operatorname{Id} - A^{-1}x_3)^2$ this equation is of degree 4 and every section with a plane $x_3 =$ constant is an ellipse as emphasized by Kummer when he presented three wire models, one corresponding to the case $\gamma = 0$ (right angles between the focal lines), one with $0 < \gamma < \delta$ (oblique angle between the focal lines) and one with $\gamma > \delta$ so that there are no real focal lines. He proved that all the cases can occur when a beam of rays emerging from a point passes through a suitable crystal.

Five years later, in 1864, Kummer returned to crystal optics. The characteristic equation of the differential equations for the propagation of light in a triclinic crystal is of the form $\sum_1^3 \xi_j^2/(\xi_0^2 - a_j^2|\xi|^2) = 0$, where $|\xi|^2 = \xi_1^2 + \xi_2^2 + \xi_3^2$, or rather, after removal of the denominators and cancellation of a factor $|\xi|^2$, the fourth degree equation

$$(5.3) \quad \xi_0^4 - \xi_0^2(\xi_1^2(a_2^2 + a_3^2) + \xi_2^2(a_1^2 + a_3^2) + \xi_3^2(a_1^2 + a_2^2))$$
$$+ (\xi_1^2 + \xi_2^2 + \xi_3^2)(\xi_1^2 a_2^2 a_3^2 + \xi_2^2 a_1^2 a_3^2 + \xi_3^2 a_1^2 a_2^2) = 0.$$

Here a_j are constants, and we assume $a_1^2 > a_2^2 > a_3^2 > 0$. The dual Fresnel wave surface, consisting of the conormals of the surface (5.3) is given by $\sum_1^3 x_j^2 a_j^2/(a_j^2 x_0^2 - |x|^2) = 0$, or

$$(5.4) \quad x_0^4 - x_0^2\left(x_1^2(a_2^{-2} + a_3^{-2}) + x_2^2(a_1^{-2} + a_3^{-2}) + x_3^2(a_1^{-2} + a_2^{-2})\right)$$
$$+ (x_1^2 + x_2^2 + x_3^2)(x_1^2 a_2^{-2} a_3^{-2} + x_2^2 a_1^{-2} a_3^{-2} + x_3^2 a_1^{-2} a_2^{-2}) = 0.$$

Physically x_0 represents time. We consider the surfaces (5.3) and (5.4) as surfaces in $P_{\mathbf{R}}^3$ or even in $P_{\mathbf{C}}^3$. The characteristic surface (5.3) has 16 conic double points with homogeneous coordinates

(5.5)
$$\left(\pm a_3\sqrt{a_1^2 - a_2^2}: \pm\sqrt{a_1^2 - a_3^2} \quad : \pm i\sqrt{a_2^2 - a_3^2} \quad : \quad 0 \qquad\right),$$
$$\left(\pm a_2\sqrt{a_1^2 - a_3^2}: \pm\sqrt{a_1^2 - a_2^2} \quad : \quad 0 \quad : \pm\sqrt{a_2^2 - a_3^2}\ \right),$$
$$\left(\pm a_1\sqrt{a_2^2 - a_3^2}: \quad 0 \quad : \pm i\sqrt{a_1^2 - a_2^2} : \pm\sqrt{a_1^2 - a_3^2}\ \right),$$
$$\left(\quad 0 \quad : \pm a_1\sqrt{a_2^2 - a_3^2}: \pm i a_2\sqrt{a_1^2 - a_3^2}: \pm a_3\sqrt{a_1^2 - a_2^2}\right),$$

with all combinations of signs. The four double points in the second row are in $P_{\mathbf{R}}^3$, the others are in $P_{\mathbf{C}}^3 \setminus P_{\mathbf{R}}^3$, and those in the last are at infinity. Similarly there are 16 conic double points of (5.4), with homogeneous coordinates

(5.6)
$$\left(\pm\sqrt{a_1^2 - a_2^2}: \pm a_2\sqrt{a_1^2 - a_3^2}: \pm i a_1\sqrt{a_2^2 - a_3^2}: \quad 0 \qquad\right),$$
$$\left(\pm\sqrt{a_1^2 - a_3^2}: \pm a_3\sqrt{a_1^2 - a_2^2}: \quad 0 \quad : \pm a_1\sqrt{a_2^2 - a_3^2}\right),$$
$$\left(\pm\sqrt{a_2^2 - a_3^2}: \quad 0 \quad : \pm i a_3\sqrt{a_1^2 - a_2^2}: \pm a_2\sqrt{a_1^2 - a_3^2}\right),$$
$$\left(\quad 0 \quad : \pm\sqrt{a_2^2 - a_3^2}: \pm i\sqrt{a_1^2 - a_3^2}: \pm\sqrt{a_1^2 - a_2^2}\ \right).$$

An element in (5.5) is orthogonal to two of the elements in (5.6) in each of the other rows and vice versa, which makes 6 points altogether, and its orthogonal plane is a tangent to (5.4) along a usually nonreal circle.

Kummer proved that a surface of degree 4 can have at most 16 conic double points and that if there are 16 conic double points (in $P_{\mathbf{C}}^3$) then there are also 16 double tangent planes, which are tangent along conics. Each of them contains 6 of the double points just as for the Fresnel surface. He gave a number of different normal forms for the equation of such surfaces, now called *Kummer surfaces*, such as $\varphi^2 = 16Kpqrs$ where p, q, r, s are four linearly independent linear forms and with constants a, b, c

$$\varphi = p^2 + q^2 + r^2 + s^2 + 2a(qr + ps) + 2b(rp + qs) + 2c(pq + rs), \quad K = a^2 + b^2 + c^2 - 2abc - 1.$$

It is clear that each of the four planes $p = 0, \ldots, s = 0$ is a tangent at the intersection with the quadratic surface $\varphi = 0$, so the four planes define a tetrahedron for which each face is a double tangent plane along a conic. An edge such as $r = s = 0$ contains two double points, defined by $\varphi = p^2 + q^2 + 2cpq = 0$, that is, $p : q = -c \pm \sqrt{c^2 - 1}$. For generic values of a, b, c this accounts for 12 double points. For the other double points one must have $\varphi \neq 0$, hence $pqrs \neq 0$, so the conditions are $2p\varphi \partial \varphi/\partial p = 16Kpqrs = \varphi^2$ and so on, that is,

$$2p\partial\varphi/\partial p = \varphi, \quad 2q\partial\varphi/\partial q = \varphi, \quad 2r\partial\varphi/\partial r = \varphi, \quad 2s\partial\varphi/\partial s = \varphi.$$

The sum of these equations is automatically fulfilled by the homogeneity of φ so only the following three differences are relevant:

$$p^2 - q^2 + a(ps - qr) + b(rp - qs) = 0, \quad q^2 - r^2 + b(qs - rp) + c(pq - rs) = 0,$$
$$r^2 - s^2 + a(qr - ps) + b(rp - qs) = 0.$$

These have a number of trivial solutions such as $p = q = r = s$ and in addition a solution which in general requires solving a fourth degree equation giving the remaining four double points; K is the common value of $\varphi^2/16pqrs$ in them. Kummer made a model for the case where p, q, r, s define the sides of a regular tetrahedron and took $a = b = c = 2$ to get tetrahedral symmetry. When $a = b = c$ the double points are

$$(p, q, r, s) = (\tfrac{1}{2}(1 + \sqrt{(a+1)/(a-1)}), \tfrac{1}{2}(1 - \sqrt{(a+1)/(a-1)}), 0, 0),$$
$$(p, q, r, s) = ((2a+1)/(2a-2), 1/(2-2a), 1/(2-2a), 1/(2-2a))$$

and permutations, if we normalize to barycentric coordinates, that is, $p + q + r + s = 1$. Thus 12 of the double points lie on the edges of the tetrahedron extended by equal amounts and four lie on the heights extended beyond the vertices. All double points and double planes are therefore real.

Further information on Kummer surfaces can be found in [FC] and in [H].

Models in the collection

H4. The model in zinc made by Klein in 1871, mentioned above, also labelled **KLEIN IV**.

H5. A plaster copy of this model [Brill #67, p. 45].

H6. A plaster model with 8 real double points and planes [Brill #68, p. 45].

H7. A plaster model with 4 real double points and planes [Brill #69, p. 45].

5.3. Quadratic line complexes. A line in the real projective space $P_{\mathbf{R}}^3$ is a two dimensional linear subspace of $\mathbf{R}^4$, thus determined by two vectors $x, y \in \mathbf{R}^4$. The exterior product $p = x \wedge y \in \bigwedge^2 \mathbf{R}^4 \cong \mathbf{R}^6$ with components given by the two row determinants

$$p_{ij} = x_i y_j - x_j y_i, \quad 0 \le i < j \le 3,$$

is independent of the choice of x and y apart from a constant factor, so it is a well defined element of $P_{\mathbf{R}}^5$. (We denote the coordinates in $\mathbf{R}^4$ by (x_0, x_1, x_2, x_3) so that the plane at infinity in $P_{\mathbf{R}}^3$ is defined by $x_0 = 0$.) One calls $p = (p_{ij})$ the Plücker coordinates of the line. They satisfy the condition

(5.7) $$p_{01}p_{23} + p_{02}p_{31} + p_{03}p_{12} = 0,$$

for $(x \wedge y) \wedge (x \wedge y) = 0$ and is equal to the left-hand side of (5.7) times a generator of $\bigwedge^4 \mathbf{R}^4 \cong \mathbf{R}$. The line is determined by p as $\{x \in \mathbf{R}^4; x \wedge p = 0\}$. It is easy to see and will be verified below that all (p_{ij}) satisfying (5.7) are Plücker coordinates of a line in $P_{\mathbf{R}}^3$. The lines determined by p and by q intersect if and only if $p \wedge q = 0$, that is,

(5.7)′ $$p_{01}q_{23} + p_{23}q_{01} + p_{02}q_{31} + p_{31}q_{02} + p_{03}q_{12} + p_{12}q_{03} = 0,$$

which means that p lies in the tangent plane of the quadric (5.7) at q. For any $X \in P_{\mathbf{R}}^3$ the lines through X form a two dimensional subspace $\sigma(X)$ contained in the quadric $G \subset P_{\mathbf{R}}^5$

defined by (5.7), and for any hyperplane $H \subset P_{\mathbf{R}}^3$ the lines contained in H also form a two dimensional subspace $\sigma^*(H)$ of G. For example, the lines through the origin $(1:0:0:0)$ are defined by $p_{12} = p_{13} = p_{23} = 0$, and the lines in the plane at infinity are defined by $p_{01} = p_{02} = p_{03} = 0$. Conversely, given a two dimensional subspace of G we can choose three elements in it corresponding to linearly independent elements $p', p'', p''' \in \mathbf{R}^6$ satisfying (5.7). The corresponding lines intersect pairwise, so either they all pass through the same point or else they lie in the same plane. This remains true for all lines intersecting the three chosen lines. Thus two dimensional spaces contained in the quadric G are either of the form $\sigma(X)$ or the form $\sigma^*(H)$. A projective transformation of $P_{\mathbf{R}}^3$ corresponds to a linear transformation in $\mathbf{R}^4$ which induces a linear transformation in $\bigwedge^2 \mathbf{R}^4$ preserving the condition (5.7) and the distinction between the two types of 3-dimensional subspaces. Conversely, a linear transformation in $\bigwedge^2 \mathbf{R}^4$ with these properties is induced by a linear transformation in $\mathbf{R}^4$. This follows easily by noting that if $X \notin H$ then the images of $\sigma(X)$ and $\sigma^*(H)$ can be brought back to $\sigma(X)$ and $\sigma^*(H)$ by such an induced transformation.

A line L in $P_{\mathbf{R}}^3$ can also be defined by two linear equations $\langle x, \xi \rangle = \langle x, \eta \rangle = 0$ in the homogeneous coordinates x; here ξ and η are linearly independent elements in (the dual of) $\mathbf{R}^4$. The line spanned by $x, y \in \mathbf{R}^4$ intersects L if and only if $\langle \lambda x + \mu y, \xi \rangle = \langle \lambda x + \mu y, \eta \rangle = 0$ for some $(\lambda, \mu) \neq (0,0)$, that is,

$$0 = \begin{vmatrix} \langle x, \xi \rangle & \langle y, \xi \rangle \\ \langle x, \eta \rangle & \langle y, \eta \rangle \end{vmatrix} = \sum_{j,k=0}^{3} (x_j \xi_j y_k \eta_k - x_j \eta_j y_k \xi_k) = \tfrac{1}{2} \sum_{j,k=0}^{3} q_{jk} \pi_{jk}$$

where $q_{jk} = x_j y_k - x_k y_j$ are Plücker coordinates of the line spanned by x and y, and $\pi_{jk} = \xi_j \eta_k - \xi_k \eta_j$ are the similarly defined dual coordinates of L. If p_{jk} are Plücker coordinates of L this means that the bilinear form $(5.7)'$ vanishes precisely when $\sum q_{jk} \pi_{jk} = 0$, so

$$p_{01} = \pi_{23}, \quad p_{23} = \pi_{01}, \quad p_{02} = \pi_{31}, \quad p_{31} = \pi_{02}, \quad p_{03} = \pi_{12}, \quad p_{12} = \pi_{03}.$$

This is the correspondence between elements $p \in \mathbf{R}^6$ and π in the dual $\mathbf{R}^6$ defined by the quadratic form (5.7).

A line complex of degree m is defined by an additional homogeneous equation of order m in the Plücker coordinates. There is a large collection of models related to quadratic complexes, defined as the intersection of the quadric G and the quadric defined by a homogeneous quadratic equation $Q(p) = 0$. Leaving now the invariant notation we write a generic line in the form

$$x = rz + \varrho, \quad y = sz + \sigma.$$

Thus the homogeneous coordinates are $(1 : x : y : z)$. The line contains the points $(1 : \varrho : \sigma : 0)$ (for $z = 0$) and $(0 : r : s : 1)$ (at infinity) so the Plücker coordinates are

$$(p_{01}, p_{02}, p_{03}, p_{12}, p_{13}, p_{23}) = (r, s, 1, \varrho s - r\sigma, \varrho, \sigma).$$

(This covers all $p \in \bigwedge^2 \mathbf{R}^4$ satisfying (5.7) with $p_{03} \neq 0$ and similarly we see that all other p satisfying (5.7) correspond to some line (which may be at infinity.) A quadratic line complex is therefore defined by the equation $Q(r, s, 1, \varrho s - r\sigma, \varrho, \sigma) = 0$.

According to Klein [Kl2, p. 171] his teacher Plücker dismissed the notion of n dimensional space as too metaphysical, and considered the study of lines in three dimensional space as the right setting for four dimensional geometry; G is of course of dimension four. To visualise a quadratic line complex he specialized it to the subset of lines intersecting a chosen line, that is, he restricted Q to a tangent plane of G. Let the line be the z axis. If the point of intersection is $(0,0,\zeta)$, then $\varrho = -r\zeta$, $\sigma = -s\zeta$, the line is defined by

$$x = r(z - \zeta), \quad y = s(z - \zeta),$$

and the restriction of the quadratic line complex is defined by $q(r,s,-r\zeta,-s\zeta) = 0$ where $q(r,s,\varrho,\sigma) = Q(r,s,1,0,\varrho,\sigma)$ is a quadratic polynomial. Hence the lines passing through $(0,0,\zeta)$ form a quadratic cone, called the complex cone at $(0,0,\zeta)$, with equation

(5.8) $q(x/(z - \zeta), y/(z - \zeta), -\zeta x/(z - \zeta), -\zeta y/(z - \zeta)) = 0,$

which may degenerate to two intersecting planes or a double plane through $(0,0,\zeta)$. The general complex surface corresponding to the z axis is defined as the envelope of these cones, so it is a surface which looks as the complex cone from any point on the z axis. The equation is obtained by eliminating ζ from (5.8) and its derivative with respect to ζ. It is more convenient to write $\tau = 1/(z - \zeta)$, thus $-\zeta/(z - \zeta) = 1 - \tau z$, for then the equations become

$$q(\tau x, \tau y, (1 - \tau z)x, (1 - \tau z)y) = 0, \quad \partial q(\tau x, \tau y, (1 - \tau z)x, (1 - \tau z)y)/\partial \tau = 0.$$

We can write

$$
\begin{aligned}
&q(\tau x, \tau y, (1 - \tau z)x, (1 - \tau z)y) \\
&= \tau^2 A(x,y) + \tau(1 - \tau z)B(x,y) + (1 - \tau z)^2 C(x,y) + \tau L(x,y) + (1 - \tau z)M(x,y) + K \\
&= \tau^2(A(x,y) - zB(x,y) + z^2 C(x,y)) \\
&\quad + \tau(B(x,y) - 2zC(x,y) + L(x,y) - zM(x,y)) + C(x,y) + M(x,y) + K
\end{aligned}
$$

where A, B, C are quadratic forms, L, M are linear forms and K is a constant. Then the equation becomes

$$
\begin{aligned}
&(B(x,y) - 2zC(x,y) + L(x,y) - zM(x,y))^2 \\
&\quad - 4(A(x,y) - zB(x,y) + z^2 C(x,y))(C(x,y) + M(x,y) + K) = 0.
\end{aligned}
$$

Expansion gives cancellations and the equivalent equation

$$
\begin{aligned}
&(M(x,y)^2 - 4C(x,y)K)z^2 + (2B(x,y)M(x,y) - 4C(x,y)L(x,y) + 4B(x,y)K \\
&\quad - 2LM)z + (B(x,y) + L(x,y))^2 - 4A(x,y)(C(x,y) + M(x,y) + K) = 0
\end{aligned}
$$

which is of degree 4. Note that the intersection with a plane π through the z axis consists of the z axis as a double line and a conic. It can also be obtained as the envelope of the lines in the complex contained in π. In general the surface has 8 double points in $P_{\mathbf{C}}^3$.

Plücker made a series of 27 models of these surfaces, 25 of which are in the local collection. (In the first 19 models the chosen line is at infinity.) Klein [Kl1 II, p. 7] observed that the series was not complete and uneven in details so that it was not in the spirit of the originator that it should be spread as a systematic collection. He writes that for this reason he made in 1871 a series of three models of the general complex surface, with 8, 4 and 2 real double points, and also the model of a Kummer surface already mentioned. It occurs in this context because, as proved by Kummer and Klein, the set of points with complex cone degenerating to two planes is a Kummer surface S. The intersection of the two planes is called a singular line of the complex. The points $X \in P^3$ where the two planes coincide are the singular points of S, and all lines through X in this plane are then called singular lines of the complex. For every singular line p there is precisely one $X \in P^3$ such that $p \in \sigma(X)$ and $\sigma(X)$ is in the tangent space of $Q^{-1}(0)$ at p. There is also precisely one hyperplane H in P^3 such that $p \in \sigma^*(H)$ and $\sigma^*(H)$ is contained in the tangent space of $Q^{-1}(0)$ at p. The maps thus defined from the set of singular lines to P^3 and its dual have the Kummer surface and its dual Kummer surface as range. (See [GH, Chapter VI] for a modern presentation.)

An example of a quadratic line complex leading to the Fresnel surface is Painvain's complex [H, p. 112], the set of lines in $\mathbf{R}^3$ which are intersections of perpendicular tangent planes of a quadric $\sum_1^3 x_j^2/\alpha_j = 1$. If the tangent points are x and y, then the normals are ξ, η where $\xi_j = x_j/\alpha_j$, $\eta_j = y_j/\alpha_j$, thus

$$\sum_1^3 \alpha_j \xi_j^2 = 1, \quad \sum_1^3 \alpha_j \eta_j^2 = 1, \quad \sum_1^3 \xi_j \eta_j = 0.$$

The dual Plücker coordinates are the two row determinants in $\begin{pmatrix} -1 & \xi_1 & \xi_2 & \xi_3 \\ -1 & \eta_1 & \eta_2 & \eta_3 \end{pmatrix}$, so they are $\pi_{0j} = \xi_j - \eta_j$ and $\pi_{jk} = \xi_j \eta_k - \xi_k \eta_j$, $j, k = 1, 2, 3$. Thus

$$\sum_{j,k=1}^3 (\alpha_j + \alpha_k)\pi_{jk}^2 = \sum_{j,k=1}^3 (\alpha_j + \alpha_k)(\xi_j \eta_k - \xi_k \eta_j)^2$$

$$= 2\sum_{j,k=1}^3 (\alpha_j + \alpha_k)\xi_j^2 \eta_k^2 - 2\sum_{j,k=1}^3 (\alpha_j + \alpha_k)\xi_j \eta_j \xi_k \eta_k = 2\sum_1^3 \eta_k^2 + 2\sum_1^3 \xi_j^2 = 2|\xi - \eta|^2,$$

which means that

$$\sum_{1 \leq j < k \leq 3} (\alpha_j + \alpha_k)\pi_{jk}^2 = \sum_1^3 \pi_{0j}^2.$$

For the Plücker coordinates this means that

$$\sum_1^3 A_j p_{0j}^2 = p_{12}^2 + p_{23}^2 + p_{31}^2, \quad A_j = \sum_1^3 \alpha_k - \alpha_j.$$

The Plücker coordinates of the ray through $x \in \mathbf{R}^3$ with direction y are the two row determinants in $\begin{pmatrix} 1 & x_1 & x_2 & x_3 \\ 0 & y_1 & y_2 & y_3 \end{pmatrix}$, so the complex cone at x is defined by

$$\sum_1^3 A_j y_j^2 = |x \times y|^2 = |x|^2|y|^2 - (x, y)^2.$$

It degenerates if $|x|^2 y - Ay = (x, y)x$ for some $y \neq 0$, where A is the diagonal matrix with diagonal elements A_1, A_2, A_3. If $|x|^2 \notin \{A_1, A_2, A_3\}$ this implies $(x, y) \neq 0$ and $y = (x, y)(|x|^2 - A)^{-1}x$, hence

$$((|x|^2 - A)^{-1}x, x) = 1 = (x, x)/|x|^2, \quad \text{that is, } (A(|x|^2 - A)^{-1}x, x) = 0,$$

which is the equation of the Fresnel surface if $A_j = a_j^2$. When $|x|^2 = A_1$, say, then $x_1 = 0$, for if $x_1 \neq 0$ we must have $(x, y) = 0$, hence $y_2 = y_3 = 0$, so $y_1 \neq 0$, which is impossible since $(x, y) = 0$. As before we now obtain $\sum_2^3 A_j x_j^2/(|x|^2 - A_j) = 0$ which is an ellipse in the Fresnel surface. Thus the singularity surface of the complex is the Fresnel wave surface when $\alpha_j = \frac{1}{2}(\sum_1^3 a_k^2 - 2a_j^2)$. (If $a_1^2 = a_2^2 + a_3^2$ then $\alpha_1 = 0$ and the quadric degenerates to an ellipse in the plane $x_1 = 0$.)

The equation $\sum_1^3 x_j^2/(|x|^2 - A_j) = 1$ can be written $\sum_1^3 x_j^2/(\alpha_j - \mu) = 1$ where $\mu = \alpha - |x|^2$, $\alpha = \sum \alpha_j$. The equation $\sum_1^3 x_j^2/(\alpha_j - \lambda) = 1$ for determining the confocal quadrics through x has three roots with sum $\alpha - |x|^2$, so the Fresnel equation means that the roots are $\alpha - |x|^2$ and $\pm\lambda$ for some λ. Thus the Fresnel surface is generated by intersections of confocal quadrics with opposite parameters $\lambda, -\lambda$. The direction of the corresponding singular line of the complex has the coordinates $x_j/(|x|^2 - A_j) = x_j/(\alpha_j - \mu)$ which is orthogonal to the normals of the confocal surfaces with parameters $\pm\lambda$.

The singularity surface for the complex

$$\sum_1^3 B_j p_{0j}^2 = C_3 p_{12}^2 + C_1 p_{23}^2 + C_2 p_{31}^2$$

with positive B_j, C_j can be obtained by a change of variables $x_j \to \gamma_j x_j$, $j = 1, 2, 3$. This changes the Plücker coordinate p_{jk} to $\gamma_j \gamma_k p_{jk}$, with $\gamma_0 = 1$, so the new equation for the complex is

$$\sum_1^3 B_j p_{0j}^2/\gamma_j^2 = C_3 p_{12}^2/\gamma_1^2\gamma_2^2 + C_1 p_{23}^2/\gamma_2^2\gamma_3^2 + C_2 p_{31}^2/\gamma_3^2\gamma_1^2.$$

If $\gamma_j^2 = \sqrt{C_1 C_2 C_3}/C_j$ this is the complex already studied, with $A_j = B_j C_j/\sqrt{C_1 C_2 C_3}$. In the expanded equation of the singularity surface

$$1 - x_1^2(A_2^{-1} + A_3^{-1}) - x_2^2(A_1^{-1} + A_3^{-1}) - x_3^2(A_1^{-1} + A_2^{-1})$$
$$+ (x_1^2 + x_2^2 + x_3^2)(x_1^2 A_2^{-1} A_3^{-1} + x_2^2 A_1^{-1} A_3^{-1} + x_3^2 A_1^{-1} A_2^{-1}) = 0$$

we return to the original coordinates by replacing x_j with $x_j \gamma_j$, which simplifies to the equation

$$1 - x_1^2(C_3/B_2 + C_2/B_3) - x_2^2(C_1/B_3 + C_3/B_1) - x_3^2(C_2/B_1 + C_1/B_2)$$
$$+ (x_1^2 C_2 C_3 + x_2^2 C_1 C_3 + x_3^2 C_1 C_2)(x_1^2/B_2 B_3 + x_2^2/B_1 B_3 + x_3^2/B_1 B_2) = 0.$$

Expressing the line complex in dual coordinates just means interchanging B_j and C_j so the equation of the dual surface is obtained by this interchange, as in (5.4), (5.5).

Klein's model of the general complex surface with 4 double points seems to have disappeared but the others are in the collection. Klein [Kl1] should also be consulted for more classical information on quadratic line complexes, in particular how to construct a quadratic line complex which gives rise to a given Kummer surface. He studied them systematically by simultaneous diagonalisation of the quadratic forms Q and (5.7), using the connection between linear transformations in $\mathbf{R}^4$ and in $\bigwedge^2 \mathbf{R}^4$ mentioned above.

Models in the collection

H8–H32. 25 of the original Plücker models, labelled also with P and roman numbers referring to the description in [YTM].

H33–H34. Two of the models made by Klein, labelled also **KLEIN I, KLEIN III**. (The model KLEIN II has disappeared, only the label was found.)

5.4. Ruled fourth degree surfaces. The *conchoid of Nicodemus* is defined as follows: Given a line L and a point P not on L, draw the line through P and an arbitrary point on L and let Q be the point at a given distance a beyond L on the line. The locus of these points is the conchoid.

If we choose coordinates (x, y) such that P is the origin and L is defined by $y = 1$, then the equation of the conchoid becomes

(5.9) $$((y-1)/y)^2(x^2 + y^2) = a^2;$$

where we are now also allowing Q to be set off on the line in the direction of P. We set $a = z$ and simplify (5.9) to

(5.10) $$(y-1)^2(x^2 + y^2) = y^2 z^2.$$

This surface in $\mathbf{R}^3$ has two skew double lines, the line $x = y = 0$ and the line $y - 1 = z = 0$. Setting

(5.11) $$x = \tau\xi, \; y = \tau, \; z = (\tau - 1)\zeta,$$

we obtain from (5.10) the equation

(5.12) $$\xi^2 + 1 = \zeta^2$$

of a a hyperbola, which we can of course rationalize, so the surface (5.10) is a rational surface, ruled by the lines (5.11). The existence of these lines in the surface is just another way of stating the definition of the conchoid. They are parallel to the generators of the light cone $\{(\xi, \eta, \zeta); \xi^2 + \eta^2 = \zeta^2\}$ and pass through the double lines. They connect the points $(0, 0, -\zeta)$ and $(\xi, 1, 0)$ on the lines, which are in $(2, 2)$ algebraic correspondence by (5.12). There are no singular points outside the double lines.

Note that on the double line $x = y = 0$ the interval $-1 < z < 1$ is isolated on the real surface (5.10), for $x^2 + y^2(1 - z^2)$ is positive definite in (x, y) when $|z| < 1$. By (5.12) there are two generators through each point on the double line with $|z| > 1$, and one through the points with $z = \pm 1$. On the other hand, since the projection of the hyperbola (5.12) on the ξ axis covers it twice, there are two generators through every point on the other double line so there are no isolated points there.

The surface can also be thought of as obtained by moving the generators of the light cone in the direction of the z axis until they pass through the double line $y - 1 = z = 0$. However, it is easily verified that the surface is *not* characteristic.

There are also some wire models. One consists of two circles in parallel planes with points running round in opposite directions connected by lines. If the circles are defined by $\xi^2 + \eta^2 = R^2$, $\zeta = \pm c$, then the surface is parametrized by

$$(x, y, z) = (\xi, \eta, c) + \lambda(2\xi, 0, 2c),$$

if (ξ, η, c) and $(-\xi, \eta, -c)$ are the points connected. Thus $x/\xi = z/c$ and $y = \eta$ which gives the fourth degree equation

$$c^2 x^2 + y^2 z^2 = R^2 z^2.$$

There is also a model where the circles have been replaced by a parabola $2a\eta = \xi^2$; then the equation is $c^2 x^2 = 2ayz^2$ which is a cubic projectively equivalent to **R21**.

There is also a somewhat more complicated badly chipped plaster model from Paris which looks like a ruled surface generated by lines intersecting two congruent ellipses in parallel planes with orthogonal major axes. It may be the envelope of planes tangent to both ellipses, so that the generators connect points in the two ellipses with parallel tangents. However, no description has been preserved. It is also conceivable that the model represents a characteristic surface through one of the ellipses, that is, the envelope of planes tangent to the ellipse and forming an angle $\pi/4$ with its plane. The intersection with a plane at distance equal to the sum of the minor and major axes approximates an ellipse quite well numerically, so a certain identification of the intention of the model cannot be made.

There is a superficial resemblance with the model outside the mathematics department. According to Lars Gårding it originated from a wire model with generators intersecting the two ellipses and the line joining their centers. This is an algebraic surface, but the architect who had it made in stainless steel misunderstood the definition. Instead he connected points at equal arc length from a starting point at the end of the major axis in one ellipse and the opposite minor axis in the other one. This gives a transcendental surface.

5.5. A quartic surface with the symmetry of a cube. The surface

$$x^4 + y^4 + z^4 - 2(y^2z^2 + z^2x^2 + x^2y^2) + 2(x^2 + y^2 + z^2) - 3 = 0$$

is a very special example of a surface with the symmetry of a cube. (A systematic study of surfaces with the symmetry of a regular polyhedron was made by Goursat [G]. This particular example is on page 188 in [G].) The 8 vertices $(\pm 1, \pm 1, \pm 1)$ of the unit cube are conic double points. The boundary planes where $z = \pm 1$ are tangents to the surface along the 4 lines $z = \pm 1, x = \pm y$ and similarly for the other four boundary planes. The 6 points at infinity on these 12 lines are conic double points, which makes 14 conic double points in all. The four planes $\pm x \pm y \pm z = 0$ intersect the surface in great circles on the sphere defined by $x^2 + y^2 + z^2 = 3/2$, which are marked on the model. They intersect the plane at infinity in four lines contained in the surface, intersecting at the 6 double points at infinity just mentioned. The model represents only the part of the surface contained in the unit cube. At infinity it would look as 6 hyperbolic cylinders with axes in the directions of the diagonals of the faces of the cube and asymptotes given by the two lines at infinity meeting in the axis direction.

5.6. The Hessian of a cubic surface. For a cubic surface S in P^3 defined by $f(x) = 0$ where $x \in \mathbf{R}^4$ and f is a cubic form, the Hessian surface H is defined by $\det f''(x) = 0$, which is a quartic surface. Every (conic) double point of S is a (conic) double point of H: If there is a double point at $(1 : 0 : 0 : 0)$ we may assume that $f(x) = \frac{1}{2}x_0 \sum_1^3 a_j x_j^2 + B(x_1, x_2, x_3)$ and obtain $\det f''(x) = -\sum a_j a_k x_j x_k C_{jk}$ where C_{jk} are algebraic complements in the matrix $(\partial^2 f/\partial x_j \partial x_k)_{j,k=1}^3$, thus $C_{jk}(1, 0, 0) = a_1 a_2 a_3 \delta_{jk}/a_j$. In addition there may be many other double points. A model in the collection [Brill #54, p. 43] is the Hessian of the cubic surface **R2**. It has 14 real double point, 3 of which are at infinity, and it contains 16 lines. (One component seems to be missing in the center.)

5.7. A rational surface. There exist four copies of a surface made in Paris, Collection Muret, for which no catalogue is available. It appears to be the surface of fourth degree with the x axis as double line

$$(x^2 + y^2 + z^2)(y^2 + z^2) = z^2.$$

(The scale is 10 cm.) We can parametrize it by

$$x = \cos\varphi \cos\theta, \quad y = \sin\varphi \sin\theta \cos\theta, \quad z = \sin\varphi \cos^2\theta.$$

There is of course also a rational parametrization. If we put $y = kz$ we get the equation

$$(1 + k^2)x^2 + (1 + k^2)^2 z^2 = 1,$$

an ellipse with a rational point $x = 0$, $z = 1/(1+k^2)$. Putting $z = tx + 1/(k^2+1)$ therefore gives a rational parametrization,

$$x = \frac{-2t}{k^2 t^2 + t^2 + 1}, \quad y = \frac{k(1 - k^2 t^2 - t^2)}{(k^2 + 1)(k^2 t^2 + t^2 + 1)}, \quad z = \frac{1 - k^2 t^2 - t^2}{(k^2 + 1)(k^2 t^2 + t^2 + 1)}.$$

Computer drawings of the surface agree well with the plaster models.

Models in the collection

H35. The conchoid of Nicodemus.
H36. Wire model of ruled surface with two director circles.
H37. Wire model of ruled surface with two director parabolas.
H38. Plaster model of a ruled surface with two director ellipses?
H39. A quartic surface with the symmetry of a cube.
H40. The Hessian of a cubic surface.
H41. A rational surface with equation $(x^2 + y^2 + z^2)(y^2 + z^2) = z^2$.

6. Function theory. A real valued function of two variables can be represented by its graph which is a surface in $\mathbf{R}^3$, usually with the independent variables in the horizontal plane. For a complex valued function of a complex variable the graph lies in $\mathbf{C} \times \mathbf{C} \cong \mathbf{R}^4$ and cannot be visualized in $\mathbf{R}^3$. For this reason the models of analytic functions display the graph of either the real part or the imaginary part; the model is marked with R or I in the two cases. Often they occur in pairs. The curves orthogonal to the level curves in one model are then level curves in the other one by the Cauchy-Riemann equations.

6.1. Elementary analytic functions. An analytic function f with a simple pole at 0 with residue a is near 0 approximately of the form $f(z) = a/z$ and the equation $\mathrm{Re}(a/z) = \gamma$ can be written $\mathrm{Re}(a\bar{z}/\gamma) = |z|^2$, which means a circle with radius $|a|/|2\gamma|$ passing through the origin. On a model this looks like a narrowing spire with circular cross section. For a double pole we get instead a cross section which is a lemniscate of Bernoulli, and for a third order pole with leading term $1/z^3$ we get in polar coordinates the curve $r^3 = \gamma^{-1}\cos(3\theta)$ which has three leaves with vertex angle $\pi/3$ separated by equally large angles.

For an essential singularity such as that of $f(z) = e^{1/z}$ the behavior at the singularity is very complicated. We have $\mathrm{Re}\, f(z) = \exp(\mathrm{Re}\, 1/z)\cos(\mathrm{Im}\, 1/z)$. On the circles where $\mathrm{Im}\, 1/z = \gamma$ we see that when $z \to 0$ then $\mathrm{Re}\, f(z) \to 0$ in one direction and $\mathrm{Re}\, f(z) \to \pm\infty$ in the other direction. There is a long discussion of this function in [FC].

One can also visualize the topology of the graph of a many valued analytic function by just displaying over its domain in $\mathbf{C}$ a surface with one sheet for each determination of the function outside the branch points where several values coalesce. The analytic continuation of a function element along a curve is displayed qualitatively by such a model.

Models in the collection

F1. The real part of $1/z = x/(x^2 + y^2)$. Note that the imaginary part is $-y/(x^2 + y^2)$ so it is obtained by turning the model $90°$.

F2. The real part of $(2\varepsilon)^{-1}\log((z - \varepsilon)/(z + \varepsilon))$ for some small ε. Note that it approaches the real part of $-1/z$ when $\varepsilon \to 0$. (The imaginary part is multivalued and would be hard to model.)

F3. The real part of $e^{1/z}$ as an example of the complexity of an essential singularity.

F4. The real and imaginary parts of $\sqrt{z^2 - 1}$ or rather the upper sheets of the graphs of the two valued functions. The square root has a singularity at ± 1, but in addition the models show singularities where $\sqrt{z^2 - 1}$ is imaginary and real respectively because of a change of branch, that is, on the intervals $i\mathbf{R}$, $[-1, 1]$ and $(-\infty, -1]$, $[1, \infty)$ respectively.

F5. The real and imaginary parts of $\sqrt{z^4 - 1}$, or rather the upper sheets of the graph of the two valued functions. The singularities are seen from the preceding model.

F6. The Riemann surface for $w^2 = z$.

F7. The Riemann surface for $w^3 = z$.

F8. The Riemann surface for $w^2 = P(z)$ where P is of degree 3 with distinct zeros. (Neighborhoods of the zeros are omitted.)

6.2. Elliptic functions. If $\omega_1, \omega_2 \in \mathbf{C}$ are linearly independent over $\mathbf{R}$ then a meromorphic function f in $\mathbf{C}$ is called elliptic (doubly-periodic) with periods ω_1 and ω_2 if $f(z) = f(z + \omega_1) = f(z + \omega_2)$, $z \in \mathbf{C}$, thus

$$f(z) = f(z + \omega) \quad \forall \omega \in \Omega = \{n_1\omega_1 + n_2\omega_2; n_1 \in \mathbf{Z}, n_2 \in \mathbf{Z}\}.$$

If f is analytic, then f is bounded, hence a constant. The sum of the residues in a parallelogram $F = \{z_0 + t_1\omega_1 + t_2\omega_2; 0 \le t_1 \le 1, 0 \le t_2 \le 1\}$ with no poles on the boundary is equal to 0 since $\int_{\partial F} f(z)\, dz = 0$; if f is not a constant it follows that f cannot just have a simple pole in F. The simplest elliptic function is therefore in a sense the Weierstrass $\wp$ function which has double poles only at Ω and for which $\wp(z) = z^{-2} + O(z)$ at 0. The difference between two such functions is a constant vanishing at 0 so $\wp$ is uniquely determined, and the existence of $\wp$ follows from the fact that

$$\wp(z) = \frac{1}{z^2} + \sum_{0 \neq \omega \in \Omega} \left(\frac{1}{(z - \omega)^2} - \frac{1}{\omega^2}\right)$$

has the required properties. Note that $\wp(z)$ is even. From the Taylor expansion

$$\frac{1}{(z - \omega)^2} - \frac{1}{\omega^2} = \frac{1}{\omega^2}\left(\left(1 - \frac{z}{\omega}\right)^{-2} - 1\right) = \sum_{1}^{5}(j + 1)\left(\frac{z^j}{\omega^{j+2}}\right) + O((z/\omega)^6)/\omega^2$$

it follows that

$$\wp(z) = \frac{1}{z^2} + c_2 z^2 + c_4 z^4 + O(z^6) \quad \text{as } z \to 0, \quad c_2 = 3\sum_{0 \neq \omega \in \Omega}\frac{1}{\omega^4}, \quad c_4 = 5\sum_{0 \neq \omega \in \Omega}\frac{1}{\omega^6}.$$

The derivative

$$\wp'(z) = -2\sum_{\omega \in \Omega}\frac{1}{(z - \omega)^3} = -\frac{2}{z^3} + 2c_2 z + 4c_4 z^3 + O(z^5)$$

is odd, and at the origin we have

$$\mathfrak{P}'(z)^2 = \frac{4}{z^6} - \frac{8c_2}{z^2} - 16c_4 + O(z^2),$$

$$\mathfrak{P}(z)^3 = \frac{1}{z^6} + \frac{3c_2}{z^2} + 3c_4 + O(z^2),$$

so $\mathfrak{P}'(z)^2 - 4\mathfrak{P}(z)^3 = -20c_2/z^2 - 28c_4 + O(z^2)$ as $z \to 0$. Hence

$$\mathfrak{P}'(z)^2 - 4\mathfrak{P}(z)^3 + 20c_2\mathfrak{P}(z) + 28c_4 = 0$$

for the left-hand side is doubly periodic, has no poles and vanishes at the origin. With the notation

$$g_2 = 20c_2 = 60 \sum_{0 \neq \omega \in \Omega} \frac{1}{\omega^4}, \quad g_3 = 28c_4 = 140 \sum_{0 \neq \omega \in \Omega} \frac{1}{\omega^6},$$

it follows that

$$\frac{\mathfrak{P}'(z)}{\sqrt{4\mathfrak{P}(z)^3 - g_2\mathfrak{P}(z) - g_3}} = 1$$

for some choice of the square root. Let E be the elliptic integral

$$E(w) = \int_w^\infty \frac{dz}{\sqrt{4z^3 - g_2 z - g_3}}$$

taken along a path avoiding the zeros of $4z^3 - g_2 z - g_3$. With the branch $E(w) = w^{-\frac{1}{2}} + O(w^{-\frac{5}{2}})$ suitably chosen at infinity we have $E(\mathfrak{P}(z)) = z + O(z^5)$ at 0, and since the derivative of $E(\mathfrak{P}(z))$ is equal to 1, we obtain $E(\mathfrak{P}(z)) \equiv z$ as long as $\mathfrak{P}(z)$ avoids the zeros of the cubic, that is, $\mathfrak{P}'(z) \neq 0$. Now $\mathfrak{P}'(z) = 0$ at $\frac{1}{2}\omega_1$, $\frac{1}{2}\omega_2$, $\frac{1}{2}(\omega_1 + \omega_2)$ by the symmetries and at no other points except those congruent modulo Ω. This follows by residue calculus applied to $\mathfrak{P}''(z)/\mathfrak{P}'(z)$ which in a suitable parallelogram F has these three poles with residue 1 and only the pole at 0 with negative residue -3. Continuing $\mathfrak{P}(E(w)) = w$ along a closed path enclosing just two zeros of the cubic we conclude that the change of E along such a path is a period.

One can also start with a cubic polynomial $q(z) = 4z^3 - g_2 z - g_3$ with simple zeros, that is, $g_2^3 - 27g_3^2 \neq 0$, and form the inverse of the elliptic integral E to obtain the $\mathfrak{P}$ function with periods given by $\int dz/\sqrt{q(z)}$ around loops enclosing precisely two zeros.

There exists a doubly periodic meromorphic function with arbitrarily prescribed singularities in a fundamental parallelogram F such that the sum of the residues is equal to 0. This can clearly be achieved by a linear combination of translates of derivatives of $\mathfrak{P}(z)$ and the observation that if $a \notin \Omega$ then $f(z) = \int_{z-a}^z \mathfrak{P}(\zeta)\, d\zeta$ is independent of the path of integration, doubly periodic, and has singularities $-1/(z - \omega)$ and $1/(z - a - \omega)$, $\omega \in \Omega$.

Models in the collection

The models are based on two different choices of cubics q, and they model both $\mathfrak{P}(z)$ and $\mathfrak{P}'(z)$.

F9. With $g_2 = 4$ and $g_3 = 0$ we take ω_1 and ω_2 as the integral of $1/\sqrt{4z(z - 1)(z + 1)}$ around $[-1, 0]$ and $[0, 1]$ respectively, which gives the periods

$$\omega_1 = \int_{-1}^0 \frac{dx}{\sqrt{x(x - 1)(x + 1)}} = 2.622\ldots \quad \text{and} \quad \omega_2 = i\omega_1$$

Note that since $i\Omega = \Omega$, it follows that $\mathfrak{P}(iz) = -\mathfrak{P}(z)$. Hence $\mathfrak{P}'(iz) = i\mathfrak{P}'(z)$ which means that in this case one obtains the imaginary part by just rotating the real part $90°$. Therefore only $\operatorname{Re}\mathfrak{P}(z)$, $\operatorname{Im}\mathfrak{P}(z)$ and $\operatorname{Re}\mathfrak{P}'(z)$ have been modelled. Unfortunately $\operatorname{Im}\mathfrak{P}(z)$ is missing from the collection.

F10. With $g_2 = 0$ and $g_3 = 4$ the zeros are the cube roots of unity. We can define ω_1 by the integral around $-\frac{1}{2} + \frac{1}{2}i[-\sqrt{3}, \sqrt{3}]$, or its rotation by $\pm\frac{2}{3}\pi$. This gives for Ω the hexagonal lattice generated by $\omega_1 = 2.4286\ldots$ and any product of it by a non-real 6th root of unity. The models for $\operatorname{Re}\mathfrak{P}(z)$, $\operatorname{Im}\mathfrak{P}(z)$, $\operatorname{Re}\mathfrak{P}'(z)$ and $\operatorname{Im}\mathfrak{P}'(z)$ are all in the collection.

F11. With ω_1 real and $\omega_2 = i\omega_1/3$, there is a model of $\operatorname{Im} f(z)$, $f(z) = \int_{z-\omega_1/2}^{z} \mathfrak{P}(\zeta)\, d\zeta$. For arbitrary Ω we have $\mathfrak{P}(-z) = \mathfrak{P}(z)$, and since in this case $\overline{\Omega} = \Omega$ we have $\mathfrak{P}(\bar{z}) = \overline{\mathfrak{P}(z)}$ also, so $\mathfrak{P}(z)$ is real on both the real and the imaginary axes. Hence $f(z)$ is real on $\mathbf{R}$, and since $f'(z) = \mathfrak{P}(z) - \mathfrak{P}(z - \omega_1/2)$ we have

$$f'(\tfrac{1}{4}\omega_1 + iy) = \mathfrak{P}(\tfrac{1}{4}\omega_1 + iy) - \mathfrak{P}(-\tfrac{1}{4}\omega_1 + iy) = \mathfrak{P}(\tfrac{1}{4}\omega_1 + iy) - \overline{\mathfrak{P}(\tfrac{1}{4}\omega_1 + iy)}, \quad y \in \mathbf{R},$$

which is purely imaginary, so $f(\tfrac{1}{4}\omega_1 + iy)$ is real valued and so is $f(-\tfrac{1}{4}\omega_1 + iy)$, for $y \in \mathbf{R}$. Thus $\operatorname{Im} f(z) = 0$ if $z \in \mathbf{R} + \omega_2\mathbf{Z}$ or $z \in \tfrac{1}{4}\omega_1 + \tfrac{1}{2}\omega_1\mathbf{Z} + i\mathbf{R}$. To make this rectangular net with size $\omega_1/2$ in the real and $\omega_1/3$ in the imaginary direction quadratic, the scale has been stretched by a factor $3/2$ in the imaginary direction which makes the cross sections at the poles elliptic instead of circular — assuming that the model has been correctly understood. No description by the designer has been found.

6.3. Elliptic integrals. The elliptic integral of the first kind in Legendre's normal form is given by

$$(6.1) \qquad F(k,\varphi) = \int_0^{\varphi} \frac{dt}{\sqrt{1 - k^2 \sin^2 t}}.$$

(See [MOS, p. 358]. In [AS, p. 589] the notation $F(\varphi|k^2)$ is used.) When φ is considered as function of $u = F(k,\varphi)$ and k one writes $\varphi = \operatorname{am}(u,k)$ which is called the *Jacobi amplitude function*. Note that F is an even function of k, an odd function of φ, and that $du = d\varphi$ for all k when $\varphi = u = 0$.

(i) When $|k| < 1$ the integral (6.1) is defined for all φ. With the standard notation

$$(6.2) \qquad K(k) = F(k, \tfrac{1}{2}\pi) = \int_0^{\frac{1}{2}\pi} \frac{dt}{\sqrt{1 - k^2 \sin^2 t}},$$

which is called a *complete elliptic integral*, we have

$$u = F(k,\varphi) = 2K(k)\varphi/\pi + O(1), \quad \text{as } \varphi \to \infty,$$

and since $du/d\varphi = 1/\sqrt{1 - k^2\varphi^2}$ it is clear that we can regard φ as a function of u defined for all $u \in \mathbf{R}$. For small k we obtain

$$\varphi = (1 - \tfrac{1}{4}k^2)u + \tfrac{1}{8}k^2 \sin(2u) + O(k^4 u).$$

(ii) When $k = 1$ we can evaluate the integral explicitly,

$$F(1,\varphi) = -\log\tan(\tfrac{1}{4}\pi - \tfrac{1}{2}\varphi), \quad |\varphi| < \tfrac{1}{2}\pi, \quad \text{thus}$$
$$\varphi = 2\arctan e^u - \tfrac{1}{2}\pi = \tfrac{1}{2}\pi - 2\arctan e^{-u} \to \tfrac{1}{2}\pi \quad \text{as } u \to +\infty.$$

(iii) When $k > 1$ the integral $F(k, \varphi)$ is only defined when φ is in the component of the origin in $\{\varphi; |\sin \varphi| < 1/k\}$. However, then we have

$$F(k, \varphi) = k^{-1} F(k^{-1}, \theta), \quad \text{if } k \sin \varphi = \sin \theta, \ |\theta| < \tfrac{1}{2}\pi.$$

In fact, since $k \cos \varphi \, d\varphi = \cos \theta \, d\theta$ and

$$dF(k, \varphi)/d\varphi = 1/\sqrt{1 - \sin^2 \theta} = 1/\cos \theta, \text{ we have}$$
$$dF(k, \varphi)/d\theta = 1/(k \cos \varphi) = 1/\sqrt{k^2 - \sin^2 \theta} = k^{-1} dF(k^{-1}, \theta)/d\theta.$$

Thus $u = F(k, \varphi)$ is equivalent to $ku = F(k^{-1}, \theta)$, that is, for $|u| < F(k, \arcsin(1/k))$ we have

$$\operatorname{am}(u, k) = \varphi = \arcsin(k^{-1} \sin \theta) \in (-\tfrac{1}{2}\pi, \tfrac{1}{2}\pi), \quad \theta = \operatorname{am}(ku, k^{-1}).$$

We can use this formula to extend the definition of $\operatorname{am}(u, k)$ to all $u \in \mathbf{R}$. Note that θ is an increasing function of u taking all values on $\mathbf{R}$ whereas φ only varies between $\pm \arcsin(1/k)$. When u increases from 0, then φ increases first to $\arcsin(1/k)$ when $\theta = \tfrac{1}{2}\pi$, then decreases to $- \arcsin(1/k)$ when $\theta = \tfrac{3}{2}\pi$ and returns to 0 when $\theta = 2\pi$, which is repeated periodically. For large k we obtain from the discussion in (i)

$$\theta = (k - \tfrac{1}{4}k^{-1})u + \tfrac{1}{8}k^{-2} \sin(2uk) + O(k^{-3}u)$$

which shows that $\varphi = \operatorname{am}(u, k) = \arcsin(k^{-1} \sin(ku)) + O(k^{-3}u^3)$. In particular we have seen that $\operatorname{am}(u, k)$ is an analytic function of (u, k) for all (u, k); the graph does not have a bijective projection on any other coordinate plane. The model describes $\operatorname{am}(u, k)$ for $u \geq 0$. The φ axis is vertical, the k axis is the upper edge of the base and the u axis is directed inward. The unit on the k axis is $3/2$ times the unit on the other axes.

Model in the collection

F12. Jacobi's amplitude function $\varphi = \operatorname{am}(u, k)$ with φ vertical and k to the right [Brill #80, p. 48].

7. Differential geometry. In subsection 3.1 we have already defined the principal curvatures, the lines of curvature and the total and mean curvatures. The models in the collection give examples particularly of surfaces of constant total or mean curvature.

7.1. Surfaces of revolution. We shall consider the surface in $\mathbf{R}^3$ obtained by letting a curve in the half plane $\{(r, z) \in \mathbf{R}^2; r > 0\}$ rotate around the z axis, that is, taking $x = r \cos \theta$, $y = r \sin \theta$ with $0 \leq \theta \leq 2\pi$. For reasons of symmetry it is clear that one direction of principal curvature will then be along the meridians while the other will be orthogonal, thus defined by $dz = 0$. The corresponding center of curvature will be the intersection of the normal of the curve with the axis of rotation.

If the curve is parametrized by the arc length s and we take the normal in the direction $(-z'(s), r'(s))$, the curvature is

$$\kappa_1(s) = z''(s)r'(s) - r''(s)z'(s) = z''(s)/r'(s) = -r''(s)/z'(s),$$

where we have used that $r'(s)^2 + z'(s)^2 = 1$, hence $r'(s)r''(s) + z'(s)z''(s) = 0$, to eliminate one of the second derivatives. The other radius of curvature R_2 is given by $r(s) - R_2 z'(s) = 0$, so the corresponding curvature is

$$\kappa_2 = z'(s)/r(s).$$

Thus the total curvature is

$$K = \kappa_1 \kappa_2 = -r''(s)/r(s).$$

and the mean curvature H is given by

$$2H = \kappa_1 + \kappa_2 = z''(s)/r'(s) + z'(s)/r(s).$$

If the *total curvature* K *is constant* then

$$r''(s) + Kr(s) = 0, \quad r'(s)^2 + z'(s)^2 = 1.$$

(i) When $K = 0$ this means that $r'(s)$ and $z'(s)$ are constant, so the surface is just a cone of revolution.

(ii) When $K > 0$ we may assume after a change of scales that $K = 1$. After a translation of the s variable we obtain $r(s) = \varrho \cos s$ where ϱ is a positive constant, hence $z'(s)^2 = 1 - \varrho^2 \sin^2 s$. When $\varrho < 1$ we have after a suitable translation in the z direction the curve defined by

$$r(s) = \varrho \cos s, \quad z(s) = \int_0^s \sqrt{1 - \varrho^2 \sin^2 t}\, dt, \quad |s| < \tfrac{1}{2}\pi,$$

which at the end points meets the z axis at an angle $\arcsin \varrho$. When $\varrho = 1$ the surface is the unit sphere with the points on the z axis removed. Note that on the surface of revolution $(\theta, s) \mapsto (r(s) \cos \theta, r(s) \sin \theta, z(s))$ the first fundamental form, that is, the square of the arc length, is $ds^2 + r(s)^2 d\theta^2 = ds^2 + \cos^2 s\, d(\varrho\theta)^2$. This proves that the surface is obtained from the unit sphere by the map $\theta \mapsto \theta/\varrho$, that is, by taking away from the unit sphere the sector between two half great circles and pulling the edges together. There is a model with a partial brass covering in the collection, and there is a sheet of brass which can be applied to it arbitrarily: by Minding's theorem surfaces of the same constant curvature are locally isometric. When $\varrho > 1$ the surface only exists when $|s| < \arcsin(1/\varrho)$, and the tangent becomes horisontal at the end points. This bulge shaped surface is also in the collection, with one and two half bulges displayed. The brass sheet just mentioned fits there. Geodesics, corresponding to great circles on the sphere under the deformation described above, are marked in the models.

(iii) If $K < 0$ we normalize to $K = -1$. The general solution of the equation $r''(s) = r(s)$ is of the form $r(s) = ae^s + be^{-s}$. By translation of s and possibly a change of sign we can reduce to one of the three cases

$$r(s) = C \sinh s, \quad r(s) = e^s, \quad r(s) = C \cosh s,$$

where $C > 0$, and $s > 0$ in the first case. The condition $r'(s)^2 \le 1$ implies the conditions

$$C \cosh s \le 1, \quad s \le 0, \quad C |\sinh s| \le 1$$

respectively, which means that $\sup r(s) = \varrho$ where

$$\varrho^2 = 1 - C^2 < 1, \quad \varrho = 1, \quad \varrho^2 = 1 + C^2 > 1$$

in the three cases. In the first case we must of course have $C < 1$.

In the second case one can write $z(s)$ down explicitly,

$$z(s) = \int_s^0 \sqrt{1 - e^{2t}}\, dt = \sqrt{1 - r(s)^2} - \log \frac{1 + \sqrt{1 - r(s)^2}}{r(s)}, \quad s \le 0.$$

The curve is *Huygens' tractrix* and the surface of revolution is called the *pseudosphere*. The usual geometric interpretation of the curve is obtained by noting that $z'(s)^2 + r'(s)^2 = 1$ implies $(dz/dr)^2 + 1 = 1/r'(s)^2 = 1/r^2$, thus $r^2(dz/dr)^2 + r^2 = 1$, which means that the distance from (r, z) to the z axis along the tangent, with direction $(1, dz/dr)$, is always equal to 1, which explains the name tractrix.

In the first case we have $0 < C < 1$, $0 < s < \operatorname{arccosh}(1/C)$, $z'(s) = \sqrt{1 - C^2 \cosh^2 s}$, hence $dz/dr = \sqrt{1/(C^2 \cosh^2 s) - 1}$, which decreases from $\sqrt{1 - C^2}/C$ to 0. The curve is concave and the surface becomes *conic* at the origin with vertex angle $2 \arcsin C$.

In the third case we have $|s| < \operatorname{arcsinh}(1/C)$, $z'(s) = \sqrt{1 - C^2 \sinh^2 s}$, hence $dr/dz = C \sinh s/\sqrt{1 - C^2 \sinh^2 s}$ which increases from $-\infty$ to $+\infty$. Thus r is an even convex function of z; the surface of revolution is said to be of *hyperbolic type* since it looks somewhat like a truncated hyperboloid with one sheet.

All three cases are represented in the collection. In the first case $C = \cos 70°$, which gives a vertex angle of $40°$. In the third case $\varrho = 1.1$ and $C = \sqrt{0.21} = 0.458....$

If the *mean curvature H is constant* we have with the notation at the beginning of this subsection

$$2Hr(s)r'(s) = r(s)z''(s) + r'(s)z'(s).$$

This means that $Hr(s)^2 - r(s)z'(s)$ is a constant.

(i) If $H = 0$ then

$$r(s)dz(s)/ds = C, \quad \text{hence } r^2(dz/ds)^2 = C^2, \ r^2(1 - (dr/ds)^2) = C^2,$$

which gives $(dr/dz)^2 = (r^2 - C^2)/C^2$. It is clear that $r \equiv C$ is a solution, which gives a circular cylinder. Otherwise we get after differentiation and division by $2dr/dz$ that $d^2r/dz^2 = r/C^2$. After a translation of the z variable this gives three possibilities:

$$r = \gamma \cosh(z/C), \quad r = \gamma e^{\pm z/C}, \quad r = \gamma \sinh(z/C).$$

However, the equation $(dr/dz)^2 = (r^2 - C^2)/C^2$ rules out the second and third possibility, for $z \to \mp\infty$ and $z = 0$ respectively, and in the first case we get $\gamma = C$. Thus

$$r = C \cosh(z/C),$$

which is the *catenary*, the equilibrium figure of a homogeneous chain under the action of gravity. The surface of revolution is the *catenoid*, which has the radii of curvature $\pm r^2/C$.

(ii) If H is a constant $\neq 0$, we have with a constant C

$$H(r^2 + C) = rdz/ds, \quad \text{hence } H^2(r^2 + C)^2 = r^2(1 - (dr/ds)^2)$$

which gives

$$\left(\frac{dz}{dr}\right)^2 = \frac{(r^2 + C)^2}{H^{-2}r^2 - (r^2 + C)^2}.$$

If $C = 0$ then $dz/dr = \pm Hr/\sqrt{1 - H^2r^2}$, which means that $Hz \pm \sqrt{1 - H^2r^2}$ is a constant. This defines a circle, so we find not surprisingly a sphere with radius $1/H$. Assume now that $C \neq 0$. In the denominator we have a quadratic polynomial in r^2 with leading coefficient -1 which is negative at the origin. No curve is obtained unless it takes positive values somewhere on the positive axis, so the roots must be positive and different. Denoting them by α_1^2 and α_2^2, where $0 < \alpha_2 < \alpha_1$, we have

$$(r^2 + C)^2 - H^{-2}r^2 = (r^2 - \alpha_1^2)(r^2 - \alpha_2^2), \quad \text{thus } C^2 = \alpha_1^2\alpha_2^2, \ H^{-2} - 2C = \alpha_1^2 + \alpha_2^2,$$

which gives $C = \pm\alpha_1\alpha_2$ and $H^{-2} = (\alpha_1 \pm \alpha_2)^2$. The curve is given by

$$\frac{dz}{dr} = \pm\frac{r^2 \pm \alpha_1\alpha_2}{\sqrt{(\alpha_1^2 - r^2)(r^2 - \alpha_2^2)}}.$$

The rotational surface is called an *onduloid* when the positive sign is taken in the numerator and a *nodoid* when the negative sign is used there. In the first case we get a monotonic curve defined for $\alpha_2 \leq r \leq \alpha_1$, with vertical tangent at the end points. In fact, it is smooth there as a function of s. Piecing together such curves with alternating signs we get a periodic analytic curve. The case of the nodoid is different, for the curve turns back when $r = \sqrt{\alpha_1\alpha_2}$. If one extends the curve then self intersections will appear when the extension is made at $r = \alpha_2$. The larger model in the collection stops before the self-intersection would appear. There is also a smaller model of the inner part up to the self-intersection. In the models $\alpha_1/\alpha_2 = 5.77$.

Delauney [De] discovered that the meridian of the onduloid is generated by the focus of an ellipse with semiaxes $\frac{1}{2}(\alpha_1 + \alpha_2)$ and $\sqrt{\alpha_1\alpha_2}$ rolling on the z axis while the nodoid is generated by the focus of a hyperbola with axes $\frac{1}{2}(\alpha_1 - \alpha_2)$ and $\sqrt{\alpha_1\alpha_2}$. Sturm [St] immediately gave the following proof. Suppose that an ellipse with axes $a > b$ and foci $F = (r, z)$, $F^* = (r^*, z^*)$ is rolling on the z axis. Since the z axis is a tangent we have $rr^* = b^2$, a consequence of the fact that the projection of a focus on a tangent is on the circumscribed circle, for r and r^* are the lengths of the intervals in which a focus divides a chord in this circle. Infinitesimally the ellipse rotates around the point K where the it touches the z axis, so the tangent of the curve described by F is orthogonal to the vector $\overrightarrow{KF}$. If α is the angle between the z axis and the vectors $\overrightarrow{KF}$, $\overrightarrow{KF^*}$, then $r = |\overrightarrow{FK}|\sin\alpha$, $r^* = |\overrightarrow{F^*K}|\sin\alpha$, hence $r + r^* = 2a\sin\alpha$ so elimination of r^* gives $r^2 + b^2 = 2ar\sin\alpha$. Since the slope $dz/dr = \pm\tan\alpha = \pm\sin\alpha/\sqrt{1 - \sin^2\alpha}$ we obtain

$$\frac{dz}{dr} = \pm\frac{r^2 + b^2}{\sqrt{4a^2r^2 - (r^2 + b^2)^2}}$$

which is the equation of the meridian of the onduloid when $\alpha_1\alpha_2 = b^2$ and $\alpha_1^2 + \alpha_2^2 = 4a^2 - 2b^2 = 4a^2 - 2\alpha_1\alpha_2$, that is, $\alpha_1 + \alpha_2 = 2a$. Note that $c = \sqrt{a^2 - b^2} = \frac{1}{2}(\alpha_1 - \alpha_2)$ so $a - c = \alpha_2$, $a + c = \alpha_1$. The case of the hyperbola with transversal axis a and conjugate axis b gives similarly the equation

$$\frac{dz}{dr} = \pm\frac{r^2 - b^2}{\sqrt{4a^2r^2 - (r^2 - b^2)^2}}$$

which is the equation of the nodoid if $b^2 = \alpha_1\alpha_2$ and $\alpha_1^2 + \alpha_2^2 = 4a^2 + 2b^2 = 4a^2 + 2\alpha_1\alpha_2$, that is, $\alpha_1 - \alpha_2 = 2a$, and $c = \sqrt{a^2 + b^2} = \frac{1}{2}(\alpha_1 + \alpha_2)$, so $\alpha_1 = c + a$ and $\alpha_2 = c - a$. Only the part where $r < b = \sqrt{\alpha_1\alpha_2}$ is obtained when the branch of hyperbola closest to the focus is rolling on the z axis; the point with $r = b$ is a limit point. The part of the curve with $b < r \leq \alpha_1$ is obtained when the branch furthest from the focus is rolling on the z axis. The curve will have a self intersection if it is continued periodically. The larger model of the nodoid uses one branch of the outer curve and two incomplete branches of the inner one. The smaller model represents the inner curve and pieces of two outer curves up to the self intersection. The dimensions are chosen so that the two axes of the hyperbola differ by less than 1%.

If the parabola rolls on a line one obtains similarly a catenary so also this case is covered by the geometrical construction.

Models in the collection

D1. Surface of revolution with positive constant total curvature which meets the axis of revolution, a piece of brass is embedded in the surface [Brill #126, p. 56].

D2. Surface of revolution with positive constant total curvature which does not meet the axis of revolution [Brill #127, p. 56].

D3. Brass sheet with positive constant total curvature which fits the preceding two models [Brill #129, p. 56].

D4. Surface of revolution with negative constant total curvature of cone type, with geodesic lines in blue [Brill #132, p. 57].

D5. Surface of revolution with negative constant total curvature of hyperbolic type, with parallel geodesic lines in green and two in red which approach the boundary [Brill #133, p. 57].

D6. Surface of revolution with negative constant total curvature, the pseudosphere, with some geodesic lines in blue [Brill #134, p. 57].

D7. An onduloid [Brill #140, p. 59].

D8. A nodoid, obtained using the full outer arc and a piece of the inner ones [Brill #141, p. 59].

D9. A nodoid obtained from the inner curves up to the point of intersection [Brill #142, p. 59].

D10. A catenoid, with some geodesic lines marked [Brill #143, p. 59].

7.2. Generalized helicoids (screws). If a curve $s \mapsto (r(s), z(s))$ in $\{(r, z) \in \mathbf{R}^2, r > 0\}$ is translated with constant speed in the z direction while it is rotated around the z axis, a (generalized) helicoid is generated. Thus it has the parametrization

$$(s, \theta) \mapsto (r(s)\cos\theta, r(s)\sin\theta, z(s) + a\theta) \in \mathbf{R}^3$$

where a is a constant. As in the preceding subsection we assume that s is the arc length, so that

$$r'(s)^2 + z'(s)^2 = 1.$$

The first fundamental form of the helicoid is then

$$ds^2 + 2az'(s)dsd\theta + (r(s)^2 + a^2)d\theta^2.$$

The unit normal is

$$-(ar'(s)\sin\theta - r(s)z'(s)\cos\theta, -ar'(s)\cos\theta - r(s)z'(s)\sin\theta, r(s)r'(s))/\sqrt{a^2r'(s)^2 + r(s)^2},$$

and the second fundamental form becomes

$$-(r(z''r' - r''z')ds^2 - 2ar'^2dsd\theta + r^2z'd\theta^2)/\sqrt{a^2r'^2 + r^2}.$$

The total curvature K and the mean curvature H are therefore

$$K = \frac{(z''r' - r''z')r^3 z' - a^2 r'^4}{(r^2 + a^2 r'^2)(r^2 + a^2 - a^2 z'^2)} = -\frac{r''r^3 + a^2 r'^4}{(r^2 + a^2 r'^2)^2},$$

$$2H = -\frac{r^2 z' + (r^2 + a^2)r(z''r' - r''z') + 2a^2 r'^2 z'}{(r^2 + a^2 - a^2 z'^2)\sqrt{r^2 + a^2 r'^2}}.$$

The total curvature K is constant if

$$r''r^3 + a^2 r'^4 + K(r^2 + a^2 r'^2)^2 = 0.$$

A special solution observed by U. Dini is $r(s) = e^s$, $s < 0$, when $K = -1/(1 + a^2)$. The curve is then again Huygens' tractrix. There is a model in the collection consisting of two arcs of the tractrix oriented in opposite directions, that is, $r(s) = e^{-|s|}$, $|s| < s_0$, which gives two surfaces of constant curvature joined at two helices.

The total curvature is identically 0 if and only if $r''r^3 + a^2 r'^4 = 0$, that is, $1/r'^2 + a^2/r^2$ is a constant C. The equations

$$\left(\frac{dr}{ds}\right)^2 = \frac{1}{C - a^2/r^2}, \quad \left(\frac{dz}{ds}\right)^2 = \frac{C - 1 - a^2/r^2}{C - a^2/r^2}$$

give the differential equation

$$\left(\frac{dz}{dr}\right)^2 = (C - 1 - a^2/r^2) = a^2(r^2/\varrho^2 - 1)/r^2$$

if we set $C - 1 = a^2/\varrho^2$. Integration gives

$$z = \pm a(\arcsin(\varrho/r) + \sqrt{r^2/\varrho^2 - 1}) + \text{const}.$$

This could also have been seen directly geometrically, for a surface with total curvature 0 is (generically) a developable surface, generated by the tangents of a curve. For the helix $\theta \mapsto (\varrho \cos \theta, \varrho \sin \theta, a\theta)$ the tangent surface is parametrized by

$$(\theta, t) \mapsto (\varrho \cos \theta, \varrho \sin \theta, a\theta) + t(-\varrho \sin \theta, \varrho \cos \theta, a) = (x, y, z).$$

If we intersect with the plane $y = 0$ and put $x = r$, we obtain

$$\varrho \cos \theta - t\varrho \sin \theta = r, \quad \varrho \sin \theta + t\varrho \cos \theta = 0, \quad a(\theta + t) = z.$$

Thus $t = -\tan \theta$, $\varrho/\cos \theta = r$, $t^2 = r^2/\varrho^2 - 1$, so

$$z = a(\arccos(\varrho/r) \pm \sqrt{r^2/\varrho^2 - 1})$$

which agrees with the previous result for suitable signs. The parametrization of the meridian curve

$$r = \varrho/\cos \theta = \varrho + \tfrac{1}{2}\varrho\theta^2 + O(\theta^4), \quad z = a(\theta - \tan \theta) = -a\theta^3/3 + O(\theta^5)$$

shows that it has a cusp when $r = \varrho$ with tangent in the radial direction. One of the models in the collection only covers the lower part of the cusp, corresponding to $\theta > 0$, that is, $t < 0$.

The classical helicoid is obtained by specializing the meridian curve to the r axis, that is, taking $r(s) = s$, $s > 0$, and $z(s) = 0$. Then we obtain

$$K = -\frac{a^2}{(r^2 + a^2)^2}, \quad H = 0,$$

so the classical helicoid is a *minimal surface* with radii of curvature $\pm(r^2+a^2)/a$. — There are many other helicoids with 0 mean curvature; the equation $H = 0$ can be written

$$(r^2 + 2a^2r'^2)(1 - r'^2) - (r^2 + a^2)rr'' = 0.$$

The classical helicoid is the singular solution with $r' = 1$, but we could instead solve the equation with arbitrary initial values such that $|r'| < 1$ and obtain a generalized helicoid which is a minimal surface. For every $r_0 > 0$ the differential equation has an even convex solution $r(s)$, $s \in \mathbf{R}$, with $r(0) = r_0$, $r'(0) = 0$, and $r'(s)^2 < 1$ for all s. It suffices to prove this for $s > 0$. Let the solution exist with $r' < 1$ for $0 \le s < S$. Then $r(s) < r_0 + s$, when $0 < s < S$, and since

$$\frac{r''r'}{1 - r'^2} = \frac{r^2 + 2a^2r'^2}{r^2 + a^2}\frac{r'}{r} < \frac{2r'}{r}$$

we have $\log(1/(1 - r'^2)) < 4\log(r/r_0) < 4\log(1 + s/r_0)$ when $0 \le s < S$. This proves that the solution can be continued beyond $s = S$, so it exists globally. Since

$$\frac{r''r'}{1 - r'^2} > \frac{r_0^2}{r_0^2 + a^2}\frac{r'}{r},$$

and since $r > \frac{1}{2}r_0 + cs$ by the convexity it follows that $r'(s) \to 1$ as $s \to \infty$, so $r(s)/s \to 1$. It is easy to see that $r'(s) - 1 = O(1/s^2)$ so $r(s) - s$ is bounded. Thus for large s we have a good approximation to the classical helicoid. No models seem to have been made of these surfaces.

By a theorem of Bour [B] every generalized helicoid is (locally) isometric to a surface of revolution. For the proof we represent the helicoid as above by

$$(s, \theta) \mapsto (r(s)\cos\theta, r(s)\sin\theta, z(s) + a\theta)$$

where $r'(s)^2 + z'(s)^2 = 1$. Recall that the first fundamental form is

$$ds^2 + 2az'(s)dsd\theta + (r(s)^2 + a^2)d\theta^2 = \frac{r(s)^2 + a^2r'(s)^2}{r(s)^2 + a^2}ds^2 + (r(s)^2 + a^2)\left(d\theta + \frac{az'(s)}{r(s)^2 + a^2}ds\right)^2.$$

For a surface of revolution

$$(\sigma, \varphi) \mapsto (\varrho(\sigma)\cos\varphi, \varrho(\sigma)\sin\varphi, \zeta(\sigma))$$

with $\varrho'(\sigma)^2 + \zeta'(\sigma)^2 = 1$, the fundamental form is $d\sigma^2 + \varrho(\sigma)^2d\varphi^2$, the special case of the helicoid with $a = 0$. They agree if

$$\varphi = \theta + \int\frac{az'(s)\,ds}{r(s)^2 + a^2}, \quad \varrho(\sigma) = \sqrt{r(s)^2 + a^2}, \quad \frac{d\sigma}{ds} = \sqrt{\frac{r(s)^2 + a^2r'(s)^2}{r(s)^2 + a^2}}.$$

The last two equations give

$$\varrho'(\sigma)^2 = \frac{r(s)^2 r'(s)^2}{r(s)^2 + a^2}\left(\frac{ds}{d\sigma}\right)^2 = \frac{r(s)^2 r'(s)^2}{r(s)^2 + a^2 r'(s)^2} \leq \frac{r(s)^2}{r(s)^2 + a^2} = \frac{\varrho(\sigma)^2 - a^2}{\varrho(\sigma)^2} < 1,$$

since $r'(s)^2 \leq 1$, so one can find $\zeta(\sigma)$ such that $\zeta'(\sigma)^2 = 1 - \varrho'(\sigma)^2$. Given the generalized helicoid one can therefore always find an isometric surface of revolution, determined up to the integration constant for σ as a function of s.

Let us now assume that the surface of revolution, that is $\varrho(\sigma)$, is given and determine the generalized helicoid, that is $r(s)$, from the preceding equations. We have already seen that this requires that

$$\varrho(\sigma) > a, \quad \varrho'(\sigma)^2 \leq (\varrho(\sigma)^2 - a^2)/\varrho(\sigma)^2.$$

The second condition means that the tangent to the meridian curve does not enter the sphere with radius a with center at the point on the axis closest to the point of tangency. Assume that the condition is fulfilled with strict inequality and set $r(s) = \sqrt{\varrho(\sigma)^2 - a^2}$ where s is determined by

$$\frac{ds}{d\sigma} = \sqrt{\frac{r(s)^2 + a^2}{r(s)^2 + a^2 r'(s)^2}}, \quad (r(s)^2 + a^2 r'(s)^2)\varrho'(\sigma)^2 = r(s)^2 r'(s)^2.$$

This gives $r(s)^2 + a^2 r'(s)^2 = r(s)^4/(r(s)^2 - a^2\varrho'(\sigma)^2)$, hence

$$\frac{ds}{d\sigma} = \frac{\varrho(\sigma)}{\varrho(\sigma)^2 - a^2}\sqrt{\varrho(\sigma)^2 - a^2(1 + \varrho'(\sigma)^2)}, \quad r'(s)^2 = \frac{(\varrho(\sigma)^2 - a^2)\varrho'(\sigma)^2}{\varrho(\sigma)^2 - a^2 - a^2\varrho'(\sigma)^2} < 1,$$

so we can determine $r(s)$ and $z(s)$, hence θ.

If we apply these results to the classical helicoid, that is, take $r(s) = s$ and $z(s) = 0$, then we obtain $\varphi = \theta$, $\varrho(\sigma) = \sqrt{s^2 + a^2}$, $d\sigma/ds = 1$, so we can take $\sigma = s$. The equation $\varrho' = \sigma/\sqrt{\sigma^2 + a^2}$ gives $\zeta' = a/\sqrt{\sigma^2 + a^2}$, that is, $\zeta = a\log(\sigma + \sqrt{\sigma^2 + a^2})$ with a suitable origin and direction of the z-axis. Thus $e^{\zeta/a} = \sigma + \sqrt{\sigma^2 + a^2}$ and $a^2 e^{-\zeta/a} = \sqrt{\sigma^2 + a^2} - \sigma$, which gives

$$\varrho = \sqrt{\sigma^2 + a^2} = \tfrac{1}{2}a(e^{\zeta/a}/a + ae^{-\zeta/a}) = \tfrac{1}{2}a\cosh((\zeta - a\log a)/a),$$

so the surface of revolution is the catenoid. This was expected for it is the only minimal surface of revolution.

Bour [B] calculated explicitly a special example of a generalized helicoid which is isometric to an ellipsoid of revolution; the equations have an elementary integral for a special choice of the parameters. His beautiful drawings are attached to this guide. In the collection there is a model of his helicoid (actually two strands of it) and the isometric ellipsoid [Brill #137, #138, p. 58]. As indicated in the drawings there is isometry only with a part of the ellipsoid. Unfortunately the brass sheet [Brill #139] which is supposed to fit the two surfaces in only one way (apart from rotation or helical motion) is missing.

Models in the collection

D11. Helicoid [Brill #146, p. 59].
D12. Another helicoid, mounted on a disk of wood.

D13. Generalized helicoid of constant negative total curvature, with meridian curve consisting of two arcs of Huygens' tractrix [Brill #135, p. 57].

D14. Generalized helicoid (two strands) which is isometric to an ellipsoid of revolution [Brill #137, p. 58].

D15. Ellipsoid of revolution isometric to the preceding model [Brill #138, p. 58].

D16. Half the tangent surface of a helix, with the central cylinder visible (model from Paris).

D17. The full tangent surface of a helix (model from Paris).

7.3. Triply periodic minimal surfaces. The much studied Plateau problem consists in finding a minimal surface with given boundary. Fairly explicit solutions (involving elliptic functions) were given by Schwarz [Sc, 1–125] for a boundary consisting of four line segments. The solution is analytic up to this boundary, which easily shows that the minimal surface can be continued across a line in the boundary by orthogonal reflection in the line. The process can be repeated, and if the original four sided boundary is chosen carefully this can give a minimal surface which is periodic in three perpendicular directions. Schwarz also constructed minimal surfaces bounded by two lines and two curves lying in planes meeting the minimal surface orthogonally. The minimal surface can then be extended both by reflection in a line in the boundary and by reflection in such a plane. This can also yield periodic minimal surfaces. (See the four plates in [Sc] and particularly the plates II. (b) and VI. (a)-(d) in [DHKW].)

As Plateau before him Schwarz liked to display minimal surfaces made by soap films on wire frames. Some of these have been preserved.

Models in the collection

D18. A number of models of Schwarz' *P*-surface, which has the periodicity of a cubic lattice.

D19. Two identical symmetric pieces of the surface which together with mirror symmetric pieces build the Schwarz *H*-surface. The surface (gray) joins two orthogonal diagonals in a cube in opposite faces (black) and intersect two other red faces orthogonally. On two of the black half squares there are red rings suggesting that they should be aligned to model reflection in a line.

D20. Wire frames for making soap films [Brill #148, p. 21 and p. 60]: a) a regular octahedron (see [Sc p. 108]), b) a pyramid with square base (see [Sc p. 98]), c) a prism with triangular base, d) a regular tetrahedron, e) four adjacent sides of a regular tetrahedron (see [Sc p. 1]), f) a prism over a hectagon (see [Sc p. 107]), g) two crossed rectangles for making the fifth Scherck surface (see [Sc, p. 102]), h) a helix, i) a Jordan curve containing two pairs of circular arcs used to give examples where there are several solutions of the Plateau problem. (See [N, p.37, p. 396 ff.])

7.4. Focal surfaces. For a surface embedded in $\mathbf{R}^3$ the centers of the two principal circles of curvature form in general two focal surfaces. For quadratic surfaces they are easy to calculate using the discussion of focal families in subsection 3.1. For example, for the elliptic paraboloid $x_1^2/A_1 + x_2^2/A_2 = 2x_3$, $0 < A_2 < A_1$, the lines of curvature are the intersections with the surfaces $\sum_1^2 x_j^2/(A_j - \lambda) = 2x_3 - \lambda$, with $A_2 < \lambda < A_1$ or $\lambda > A_1$ for the two families. Elimination gives

$$x_1^2(A_1 - A_2)/(A_1(A_1 - \lambda)) = 2x_3 + A_2 - \lambda, \quad x_2^2(A_2 - A_1)/(A_2(A_2 - \lambda)) = 2x_3 + A_1 - \lambda$$

as parametrization of the lines of curvature. The normal of the surface has the direction $(x_1/A_1, x_2/A_2, -1)$, so $y_j = x_j + (x_3 - y_3)x_j/A_j$, $j = 1, 2$, if y is on the normal. The locus of the center of curvature is the envelope, defined by

$$(1 + (x_3 - y_3)/A_1)dx_1 + x_1 dx_3/A_1 = 0 \quad \text{if } x_1(A_1 - A_2)/(A_1(A_1 - \lambda))dx_1 = dx_3,$$

which gives $y_3 = A_1 + A_2 + 3x_3 - \lambda$ and the parametrization of the focal surface

$$A_1 y_1^2 = \frac{A_1 - \lambda}{A_1 - A_2}(2x_3 + A_2 - \lambda)^3, \quad A_2 y_2^2 = \frac{\lambda - A_2}{A_1 - A_2}(2x_3 + A_1 - \lambda)^3, \quad y_3 = A_1 + A_2 + 3x_3 - \lambda.$$

A similar computation works for quadratic surfaces with center.

Models in the collection

D21. One of the focal surfaces for the paraboloid $y^2/12 + z^2/10 = 2x$ rescaled by $z \mapsto \sqrt{1.2}z$ [Brill #149, p. 60].

D22. The preceding focal surface joined to the other focal surface [Brill #151, p. 60].

D23. A focal surface of a hyperboloid with one sheet [Brill #152, p. 60].

D24. The other focal surface of the same hyperboloid [Brill #153, p. 60].

D25. The preceding two surfaces joined together [Brill #154, p. 60].

References

[AS]   M. Abramowitz and I. A. Stegun, *Handbook of Mathematical Functions*, National Bureau of Standards, 1964.

[A]   V. I. Arnol'd, *Normal forms for functions near degenerate critical points, the Weyl groups of A_k, D_k, E_k and Lagrangian singularities*, Functional Anal. Appl. **6** (1972), 254–272.

[B]   E. Bour, *Théorie de la déformation des surfaces*, Journal de l'École Polytechnique **22** (1861–62), 1–148.

[De]   Ch. Delauney, *Sur la surface de révolution dont la courbure moyenne est constante*, J. Math. Pures Appl. **6** (1841), 309–315.

[Dy]   W. Dyck, *Einleitender Bericht über die Mathematische Ausstellung in München*, Jahresber. Deutsch. Math.-Verein. **3** (1892–93), 39–56.

[DHKW]   U. Dierkes, S. Hildebrandt, A. Küster, O. Wohlrab, *Minimal Surfaces I*, Springer Verlag, 1991.

[FC]   G. Fischer, *Mathematical Models. Commentary*, Vieweg Verlag, 1986.

[FP]   ——, *Mathematische Modelle. Mathematical Models*, Vieweg Verlag, 1986.

[F]   F. G. Friedlander, *Simple progressive solutions of the wave equation*, Proc. Cambridge Phil. Soc. **43** (1947), 360–373.

[G]   E. Goursat, *Étude des surfaces qui admettent tous les plans de symétrie d'un polyèdre régulier*, Ann. Sci. École Norm. Sup. **4** (1887), 159–200.

[GH]   P. Griffiths and J. Harris, *Principles of algebraic geometry*, John Wiley & Sons, 1978.

[H]   R. W. H. T. Hudson, *Kummer's quartic surface*, Cambridge University Press, 1905.

[Kl1]   F. Klein, *Gesammelte Mathematische Abhandlungen I, II*, Springer Verlag, 1921–1922.

[Kl2]   ——, *Vorlesungen über die Entwicklung der Mathematik im 19. Jahrhundert I*, Springer Verlag, 1926.

[Ku]   E. E. Kummer, *Gesammelte Abhandlungen II*, Springer Verlag, 1975.

[M]   J. Mather, *Stability of C^∞ mappings, III: Finitely determined map-germs*, Publ. Math. IHES No **35** (1968), 127–156.

[Me]   W. Fr. Meyer, *Flächen Dritter Ordnung*, Enc. d. math. Wiss. III 2.II.B, pp. 1437–1531; *Spezielle algebraische Flächen*, pp. 1533–1779.

[MOS]   W. Magnus, F. Oberhettinger and R. P. Soni, *Formulas and Theorems for the Special Functions of Mathematical Physics*, Springer Verlag, 1966.

[N]   J. C. C. Nitsche, *Vorlesungen über Minimalflächen*, Springer Verlag, 1975.

[R]   M. Riesz, *A special characteristic surface — a new relativistic model for a particle?*, Collected works, Springer Verlag, 1988, pp. 848–858.

[Ro]   K. Rohn, *Die verschiedenen Gestalten der Kummer'schen Fläche*, Math. Ann. **18** (1881), 99–159.

[Sc]   H. A. Schwarz, *Gesammelte Mathematische Abhandlungen*, Springer Verlag, 1890.

[St]   M. Sturm, *Note à l'occasion de l'article précédent*, J. Math. Pures Appl. **6** (1841), 315–320.

[T]   J. C. Tougeron, *Idéaux de fonctions différentiables. I.*, Ann. Inst. Fourier **18** (1968), 177–240.

[YTM]   R. Diesel et al., *Ytmodellerna*, A collection of catalogs and descriptions of the models in the section "Allmänt" of the library.

[Z]   Konrad Zindler, *Algebraische Liniengeometrie*, Enc. d. math. Wiss. III 2.IIA, pp. 973–1228.

THE PROOF OF THE NIRENBERG-TREVES CONJECTURE
ACCORDING TO N. DENCKER AND N. LERNER

LARS HÖRMANDER

1. Introduction. Since 1980 the remaining part of this conjecture has been to prove that if P is a pseudodifferential operator of principal type, then the equation $Pu = f$ has a distribution solution locally for every $f \in C^\infty$ if the principal symbol p of P satisfies a condition called (Ψ). It means roughly speaking that $\operatorname{Im} p$ does not change sign from $-$ to $+$ when one moves in the positive direction along a bicharacteristic of $\operatorname{Re} p$, where $\operatorname{Re} p = 0$. (See Definition 26.4.6 in [H1] for a precise formulation and Theorem 26.4.7 for a proof of the necessity.) When P is a differential operator and in a number of other cases a positive answer is known, stating that if f is in the Sobolev space $H_{(s)}$ and P is of order m then a local solution in $H_{(s+m-1)}$ exists for these operators. N. Lerner has proved that this is true for all operators of principal type satisfying condition (Ψ) in the two dimensional case but for no dimension greater than two. (See [L5], [L6] and the survey paper [H2].)

Nils Dencker [D1] has recently completed a proof of the Nirenberg-Treves conjecture, with local solutions $u \in H_{(s+m-2)}$ when $f \in H_{(s)}$. His work contains new geometrical ideas and very subtle estimates, to a large extent present already in the early versions [D2] several years ago, but much effort has been required to fill in the gaps there as indicated by the large number of successive manuscript versions.

Very recently Lerner [L7] has proved that in fact there exist local solutions $u \in H_{(s+m-\frac{3}{2})}$ when $f \in H_{(s)}$. His proofs follow the same main lines as those in [D1]. We shall indicate below where the decisive improvements lie. The purpose of this manuscript is just to try to give a simple and self-contained presentation of the result using arguments from both [D1] and [L7] freely.

As in most earlier work the major part of [D1] and of [L7] is devoted to operators of the form

$$(1.1) \qquad P = D_n + iF(x, D')$$

where $D_n = -i\partial/\partial x_n$ and $F(x, D')$ is a pseudodifferential operator in the variables $x' = (x_1, \dots, x_{n-1})$ with real principal symbol depending smoothly on the parameter x_n, for the general case can be reduced microlocally to this situation, at least formally. Partitions of unity of the Littlewood-Paley type have often been used to reduce the study of operators of the form (1.1) to the case where F is real valued, the operator is defined as in the Weyl calculus, and for some small $h > 0$

$$(1.2) \qquad |\partial_{x',\xi'}^k F(x', x_n, \xi')| \le C_k h^{\frac{1}{2}k-1}.$$

However, the passage from operators (1.1) with such symbols to more conventional pseudo-differential operators is very delicate for the weak estimates involved here and will occupy both Subsection 4.4 and a major part of Section 5.

© Springer International Publishing AG, part of Springer Nature 2018

L. Hörmander, *Unpublished Manuscripts*,

https://doi.org/10.1007/978-3-319-69850-2_23

The condition (Ψ) means that $x_n \mapsto F(x', x_n, \xi')$ does not change sign from $-$ to $+$ for increasing x_n, at least if one ignores terms of lower order. The geometric properties of the zeros of F are of central importance to the problem. In Section 2 we shall present Dencker's analysis of this geometry beginning with a study of functions satisfying (1.2) which do not depend on x_n. The relevant properties of F are encoded in a number of metrics generalizing the Beals-Fefferman metric which was introduced for the case of differential operators. In Subsection 2.4 we allow functions depending on a parameter and satisfying condition (Ψ). The main result where this condition enters is Proposition 2.4.2 which is due to Lerner [L7]. It improves a key estimate of Dencker [D1] and the proof is actually simpler and more natural. However, the improvement is not essential for the proofs in [L7]. Altogether Section 2 relies on [D1], [L1], [L3], [L4], [L7] in a way which would be hard to disentangle here.

In Section 3 we recall the results from the theory of pseudodifferential operators which will be used. In particular we give a short presentations of what Lerner calls the "Wick quantization" which is needed to regularize the symbols which we have to consider.

Section 4 is devoted to the proof of the relevant a priori estimates for operators of the form (1.1). The main lines will be indicated below. In Section 5 this leads to the proof of Lerner's improved version of the sufficiency of condition (Ψ). In a final Section 6 we list some open problems concerning solvability, and take this opportunity to correct an error in [H2].

We shall end this introduction by some hints on the approach of Dencker. To prove an existence theorem for the operator (1.1) one must obtain some bound for the inverse of the adjoint operator acting on functions with small compact support. In the very special case where $F(x, D') = -q(x_n)|D'|$ this means proving an estimate of the form

$$(1.3) \qquad \int |u(t)|^2 \, dt \leq C \int |u'(t) - \lambda q(t) u(t)|^2 \, dt, \quad u \in C_0^\infty(-T, T), \ \lambda \geq 0,$$

where we have now written t instead of x_n, and T is some positive number. If we set $u = v e^{\lambda Q}$ where $Q' = q$, then (1.3) is equivalent to

$$(1.4) \qquad \int |v|^2 e^{2\lambda Q} \, dt \leq C \int |v'|^2 e^{2\lambda Q} \, dt, \quad v \in C_0^\infty(-T, T), \ \lambda \geq 0.$$

If $I \subset (-T, T)$ is a closed interval then (1.4) implies that $\max_I Q \leq \max_{\partial I} Q$, for if this were not true we could choose $v \in C_0^\infty(I)$ with $\operatorname{supp} v = I$ but $\operatorname{supp} v'$ so close to ∂I that the maximum of Q in I exceeds the maximum in $\operatorname{supp} v'$, and then the inequality cannot hold when $\lambda \to +\infty$. Thus Q must be *quasiconvex*, which means that there is a point $t_0 \in [-T, T]$ such that Q is decreasing when $-T < t \leq t_0$ and increasing when $t_0 \leq t < T$, or equivalently that q *does not change sign from $+$ to $-$ for increasing t*. (See e.g. [H3, p. 27].) This condition will be denoted by ($\bar\Psi$); an operator satisfies condition (Ψ) precisely when the adjoint satisfies ($\bar\Psi$). Conversely, if Q is quasiconvex and $b(t) = t - t_0$, then

$$2\operatorname{Re}(\lambda q u - u', bu) = \int |u|^2 \, dt + 2\operatorname{Re}(\lambda q u, bu), \quad u \in C_0^1((-T, T)), \ \lambda \geq 0,$$

and since $bq \geq 0$ and $|b(t)| \leq 2T$, it follows that

$$\|u\|^2 \leq 4T \|u' - \lambda q u\| \|u\|, \quad u \in C_0^1((-T, T)), \ \lambda \geq 0,$$

which proves (1.3) with $C = 16T^2$. (Actually,

$$(1.5) \qquad \sup |u| \leq \int |u'(t) - \lambda q(t)u(t)| \, dt, \quad u \in C_0^1((-T, T)), \ \lambda \geq 0,$$

which implies (1.3) with $C = 4T^2$. In fact, if $-T < t \leq t_0$ then

$$|u(t)e^{-\lambda Q(t)}| \leq \int_{-T}^{t} |u'(s) - \lambda q(s)u(s)|e^{-\lambda Q(s)} \, ds \leq e^{-\lambda Q(t)} \int_{-T}^{T} |u'(s) - \lambda q(s)| \, ds,$$

and when $t_0 \leq t \leq T$ the same result is obtained by integrating from T instead.) Also in general the condition (Ψ) will by Proposition 2.4.2 lead to a function satisfying a modified quasiconvexity condition.

Dencker's approach to proving an existence theorem for (1.1) follows the same traditional pattern. With $-f$ equal to the adjoint of F and a selfadjoint pseudodifferential operator $B(x, D')$ he forms

$$2 \operatorname{Re}(f(x, D')u - \partial u/\partial x_n, B(x, D')u) = (B_n(x, D')u, u) + 2 \operatorname{Re}(f(x, D')u, B(x, D')u).$$

Here $B_n(x, \xi') = \partial B(x, \xi')/\partial x_n$. If B can be chosen so that

$$B_n(x, D') + B(x, D')f(x, D') + f(x, D')B(x, D')$$

has a good lower bound and an upper bound for $B(x, D')$ is also available, then a bound for u in terms of $\partial_n u - f(x, D')u$ will be obtained. The construction of B is based on a detailed examination of the symbol of f; the main part is roughly a regularization of a truncated signed distance δ_0 to the sign changes (for fixed t). Since B will belong to a quite exotic class of pseudodifferential operators with a very weak calculus the execution of this program is rather technical. The crucial improvement made by Lerner [L7] is to multiply δ_0 by a factor with values in $[1, 2]$ which gives rise to an important positive contribution to B_n. (For the precise construction we refer to (4.1).)

2. Metrics and weights. When studying symbols satisfying (1.2) it is convenient to change notation and write t instead of x_n, and X instead of (x', ξ') so that the function becomes $F(t, X)$. For typographical reasons we shall in fact use lower case x instead of X. Before coming to statements depending on condition (Ψ) we shall also omit any dependence on t and just consider C^∞ functions of $x \in \mathbf{R}^N$ for some N (which will ultimately become $2(n-1)$).

2.1. Generalities on symbols and weights. A Riemannian metric $\gamma(x)|dx|^2$, $x \in \mathbf{R}^N$, conformal to the Euclidean metric, is said to be *slowly varying* if there are positive constants c and C such that

$$(2.1.1) \qquad \gamma(x + y) \leq C\gamma(x) \quad \text{when } \gamma(x)|y|^2 \leq c, \ x, y \in \mathbf{R}^N.$$

This implies $C \geq 1$, and since $\gamma(x+y)|y|^2 \leq C\gamma(x)|y|^2$ when $\gamma(x)|y|^2 \leq c$, we obtain

$$(2.1.1)' \qquad C^{-1}\gamma(x) \leq \gamma(x+y) \leq C\gamma(x) \quad \text{when } \gamma(x)|y|^2 \leq c/C, \ x, y \in \mathbf{R}^N.$$

If $\gamma \leq 1$ so that the metric is bounded above by the Euclidean one, then

$$(2.1.2) \qquad \gamma(x+y) \leq \gamma(x)(C + |y|^2/c), \quad x, y \in \mathbf{R}^N.$$

This follows from (2.1.1) if $\gamma(x)|y|^2 \leq c$, and if $\gamma(x)|y|^2 \geq c$ it follows from the assumption that $\gamma(x+y) \leq 1$.

Using the partition of unity in [H1, Lemma 18.4.4] it is easy to see that to every slowly varying metric there is an equivalent one $\gamma(x)|dx|^2$ with $\gamma \in C^\infty$ such that

$$(2.1.3) \qquad |\gamma^{(k)}(x)| \leq C_k \gamma(x)^{1+\frac{1}{2}k}, \quad x \in \mathbf{R}^N, \; k = 1, 2, \ldots,$$

where the left-hand side is the norm of the multilinear form $\gamma^{(k)}(x)$ with respect to the Euclidean metric. In particular, $|\gamma'(x)\gamma(x)^{-\frac{3}{2}}| \leq C_1$, that is, $\gamma(x)^{-\frac{1}{2}}$ is Lipschitz continuous with Lipschitz constant $C_1/2$. Conversely:

LEMMA 2.1.1. *If γ is a positive function in $\mathbf{R}^N$ such that $\gamma^{-\frac{1}{2}}$ is Lipschitz continuous with Lipschitz constant K, then*

$$(2.1.1)'' \qquad \tfrac{4}{9}\gamma(x) \leq \gamma(x+y) \leq 4\gamma(x), \quad \text{if } \gamma(x)|y|^2 \leq (2K)^{-2}.$$

If in addition $\gamma \leq 1$ then

$$(2.1.2)' \qquad \gamma(x)(1 + K|y|)^{-2} \leq \gamma(x+y) \leq \gamma(x)(1 + K|y|)^2.$$

PROOF. Since $|\gamma(x+y)^{-\frac{1}{2}} - \gamma(x)^{-\frac{1}{2}}| \leq K|y| \leq \frac{1}{2}\gamma(x)^{-\frac{1}{2}}$ if $\gamma(x)|y|^2 \leq (2K)^{-2}$ we obtain $(2.1.1)''$. If $\gamma(x) \leq 1$ it follows that $\gamma(x+y)^{-\frac{1}{2}} \leq \gamma(x)^{-\frac{1}{2}}(1 + K|y|)$ which proves $(2.1.2)'$.

For the many metrics which will be introduced below the Lipschitz continuity of $\gamma^{-\frac{1}{2}}$ will be verified with explicit constants making $(2.1.1)''$ and $(2.1.2)'$ applicable.

The regularization of γ can also be made as follows. Choose $\chi \in C_0^\infty(\mathbf{R}^N)$ with $\chi \geq 0$, $\int_{\mathbf{R}^N} \chi(y)\,dy = 1$, so that $\chi(y)$ is a decreasing function of $|y|$ vanishing when $|y|^2 \geq c/C$. Then

$$\tilde{\gamma}(x) = \int \chi((x-y)\sqrt{\gamma(y)})\gamma(y)^{1+\frac{1}{2}N}\,dy$$

is equivalent to γ and (2.1.3) is valid with γ replaced by $\tilde{\gamma}$ and constants depending only on the constants in the hypothesis (2.1.1) of slow variation. For the proof we note that in the support of the integrand we have $\gamma(y)/C \leq \gamma(x) \leq C\gamma(y)$ by $(2.1.1)'$, hence

$$\tilde{\gamma}(x) \leq \int \chi((x-y)\sqrt{\gamma(x)/C})(C\gamma(x))^{1+\frac{1}{2}N}\,dy = C^{1+N}\gamma(x),$$

$$\tilde{\gamma}(x) \geq \int \chi((x-y)\sqrt{C\gamma(x)})(\gamma(x)/C)^{1+\frac{1}{2}N}\,dy = C^{-1-N}\gamma(x),$$

$$|\tilde{\gamma}^{(j)}(x)| \leq \sup|\chi^{(j)}| \int_{|x-y|^2\gamma(x)<c} (C\gamma(x))^{1+\frac{1}{2}(N+j)}\,dy = C_j\gamma(x)^{1+\frac{1}{2}j}.$$

A positive real-valued function m in $\mathbf{R}^N$ is said to be continuous with respect to the metric $\Gamma = \gamma(x)|dx|^2$ if there are positive constants c and C such that

$$(2.1.4) \qquad m(x)/C \leq m(x+y) \leq Cm(x) \quad \text{when } \gamma(x)|y|^2 < c.$$

Then $S(m, \Gamma)$ is defined to be the set of functions $u \in C^\infty(\mathbf{R}^N)$ such that for every integer $j \geq 0$

$$(2.1.5) \qquad |u^{(j)}(x)| \leq C_j m(x)\gamma(x)^{\frac{1}{2}j}$$

where $|u^{(j)}(x)|$ is the norm of the multilinear form $u^{(j)}(x)$ with respect to the Euclidean metric. With the best constants C_j in (2.1.5) as seminorms, the symbol space $S(m,g)$ is a Fréchet space. With this notation (2.1.3) means that $\gamma \in S(\gamma, \Gamma)$. Quite generally, if m satisfies (2.1.4) and γ is chosen so that (2.1.3) is valid then

$$\tilde{m}(x) = \int m(y)\chi((x-y)\gamma(y)^{\frac{1}{2}})\gamma(y)^{\frac{N}{2}} \, dy$$

is in $S(m, \Gamma)$ and is equivalent to m if $0 \leq \chi \in C_0^\infty(\mathbf{R}^N)$, $\int \chi(y) \, dy = 1$, and $\chi(y)$ is a decreasing function of $|y|$ vanishing when $|x| > r$ for a sufficiently small r. In fact, if C' is a Lipschitz constant for $S = \gamma^{-\frac{1}{2}}$ and $2C'r < 1$, then $|S(x) - S(y)| \leq S(y)/2$ in the support of the integrand since $|x - y| \leq rS(y)$, hence $S(y)/2 \leq S(x) \leq 3S(y)/2$, and $m(x)/C \leq m(y) \leq Cm(x)$ by (2.1.4) if $r^2 < c$. Since

$$\chi(\tfrac{3}{2}(x-y)/S(x))2^{-N}S(x)^{-N} \leq \chi((x-y)/S(y))S(y)^{-N} \leq \chi(\tfrac{1}{2}(x-y)/S(x))(3/2)^N S(x)^{-N}$$

it follows that

$$3^{-N}C^{-1}m(x) \leq \tilde{m}(x) \leq 3^N Cm(x),$$

so $\tilde{m}$ is equivalent to m. In the same way we obtain

$$|\tilde{m}^{(j)}(x)| \leq 3^{N+j}Cm(x) \int \chi_j(y) \, dy S(x)^{-j}, \quad \chi_j(y) = \max_{|z| \geq |y|} |\chi^{(j)}(z)|,$$

which proves that $\tilde{m} \in S(m, \Gamma) = S(\tilde{m}, \Gamma)$.

In Subsection 4.3 we shall need to choose cutoff functions using the following

LEMMA 2.1.2. *If F is a closed set in $\mathbf{R}^N$ and Ω is a Γ neighborhood of F in the sense that $y \in F$ and $|x - y|^2\gamma(y) < \kappa$ implies $x \in \Omega$ for some $\kappa > 0$, then one can find $\varphi \in S(1, \Gamma)$ with $\operatorname{supp}\varphi \subset \Omega$, $0 \leq \varphi \leq 1$, and $\varphi = 1$ in a neighborhood of F.*

PROOF. We may again assume that γ satisfies (2.1.3), and we denote by C' a Lipschitz constant for $S = \gamma^{-\frac{1}{2}}$. Let ε_1 be a small positive number and set

$$F_1 = \{x; |x - y| < \varepsilon_1 S(y) \text{ for some } y \in F\}.$$

If $|x - y_1| \leq \varepsilon_2 S(y_1)$ for some $y_1 \in F_1$ then we can find $y \in F$ such that $|y_1 - y| < \varepsilon_1 S(y)$, hence $|S(y_1) - S(y)| \leq C'\varepsilon_1 S(y_1)$ and $S(y_1) \leq S(y)/(1 - C'\varepsilon_1)$, if $C'\varepsilon_1 < 1$, so

$$|x - y| \leq \varepsilon_2 S(y_1) + \varepsilon_1 S(y) \leq S(y)(\varepsilon_2/(1 - C'\varepsilon_1) + \varepsilon_1) < \sqrt{\kappa}S(y),$$

if ε_1 and ε_2 are small enough. Hence $x \in \Omega$ then and we conclude that Ω contains the support of

$$\psi(x) = \int_{F_1} \chi((x-y)/S(y))S(y)^{-N} \, dy,$$

if $|y| < \varepsilon_2$ when $|y| \in \operatorname{supp}\chi$. As above we conclude that $\psi \in S(1, \Gamma)$ and that ψ has a positive lower bound c in F. If $\Phi \in C^\infty(\mathbf{R})$ is increasing and $\Phi(0) = 0$, $\Phi(t) = 1$ when $t \geq c/2$, then $\varphi = \Phi(\psi) \in S(1, \Gamma)$, $0 \leq \varphi \leq 1$, $\varphi = 1$ in a neighborhood of F, and $\operatorname{supp}\varphi \subset \Omega$.

2.2. The Beals-Fefferman metric and modifications by Dencker and Lerner. Let $f \in C^\infty(\mathbf{R}^N)$ be a real valued function such that for some $h \in (0,1)$ and constants γ_j

(2.2.1) $$|f^{(j)}(x)| \leq \gamma_j h^{\frac{1}{2}j-1}, \quad x \in \mathbf{R}^N, \ j = 0,1,2,\ldots.$$

This means that $f \in S(h^{-1}, h|dx|^2)$. It is often convenient and no essential loss of generality to assume that $\gamma_j = 1$ for $j \leq 3$, for example. In the beginning this will not be assumed though.

In (2.2.1) h is a constant which as indicated in (1.2) will be assumed in a first step toward the desired estimates in Section 4, but it is essential to maintain uniformity in $h \in (0,1)$ in the following discussion. However, a key idea of Beals and Fefferman in their proof of solvability for operators satisfying both conditions (Ψ) and ($\bar{\Psi}$) was to introduce estimates of the form (2.2.1) with h replaced by a function $H(x)$ with $h \leq H(x) \leq 1$ which are optimal in the sense that H is as large as possible, and the corresponding balls therefore as small as possible. This requires that $|f(x)/\gamma_0| \leq H(x)^{-1}$ and that $|f'(x)/\gamma_1| \leq H(x)^{-\frac{1}{2}}$, so the optimal choice is

(2.2.2) $$H(x)^{-1} = \max(1, |f(x)/\gamma_0|, |f'(x)/\gamma_1|^2),$$

which does imply $h \leq H(x) \leq 1$ and that

(2.2.3) $$|f^{(j)}(x)| \leq \gamma_j H(x)^{\frac{1}{2}j-1}, \quad x \in \mathbf{R}^N, \ j = 0,1,2,\ldots.$$

In fact, (2.2.3) follows from (2.2.2) when $j < 2$ and from (2.2.1) when $j \geq 2$ since $h \leq H(x)$. The metric $H(x)|dx|^2$ in $\mathbf{R}^N$, called the *Beals-Fefferman metric*, is slowly varying, for

$$H(x)^{-\frac{1}{2}} = \max(1, |f(x)|^{\frac{1}{2}}, |f'(x)|)$$

is Lipschitz continuous. More generally:

PROPOSITION 2.2.1. *If $f \in C^\infty(\mathbf{R}^N$ is a symbol satisfying*

(2.2.4) $$|f^{(j)}(x)| \leq \gamma_j h^{\frac{1}{2}j-m}, \quad x \in \mathbf{R}^N, \ j = 0,1,2,\ldots,$$

for some positive half integer m then

(2.2.5) $$H(x)^{-1} = \max\left(1, \max_{0 \leq j < 2m} |f^{(j)}(x)/\gamma_j|^{1/(m-\frac{1}{2}j)}\right)$$

is the largest number ≤ 1 such that

(2.2.6) $$|f^{(j)}(x)| \leq \gamma_j H(x)^{\frac{1}{2}j-m}, \quad j = 0,1,\ldots.$$

We have $h \leq H \leq 1$, and $H^{-\frac{1}{2}}$ is Lipschitz continuous with Lipschitz constant

(2.2.7) $$\max_{0 \leq j < 2m} \gamma_{j+1}/(\gamma_j(2m-j)).$$

PROOF. By (2.2.4) we have $h \leq H(x)$ so (2.2.6) follows from (2.2.4) when $j \geq 2m$ and from (2.2.5) when $j < 2m$. It is clear that $H(x)^{-\frac{1}{2}}$ is locally Lipschitz continuous. At a differentiable point we have if the maximum in (2.2.5) is taken for the index $j < 2m$

$$|(H(x)^{-\frac{1}{2}})'| \leq |f^{(j)}(x)/\gamma_j|^{1/(2m-j)-1}|f^{(j+1)}(x)/\gamma_j|/(2m-j)$$

$$\leq H(x)^{-(1/(2m-j)-1)(2m-j)/2}H(x)^{-(2m-j-1)/2}\gamma_{j+1}/(\gamma_j(2m-j)) = \gamma_{j+1}/(\gamma_j(2m-j)),$$

which completes the proof.

For $m = 2$ the metric (2.2.5) was already used in the proof of Theorem 18.1.8 in [H1], so the idea is not new. However, as observed by Lerner [L7], if f satisfies (2.2.1) then $h^{1-m}f$ satisfies (2.2.4), so we have:

PROPOSITION 2.2.2. *If f satisfies (2.2.1) and m is a positive half integer, then*

$$(2.2.8) \qquad |f^{(j)}(x)| \le \gamma_j h^{m-1} H_{(m)}(x)^{\frac{1}{2}j-m}, \quad j = 0, 1, \ldots,$$

$$(2.2.9) \qquad where \quad H_{(m)}(x)^{-\frac{1}{2}} = \max(1, \max_{0 \le j < 2m} |h^{1-m} f^{(j)}(x)/\gamma_j|^{1/(2m-j)}.$$

We have $h \le H_{(m)} \le 1$, and $H_{(m)}$ decreases to h when $m \to \infty$. If we allow h to be a function of x such that $h^{-\frac{1}{2}}$ is Lipschitz continuous with Lipschitz constant K, then $H_{(m)}^{-\frac{1}{2}}$ has Lipschitz constant

$$(2.2.10) \qquad \max_{0 \le j < 2m} (\gamma_{j+1}/\gamma_j + |2m - 2|K)/(2m - j).$$

PROOF. Most of this follows from Proposition 2.2.1, but in proving the Lipschitz continuity of $H_m^{-\frac{1}{2}}$ as there we must also examine the term

$$|f^{(j)}(x)/\gamma_j|^{1/(2m-j)}|((h(x)^{(1-m)/(2m-j)})'|.$$

The first factor is at most $h^{(\frac{1}{2}j-1)/(2m-j)}$, and the second factor is the norm of the differential of $(h^{-\frac{1}{2}})^{(2m-2)/(2m-j)}$ which is $\le K|(2m - 2)/(2m - j)|h^{(1-m)/(2m-j)+\frac{1}{2}}$. Since $(\frac{1}{2}j-1)/(2m-j) = -\frac{1}{2}+(m-1)/(2m-j)$, the two powers of h cancel each other. To prove that $H_{(m)}$ is a decreasing function of m assume that $H_{(m)}(x) < 1$ and choose $j \in [0, 2m)$ so that $H_{(m)}(x)^{-\frac{1}{2}} = |h^{1-m} f^{(j)}(x)/\gamma_j|^{1/(2m-j)}$. Then

$$H_{(m+\frac{1}{2})}(x)^{(j-1-2m)/2} \ge |h^{1-m-\frac{1}{2}} f^{(j)}(x)/\gamma_j| = H_{(m)}(x)^{\frac{1}{2}j-m} h^{-\frac{1}{2}} \ge H_{(m)}(x)^{(j-1-2m)/2}$$

which proves that $H_{(m+\frac{1}{2})} \le H_{(m)}$.

Taking $m = 1$ gives of course the Beals-Fefferman metric back.

With $m = \frac{1}{2}$ we get if we let h be any metric H such that (2.2.3) is valid that

$$(2.2.11) \qquad |f^{(j)}(x)| \le \gamma_j H(x)^{-\frac{1}{2}} H_{(\frac{1}{2})}(x)^{\frac{1}{2}j-\frac{1}{2}},$$

$$(2.2.12) \qquad where \quad H_{(\frac{1}{2})}(x)^{-1} = \max(1, |f(x)/\gamma_0|^2 H(x)),$$

and $H \le H_{(\frac{1}{2})} \le 1$. If H is the Beals-Fefferman metric then $H^{-\frac{1}{2}}$ has Lipschitz constant $\max(\gamma_1/2\gamma_0, \gamma_2/\gamma_1)$ and the Lipschitz constant of $H_{(\frac{1}{2})}^{-\frac{1}{2}}$ is at most γ_1/γ_0 larger.

When $m = \frac{3}{2}$ we have, *assuming from now on that $\gamma_j = 1$ for $j \le 3$,*

$$|h^{-\frac{1}{2}} f^{(j)}(x)| \le \gamma_j H_{(\frac{3}{2})}(x)^{\frac{1}{2}j-\frac{3}{2}}, \quad j = 0, 1, \ldots, \quad where$$

$$H_{(\frac{3}{2})}(x)^{-1} = \max(1, |h^{-\frac{1}{2}} f(x)|^{\frac{2}{3}}, |h^{-\frac{1}{2}} f'(x)|, |h^{-\frac{1}{2}} f''(x)|^2) = \max(H_3(x)^{-1}, |h^{-\frac{1}{2}} f(x)|^{\frac{2}{3}}),$$

$$(2.2.13) \qquad H_3(x)^{-1} = \max(1, |h^{-\frac{1}{2}} f'(x)|, |h^{-\frac{1}{2}} f''(x)|^2).$$

This means that $H_3(x)|dx|^2$ can be regarded as the Beals-Fefferman metric for the vector valued function $x \mapsto h^{-\frac{1}{2}} f'(x)$ which also satisfies (2.2.1) with a left shift of the sequence (γ_j). Thus $H_3^{-\frac{1}{2}}$ is Lipschitz continuous with constant 1 and we have

$$(2.2.14) \qquad |f^{(j)}(x)| \le \gamma_j h^{\frac{1}{2}} H_3(x)^{(j-3)/2}, \quad j = 1, 2, \ldots.$$

Of course this cannot be expected when $j = 0$ for $f(x)$ played no role in the definition of H_3.[18] When $H_{(\frac{3}{2})}(x)^{-1} > H_3(x)^{-1}$ then $H_{(\frac{3}{2})}(x)^{-1} = |h^{-\frac{1}{2}} f(x)|^{\frac{2}{3}}$, and Taylor's formula gives

$$|f(x+y)| \geq |f(x)| - |f'(x)||y| - |f''(x)||y|^2/2 - h^{\frac{1}{2}}|y|^3/6$$

$$\geq h^{\frac{1}{2}}(H_{(\frac{3}{2})}(x)^{-\frac{3}{2}} - H_{(\frac{3}{2})}(x)^{-1}|y| - H_{(\frac{3}{2})}^{-\frac{1}{2}}|y|^2/2 - |y|^3/6) > 0$$

if $|y| \leq \frac{1}{2} H_{(\frac{3}{2})}(x)^{-\frac{1}{2}}$. Thus $H_{(\frac{3}{2})}(x)^{-\frac{1}{2}} \leq 2S(x)$ where

(2.2.15) $$S(x) = \sup\{R; 0 \leq R \leq \tfrac{1}{2} h^{-\frac{1}{2}}, f(x+y) \neq 0 \text{ when } |y| < R\};$$

the bound $2R \leq h^{-\frac{1}{2}}$ is permissible since $H_{(\frac{3}{2})} \geq h$. Hence we have proved that

(2.2.16) $$H_{(\frac{3}{2})}(x)^{-1} \leq H_4(x)^{-1} = \max\left(H_3(x)^{-1}, (2S(x))^2\right),$$

(2.2.17) $$|f^{(j)}(x)| \leq \gamma_j h^{\frac{1}{2}} H_4(x)^{(j-3)/2}, \quad j = 0, 1, \ldots.$$

The last bounds follow from the properties of $H_{(\frac{3}{2})} \geq H_4$ when $j \leq 3$ and when $j > 3$ since $h \leq H_4$.

Recalling that H_3 can be regarded as the Beals-Fefferman metric for the vector valued function $x \mapsto h^{-\frac{1}{2}} f'(x)$ we can obtain another metric $H_2(x)|dx|^2$ by applying to it the case $m = \frac{1}{2}$ of Proposition 2.2.2. We define following (2.2.12)

(2.2.18) $$H_2(x)^{-1} = \max(1, h^{-1}|f'(x)|^2 H_3(x)),$$

noting that $H_2^{-\frac{1}{2}}$ has Lipschitz constant 2, and get from (2.2.11)

(2.2.19) $$|f^{(j)}(x)| \leq \gamma_j h^{\frac{1}{2}} H_3(x)^{-\frac{1}{2}} H_2(x)^{\frac{1}{2}j-1}, \quad j = 1, 2, \ldots.$$

Defining

(2.2.20) $$H_1(x)^{-1} = \max(H_2(x)^{-1}, (2S(x))^2),$$

in analogy to the passage from H_3 to H_4 in (2.2.16), we shall prove that

(2.2.21) $$|f^{(j)}(x)| \leq \gamma_j h^{\frac{1}{2}} H_4(x)^{-\frac{1}{2}} H_1(x)^{\frac{1}{2}j-1}, \quad j = 0, 1, \ldots.$$

When $j = 1$ or $j = 2$ this follows from (2.2.19), for $H_4^{-1} \geq H_3^{-1}$ and $H_1^{-1} \geq H_2^{-1}$. Since $H_2 \geq H_3$ we have $H_1 \geq H_4$ which gives (2.2.21) when $j \geq 3$, since $H_1^{(j-3)/2} \geq h^{(j-3)/2}$. If $f(x) = h^{\frac{1}{2}} H_4(x)^{-\frac{1}{2}} H_1(x)^{-1} \mu$ and $|\mu| > 1$, then by Taylor's formula

$$h^{-\frac{1}{2}} H_4(x)^{\frac{1}{2}}|f(x+y)| \geq H_1(x)^{-1}|\mu| - H_1(x)^{-\frac{1}{2}}|y| - |y|^2/2 - H_4(x)^{\frac{1}{2}}|y|^3/6 > 0$$

when $|y| \leq \frac{1}{2} H_1(x)^{-1}$, for $H_4 \leq H_1$. Hence $2S(x) > H_1(x)^{-\frac{1}{2}}$, which is a contradiction, unless $H_1(x) = h$, $2S(x) = h^{-\frac{1}{2}}$, which implies $H_4(x) = h$ too and makes (2.2.21) obvious.

Summing up, we have four metrics $G_j = H_j(x)|dx|^2$, $j = 1, \ldots, 4$, such that $H_j^{-\frac{1}{2}}$ is Lipschitz continuous with constant 2, $H_4 \leq H_3$, $H_1 \leq H_2$, and $H_2 \geq H_3$, $H_1 \geq H_4$. They are defined in (2.2.13), (2.2.16), (2.2.15), (2.2.18) and (2.2.20). In terms of these metrics f and its derivatives have the bounds (2.2.14), (2.2.17), (2.2.19) and (2.2.21).

REMARK. H_1, H_2, H_3 all occur in [D1] with the same notation, H_1^{-1} is denoted by ν and H_4^{-1} is denoted by μ in [L7].

---

[18]Note the distinction between H_3 and $H_{(3)}$. The notation H_3 is used here to conform with that of Dencker while the notation with parenthesis around the subscript refers to the construction in Proposition 2.2.2.

2.3. The distance to the regular zeros. It is an immediate consequence of the implicit function theorem that a zero of f where $f' \neq 0$ has a neighborhood where the distance to the zeros of f is a C^∞ function if it is provided with the sign of f. We shall now make this remark quantitative at a zero y where $H_2(y) < 1$. By (2.2.19) we have

$$|f^{(j)}(y)| \leq \gamma_j h^{\frac{1}{2}} H_3(y)^{-\frac{1}{2}} H_2(y)^{\frac{1}{2}j-1}, \quad j = 0, 1, \dots,$$

with equality when $j = 1$. (Recall that we have asssumed $\gamma_j = 1$ when $j \leq 2$.) Moreover,

$$|f''(y + x)| \leq h^{\frac{1}{2}} H_3(y + x)^{-\frac{1}{2}} \leq h^{\frac{1}{2}} (H_3(y)^{-\frac{1}{2}} + |x|), \quad x \in \mathbf{R}^N,$$

for $H_3^{-\frac{1}{2}}$ has Lipschitz constant 1. If we set

$$\Phi(x) = f(y + xH_2(y)^{-\frac{1}{2}}) h^{-\frac{1}{2}} H_3(y)^{\frac{1}{2}} H_2(y)$$

it follows that

(2.3.1) $\Phi(0) = 0, \ |\Phi'(0)| = 1, \ |\Phi^{(j)}(0)| \leq \gamma_j, \ \forall j; \ |\Phi''(x)| \leq 1 + |x|, \ x \in \mathbf{R}^N,$

for $H_3(y) \leq 1$.

LEMMA 2.3.1. *If $\Phi \in C^\infty(\mathbf{R}^N)$ satisfies (2.3.1) and $x \in \mathbf{R}\Phi'(0)$, $|x| < \frac{2}{3}$, then $\Phi(y) \neq 0$ when $|x - y| \leq |x|$ and $y \neq 0$. The signed distance $\delta(z) = \mathrm{sgn}\, f(z) \inf_{f(y)=0} |z - y|$ to the zeros of f is a C^∞ function in a neighborhood of x. For every j there is a bound for $\delta^{(j)}(x)$ which depends only on the constants γ_k with $k \leq j$; we have $|\delta'| = 1$ and $|\delta''| \leq 3$. Also Φ/δ is a positive C^∞ function at x, and the derivatives of order j have bounds depending only on γ_k for $k \leq j + 1$.*

PROOF. We may assume that $\partial_j \Phi(0) = 0$ when $j < N$ and that $\partial_N \Phi(0) = 1$. When $|x|^2 \leq 2r x_N$ Taylor's formula gives

$$\Phi(x) \geq x_N - \tfrac{1}{2}|x|^2 - \tfrac{1}{6}|x|^3 \geq \tfrac{1}{2}|x|^2(r^{-1} - 1 - \tfrac{1}{3}|x|) \geq \tfrac{1}{2}|x|^2(r^{-1} - 1 - \tfrac{2}{3}r) > 0$$

if $x \neq 0$ and $2r^2 + 3r < 3$, hence if $r \leq \frac{2}{3}$. If $0 \neq |x|^2 \leq -2r x_N$ we obtain $\Phi(x) < 0$ in the same way, so the origin is the only zero of Φ at distance $\leq |x|$ from x if x is on the x_N axis $\mathbf{R}\Phi'(0)$ and $|x| \leq \frac{2}{3}$. From the implicit function theorem it follows that δ is then a C^∞ function in a neighborhood of x. First it shows that the zeros of Φ near the origin are given by $y_N = \psi(y')$ where $y' = (y_1, \dots, y_{N-1})$, and $\psi \in C^\infty$, $\psi(0) = 0$ and $\psi'(0) = 0$. Successive differentiations of the equation $\Phi(y', \psi(y')) = 0$ give that every derivative $\partial^\alpha \psi$ at 0 is a polynomial in the derivatives of Φ of order $\leq |\alpha|$, for the equation $\partial^\alpha \Phi(y', \psi(y')) = 0$ contains when $y' = 0$ besides the term $\partial^\alpha \psi(0)$ only terms involving derivatives of ψ of lower order. Thus $\partial^\alpha \psi(0)$ can be estimated by a polynomial in γ_j for $j \leq |\alpha|$. In particular, $\partial^2 \psi(0)/\partial y_j \partial y_k = -\partial^2 \Phi(0)/\partial y_j \partial y_k$ when $j, k < N$.

Writing $x = X(y', t) = (y', \psi(y')) + t(-\psi'(y'), 1)/\sqrt{1 + |\psi'(y')|^2}$ we have $\delta(X(y', t)) = t$ and $\delta'(X(y', t)) = (-\psi'(y'), 1)/\sqrt{1 + |\psi'(y')|^2} = (-\psi'(y'), 1) + O(|y'|^2)$. Since

$$dx = dX(y', t) = (dy', 0) + (0, dt) - t(\psi''(0)dy', 0) \quad \text{when } y' = 0$$

it follows that $dy' = (\mathrm{Id} - t\psi''(0))^{-1} dx'$ then, and

$$\delta''(x)\, dx = (-\psi''(0)\, dy', 0) = (-\psi''(0)(\mathrm{Id} - t\psi''(0))^{-1} dx', 0).$$

When $|t| \leq \frac{2}{3}$ we have $|t\psi''(0)| \leq \frac{2}{3}$, so $\mathrm{Id} - t\psi''(0)$ is invertible and $\delta \in C^\infty$ by the implicit function theorem. The inverse $\partial(y', t)/\partial x$ of the Jacobian matrix $\partial X(y', t)/\partial(y', t)$ contains second derivatives of $\psi(y')$, but in calculating

$$(\partial/\partial x)^{j-1}\delta'(x) = \left(\frac{\partial y'}{\partial x}\frac{\partial}{\partial y'}\right)^{j-1}((-\psi'(y'), 1)/\sqrt{1 + |\psi'(y')|^2})$$

only derivatives of order $\leq j$ of ψ will appear since the Jacobian is never differentiated more than $j - 2$ times. Thus we obtain bounds for $\delta^{(j)}$ when $x' = 0$ and $|x_N| \leq \frac{2}{3}$ in terms of the constants γ_k for $k \leq j$; in particular $\partial^2\delta(x)/\partial x_j\partial x_k = 0$ if $x' = 0$ and $j = N$ or $k = N$, and

$$(\partial^2\delta(x)/\partial x_j\partial x_k)_{j,k=1}^{N-1} = -\psi''(0)(\mathrm{Id} - x_N\psi''(0))^{-1}, \quad \text{when } x' = 0.$$

Since $|\psi''(0)| \leq 1$ and $|x_N| \leq \frac{2}{3}$ it follows that

$$|\partial^2\delta(x)/\partial x^2| \leq 3.$$

What remains is to study Φ/δ. Returning to the coordinates used in studying the differentiability of δ, we note that

$$\Phi(X(y', t))/t = \int_0^1 \left(\Phi'(X(y', \tau t)), (-\psi'(y'), 1)\right)/\sqrt{1 + |\psi'(y')|^2}\, d\tau,$$

for $\Phi(X(y', 0)) = 0$. Since $t = \delta(X(y', t))$ this gives an estimate for the derivatives of Φ/δ of order j with respect to the (y', t) variables in terms of those of Φ of order $\leq j + 1$. Returning to the variables x we conclude that $\Phi(x)/\delta(x)$ is in C^∞ in a neighborhood of the x_N axis when $|x| \leq \frac{1}{5}$, with the derivatives of order j bounded in terms of γ_k, $k \leq j+1$. The proof is complete.

With the original variables we have now proved:

PROPOSITION 2.3.2. *If f satisfies* (2.2.1), $f(y) = 0$, $H_2(y) < 1$, *and* $x - y \in \mathbf{R}f'(y)$, $|x - y| < \frac{2}{3}H_2(y)^{-\frac{1}{2}}$, *then the signed distance δ to the (regular) zeros of f is a C^∞ function at x,*

$$(2.3.2) \qquad |\delta'(x)| = 1, \quad |\delta''(x)| \leq 3H_2(y)^{\frac{1}{2}}, \quad |\delta^{(j)}(x)| \leq \Gamma_j H_2(y)^{\frac{1}{2}(j-1)}, \quad j > 2,$$

where Γ_j only depends on γ_k for $k \leq j$. We have $f = a^2\delta$ where $a \in C^\infty$, and

$$(2.3.3) \qquad |a^{(j)}(x)| \leq \tilde{\Gamma}_j|f'(y)|^{\frac{1}{2}}H_2(y)^{\frac{1}{2}j}, \quad |f'(y)| = h^{\frac{1}{2}}H_3(y)^{-\frac{1}{2}}H_2(y)^{-\frac{1}{2}},$$

where $\tilde{\Gamma}_j$ only depends on γ_k for $k \leq j + 1$.

The following is a much more useful but essentially equivalent reformulation of Proposition 2.3.2:

COROLLARY 2.3.3. *The signed distance $\delta(x)$ from x to the regular zeros of f is a C^∞ function of x in $\{x \in \mathbf{R}^N; 4S(x) < H_1(x)^{-\frac{1}{2}}, 4H_1(x) < 1\}$, and $|\delta(x)| = S(x)$. In this open set we have $f = a^2\delta$, where $0 < a \in C^\infty$,*

$$(2.3.2)' \quad |\delta'(x)| = 1, \quad |\delta''(x)| \leq 6H_2(x)^{\frac{1}{2}}, \quad |\delta^{(j)}(x)| \leq \Gamma_j(4H_2(x))^{\frac{1}{2}(j-1)}, \quad j > 2,$$

$$(2.3.3)' \qquad |a^{(j)}(x)| \leq 2^{j+1}\tilde{\Gamma}_j h^{\frac{1}{4}}H_3(x)^{-\frac{1}{4}}H_2(x)^{\frac{1}{2}j-\frac{1}{4}}, \quad j = 0, 1, \ldots.$$

with Γ_j and $\tilde{\Gamma}_j$ as in Proposition 2.3.2, and $H_1(x) = H_2(x)$, $H_3(x) = H_4(x)$.

PROOF. Since $4S(x) < H_1(x)^{-\frac{1}{2}}$ we have $H_1(x) = H_2(x)$ and can find y with $|x - y| = S(x) < \frac{1}{4}H_1(x)^{-\frac{1}{2}}$ such that $f(y) = 0$. Then

$$|H_2(y)^{-\frac{1}{2}} - H_2(x)^{-\frac{1}{2}}| \leq 2|x - y| < \frac{1}{2}H_2(x)^{-\frac{1}{2}}, \quad \text{hence}$$
$$\tfrac{1}{2}H_2(x)^{-\frac{1}{2}} \leq H_2(y)^{-\frac{1}{2}} \leq \tfrac{3}{2}H_2(x)^{-\frac{1}{2}}.$$

If $4H_2(x) < 1$ it follows that $H_2(y) < 1$ so that $H_2(y)^{-1} = h^{-1}|f'(y)|^2 H_3(y)$, and

$$|H_3(y)^{-\frac{1}{2}} - H_3(x)^{-\frac{1}{2}}| \leq |x - y| < \tfrac{1}{4}H_2(x)^{-\frac{1}{2}} \leq \tfrac{1}{4}H_3(x)^{-\frac{1}{2}},$$

for $H_3 \leq H_2$. This implies $H_3(y)^{-\frac{1}{2}} \leq \frac{5}{4}H_3(x)^{-\frac{1}{2}}$. The estimates $(2.3.2)'$ and $(2.3.3)'$ are now immediate consequences of $(2.3.2)$ and $(2.3.3)$, and $H_3(x) = H_4(x)$ since $4S(x) < H_2(x)^{-\frac{1}{2}} \leq H_3(x)^{-\frac{1}{2}}$.

Since δ is not smooth everywhere we shall later on have to regularize δ by convolution with a function $\psi \in \mathcal{S}$, in fact a Gaussian. By the following proposition this will not change δ very much near the regular zeros; the statement leaves some freedom to redefine δ at some distance from them.

PROPOSITION 2.3.4. Let $\psi \in \mathcal{S}(\mathbf{R}^N)$ be an even function, $\int_{\mathbf{R}^N} \psi(x)\,dx = 1$. If f satisfies $(2.2.1)$, and δ_0 is a function with Lipschitz constant 1 which is equal to the signed distance $\delta(x)$ from x to the regular zeros of f when $4S(x) < H_1(x)^{-\frac{1}{2}}$ and $4H_1(x) < 1$, and $\delta_\psi = \delta_0 * \psi$, then

$(2.3.4) \quad |\delta_\psi - \delta_0| \leq \int_{\mathbf{R}^N} |\psi(z)|\,dz, \quad |\delta_\psi^{(j)}(x)| \leq \int_{\mathbf{R}^N} |\psi^{(j-1)}(z)|\,dz, \ \text{if } j \geq 1, \ x \in \mathbf{R}^N,$

$(2.3.5)$
$$|\delta_\psi^{(j)}(x) - \delta^{(j)}(x)| \leq C_j H_1(x)^{\frac{1}{2}(j+1)}, \ \text{if } j \geq 0, \ 8S(x) < H_1(x_0)^{-\frac{1}{2}}, \ 6H_1(x) < 1,$$

where C_j only depends on ψ and the constants γ_k with $k \leq j + 2$.

PROOF. Since $\delta'_\psi = \delta'_0 * \psi$ and $|\delta'_0| \leq 1$, we obtain $(2.3.4)$ for $j \geq 1$ if we let the remaining $j - 1$ derivatives act on the factor ψ in the convolution. The first inequality also follows at once from the Lipschitz condition.

To prove $(2.3.5)$ we consider a point x_0 with $8S(x_0) < H_1(x_0)^{-\frac{1}{2}}$ and $H_1(x_0)^{-\frac{1}{2}} \geq 8$. If $|x - x_0| < \kappa H_1(x_0)^{-\frac{1}{2}}$ then $4S(x) \leq 4S(x_0) + 4\kappa H_1(x_0)^{-\frac{1}{2}}$, which implies $H_1(x)^{-\frac{1}{2}} \geq (1 - 2\kappa)H_1(x_0)^{-\frac{1}{2}} > 4S(x)$ if $H_1(x_0)^{-\frac{1}{2}}(1 - 2\kappa) > 4S(x_0) + 4\kappa H_1(x_0)^{-\frac{1}{2}}$. This is true if $\kappa = \frac{1}{12}$, and then we have $H_1(x)^{-1} \geq 6(1 - \frac{1}{6})^2 > 4$.

To verify $(2.3.5)$ at x_0 we choose $\chi \in C_0^\infty(\mathbf{R}^N)$ with $0 \leq \chi \leq 1$ so that $\chi(x) = 1$ when $|x| \leq \frac{1}{2}$ and $|x| < 1$ when $x \in \operatorname{supp}\chi$, and we write $\delta_0 = \delta_1 + \delta_2$ where

$$\delta_1(x) = \chi((x - x_0)/R)\delta_0(x) = \chi((x - x_0)/R)\delta(x), \quad R = \tfrac{1}{12}H_1(x_0)^{-\frac{1}{2}},$$

so that $|x - x_0| < \frac{1}{12}H_1(x_0)^{-\frac{1}{2}}$ in the support. By Corollary 2.3.3

$$|\delta_1^{(j)}| \leq C_j' R^{1-j}, \quad j \geq 0,$$

for some constants depending only on ψ and γ_k for $k \leq j$. Taylor's formula gives

$$\delta_1 * \psi(x) - \delta_1(x) = \int (\delta_1(x - z) - \delta_1(x))\psi(z) \, dz = \int_0^1 \int_{\mathbf{R}^N} \delta_1''(x - tz; z, z)(1 - t)\psi(z) \, dt \, dz,$$

and after j differentiations with respect to x we obtain

$$|\delta_1^{(j)} * \psi(x) - \delta_1^{(j)}(x)| \leq C_{j+2}' R^{-1-j} \int_0^1 \int_{\mathbf{R}^N} |z|^2 (1 - t)|\psi(z)| \, dt \, dz.$$

Since $\delta_2(x) = 0$ when $|x - x_0| < R/2$ and $|\delta_2(x)| \leq |\delta_0(x)| \leq |x - x_0| + S(x_0)$, where $S(x_0) < \frac{1}{8} H_1(x_0)^{-\frac{1}{2}} = \frac{3}{2}R \leq 3|x - x_0|$ in $\operatorname{supp} \delta_2$, we obtain

$$|\delta_2 * \psi^{(j)}(x_0)| \leq 4 \int_{|z| > R/2} |z||\psi^{(j)}(z)| \, dz.$$

Since $\psi \in \mathcal{S}(\mathbf{R}^N)$ and $\delta_2 = 0$ in a neighborhood of x_0, it follows that

$$|\delta_\psi^{(j)}(x_0) - \delta^{(j)}(x_0)| \leq C_j''' R^{-1-j}$$

which proves (2.3.5).

2.4. Parameter dependent symbols satisfying condition ($\bar{\Psi}$). In this section f will be a function of $(t, x) \in [-1, 1] \times \mathbf{R}^N$ which is a continuous function of t with values in $C^\infty(\mathbf{R}^N)$ satisfying for some $h \in (0, 1)$

(2.4.1) $\qquad |f^{(k)}(t, x)| \leq \gamma_k h^{\frac{1}{2}k-1}, \quad x \in \mathbf{R}^N, \ t \in [-1, 1], \ k = 0, 1, 2, \ldots$

where $f^{(k)}(t, x)$ denotes the differential of order k for fixed t. As in Section 2 we shall assume that $\gamma_k = 1$ when $k \leq 3$. (The interval $[-1, 1]$ could of course be any compact interval on $\mathbf{R}$.) We shall also assume that f satisfies the condition ($\bar{\Psi}$):

(2.4.2) $\qquad f$ does not change sign from $+$ to $-$ for increasing $t \in [-1, 1]$.

This implies that the open subsets $X_\pm$ of $[-1, 1] \times \mathbf{R}^N$ defined by

(2.4.3) $\qquad X_+ = \{(t, x) \in [-1, 1] \times \mathbf{R}^N; f(s, x) > 0 \text{ for some } s \in [-1, t]\},$

(2.4.4) $\qquad X_- = \{(t, x) \in [-1, 1] \times \mathbf{R}^N; f(s, x) < 0 \text{ for some } s \in [t, 1]\},$

are disjoint; $\pm f \geq 0$ in $X_\pm$ and $\pm f \leq 0$ in the complement of $X_\pm$, thus $f = 0$ in the closed set $X_0 = [-1, 1] \times \mathbf{R}^N \setminus (X_+ \cup X_-)$. With the notation $X_\pm(t) = \{x \in \mathbf{R}^N; (t, x) \in X_\pm\}$, $t \in [-1, 1]$, it is clear that $X_+(t)$ increases with t while $X_-(t)$ decreases. If $X_0(t) = \emptyset$ then $X_+(t) = \mathbf{R}^N$ or $X_-(t) = \mathbf{R}^N$. If we define the signed truncated distance function $\delta_0(t, x)$ by

$$\delta_0(t, x) = \begin{cases} \min(\frac{1}{2}h^{-\frac{1}{2}}, \sup\{R; (t, x + y) \in X_+ \text{ when } |y| < R\}), & \text{when } x \in X_+(t), \\ -\min(\frac{1}{2}h^{-\frac{1}{2}}, \sup\{R; (t, x + y) \in X_- \text{ when } |y| < R\}, & \text{when } x \in X_-(t)), \\ 0, & \text{when } x \in X_0(t), \end{cases}$$

it follows that $\delta_0(t,x)$ *is an increasing function of t and that* $\delta_0(t,x)f(t,x) \geq 0$. When $X_0(t) \neq \emptyset$ then δ_0 is the signed truncated distance to $X_0(t)$. From the triangle inequality it follows that $\delta_0(t,x)$ is Lipschitz continuous with Lipschitz constant 1 (for fixed t), for $|\delta_0(t,x) - \delta_0(t,y)| \leq |x - y|$ if the open line segment (x,y) lies in $X_+(t)$ or in $X_-(t)$, or if $x,y \in X_0(t)$, which contains the boundary of $X_\pm(t)$. Since $2|\delta_0(t,x)| \geq \min(h^{-\frac{1}{2}}, \inf_{f(t,y)=0} 2|x - y|)$ we can modify (2.2.16) and (2.2.20) and define

$$H_1(t,x)^{-1} = \max(4\delta_0(t,x)^2, H_2(t,x)^{-1}), \quad H_4(t,x)^{-1} = \max(4\delta_0(t,x)^2, H_3(t,x)^{-1})$$

where $H_2(t,x)$ and $H_3(t,x)$ are defined by (2.2.18) and (2.2.13) for fixed t. This gives for fixed t the estimates (2.2.17) and (2.2.21), for $|\delta_0(t,x)|$ is at least as large as the function $S(x)$ there so the inequalities remain valid when $j \leq 2$ and are trivial when $j > 2$. With the convenient notation $\langle \delta_0 \rangle = \max(1, 2|\delta_0|)$ these definitions do not change if we replace $4\delta_0^2$ by $\langle \delta_0 \rangle^2$.

On a very formal level $\delta_0(t,x)$ has the properties which the symbol of the operator $B(x,D')$ in the introduction should have (with $t = x_n$ and $(x',\xi') = x$). However, it is not sufficiently regular to allow any symbolic calculus. This section will be devoted to establishing properties of δ_0 which allow a regularisation sufficient for a weak pseudodifferential calculus. To extract further information from the condition $(\bar{\Psi})$ we shall first prove an elementary lemma on functions with sign conditions like those of $f(t,x)$ for three values of t.

As a motivation for the lemma, which is essentially due to Lerner [L7], he observed that if $F_1 \in C^1$ and $F_2 \in C^2$ are real valued functions in a neighborhood of $0 \in \mathbf{R}^N$ and $F_1(0) = F_2(0) = 0$, $F_1'(0) \neq 0$, $F_2'(0) = 0$, then $F_2''(0) \geq 0$ if $F_1(x) > 0$ implies $F_2(x) \geq 0$. For the proof it suffices to note that if $\langle F_1'(0), x \rangle > 0$ then $F_1(tx) > 0$ for small $t > 0$, hence $0 \leq F_2(tx)/t^2 \to \frac{1}{2}\langle F_2''(0)x, x \rangle$ when $t \to +0$, which proves that $\langle F_2''(0)x, x \rangle \geq 0$ when $\langle F_1'(0), x \rangle \neq 0$ and so for all x. Here is the lemma:

LEMMA 2.4.1. *Let* $F_1 \in C^2(B_R)$ *and* $F_3 \in C^3(B_R)$ *where* $B_R = \{x \in \mathbf{R}^N; |x| \leq R\}$, *and assume that*

$$0 < \varrho_1 = |F_1'(0)| \leq R; \quad |F_1''| \leq M_1 \text{ in } B_R,$$
$$\varrho_2 = |F_2''(0)| \leq \varrho_1, \quad |F_2'(0)| < \kappa\varrho_2^2; \quad |F_2'''| \leq M_2 \text{ in } B_R.$$

Assume that $F_1(z) \geq 0$ *for some* z *with* $|z| \leq \kappa\varrho_2$ *and* $F_2(w) \leq 0$ *for some* w *with* $|w| \leq \kappa\varrho_2$ *where* $0 < 8\kappa \leq \min(1/(2M_1), 1/(2M_2), 1)$. *If* $F_1(x) > 0$ *implies* $F_2(x) \geq 0$ *when* $x \in B_R$, *it follows that*

$$\langle F_2''(0)x, x \rangle \geq -3\varrho_2|x|^2/4 \quad \text{when } x \in \mathbf{R}^N.$$

PROOF. We may assume that $F_1'(0) = (\varrho_1, 0, \ldots, 0)$. Then

$$F_1(x) \geq F_1(0) + \varrho_1 x_1 - \tfrac{1}{2}M_1|x|^2, \quad x \in B_R,$$
$$0 \leq F_1(z) \leq F_1(0) + \varrho_1 z_1 + \tfrac{1}{2}M_1|z|^2, \quad \text{hence}$$
$$F_1(x) \geq \varrho_1 x_1 - \varrho_1 z_1 - \tfrac{1}{2}M_1(|x|^2 + |z|^2), \quad x \in B_R.$$

Similarly,

$$F_2(x) \leq F_2(0) + \langle F_2'(0), x \rangle + \tfrac{1}{2}\langle F_2''(0)x, x \rangle + M_2|x|^3/6, \quad x \in B_R,$$
$$0 \geq F_2(w) \geq F_2(0) + \langle F_2'(0), w \rangle + \tfrac{1}{2}\langle F_2''(0)w, w \rangle - M_2|w|^3/6, \quad \text{hence}$$
$$F_2(x) \leq \langle F_2'(0), x - w \rangle + \tfrac{1}{2}\langle F_2''(0)x, x \rangle - \tfrac{1}{2}\langle F_2''(0)w, w \rangle + M_2(|x|^3 + |w|^3)/6, \quad x \in B_R.$$

By the assumed sign condition we have if $x \in B_R$

$$x_1 > z_1 + \tfrac{1}{2}M_1(|x|^2 + |z|^2)/\varrho_1 \implies F_1(x) > 0 \implies F_2(x) \geq 0$$
$$\implies -\langle F_2''(0)x, x\rangle \leq 2\langle F_2'(0), x - w\rangle - \langle F_2''(0)w, w\rangle + M_2(|x|^3 + |w|^3)/3.$$

If we set $\tilde{\kappa} = 8\kappa$ and $x = \tilde{\kappa}\varrho_2 X$ where $|X| = 1$, it follows that

$$X_1 > c \implies -\langle F_2''(0)X, X\rangle \leq \varrho_2 b,$$

if $c \geq \kappa/\tilde{\kappa} + \tfrac{1}{2}M_1\tilde{\kappa}(1 + (\kappa/\tilde{\kappa})^2)\varrho_2/\varrho_1$ and $b \geq 2\kappa(\kappa + \tilde{\kappa})/\tilde{\kappa}^2 + (\kappa/\tilde{\kappa})^2 + M_2\tilde{\kappa}(1 + (\kappa/\tilde{\kappa})^3)/3$.
By the hypotheses we can take

$$c = \tfrac{1}{8} + \tfrac{1}{4}(1 + \tfrac{1}{64}), \quad b = \tfrac{1}{4}(1 + \tfrac{1}{8}) + \tfrac{1}{64} + \tfrac{1}{6}(1 + \tfrac{1}{512}).$$

If we homogenize this means that $X_1 > c|X|$ implies $-\langle F_2''(0)X, X\rangle \leq \varrho_2 b|X|^2$. Set $X = (1, tY)$ where $Y \in \mathbf{R}^{N-1}$ is a unit vector, and set $g(t) = \langle F_2''(0)X, X\rangle/\varrho_2$. Then we have $g(t) \leq 1 + t^2$ by the definition of ϱ_2, and

$$c^2(1 + t^2) < 1 \implies g(t) \geq -b(1 + t^2).$$

Here $0 < c < 1$, and we can conclude that $g(t) \geq 1 - Kt^2$ when $c^2(1 + t^2) > 1$ if $1 - Kt^2 = -b(1 + t^2)$ when $c^2(1 + t^2) = 1$, that is, $K = (b + c^2)/(1 - c^2)$. In fact, if $g(0) = 1$ then $g'(0) = 0$ and $g(t) = 1 - kt^2$ for some $k \leq K$. Otherwise the quadratic polynomial $g(t) - 1 + Kt^2$ is negative at the origin and non-negative when $c^2(1 + t^2) = 1$, hence non-negative when $c(1 + t^2) \geq 1$. Thus $g(t) \geq -(1 + t^2)(b + c^2)/(1 - c^2)$, $t \in \mathbf{R}$, which means that $\langle F_2''(0)X, X\rangle \geq -C\varrho_2|X|^2$ where $C = (b + c^2)/(1 - c^2) < 3/4$ by a simple numerical calculation.

PROPOSITION 2.4.2. *If f satisfies (2.4.1) and (2.4.2) then*

$$(2.4.5) \quad h^{\frac{1}{2}}H_4(t, x)^{-\frac{1}{2}}H_1(t, x)^{\frac{1}{2}}\langle\delta_0(t, x)\rangle \leq C \max_{j=1,2}(\langle\delta_0(t_j, x)\rangle + \langle\delta_0(t, x)\rangle)H_1(t_j, x)^{\frac{1}{2}},$$

if $-1 \leq t_1 \leq t \leq t_2$, for some constant C independent of t_1, t, t_2, x and f, in fact for $C = 32$.

PROOF. To prove this we shall assume that for some t, t_1, t_2, x we have

$$(2.4.6) \quad \max_{j=1,2}(\langle\delta_0(t_j, x)\rangle + \langle\delta_0(t, x)\rangle)H_1(t_j, x)^{\frac{1}{2}} < \kappa h^{\frac{1}{2}}H_4(t, x)^{-\frac{1}{2}}H_1(t, x)^{\frac{1}{2}}\langle\delta_0(t, x)\rangle$$

and derive a contradiction if κ is small enough. At first we shall only use the part of (2.4.6) where $j = 1$ in the left-hand side. From the second term on the left we obtain

$$H_1(t_1, x)^{\frac{1}{2}} < \kappa H_1(t, x)^{\frac{1}{2}}, \quad H_4(t, x)^{\frac{1}{2}} < \kappa H_1(t, x)^{\frac{1}{2}},$$

for $h \leq H_4(t, x)$ and $h \leq H_1(t_1, x)$. This implies, if $\kappa < 1$, that

$$\langle\delta_0(t, x)\rangle \leq H_1(t, x)^{-\frac{1}{2}} \leq \kappa H_4(t, x)^{-\frac{1}{2}},$$
$$|h^{-\frac{1}{2}}f'(t, x)| \leq H_4(t, x)^{-\frac{1}{2}}H_1(t, x)^{-\frac{1}{2}} < \kappa H_4(t, x)^{-1}, \quad \text{hence}$$
$$H_4(t, x)^{-1} = H_3(t, x)^{-1} = h^{-1}|f''(t, x)|^2,$$

so the right-hand side of (2.4.6) becomes

$$\kappa|f''(t,x)|H_1(t,x)^{\frac{1}{2}}\langle\delta_0(t,x)\rangle \le \kappa|f''(t,x)|,$$

Hence

$$H_1(t_1,x)^{\frac{1}{2}} < \kappa|f''(t,x)|H_1(t,x)^{\frac{1}{2}}, \quad \langle\delta_0(t,x)\rangle < \kappa H_1(t_1,x)^{-\frac{1}{2}}|f''(t,x)|,$$

and since $H_1(t,x)^{-1} \ge h^{-1}|f'(t,x)|^2/(h^{-1}|f''(t,x)|^2) = |f'(t,x)|^2/|f''(t,x)|^2$, we conclude that

$$H_1(t_1,x)^{\frac{1}{2}} \le \kappa|f''(t,x)|^2/|f'(t,x)|, \quad\text{or equivalently}$$

$$|H_1(t_1,x)^{-\frac{1}{2}}f'(t,x)| \le \kappa|H_1(t_1,x)^{-\frac{1}{2}}f''(t,x)|^2.$$

Now consider the rescaled functions

$$F_1(y) = h^{-\frac{1}{2}}H_4(t_1,x)^{\frac{1}{2}}f(t_1,x+y), \quad F_2(y) = H_1(t_1,x)^{-\frac{1}{2}}f(t,x+y).$$

Here $|F_1'(0)| = h^{-\frac{1}{2}}H_4(t_1,x)^{\frac{1}{2}}|f'(t_1,x)|$. Since $\langle\delta_0(t_1,x)\rangle < \kappa H_1(t_1,x)^{-\frac{1}{2}}$ by (2.4.6), we have $H_1(t_1,x)^{-1} = H_2(t_1,x)^{-1} = h^{-1}|f'(t_1,x)|^2 H_3(t_1,x)$, and $H_3(t_1,x) = H_4(t_1,x)$ since $H_1^{-1} \le H_4^{-1}$, which means that $|F_1'(0)| = H_1(t_1,x)^{-\frac{1}{2}} = \varrho_1$. Moreover, $|F_1''(0)| \le 1$, and $|F_1'''| \le H_4(t_1,x)^{\frac{1}{2}} \le H_1(t_1,x)^{\frac{1}{2}}$ everywhere. Summing up, we have with the notation $\varrho_1 = H_1(t_1,x)^{-\frac{1}{2}}$

$$|F_1'(0)| = \varrho_1, \ |F_1''(0)| \le 1; \ |F_1'''(y)| \le 1/\varrho_1, \ y \in \mathbf{R}^N.$$

We have $|F_2'''| \le 1$ in $\mathbf{R}^N$ and

$$\varrho_2 = |F_2''(0)| = |H_1(t_1,x)^{-\frac{1}{2}}f''(t,x)| \le H_1(t_1,x)^{-\frac{1}{2}} = \varrho_1,$$

$$|F_2'(0)| = |H_1(t_1,x)^{-\frac{1}{2}}f'(t,x)| \le \kappa\varrho_2^2.$$

The first bound is obvious, and the second was proved above. Since

$$\langle\delta_0(t,x)\rangle < \kappa H_1(t_1,x)^{-\frac{1}{2}}|f''(t,x)| = \kappa\varrho_2,$$

there is some w with $|w| < \kappa\varrho_2$ where $F_2(w) \le 0$. By (2.4.6)

$$\langle\delta_0(t_1,x)\rangle \le \kappa H_1(t_1,x)^{-\frac{1}{2}}|f''(t,x)| = \kappa\varrho_2,$$

so $F_1(z) \ge 0$ for some z with $|z| \le \kappa\varrho_2$. We can now apply Lemma 2.4.1 with $R = \varrho_1$ and $M_1 = 2$, $M_2 = 1$ and conclude that $\langle F_2''(0)y,y\rangle \ge -\frac{3}{4}\varrho_2|y|^2$ when $y \in \mathbf{R}^N$, that is, $\langle f''(t,x)y,y\rangle \ge -\frac{3}{4}|f''(t,x)||y|^2$, if $\kappa < 1/32$.

Replacing f by $-f$ and t_1 by t_2 we can argue in the same way and obtain $\langle f''(t,x)y,y\rangle \le \frac{3}{4}|f''(t,x)||y|^2$, which implies that $|f''(t,x)| \le \frac{3}{4}|f''(t,x)|$ so $f''(t,x) = 0$ which is a contradiction. Thus (2.4.5) is valid with $C = 32$.

Proposition 2.4.2 is due to Lerner [L7]. From (2.4.5) it follows that

$$(2.4.5)' \quad h^{\frac{1}{2}}H_4(t,x)^{-\frac{1}{2}}H_1(t,x)^{\frac{1}{2}}\langle\delta_0(t,x)\rangle$$
$$\le C(2\delta_0(t_2,x) - 2\delta_0(t_1,x) + \max_{j=1,2}2\langle\delta_0(t_j,x)H_1(t_j,x)^{\frac{1}{2}}\rangle).$$

By (2.2.21) $M(t,x) = h^{\frac{1}{2}}H_4(t,x)^{-\frac{1}{2}}H_1(t,x)^{-1}$ is a weight for $f(t,x)$ with respect to the metric $H_1(t,x)|dx|^2$. The left-hand side of (2.4.5)' is equal to $M(t,x)H_1(t,x)^{\frac{3}{2}}\langle\delta_0(t,x)\rangle$. Dencker's key estimate states precisely that there is a weight M for f such that this product can be estimated by the right-hand side of (2.4.5)', so Proposition 2.4.2 strengthens his result. It is not obvious though how this improvement can be exploited, but the proof of the stronger estimate is actually simpler since it allows separate study of the situation at t_1 and at t_2.

PROPOSITION 2.4.3. *If f satisfies (2.4.1) and (2.4.2), and*

$$(2.4.7) \qquad m(t,x) = \inf_{-1 \le t_1 \le t \le t_2 \le 1} \max_{j=1,2} ((\langle \delta_0(t_j,x) \rangle + \langle \delta_0(t,x) \rangle) H_1(t_j,x)^{\frac{1}{2}}$$

is the minimum with respect to t_1, t_2 of the right-hand side of (2.4.5), then

$$(2.4.8) \qquad m(t,x) \le \max_{j=1,2} \left(m(t_j,x) + 4(\delta_0(t_2,x) - \delta_0(t_1,x)) H_1(t_j,x)^{\frac{1}{2}} \right),$$

when $-1 \le t_1 \le t \le t_2 \le 1$.

PROOF. Since x plays no role in the proof we omit the dependence on x from the notation. If $\varepsilon > 0$ we can by the definition of $m(t_1)$ find $t' \in [-1, t_1]$ such that

$$(\langle \delta_0(t') \rangle + \langle \delta_0(t_1) \rangle) H_1(t')^{\frac{1}{2}} \le m(t_1) + \varepsilon,$$

and by the definition of $m(t_2)$ we can find $t'' \in [t_2, 1]$ such that

$$(\langle \delta_0(t'') \rangle + \langle \delta_0(t_2) \rangle) H_1(t'')^{\frac{1}{2}} \le m(t_2) + \varepsilon.$$

Since $t' \le t \le t''$ it follows that

$$\begin{aligned}
m(t) &\le \max \left((\langle \delta_0(t') \rangle + \langle \delta_0(t) \rangle) H_1(t')^{\frac{1}{2}}, (\langle \delta_0(t'') \rangle + \langle \delta_0(t) \rangle) H_1(t'')^{\frac{1}{2}} \right) \\
&\le \varepsilon + \max \left(m(t_1) + (\langle \delta_0(t) \rangle - \langle \delta_0(t_1) \rangle) H_1(t')^{\frac{1}{2}}, m(t_2) + (\langle \delta_0(t) \rangle - \langle \delta_0(t_2) \rangle) H_1(t'')^{\frac{1}{2}} \right) \\
&= \varepsilon + \max \left(m(t_1) + (|s| - |s_1|) H_1(t')^{\frac{1}{2}}, m(t_2) + (|s| - |s_2|) H_1(t'')^{\frac{1}{2}} \right),
\end{aligned}$$

where $s = 2\delta_0(t)$, $s_j = 2\delta_0(t_j)$, thus $s_1 \le s \le s_2$. The maximum of this convex function of $s \in [s_1, s_2]$ is taken at an end point, and $||s_2| - |s_1|| \le s_2 - s_1$. Since

$$\langle \delta_0(t_1) \rangle H_1(t')^{\frac{1}{2}} \le m(t_1) + \varepsilon \le 2\langle \delta_0(t_1) \rangle H_1(t_1)^{\frac{1}{2}} + \varepsilon$$

we have $H_1(t')^{\frac{1}{2}} \le 2H_1(t_1)^{\frac{1}{2}} + \varepsilon$ and similarly $H_1(t'')^{\frac{1}{2}} \le 2H_1(t_2)^{\frac{1}{2}} + \varepsilon$, which proves (2.4.8) with 2ε added on the right-hand side.

REMARK. In the proof we could just have estimated $H_1(t')$ and $H_1(t'')$ by 1, which gives

$$(2.4.8)' \quad m(t,x) \le \max \left(m(t_1,x), m(t_2,x) \right) + 2(\delta_0(t_2,x) - \delta_0(t_1,x)), \quad -1 \le t_1 \le t \le t_2 \le 1.$$

Only this weaker result will be used below.

The estimate $(2.4.8)'$ can be used to modify the increasing function $\delta_0(t,x)$ so that we get a lower bound for the rate of increase and a bound for the modifying term.

PROPOSITION 2.4.4. *If f satisfies (2.4.1) and (2.4.2), and m is defined by (2.4.7) then*

$$(2.4.9) \qquad 4T\partial(\delta_0(t,x) + \Theta_T(t,x)) \ge m(t,x), \quad 2|\Theta_T(t,x)| \le m(t,x), \quad if$$

$$(2.4.10) \quad \Theta_T(t,x) = \sup_{-T \le s \le t} \left(\delta_0(s,x) - \delta_0(t,x) - \tfrac{1}{2}m(s,x) + \frac{1}{4T} \int_s^t m(\tau,x) \, d\tau \right),$$

where $|t| < T \leq 1$ and $x \in \mathbf{R}^N$. The derivative is taken in the sense of distribution theory.

PROOF. Taking $s = t$ in (2.4.10) gives $2\Theta_T(t,x) \geq -m(t,x)$, and $\Theta_T(t,x) \leq \frac{1}{2}m(t,x)$ for

$$\frac{1}{4T}\int_s^t m(\tau,x)\,d\tau \leq \frac{1}{2}\sup_{s\leq\tau\leq t} m(\tau,x) \leq \frac{1}{2}\max(m(s,x),m(t,x)) + \delta_0(t,x) - \delta_0(s,x)$$

by (2.4.8)′. Since

$$\delta_0(t,x) + \Theta_T(t,x) - \frac{1}{4T}\int_0^t m(\tau,x)\,d\tau = \sup_{-T\leq s\leq t}\left(\delta_0(s,x) - \tfrac{1}{2}m(s,x) + \frac{1}{4T}\int_s^0 m(\tau,x)\,d\tau\right)$$

and the right-hand side is an increasing function of t, the first part of (2.4.9) is also proved.

Taking $t_1 = t_2 = t$ in (2.4.7) we obtain

$$(2.4.11) \qquad\qquad m(t,x) \leq 2\langle\delta_0(t,x)\rangle H_1(t,x)^{\frac{1}{2}} \leq 2.$$

Since $\langle\delta_0(t,x+y)\rangle \leq \langle\delta_0(t,x)\rangle(1+2|y|)$ by the triangle inequality, and $H_1(t,x+y)^{\frac{1}{2}} \leq H_1(t,x)^{\frac{1}{2}}(1+2|y|)$ by (2.1.2)′ since $H_1^{-\frac{1}{2}}$ has Lipschitz constant 2, it follows that

$$(2.4.12) \qquad\qquad m(t,x+y) \leq m(t,x)(1+2|y|)^2.$$

Dencker's energy estimates relied on an operator with symbol derived from the increasing function $B = \delta_0(t,x) + \Theta_T(t,x)$. Lerner has modified this to

$$(2.4.13) \qquad\qquad B_T(t,x) = \delta_0(t,x) + \Theta_T(t,x) + \delta_0(t,x)\eta_T(t,x),$$

$$(2.4.14) \qquad\qquad \eta_T(t,x) = h^{\frac{1}{2}}T^{-1}\int_0^t \delta_0(s,x)\,ds + \tfrac{1}{2}, \quad |t| < T.$$

Since $|\delta_0(s,x)| \leq \frac{1}{2}h^{-\frac{1}{2}}$ we have

$$0 \leq \eta_T(t,x) \leq 1, \partial\eta_T(t,x)/\partial t = h^{\frac{1}{2}}\delta_0(t,x)/T,$$

so the size of B_T does not change very much but

$$(2.4.15) \quad T\partial B_T(t,x)/\partial t \geq m(t,x)/4 + h^{\frac{1}{2}}\delta_0(t,x)^2 \geq \tfrac{1}{6}\left(m(t,x) + h^{\frac{1}{2}}(1+\delta_0(t,x)^2)\right),$$

for $m(t,x) \geq 2h^{\frac{1}{2}}$.

3. Pseudodifferential calculus. In this section we shall review the techniques required for the proof of the key estimates in Sections 4 and 5. Let us first recall the basic definitions and results of the Weyl calculus while referring to [H1, Sections 18.4 – 18.6] for details and proofs. A Riemannian metric g in $W = T^*(\mathbf{R}^n) = \mathbf{R}^n \oplus \mathbf{R}^n$ is called slowly varying if there are positive constants c and C such that

$$g_{w+w_1} \leq Cg_w \quad \text{if } g_w(w_1) \leq c, \quad w,w_1 \in W.$$

Here g_w is the quadratic form on W defined by the metric at w. This is just an extension of (2.1.1), and as before there is an analogue of (2.1.1)'. The dual g_w^σ of this quadratic form with respect to the symplectic form σ is defined by

$$g_w^\sigma(t) = \sup_{t'} |\sigma(t,t')|^2/g_w(t'), \quad w,t,t' \in W; \quad \sigma((x,\xi),(y,\eta)) = \langle \xi,y \rangle - \langle \eta,x \rangle.$$

The metric g is called σ temperate if it is slowly varying and for some constants C and N

$$g_{w_1}(t) \le Cg_w(t)(1 + g_{w_1}^\sigma(w - w_1))^N, \quad w,w_1 \in W.$$

A positive function M in W is called σ, g temperate if

$$M(w + w_1) \le CM(w) \quad \text{when } g_w(w_1) \le c, \quad w,w_1 \in W,$$
$$M(w_1) \le CM(w)(1 + g_{w_1}^\sigma(w - w_1))^N, \quad w,w_1 \in W.$$

The space $S(M,g)$ of symbols with weight M with respect to g is then defined as the set of all $a \in C^\infty(W)$ such that $|a^{(k)}|^g \le C_k M$ for all integers $k \ge 0$; here $|a^{(k)}(w)|^g$ is the norm of the kth differential of a at w with respect to the quadratic form g_w. A corresponding Weyl operator $a^w = \mathrm{Op}^w a$ is then defined as in [H1, Section 18.5], formally as

$$a^w(x,D)u(x) = (2\pi)^{-n} \iint a((x+y)/2,\xi)e^{i\langle x-y,\xi \rangle}u(y)\,dy\,d\xi, \quad u \in \mathcal{S}(\mathbf{R}^n).$$

It is a continuous map on the Schwartz space $\mathcal{S}(\mathbf{R}^n)$ and on its dual $\mathcal{S}'(\mathbf{R}^n)$. If a is real valued then a^w is symmetric, if M is bounded and $g \le g^\sigma$ then a^w is bounded in $L^2(\mathbf{R}^n)$.

3.1. Some (lower) bounds. If $g \le g^\sigma$, as we shall always assume, then there is a good calculus: If M_1 and M_2 are σ, g temperate and $a_j \in S(M_j,g)$, $j = 1,2$, then $a_1^w(x,D)a_2^w(x,D) = (a_1 \# a_2)^w(x,D)$ where $a_1 \# a_2 \in S(M_1M_2,g)$ and the bilinear map $(a_1,a_2) \mapsto a_1 \# a_2$ is continuous. If $h(w)^2 = \sup g_w/g_w^\sigma$, thus $0 < h \le 1$, then there is an asymptotic expansion of $a_1 \# a_2$ with terms in $S(h^j M_1 M_2,g)$, in particular

$$a_1 \# a_2 - a_1 a_2 - \{a_1,a_2\}/(2i) \in S(h^2 M_1 M_2,g),$$

where $\{a_1,a_2\}$ is the Poisson bracket. (See [H1, Theorem 18.5.4].) In fact, *this only requires that* $|a_j^{(k)}|^g \le C_k M_j$ *when* $k \ge 2$. (See Bony [B] for the required additions to the proof of [H1, Theorem 18.5.4].) If $a \in S(M,g)$ is real valued and we take $a_1 = a_2 = a$ then the Poisson bracket vanishes and we get

(3.1.1) $$(a^2)^w(x,D) - a^w(x,D)^2 = R^w(x,D), \quad R \in S(h^2 M^2,g);$$

thus $R^w(x,D)$ is a lower bound for $(a^2)^w(x,D)$. A much deeper result of this form is the Fefferman-Phong bound which we need in a somewhat more general version than stated in [H1]:

PROPOSITION 3.1.1. *If g is σ temperate and $h(w)^2 = \sup g_w/g_w^\sigma \le 1$, and M is σ, g temperate, $0 \le a \in S(M,g)$, then $a^w(x,D) \ge R^w(x,D)$ for some $R \in S(h^2 M,g)$.*

PROOF. When $M = h^{-2}$ this is precisely [H1, Theorem 18.6.8], and we shall reduce the general statement to that case. Choose a real valued $c \in S(M^{\frac{1}{2}}h,g)$ so that $b = c^{-1} \in S(M^{-\frac{1}{2}}h^{-1},g)$. Let $u \in \mathcal{S}(\mathbf{R}^n)$ and set $v = c^w(x,D)u$. Then

$$b^w v = b^w c^w u = u - r^w u, \quad r \in S(h^2,g),$$

hence with scalar products in L^2

$$(a^w u, u) = (a^w b^w v, b^w v) + 2\operatorname{Re}(a^w b^w v, r^w u) + (a^w r^w u, r^w u).$$

Since $b^w a^w b^w = (b^2 a)^w + r_1^w$ where $r_1 \in S(1, g)$ and $0 \leq b^2 a \in S(h^{-2}, g)$, the first term on the right is bounded below by $-C\|v\|^2$ for some constant C by [H1, Theorem 18.6.8], and $\|v\|^2 = ((c^w)^2 u, u)$; the symbol of $(c^w)^2$ is in $S(h^2 M, g)$. The symbol of $\bar{r}^w a^w r^w$ is in $S(h^4 M, g) \subset S(h^2 M, g)$. We have $(a^w b^w v, r^w u) = (\bar{r}^w a^w b^w c^w u, u)$, and the symbol of $\bar{r}^w a^w b^w c^w$ is in $S(h^2 M, g)$, which completes the proof.

In carrying out the program outlined in the introduction we must work with Weyl operators associated with different metrics. The composition of such operators was discussed quite generally in Proposition 18.5.3 and in Theorem 18.5.5 in [H1] but we shall confine ourselves here to the case of two conformal metrics as in [H1, Proposition 18.5.7]. With g as before we let $g_0 = \gamma g$ be a conformal metric where $\gamma \geq 1$. Then $h_0^2(w) = \sup g_{0,w}/g_{0,w}^\sigma = \gamma^2 h(w)^2$, and we require that $h_0 \leq 1$, thus $\gamma h \leq 1$ and $\gamma^2 g \leq g^\sigma$. If g_0 is slowly varying then g_0 is σ temperate by [H1, Proposition 18.5.6], and the same proof shows that if M is σ, g temperate then M is also σ, g_0 temperate. In fact, if

$$M(w_1) \leq CM(w)(1 + g_{w_1}^\sigma(w - w_1))^N, \quad w, w_1 \in W,$$

it follows that

$$M(w_1) \leq C'M(w)(1 + \gamma^{-1}g_{w_1}^\sigma(w - w_1))^{2N}, \quad w, w_1 \in W;$$

this is obvious if $g_{w_1}(w - w_1) \leq c$ and c is small since M is slowly varying with respect to g, and if $g_{w_1}(w - w_1) \geq c$ then $g_{w_1}^\sigma(w - w_1) \geq \gamma^2 c$ and $g_{w_1}^\sigma(w - w_1)/\gamma \geq \sqrt{cg_{w_1}^\sigma(w - w_1)}$.

If $a \in S(M, g)$ and $a_0 \in S(M_0, g_0)$ then $a \# a_0 \in S(MM_0, g_0)$ and there is an asymptotic expansion with the jth term in $S((hh_0)^{j/2} MM_0, g_0)$. In particular,

$$(3.1.2) \quad \begin{aligned} a \# a_0 - aa_0 - \{a, a_0\}/(2i) &\in S(hh_0 MM_0, g_0), \\ a_0 \# a - aa_0 - \{a_0, a\}/(2i) &\in S(hh_0 MM_0, g_0). \end{aligned}$$

As for (3.1.1) this remains true if $|a^{(k)}|^g \leq C_k M$ and $|a_0^{(k)}|^{g_0} \leq C_k M_0$ for $k \geq 2$ only.

By G_0 we shall denote the self dual metric $|dw|^2$ in W. That the calculus is useful even for such a large metric is shown by the proof of the following proposition which will be needed later on.

PROPOSITION 3.1.2. Let $\mu \in C^\infty(T^*\mathbf{R}^n)$ be a positive weight function such that

$$(3.1.3) \qquad \mu(w + w') \leq C\mu(w)(1 + \varepsilon^2|w'|^2)^N, \quad w, w' \in T^*\mathbf{R}^n,$$

and assume that $\mu \in S(\mu, \varepsilon^2 G_0)$, that is, that for all multiindices α

$$(3.1.4) \qquad |\mu^{(\alpha)}(w)| \leq C_\alpha \varepsilon^{|\alpha|} \mu(w), \quad w \in T^*\mathbf{R}^n.$$

If $0 < \varepsilon < \varepsilon_0$, where ε_0 only depends on the constants C and finitely many C_α, and if $a \in S(\mu, G_0)$, then

$$(3.1.5) \qquad |(a^w(x, D)u, u)| \leq A(a)(\mu^w(x, D)u, u), \quad u \in \mathcal{S}(\mathbf{R}^n),$$

where $A(a)$ is a seminorm in the Fréchet space $S(\mu, G_0)$.

PROOF. If $b = \sqrt{\mu}$ and $c = 1/\sqrt{\mu}$, then

$$|b^{(\alpha)}(w)| \leq C'_\alpha \varepsilon^{|\alpha|} b(w), \quad |c^{(\alpha)}(w)| \leq C'_\alpha \varepsilon^{|\alpha|} c(w), \quad w \in T^*\mathbf{R}^n,$$

where the constants C'_α only depend on the constants C_β with $|\beta| \leq |\alpha|$. In fact, a derivative of b of order α is a linear combination of products of b and factors $\mu^{(\beta)}/\mu$ of orders β adding up to α, and similarly for c. Hence

$$c^w \mu^w c^w = \mathrm{Id} + R_1^w, \quad c^w b^w = \mathrm{Id} + R_2^w$$

where R_j/ε^2 is bounded in $S(1, G_0)$, which implies that the norm of R_j^w as operator in L^2 is $< \frac{1}{2}$ if ε is small enough, $j = 1, 2$. Hence

$$(\mu^w c^w v, c^w v) = (c^w \mu^w c^w v, v) = (v, v) + (R_1^w v, v) \geq \tfrac{1}{2}\|v\|^2, \quad v \in \mathcal{S}(\mathbf{R}^n).$$

Since the Weyl symbol of $c^w a^w c^w$ is in $S(1, G_0)$ we obtain

$$|(a^w c^w v, c^w v)| = |(c^w a^w c^w v, v)| \leq \tfrac{1}{2} A(a)\|v\|^2 \leq A(a)(\mu^w c^w v, c^w v), \quad v \in \mathcal{S}(\mathbf{R}^n),$$

where $A(a)$ is a seminorm in $S(\mu, g_0)$. Now $\{c^w v; v \in \mathcal{S}(\mathbf{R}^n)\}$ is dense in L^2, for if we take $v = b^w u$, $u \in \mathcal{S}(\mathbf{R}^n)$, then $v \in \mathcal{S}(\mathbf{R}^n)$ and $c^w v = c^w b^w u = u + R_2^w u$, and $\mathrm{Id} + R_2^w$ is an invertible operator in L^2. If μ is bounded then both a^w and μ^w are L^2 continuous and the inequality (3.1.5) follows even for all $u \in L^2(\mathbf{R}^n)$. Otherwise we replace $a(w)$ by $a(w)/(1 + \delta^2|w|^2)^N$ and $\mu(w)$ by $\mu(w)/(1 + \delta^2|w|^2)^N$ and obtain (3.1.5) when $\delta \to 0$.

3.2. Regularization of symbols and Wick operators. When proving estimates for the operator (1.1) along the lines indicated in the introduction, the symbol of the auxiliary operator B will be closely connected to the truncated signed distance function δ_0 discussed in Section 2.4. It is an increasing function of t but even for fixed t it is not in any acceptable symbol class so it must be regularized before pseudodifferential calculus can be applied. If $a \in L^\infty(W)$ we shall write

$$(3.2.1) \qquad a^b(X) = \pi^{-n} \int_W e^{-|X-Y|^2} a(Y)\, dY, \quad X \in W,$$

for the Gaussian regularization. We have $a^b \in S(1, G_0)$, and following Lerner the operator with Weyl symbol a^b will be called the *Wick quantization of a* and denoted by a^{Wick}. We have

$$(3.2.2) \quad a^{\mathrm{Wick}} = \int_W a(Z)\Sigma_Z\, dZ, \quad \Sigma_Z = \mathrm{Op}^w \exp(-|x - z|^2 - |\xi - \zeta|^2), \quad Z = (z, \zeta) \in W.$$

The kernel of the Gaussian Weyl operator centered at Z is

$$(x, y) \mapsto (2\pi)^{-n} \pi^{-n} \int \exp\left(-|\tfrac{1}{2}(x + y) - z|^2 - |\xi - \zeta|^2 + i\langle x - y, \xi\rangle\right) d\xi$$

$$= (2\pi)^{-n} \pi^{-\frac{n}{2}} \exp\left(-|\tfrac{1}{2}(x + y) - z|^2 + i\langle x - y, \zeta\rangle - \tfrac{1}{4}|x - y|^2\right)$$

$$= (2\pi)^{-n} \pi^{-\frac{n}{2}} \exp\left(-\tfrac{1}{2}(|x|^2 + |y|^2) - |z|^2 + \langle x + y, z\rangle + i\langle x - y, \zeta\rangle\right).$$

With the notation

$$G_Z(x) = \pi^{-\frac{n}{4}} \exp\left(-\tfrac{1}{2}|x|^2 + \langle x, z + i\zeta\rangle - \tfrac{1}{2}|z|^2\right), \quad x \in \mathbf{R}^n,$$

we have $\|G_Z(x)\|_{L^2} = 1$, and the kernel of Σ_Z is $(2\pi)^{-n} G_Z(x)\overline{G_Z(y)}$ so Σ_Z is $(2\pi)^{-n}$ times the orthogonal projection on G_Z, and

$$a^{\text{Wick}} = (2\pi)^n \int_W a(Y)\Sigma_Y^2 \, dY = (2\pi)^n \int_W a(Y)\Sigma_Y^* \Sigma_Y \, dY$$

which shows that $a^{\text{Wick}} \geq 0$ when $a \geq 0$. When $u \in \mathcal{S}(\mathbf{R}^n)$ we have

$$|(a^{\text{Wick}}u, u)| = |(2\pi)^n \int_W a(Y)\|\Sigma_Y u\|^2 \, dY| \leq \|a\|_{L^\infty} (2\pi)^n \int_W (\Sigma_Y^* \Sigma_Y u, u) \, dY,$$

so a^{Wick} is bounded in L^2 with norm $\leq \|a\|_{L^\infty}$ since 1^{Wick} is the identity operator.

The Wick quantization is well defined for every $a \in \mathcal{S}'(W)$, for then we have for some N and C

$$|a(\varphi)| \leq C \sum_{|\alpha+\beta|\leq N} \sup |Y^\alpha D_Y^\beta \varphi(Y)|, \quad \varphi \in \mathcal{S}(W),$$

hence

$$|D_X^\gamma a^\flat(X)| = \pi^{-n}|a_Y(D_X^\gamma e^{-|X-Y|^2})| \leq C' \sup_Y \sum_{\nu+\mu\leq N} (1+|Y|)^\nu (1+|X-Y|)^{\mu+|\gamma|} e^{-|X-Y|^2}$$

$$\leq C'(1+|X|)^N \sup_Y (1+|X-Y|)^{N+|\gamma|} e^{-|X-Y|^2},$$

which proves that $a^\flat \in S((1+|X|)^N, |dX|^2)$. This implies that a^{Wick} defines a continuous map from $\mathcal{S}$ to $\mathcal{S}$ and from $\mathcal{S}'$ to $\mathcal{S}'$ for every $a \in \mathcal{S}'(W)$.

PROPOSITION 3.2.1. *If M is a positive measurable function in W with $M(X+Y) \leq CM(X)(1+|Y|)^N$ for all $X, Y \in W$, then $M^\flat \in S(M, G_0)$, and if $a \in S(M, G_0)$ then*

$$(3.2.3) \qquad |(a^w(x, D)u, u)| \leq A(a)(M^{\text{Wick}}u, u), \quad u \in \mathcal{S}(\mathbf{R}^n),$$

where $A(a)$ is a seminorm in the Fréchet space $S(M, G_0)$.

PROOF. The weight M is equivalent to $M^\flat$, for

$$\pi^{-n} M(X) \int_W e^{-|Y|^2}(1+|Y|)^{-N} \, dY/C \leq M^\flat(X) \leq \pi^{-n} M(X) \int_W e^{-|Y|^2}(1+|Y|)^N C \, dY,$$

and $M^\flat \in S(M, G_0)$ since

$$|D^\alpha M^\flat(X)| \leq M(X)\pi^{-n} \int_W C(1+|Y|)^N |D^\alpha e^{-|Y|^2}| \, dY.$$

To prove (3.2.3) we shall apply Proposition 3.1.2. First note that if $0 < \varepsilon < 1$ and $M_\varepsilon(X) = \sup_Y M(X+Y)/(1+\varepsilon|Y|)^N$, then

$$M(X) \leq M_\varepsilon(X) \leq C\varepsilon^{-N} M(X), \quad M_\varepsilon(X+Y) \leq M_\varepsilon(X)(1+\varepsilon|Y|)^N, \quad X, Y \in W,$$

for $(1 + |Y|)/(1 + \varepsilon|Y|) \leq 1/\varepsilon$ and $1 + \varepsilon|Y + Z| \leq (1 + \varepsilon|Y|)(1 + \varepsilon|Z|)$. If $0 \leq \chi \in C_0^\infty(W)$ and χ has suppport in the unit ball and integral 1, $\chi_\varepsilon(y) = \varepsilon^{2n}\chi(\varepsilon Y)$, and $m_\varepsilon = \chi_\varepsilon * M_\varepsilon$, then

$$2^{-N} M_\varepsilon \leq m_\varepsilon \leq 2^N M_\varepsilon, \quad |D^\alpha m_\varepsilon| \leq C_\alpha m_\varepsilon \varepsilon^{|\alpha|}, \quad m_\varepsilon(X + Y) \leq m_\varepsilon(X)(1 + \varepsilon|Y|)^N,$$

where $C_\alpha = 2^N \int_W |D^\alpha \chi(Y)|\, dY$. The last two inequalities remain valid with m_ε replaced by $m_\varepsilon{}^\flat$, which is also equivalent to M. Hence it follows from Proposition 3.1.2 with $\mu = m_\varepsilon{}^\flat$ that for fixed small $\varepsilon > 0$

$$|(a^w(x, D)u, u) \leq A(a)(m_\varepsilon{}^{\flat w}(x, D)u, u) = A(a)(m_\varepsilon^{\text{Wick}}u, u), \quad u \in \mathcal{S}(\mathbf{R}^n),$$

if $a \in S(M, G_0)$. Here $A(a)$ is a seminorm in the Fréchet space $S(M, G_0) = S(m_\varepsilon{}^\flat, G_0)$. Since $m_\varepsilon \leq CM$ implies $m_\varepsilon^{\text{Wick}} \leq CM^{\text{Wick}}$, we have proved (3.2.3).

4. A priori estimates for model operators. We have now presented all the tools required to study the adjoints of the model operator (1.1), (1.2) of the introduction. Since the x_n variable plays a special role there we shall now denote it by t, and replacing n by $n + 1$ we denote the others by $x \in \mathbf{R}^n$. We want to prove a bound for the inverse of the operator $\partial/\partial t - f^w(t, x, D)$ where $f(t, x, \xi)$ is a continuous function of t with values in $C^\infty(\mathbf{R}^{2n})$ satisfying (2.4.1) with x replaced by $w = (x, \xi)$, $N = 2n$, and also condition $(\bar{\Psi})$ as in (2.4.2). We may assume that $\gamma_k = 1$ when $k \leq 3$, for this can be attained by a change of scale for the variable t, so all the results of Section 2 can be applied to f. We define $m(t, w)$ and $\Theta_T(t, w)$, $0 < T \leq 1$, by (2.4.7) and Proposition 2.4.4 respectively, recalling that $\delta_0(t, w)$ is the signed truncated distance function to the zeros of f defined in Subsection 2.4. Set

(4.1)
$$B_T(t, w) = \delta_0(t, w) + \Theta_T(t, w) + \delta_0(t, w)\eta_T(t, w), \quad |t| < T,$$
$$\eta_T(t, w) = h^{\frac{1}{2}}T^{-1} \int_0^t \delta_0(s)\, ds + \tfrac{1}{2}, \quad |t| < T,$$

as in (2.4.13), (2.4.14), and

(4.2)
$$\tilde{m}(t, w) = m(t, w) + h^{\frac{1}{2}}(1 + \delta_0(t, w)^2),$$

as in (2.4.15). Both $B_T(t, w)$ and $\tilde{m}(t, w)$ are increasing functions of t and satisfy the analogue of (2.4.12) as functions of w, uniformly with respect to t, and are thus continuous apart from countably many values of t. To simplify notation we drop the subscript T in what follows. In the definition of the Wick operators t is considered as a parameter.

THEOREM 4.1. *If T is sufficiently small then*

(4.3)
$$(\tilde{m}^{\text{Wick}}u, u) \leq CT \operatorname{Re}(f^w(t, x, D) - \partial/\partial t)u, B^{\text{Wick}}u),$$

when $u \in \mathcal{S}(\mathbf{R}^{1+n})$ and $\sup\{|t|; (t, w) \in \operatorname{supp} u\} < T$.

The positivity of the Wick quantization allows us to translate (2.4.15) to a lower bound for the term in (4.3) involving $\partial u/\partial t$:

LEMMA 4.2. *If $u \in \mathcal{S}(\mathbf{R}^{1+n})$ and $\sup\{|t|; (t,w) \in \operatorname{supp} u\} < T$ then*

$$(4.4) \qquad (\tilde{m}^{\mathrm{Wick}}u, u) \le -12\operatorname{Re}(\partial u/\partial t, B^{\mathrm{Wick}}u).$$

PROOF. By (2.4.15) we have

$$\int \tilde{m}(t,w)\chi(t)\, dt + 6T \int B(t,w)\chi'(t)\, dt \le 0$$

if $0 \le \chi \in C_0^\infty(-T, T)$. Hence the Wick quantization of

$$w \mapsto \int \tilde{m}(t,w)\chi(t)\, dt + 6T \int B(t,w)\chi'(t)\, dt$$

is ≤ 0, that is, if we write $\tilde{m}_t(w) = \tilde{m}(t,w)$, $B_t(w) = B(t,w)$

$$(4.5) \qquad \int (\tilde{m}_t^{\mathrm{Wick}}v, v)\chi(t)\, dt + 6T \int (B_t^{\mathrm{Wick}}v, v)\chi'(t)\, dt \le 0, \quad v \in \mathcal{S}(\mathbf{R}^n).$$

Choose a fixed $\chi \in C_0^\infty(-1, 1)$ with $\chi \ge 0$ and $\int \chi(t)\, dt = 1$, and set $\chi_\varepsilon(t) = \chi(t/\varepsilon)/\varepsilon$ where $\varepsilon > 0$ is small. If we apply (4.5) with $v = u(s, \cdot)$ and u as in the lemma, replacing χ by $\chi_\varepsilon(t-s)$ and integrating with respect to s also, we obtain

$$\iint_{|s|<T-\varepsilon} (\tilde{m}_t^{\mathrm{Wick}}u(s,\cdot), u(s,\cdot))\chi_\varepsilon(t-s)\, dt\, ds$$

$$+ 6T \iint_{|s|<T-\varepsilon} (B_t^{\mathrm{Wick}}u(s,\cdot), u(s,\cdot))\chi_\varepsilon'(t-s)\, dt\, ds \le 0.$$

Since $\chi_\varepsilon'(t-s) = -d\chi_\varepsilon(t-s)/ds$ an integration by parts in the second term gives

$$\iint_{|s|<T-\varepsilon} ((\tilde{m}_t^{\mathrm{Wick}}u(s,\cdot), u(s,\cdot)) + 12T\operatorname{Re}(B_t^{\mathrm{Wick}}\partial u(s,\cdot)/\partial s, u(s,\cdot)))\chi_\varepsilon(t-s)\, dt\, ds \le 0,$$

which gives (4.4) when $\varepsilon \to 0$ since $u(s, \cdot)$ is a continuous function of s with values in $\mathcal{S}(\mathbf{R}^n)$.

To prove (4.3) it remains to give good lower bounds for the term $\operatorname{Re}(f^w(t, x, D)u, u)$ in the right-hand side of (4.4). This will be done in the next three subsections by studying separately the contributions near the sign changes of f, those far from the sign changes and those where f is bounded. Since t only occurs as a parameter in these estimates, we shall omit t from the notation in this discussion.

4.1. Lower bounds near the sign changes. By Corollary 2.3.3 $\delta_0 \in C^\infty$ and δ_0 has the estimates of a symbol belonging to $S(H_1^{-\frac{1}{2}}, G_1)$, $G_1 = H_1 G_0$, $G_0 = |dw|^2$, in

$$(4.1.1) \qquad W_0 = \{w \in W; 8|\delta_0(w)| < H_1(w)^{-\frac{1}{2}}, 6H_1(w) < 1\},$$

and $f = \delta_0 a^2$ in W_0, where $0 < a \in C^\infty(W_0)$ and a has the estimates of a symbol in $S(h^{\frac{1}{4}}H_4^{-\frac{1}{4}}H_1^{-\frac{1}{4}}, G_1)$ in W_0. (Such statements will tacitly assert uniformity in parameters such as t and h.) In this subsection we shall prove:

PROPOSITION 4.1.1. *If $\varphi_0 \in S(1, G_1)$ and supp $\varphi_0 \subset W_0$, then*

$$(4.1.2) \qquad \|q^w(x, D)u\|^2 \leq 2 \operatorname{Re}\left((\varphi_0^2 f)^w(x, D)u, B^{\mathrm{Wick}}u\right) + (R_0^w(x, D)u, u),$$

when $u \in S(\mathbf{R}^n)$. Here $q = \varphi_0 \delta_0 a \in S(h^{\frac{1}{4}} H_4^{-\frac{1}{4}} H_1^{-\frac{3}{4}}, G_1)$ and $R_0 \in S(h^{\frac{1}{2}} H_4^{-\frac{1}{2}} H_1^{\frac{1}{2}}, G_0)$.

Recall that $B(w) = \delta_0(w)(1 + \eta(w)) + \Theta(w)$. Since $\delta_0 \varphi_0^2 f = q^2$ where $q = \varphi_0 \delta_0 a \in S(h^{\frac{1}{4}} H_4^{-\frac{1}{4}} H_1^{-\frac{3}{4}}, G_1)$ we shall use the following lemma to bring out a factor $\delta_0^\flat$ from the Gaussian regularization $(\delta_0(1+\eta))^\flat$.

LEMMA 4.1.2. *By (2.4.14) we have $0 \leq \eta \leq 1$, $|\eta'| \leq h^{\frac{1}{2}}$, which gives $0 \leq \eta^\flat \leq 1$ and*

$$(4.1.3) \qquad \delta_0^\flat \eta^\flat - (\delta_0 \eta)^\flat \in S(h^{\frac{1}{2}}, G_0), \quad \delta_0^\flat \# \eta^\flat - (\delta_0 \eta)^\flat \in S(h^{\frac{1}{2}}, G_0),$$

$$(4.1.4) \qquad |\delta_0^\flat - \delta_0| \leq \sqrt{n}, \; \delta_0^{\flat\prime} \in S(1, G_0), \; |\eta^\flat - \eta| \leq \sqrt{n} h^{\frac{1}{2}}, \; \eta^{\flat\prime} \in S(h^{\frac{1}{2}}, G_0).$$

PROOF. (4.1.4) follows at once from the definition (3.2.1), for $|\delta_0'| \leq 1$ and $|\eta'| \leq h^{\frac{1}{2}}$, and we have $\pi^{-n} \int_W |Y| e^{-|Y|^2} dY \leq \left(\pi^{-n} \int_W |Y|^2 e^{-|Y|^2} dY\right)^{\frac{1}{2}} = \sqrt{n}$. We have

$$\pi^{-n} \int_W (\delta_0(X - Y) - \delta_0(X))(\eta(X - Y) - \eta(X)) e^{-|Y|^2} dY$$

$$= (\delta_0 \eta)^\flat(X) - \delta_0(X)\eta^\flat(X) - \delta_0^\flat(X)\eta(X) + \delta_0(X)\eta(X)$$

$$= (\delta_0 \eta)^\flat(X) - \delta_0^\flat(X)\eta^\flat(X) + (\delta_0(X) - \delta_0^\flat(X))(\eta(X) - \eta^\flat(X)).$$

The integral on the left is $\leq n h^{\frac{1}{2}}$ so $|(\delta_0 \eta)^\flat(X) - \delta_0^\flat(X)\eta^\flat(X)| \leq 2n h^{\frac{1}{2}}$. If α is a multiindex $\neq 0$ we have similarly

$$\pi^{-n} \int_W (\delta_0(X - Y) - \delta_0(X))(\eta(X - Y) - \eta(X)) D_Y^\alpha e^{-|Y|^2} dY$$

$$= D^\alpha(\delta_0 \eta)^\flat(X) - \delta_0(X)D^\alpha \eta^\flat(X) - \eta(X)D^\alpha \delta_0^\flat(X) = D^\alpha(\delta_0 \eta)^\flat(X) - \delta_0^\flat(X)D^\alpha \eta^\flat(X)$$

$$- \eta^\flat(X)D^\alpha \delta_0^\flat(X) + (\delta_0^\flat(X) - \delta_0(X))D^\alpha \eta^\flat(X) + (\eta^\flat(X) - \eta(X))D^\alpha \delta_0^\flat(X).$$

The left-hand side is $\leq h^{\frac{1}{2}} \pi^{-n} \int_W |Y|^2 |D_Y^\alpha e^{-|Y|^2}| dY$. By (4.1.4) the last two terms can also be estimated by constants times $h^{\frac{1}{2}}$, and so can $D^\alpha(\delta_0^\flat(X)\eta^\flat(X)) - \delta_0^\flat(X)D^\alpha \eta^\flat(X) - \eta^\flat(X)D^\alpha \delta_0^\flat(X)$ for this is a sum of products of derivatives of $\delta_0^\flat$ and of $\eta^\flat$ of non-zero orders. This proves the first part of (4.1.3), and it implies the second part by the calculus and (4.1.4).

The error committed when $(\delta_0(1 + \eta))^{\mathrm{Wick}}$ is replaced by $\delta_0^{\mathrm{Wick}}(1 + \eta)^{\mathrm{Wick}}$ will be examined together with the term coming from Θ^{Wick}, but first we shall prove:

LEMMA 4.1.3. *With $q_0 = (\delta_0^\flat - \delta_0)\varphi_0 a \in S(h^{\frac{1}{4}} H_4^{-\frac{1}{4}} H_1^{\frac{1}{4}}, G_1)$ and $f_1 = \frac{1}{2}\{\varphi_0^2 f, \delta_0^\flat\} \in S(h^{\frac{1}{2}} H_4^{-\frac{1}{2}} H_1^{-\frac{1}{2}}, G_1)$ we have*

$$(4.1.5) \qquad \delta_0^{\mathrm{Wick}}(\varphi_0^2 f)^w(x, D) = q^w(x, D)q^w(x, D) + q_0^w(x, D)q^w(x, D)$$

$$+ if_1^w(x, D) + R_1^w(x, D),$$

where $R_1 \in S(h^{\frac{1}{2}}H_4^{-\frac{1}{2}}H_1^{\frac{1}{2}}, G_0)$.

PROOF. Since the derivatives of $\delta_0^{\flat}$ are bounded and $\varphi_0^2 f \in S(h^{\frac{1}{2}}H_4^{-\frac{1}{2}}H_1^{-1}, G_0)$ the error in the expansion of $\delta_0^{\flat} \# \varphi_0^2 f$ after j terms is in $S(h^{\frac{1}{2}}H_4^{-\frac{1}{2}}H_1^{-1+\frac{1}{2}j}, G_0)$ and can be included in R_1 if $j \geq 3$. Since $\delta_0^{\flat}$ behaves as a symbol in $S(H_1^{-\frac{1}{2}}, G_1)$ in the support of φ_0, the jth term is actually in $S(h^{\frac{1}{2}}H_4^{-\frac{1}{2}}H_1^{-\frac{3}{2}+j}, G_1)$ and can be included in R_1 when $j = 2$ also. When $j = 1$ it is if_1, where we could also replace $\delta_0^{\flat}$ by δ_0. The first term is $\delta_0^{\flat}\varphi_0^2 f = q^2 + q_0 q$ since $(\delta_0^{\flat} - \delta_0)\varphi_0^2 f = q_0 q$, and $q_0 \in S(h^{\frac{1}{4}}H_4^{-\frac{1}{4}}H_1^{\frac{1}{4}}, G_1)$ by Proposition 2.3.4. We have $q \# q - q^2 \in S(h^{\frac{1}{2}}H_4^{-\frac{1}{2}}H_1^{\frac{1}{2}}, G_1)$ and $q_0 \# q - q_0 q \in S(h^{\frac{1}{2}}H_4^{-\frac{1}{2}}H_1^{\frac{1}{2}}, G_1)$ by the pseudodifferential calculus, so these remainder terms can be included in the error term R_1, which completes the proof of the lemma.

PROOF OF PROPOSITION 4.1.1. By Lemma 4.1.3 we can write

$$(4.1.6) \quad \mathrm{Re}\left(((\varphi_0^2 f)^w(x, D)u, \delta_0^{\mathrm{Wick}}(1 + \eta^{\mathrm{Wick}})u\right) = \left((1 + \eta^{\mathrm{Wick}})q^w(x, D)u, q^w(x, D)u\right)$$
$$+ \mathrm{Re}\left(q^w(x, D)u, f_2^w(x, D)u\right) + (i[\eta^{\mathrm{Wick}}, f_1^w(x, D)]u, u) + \mathrm{Re}\left((1 + \eta^{\mathrm{Wick}})R_1^w(x, D)u, u\right),$$

where

$$(4.1.7) \qquad\qquad f_2 = q \# \eta^{\flat} - \eta^{\flat} \# q + q_0 \in S(h^{\frac{1}{4}}H_4^{-\frac{1}{4}}H_1^{\frac{1}{4}}, G_0).$$

In fact, the symbol $q \# \eta^{\flat} - \eta^{\flat} \# q$ of $[q^w, \eta^{\mathrm{Wick}}]$ is in $S(h^{\frac{1}{4}}H_4^{-\frac{1}{4}}H_1^{-\frac{3}{4}}H_1^{\frac{1}{2}}h^{\frac{1}{2}}, G_0)$ by (4.1.4). Since $\eta \geq 0$ the first term on the right is $\geq \|q^w(x, D)u\|^2$, and the second can be estimated by

$$|(q^w(x, D)u, f_2^w(x, D)u)| \leq \tfrac{1}{4}\|q^w(x, D)u\|^2 + \|f_2^w(x, D)u\|^2.$$

Since $\eta^{\flat} \# f_1 - f_1 \# \eta^{\flat} \in S(h^{\frac{1}{2}}H_4^{-\frac{1}{2}}H_1^{-\frac{1}{2}}H_1^{\frac{1}{2}}h^{\frac{1}{2}}, G_0)$, we can now conclude that

$$(4.1.8) \quad \tfrac{3}{4}\|q^w(x, D)u\|^2 \leq \mathrm{Re}\left((\varphi_0^2 f)^w(x, D)u, \delta_0^{\mathrm{Wick}}(1 + \eta^{\mathrm{Wick}})u\right) + (R_2^w(x, D)u, u),$$

$$(4.1.9) \quad R_2 = f_2 \# f_2 - \mathrm{Re}(1 + \eta^{\flat}) \# R_1 - i(\eta^{\flat} \# f_1 - f_1 \# \eta^{\flat}) \in S(h^{\frac{1}{2}}H_4^{-\frac{1}{2}}H_1^{\frac{1}{2}}, G_0).$$

What remains is to estimate $\mathrm{Re}\left((\varphi_0^2 f)^w(x, D)u, g^w(x, D)u\right)$ where

$$(4.1.10) \qquad\qquad g = (\delta_0 \eta)^{\flat} - \delta_0^{\flat} \# \eta^{\flat} + \Theta^{\flat} \in S(\langle \delta_0 \rangle H_1^{\frac{1}{2}}, G_0),$$

by (4.1.3) and by (2.4.9), (2.4.7). To do so we shall split g^w into two terms where one contains a factor δ_0^{Wick} which can be handled using Lemma 4.1.3, and the other is in $S(H_1^{\frac{1}{2}}, G_0)$. With a small $\varepsilon \in (0, 1)$ we write

$$g_1 = \varepsilon^2 \delta_0^{\flat} g / (1 + \varepsilon^2 (\delta_0^{\flat})^2), \quad g_2 = \delta_0^{\flat} g_1 - \delta_0^{\flat} \# g_1 + g / (1 + \varepsilon^2 (\delta_0^{\flat})^2),$$

so that $g^w = \delta_0^{\mathrm{Wick}} g_1^w + g_2^w$. If

$$\psi_\varepsilon(t) = \frac{\varepsilon t}{1 + \varepsilon^2 t^2} = \frac{1}{2i}\left(\frac{1}{1 - i\varepsilon t} - \frac{1}{1 + i\varepsilon t}\right)$$

then $|\psi_\varepsilon^{(j)}| \leq j!\varepsilon^j$, and in view of (4.1.4) it follows that $\psi_\varepsilon(\delta_0^\flat) \in S(1, G_0)$ uniformly in ε, so g_1/ε is uniformly bounded in $S(1, G_0)$ since $\langle\delta_0\rangle H_1^{\frac{1}{2}} \leq 2$. On the other hand,

$$|\varepsilon\psi_\varepsilon(t)| = |\varepsilon^2 t|/(1+\varepsilon^2 t^2) \leq 1/\sqrt{1+t^2}, \quad |\varepsilon\psi_\varepsilon^{(j)}(t)| \leq j!\varepsilon^{j+1}/(1+\varepsilon^2 t^2) \leq j!/(1+t^2), \quad j \geq 1,$$

so it follows that $\varepsilon\psi_\varepsilon(\delta_0^\flat) \in S(\langle\delta_0\rangle^{-1}, G_0)$, so $g_1 \in S(H_1^{\frac{1}{2}}, G_0)$ with bounds independent of ε. For

$$\tilde{\psi}_\varepsilon(t) = \frac{1}{1+\varepsilon^2 t^2} = \frac{1}{2}\left(\frac{1}{1-i\varepsilon t} + \frac{1}{1+i\varepsilon t}\right)$$

we have $\varepsilon|\tilde{\psi}_\varepsilon(t)| \leq 1/\sqrt{1+t^2}$ and $|\varepsilon\tilde{\psi}_\varepsilon^{(j)}(t)| \leq j!\varepsilon^{j+1}/(1+\varepsilon^2 t^2) \leq j!/(1+t^2)$ when $j \geq 1$, so $\varepsilon\tilde{\psi}_\varepsilon(\delta_0^\flat)$ is bounded in $S(\langle\delta_0\rangle^{-1}, G_0)$. Hence it follows that $\varepsilon g/(1+\varepsilon^2(\delta_0^\flat)^2)$ is uniformly bounded in $S(H_1^{\frac{1}{2}}, G_0)$, and that εg_2 is uniformly bounded in $S(H_1^{\frac{1}{2}}, G_0)$.

In analogy to (4.1.6) we have

$$\mathrm{Re}\,\left((\varphi_0^2 f)^w(x, D)u, \delta_0^{\mathrm{Wick}} g_1^w(x, D)u\right) = \left(g_1^w(x, D)q^w(x, D)u, q^w(x, D)u\right)$$
$$+\mathrm{Re}\,\left(q^w(x, D)u, f_3^w(x, D)u\right) + \left(i[g_1^w(x, D), f_1^w(x, D)]u, u\right) + \mathrm{Re}\,\left(g_1^w(x, D)R_1^w(x, D)u, u\right),$$

where by the calculus

$$f_3 = q\#g_1 - g_1\#q + q_0 \in S(h^{\frac{1}{4}}H_4^{-\frac{1}{4}}H_1^{-\frac{3}{4}}H_1^{\frac{1}{2}}H_1^{\frac{1}{2}}, G_0) = S(h^{\frac{1}{4}}H_4^{-\frac{1}{4}}H_1^{\frac{1}{4}}, G_0),$$
$$g_1\#f_1 - f_1\#g_1 \in S(h^{\frac{1}{2}}H_4^{-\frac{1}{2}}H_1^{-\frac{1}{2}}H_1^{\frac{1}{2}}H_1^{\frac{1}{2}}, G_0) = S(h^{\frac{1}{2}}H_4^{-\frac{1}{2}}H_1^{\frac{1}{2}}, G_0).$$

The L^2 norm of g_1^w is $\leq C\varepsilon$, hence $\leq 1/16$ if ε is small enough, so we obtain

$$\mathrm{Re}\,\left((\varphi_0^2 f)^w(x, D)u, \delta_0^{\mathrm{Wick}} g_1^w(x, D)u\right) \geq -\tfrac{1}{8}\|q^w(x, D)u\|^2 + (R_3^w(x, D)u, u)$$

where $R_3 = -4f_3\#f_3 + i(g_1\#f_1 - f_1\#g_1) + \mathrm{Re}\,g_1\#R_1 \in S(h^{\frac{1}{2}}H_4^{-\frac{1}{2}}H_1^{\frac{1}{2}}, G_0)$. Here we have used that $\|q^w(x, D)u\|\|f_3^w(x, D)u\| \leq \frac{1}{16}\|q^w(x, D)u\|^2 + 4\|f_3^w(x, D)u\|^2$.

Finally we shall estimate $\mathrm{Re}\,\left((\varphi_0^2 f)^w(x, D)u, g_2^w(x, D)u\right)$ by observing that since $\varphi_0^2 f = a\varphi_0 q$ we have

$$(\varphi_0^2 f)^w(x, D) = (a\varphi_0)^w(x, D)q^w(x, D) + f_4^w(x, D), \quad f_4 \in S(h^{\frac{1}{2}}H_4^{-\frac{1}{2}}, G_1).$$

We have $g_2\#f_4 \in S(h^{\frac{1}{2}}H_4^{-\frac{1}{2}}H_1^{\frac{1}{2}}, G_0)$, and $f_5 = (a\varphi_0)\#g_2 \in S(h^{\frac{1}{4}}H_4^{-\frac{1}{4}}H_1^{\frac{1}{4}}, G_0)$, which implies that

$$|(q^w(x, D)u, f_5^w(x, D)u)| \leq \tfrac{1}{8}\|q^w(x, D)u\|^2 + 2\|f_5^w(x, D)u\|^2,$$

and it follows that

$$\mathrm{Re}\,\left((\varphi_0^2 f)^w(x, D)u, g_2^w(x, D)u\right) \geq -\tfrac{1}{8}\|q^w(x, D)u\|^2 + (R_4^w(x, D)u, u)$$

where $R_4 = -2f_5\#f_5 + \mathrm{Re}\,g_2\#f_4 \in S(h^{\frac{1}{2}}H_4^{-\frac{1}{2}}H_1^{\frac{1}{2}}, G_0)$. Summing up we have proved (4.1.2) with $R_0 = 2R_2 - 2R_3 - 2R_4$.

4.2. Lower bounds far from the sign changes. In Subsection 4.1 we have studied the set where $8|\delta_0(w)| < H_1(w)^{\frac{1}{2}}$ and $6H_1(w) < 1$. Here we shall examine the sets

$$(4.2.1) \qquad W_\pm = \{w \in W; \pm 16\delta_0(w) > H_1(w)^{-\frac{1}{2}}, \; H_1(w) < c^2\},$$

where c is a small positive number to be chosen in a moment. It is clearly sufficient to study W_+. Since $B = \delta_0(1+\eta) + \Theta$ and $|\Theta| \leq 2$, $0 \leq \eta \leq 1$, we have $\delta_0 - 2 \leq B \leq 2\delta_0 + 2$, and $\delta_0 - \sqrt{n} - 2 \leq B^\delta \leq 2(\delta_0 + \sqrt{n}) + 2$ by (4.1.4). When $w \in W_+$ it follows that $B^\flat(w) > H_1(w)^{-\frac{1}{2}}/32$ if $\sqrt{n} + 2 < \delta_0(w)/2$, thus if $\sqrt{n} + 2 \leq H_1(w)^{-\frac{1}{2}}/32$, hence if $c = 1/(32(\sqrt{n} + 2))$, which we assume from now on. Note that $H_1(w)^{-\frac{1}{2}} > c^{-1} > 32$ in W_+, hence $\delta_0(w) > 2$.

Our first step is to find an approximate square root of B^{Wick} in W_+. When doing so we shall use that $\frac{1}{16} \leq H_1(w)^{\frac{1}{2}} \langle \delta_0 \rangle \leq 1$, and that $\frac{1}{2}\delta_0 \leq B^\flat \leq 2\delta_0 + 2\sqrt{n} + 2 < 3\delta_0$ in W_+. Since $\delta_0(1+\eta)$ has Lipschitz constant 3 and $|\Theta| \leq 2$ we have $B^{\flat\prime} \in S(1, G_0)$.

LEMMA 4.2.1. *If $\psi \subset S(1, G_1)$ and $\operatorname{supp}\psi \subset W_+$, one can find $E \in S(H_1^{-\frac{1}{4}}, G_0)$ such that $E' \in S(H_1^{\frac{1}{4}}, G_0)$, $\operatorname{supp} E \subset \operatorname{supp}\psi$, and*

$$(4.2.2) \qquad \psi^2 B^\flat = E\#E + (\psi^2 - 1)r_0 + r_1,$$

where $r_0 \in S(H_1^{\frac{1}{2}}, G_0)$ and $r_1 \in S(H_1, G_0)$.

PROOF. Set $B^\flat = b$. As a first approximation we take $e_0 = \psi b^{\frac{1}{2}}$. Since $1/b \leq CH_1^{\frac{1}{2}}$ in $\operatorname{supp}\psi$ and the derivatives of b are bounded, it follows that $D^\alpha b^{\frac{1}{2}}$ can be estimated by a constant times $H_1^{\frac{1}{4}}$ if $\alpha \neq 0$, and so can $(D^\alpha \psi)b^{\frac{1}{2}}$. Hence

$$|e_0| \leq CH_1^{-\frac{1}{4}}, \quad e_0' \in S(H_1^{\frac{1}{4}}, G_0).$$

By the calculus this implies that $e_0\#e_0 - \psi^2 b = r_0 \in S(H_1^{\frac{1}{2}}, G_0)$. To improve the error term we introduce $e_1 = \frac{1}{2}r_0\psi b^{-\frac{1}{2}} \in S(H_1^{\frac{3}{4}}, G_0)$, for $D^\alpha b^{-\frac{1}{2}}$ can be estimated by a constant times $H_1^{\frac{1}{4}}$ in W_+. If $E = e_0 - e_1$ then $E \in S(H_1^{-\frac{1}{4}}, G_0)$ and $E' \in S(H_1^{\frac{1}{4}}, G_0)$, and

$$E\#E = e_0\#e_0 - e_0\#e_1 - e_1\#e_0 + e_1\#e_1 = \psi^2 b + r_0 - \psi^2 r_0 - r_1$$

where $-r_1 = 2e_0 e_1 - e_0\#e_1 - e_1\#e_0 + e_1\#e_1 \in S(H_1, G_0)$ by the calculus.

We can now prove a somewhat weaker analogue of Proposition 4.1.1:

PROPOSITION 4.2.2. *If $\varphi_\pm, \psi \in S(1, G_1)$, $0 \leq \varphi_\pm \leq 1$, $\psi = 1$ in a neighborhood of $\operatorname{supp}\varphi_\pm$, and $\operatorname{supp}\psi \subset W_\pm$, then*

$$(4.2.3) \qquad 2\operatorname{Re}\left((\varphi_\pm f)^w(x, D)u, B^{\mathrm{Wick}}u\right) \geq (R_\pm^w(x, D)u, u),$$

when $u \in \mathcal{S}(\mathbf{R}^n)$, for some $R_\pm \in S(h^{\frac{1}{2}} H_4^{-\frac{1}{2}} H_1^{\frac{1}{2}} \langle \delta_0 \rangle, G_0)$.

PROOF. It is sufficient to prove this for φ_+. With E as in Lemma 4.2.1 we have

$$B^\flat = E\#E + (1-\psi^2)B^\flat + (\psi^2 - 1)r_0 + r_1 = E\#E + (1-\psi^2)\#(B^\flat - r_0) + \tfrac{i}{2}\{\psi^2, B^\flat\} + r_2,$$

where $r_2 \in S(H_1, G_0)$. We insert this in the first term in (4.2.3). Since $(1 - \psi^2) \# (\varphi_+ f) \in S(h^{\frac{1}{2}} H_4^{-\frac{1}{2}} H_1^N, G_0)$ for every N, the term containing a factor $(1 - \psi^2)^w$ can be included in $R_+^w(x, D)$. Since $\{\psi^2, B^b\}$ vanishes in a neighborhood of $\mathrm{supp}(\varphi_+ f)$, the corresponding term can also be included in $R_+^w(x, D)$. We have $r_2 \# (\varphi_+ f) \in S(h^{\frac{1}{2}} H_4^{-\frac{1}{2}} H_1^{\frac{1}{2}} \langle \delta_0 \rangle, G_0)$ since $H^{\frac{1}{2}} \langle \delta_0 \rangle$ has a positive lower bound in $\mathrm{supp}(\varphi_+ f)$, so all that remains is to estimate

$$\mathrm{Re}\,\big((\varphi_+ f)^w(x, D)u, E^w(x, D)E^w(x, D)u\big)$$
$$= \big((\varphi_+ f)^w(x, D)E^w(x, D)u, E^w(x, D)u\big) + \big([E^w(x, D), [E^w(x, D), (\varphi_+ f)^w(x, D)]]u, u\big)$$

from below. The symbol of $[E^w(x, D), (\varphi_+ f)^w(x, D)]$ is in $S(h^{\frac{1}{2}} H_4^{-\frac{1}{2}} H_1^{-\frac{1}{2}} H_1^{\frac{1}{4}}, G_0)$, and that of the second commutator is in $S(h^{\frac{1}{2}} H_4^{-\frac{1}{2}} H_1^{\frac{1}{2}} \langle \delta_0 \rangle, G_0)$ where we have used that $E' \in S(H_1^{\frac{1}{4}} H_1^{\frac{1}{2}} \langle \delta_0 \rangle, G_0)$ since $H_1^{\frac{1}{2}} \langle \delta_0 \rangle$ has a positive lower bound in the support. By the Fefferman-Phong inequality, as stated in Proposition 3.1.1, we have $(\varphi_+ f)^w(x, D) \geq R^w(x, D)$ for some $R \in S(h^{\frac{1}{2}} H_4^{-\frac{1}{2}} H_1, G_1)$. Hence

$$\big((\varphi_+ f)^w(x, D)E^w(x, D)u, E^w(x, D)u\big) \geq (E^w(x, D)R^w(x, D)E^w(x, D)u, u),$$

and $E \# R \# E \in S(h^{\frac{1}{2}} H_4^{-\frac{1}{2}} H_1^{\frac{1}{2}}, G_0)$ which can also be included in R_+. This completes the proof of Proposition 4.2.2.

4.3. Proof of Theorem 4.1. By Lemma 2.1.2 we can choose φ_0 as in Proposition 4.1.1 with $\varphi_0 \leq 1$ so that

$$\varphi_0(w) = 1 \quad \text{if } 10|\delta_0(w)| < H_1(w)^{-\frac{1}{2}}, \text{ and } 8H_1(w) < 1.$$

We can then choose $\varphi_\pm$ and ψ as in Proposition 4.2.2 so that

$$\varphi_\pm(w) = 1 \quad \text{if } \pm 12\delta_0(w) > H_1(w)^{-\frac{1}{2}} \text{ and } H_1(w)^{-\frac{1}{2}} > 2/c = 64(\sqrt{n} + 2).$$

If we define $\varphi_b = (1 - \varphi_0^2)(1 - \varphi_+ - \varphi_-)$, it follows that $\varphi_0^2 + (1 - \varphi_0^2)(\varphi_+ + \varphi_-) + \varphi_b = 1$, that $\varphi_b \in S(1, G_1)$, and that $H_1(w)^{-\frac{1}{2}} \leq 2/c$ if $w \in \mathrm{supp}\,\varphi_b$. This implies that

$$(4.3.1) \qquad \mathrm{Re}\,\big((\varphi_b f)^w(x, D)u, B^{\mathrm{Wick}}u\big) = (R_b^w(x, D)u, u),$$

where $R_b = \mathrm{Re}\, B^b \# (\varphi_b f) \in S(h^{\frac{1}{2}} H_4^{-\frac{1}{2}} H_1^N, G_0)$ for any N since $\varphi_b f \in S(h^{\frac{1}{2}} H_4^{-\frac{1}{2}} H_1^N, G_1)$ for every N. Summing up (4.1.2), (4.2.3) with $\varphi_\pm$ replaced by $(1 - \varphi_0^2)\varphi_\pm$, and (4.3.1) we have proved that

$$(4.3.2) \qquad 2\,\mathrm{Re}((f^w(x, D)u, B^{\mathrm{Wick}}u) + (R^w(x, D)u, u) \geq 0,$$

where $R = R_0 - R_+ - R_- - 2R_b \in S(h^{\frac{1}{2}} H_4^{-\frac{1}{2}} H_1^{\frac{1}{2}} \langle \delta_0 \rangle, G_0)$.

By Proposition 2.4.2 we have $h^{\frac{1}{2}} H_4^{-\frac{1}{2}} H_1^{\frac{1}{2}} \langle \delta_0 \rangle \leq Cm \leq C\tilde{m}$, so it follows from Proposition 3.2.1 that

$$(4.3.3) \qquad (R^w(x, D)u, u) \leq C'(\tilde{m}^{\mathrm{Wick}}u, u), \quad u \in \mathcal{S}(\mathbf{R}^n), \quad \text{hence}$$

$$(4.3.4) \qquad 2\,\mathrm{Re}(f^w(x, D)u, B^{\mathrm{Wick}}u) + C'(\tilde{m}^{\mathrm{Wick}}u, u) \geq 0.$$

We recall that implicitly there has been a parameter $t \in (-T, T)$ present in all our arguments. Combining (4.4) with (4.3.4) we now obtain

$$(\tilde{m}^{\text{Wick}} u, u) \leq 12T \operatorname{Re} \left((f^w(t, x, D) - \partial/\partial t)u, B^{\text{Wick}} u \right) + 6T C' (\tilde{m}^{\text{Wick}} u, u),$$

when $u \in \mathcal{S}(\mathbf{R}^{1+n})$ and $\sup\{|t|; (t, w) \in \operatorname{supp} \varphi\} < T$; the scalar products are of course taken in $L^2(\mathbf{R}^{1+n})$. When T is so small that $12T C' \leq 1$, it follows that (4.3) is valid with $C = 24$, and the proof of Theorem 4.1 is complete.

Since $B^\flat \# B^\flat \in S(\langle \delta_0 \rangle^2, G_0)$ it follows from Proposition 3.2.1 and (4.2) that

$$\|u\|^2 + \|B^{\text{Wick}} u\|^2 \leq C_1^2 (h^{-\frac{1}{2}} \tilde{m}^{\text{Wick}} u, u),$$

so (4.3) implies

(4.3.5) $h^{\frac{1}{2}} (\|u\|^2 + \|B^{\text{Wick}} u\|^2) \leq C_2 \operatorname{Re} \left((f^w(t, x, D) - \partial/\partial t)u, B^{\text{Wick}} u \right).$

From (4.3.5) we obtain by two applications of Cauchy-Schwarz' inequality

$$h^{\frac{1}{2}} \|B^{\text{Wick}} u\| \leq C_2 \|(f^w(t, x, D) - \partial/\partial t)u\|, \quad h^{\frac{1}{2}} \|u\| \leq C_2 \|(f^w(t, x, D) - \partial/\partial t)u\|.$$

The second inequality suggests a loss of $\frac{3}{2}$ derivatives compared to the elliptic case. However to continue we must rely on the stronger and much more stable estimate (4.3.5).

4.4. A first step towards classical pseudodifferential operators. In this subsection we shall prove that the estimate (4.3.5) which was a consequence of Theorem 4.1 has an analogue involving more conventional pseudodifferential operators. Recall the notation $S_{\varrho, \delta}^m = S_{\varrho, \delta}^m(T^* \mathbf{R}^n)$ for the space of symbols in $T^* \mathbf{R}^n = \mathbf{R}^n \times \mathbf{R}^n$ such that for all multiindices α and β

(4.4.1) $|\partial_\xi^\alpha \partial_x^\beta a(x, \xi)| \leq C_{\alpha, \beta} (1 + |\xi|)^{m - \varrho|\alpha| + \delta|\beta|}, \quad x, \xi \in \mathbf{R}^n,$

that is, this is the space of symbols of weight $(1 + |\xi|)^m$ with respect to the metric $g_{\varrho, \delta} = (1 + |\xi|)^{2\delta} |dx|^2 + (1 + |\xi|)^{-2\varrho} |d\xi|^2$. The corresponding operator $a^w(x, D)$ is continuous from $H_{(s)}(\mathbf{R}^n)$ to $H_{(s-m)}(\mathbf{R}^n)$ if $0 \leq \delta \leq \varrho \leq 1$ and $\delta < 1$. We shall only use $S_{1,0}^m$ and $S_{\frac{1}{2}, \frac{1}{2}}^m$. By $\dot{S}_{\varrho, \delta}^m$ we shall denote the set of symbols satisfying (4.4.1) when $|\alpha| + |\beta| \neq 0$.

THEOREM 4.4.1. *Let $f(t, x, \xi)$ be a real valued continuous function of $t \in [-1, 1]$ with values in $S_{1,0}^1(T^* \mathbf{R}^n)$ satisfying (2.4.2), the condition $(\bar{\Psi})$. Then it follows for small $T > 0$ that, with norms in $L^2(\mathbf{R}^{1+n})$,*

(4.4.2) $\|u\|^2 + \|(1 + |D|^2)^{-\frac{1}{4}} B^w(t, x, D)u\|^2$

$$\leq CT \operatorname{Re} \left((f^w(t, x, D) - \partial/\partial t)u, B^w(t, x, D)u \right),$$

when $u \in \mathcal{S}(\mathbf{R}^{1+n})$ and $\sup\{|t|; (t, w) \in \operatorname{supp} u\} < T$. Here B is real valued, and B is bounded as a function of $t \in [-T, T]$ with values in $S_{\frac{1}{2}, \frac{1}{2}}^1 \cap \dot{S}_{\frac{1}{2}, \frac{1}{2}}^{\frac{1}{2}}$. Although B depends on T, both C and the bounds for B are independent of T.

PROOF. We shall use a Littlewood-Paley (dyadic) decomposition in the frequency variables ξ to deduce (4.4.2) from (4.3.5). Choose non-negative $\varphi, \psi \in C_0^\infty(\{\xi \in \mathbf{R}^n; \frac{1}{2} < |\xi| < 2\})$ such that $\psi = 1$ in a neighborhood of the support of φ and $\sum_{-\infty}^\infty \varphi(\xi/2^j)^2 = 1$ when

$\xi \neq 0$. (It suffices to choose φ strictly positive when $2^{-\frac{1}{2}} \leq |\xi| \leq 2^{\frac{1}{2}}$ and then replace $\varphi(\xi)$ by $\varphi(\xi)(\varphi(\xi)^2 + \varphi(2\xi)^2 + \varphi(\xi/2)^2)^{-\frac{1}{2}}$. Set $\varphi_j(\xi) = \varphi(\xi/2^j)$, $\psi_j(\xi) = \psi(\xi/2^j)$ when $0 < j \in \mathbf{Z}$, and define $\varphi_0(\xi) = \varphi(\xi)$, $\psi_0(\xi) = \psi(\xi)$ when $|\xi| \geq 1$ and $\varphi_0(\xi) = \psi_0(\xi) = 1$ when $|\xi| \leq 1$. Also φ_0 and ψ_0 are in C_0^∞, for $\varphi(\xi) = 1$ when ξ is in a neighborhood of the unit sphere, and we have $\sum_0^\infty \varphi_j(\xi)^2 = 1$, $\xi \in \mathbf{R}^n$, hence

$$(4.4.3) \qquad \|u\|^2 = \sum_0^\infty \|\varphi_j(D)u\|^2, \quad u \in \mathcal{S}(\mathbf{R}^{1+n}).$$

Set $f_j(t,x,\xi) = f(t,x,\xi)\psi_j(\xi)$ and $F_j(t,x,\xi) = f_j(t,x\sqrt{h_j},\xi/\sqrt{h_j})$, where $h_j = 2^{-j}$. Since $\frac{1}{2} < |\xi|h_j < 2$ in $\operatorname{supp} f_j$ when $j \neq 0$ and $f \in S_{1,0}^1$, it follows that

$$|\partial_{x,\xi}^k F_j(t,x,\xi)| \leq \gamma_k h_j^{\frac{1}{2}k-1}, \quad x,\xi \in \mathbf{R}^n,\ t \in [-1,1],\ k = 0,1,2,\dots,$$

where the constants γ_k are independent of j. The condition $(\bar{\Psi})$ is inherited by F_j, so it follows from (4.3.5) when T is small and u is as in Theorem 4.4.1 that

$$h_j^{\frac{1}{2}}(\|u\|^2 + \|b_j^w(t,x,D)u\|^2) \leq C_2 T \operatorname{Re}\left((F_j^w(t,x,D) - \partial/\partial t)u, b_j^w(t,x,D)u\right),$$

where b_j is a bounded function of t with values in $S(h_j^{-\frac{1}{2}}, G_0) \cap \dot{S}(1, G_0)$, depending on T but with bounds independent of T. Since the passage from f_j to F_j was made by a symplectic dilation, this is equivalent to the estimate

$$(4.4.4) \qquad \|u\|^2 + h_j\|B_j^w(t,x,D)u\|^2 \leq C_2 T \operatorname{Re}\left((f_j^w(t,x,D) - \partial/\partial t)u, B_j^w(t,x,D)u\right)$$

where $B_j(t,x,\xi) = h_j^{-\frac{1}{2}}b_j(t,x/\sqrt{h_j},\xi\sqrt{h_j})$. (Cf. [H1, Theorem 18.5.9].) We have uniform bounds for $B_j(t,x,\xi)$ in $S(h_j^{-1}, g_j) \cap \dot{S}(h_j^{-\frac{1}{2}}, g_j)$ where

$$(4.4.5) \qquad g_j = |dx|^2/h_j + |d\xi|^2 h_j.$$

The metric g_j is equivalent to $g_{\frac{1}{2},\frac{1}{2}}$ when $2^{j-1} < |\xi| < 2^{j+1}$, and in general we have

$$(4.4.6) \qquad m_j g_j \leq g_{\frac{1}{2},\frac{1}{2}} \leq g_j/m_j, \quad m_j(\xi) = \min\left((1+|\xi|)/2^j, 2^j/(1+|\xi|)\right).$$

For later reference we observe that

$$(4.4.7) \qquad \sum_0^\infty m_j(\xi) \leq (1+|\xi|) \sum_{2^j \geq 1+|\xi|} 2^{-j} + (1+|\xi|)^{-1} \sum_{2^j \leq 1+|\xi|} 2^j \leq 4.$$

If we apply (4.4.4) to $\varphi_j(D)u$ we obtain

$$(4.4.8) \quad \|\varphi_j(D)u\|^2 + h_j\|B_j^w(t,x,D)\varphi_j(D)u\|^2$$
$$\leq C_2 T \operatorname{Re}\left((f^w(t,x,D) - \partial/\partial t)u, \varphi_j(D)B_j^w(t,x,D)\varphi_j(D)u\right) - C_2 T R_j,$$
$$R_j = \operatorname{Re}\left((\varphi_j(D)f^w(t,x,D) - f_j^w(t,x,D)\varphi_j(D))u, B_j^w(t,x,D)\varphi_j(D)u\right).$$

To estimate R_j we note that φ_j is uniformly bounded in $S(m_j^N, g_{1,0})$ for every N, and that m_j is uniformly σ, $g_{1,0}$ temperate. Since $\psi_j = 1$ in a neighborhood of $\operatorname{supp}\varphi_j$ we conclude that

$$\varphi_j(D)f^w(t,x,D) - f_j^w(t,x,D)\varphi_j(D) = ir_j^w(t,x,D) + s_j^w(t,x,D),$$

where $r_j = \{f, \varphi_j\} = -\langle \partial f/\partial x, \partial \varphi_j/\partial \xi \rangle$ is uniformly bounded in $S(m_j^N, g_{1,0})$ and s_j is uniformly bounded in $S((1 + |\xi|)^{-2} m_j^N, g_{1,0})$, for any N. Since $m_j(\xi) \leq (1 + |\xi|) h_j$, it follows that s_j is uniformly bounded in $S(h_j^2 m_j^N, g_{1,0})$ for any N, which implies that the symbol of $\varphi_j(D) B_j^w(t, x, D) s_j^w(t, x, D)$ is uniformly bounded in $S(h_j m_j^N, g_j)$ and in view of (4.4.6) also in $S(h_j m_j^N, g_{\frac{1}{2}, \frac{1}{2}})$. We have $R_j = R_j' + R_j''$ where

$$R_j' = \mathrm{Re}(\varphi_j(D) B_j^w(t, x, D) s_j^w(t, x, D) u, u),$$
$$R_j'' = \mathrm{Re}(i r_j^w(t, x, D) u, B_j^w(t, x, D) \varphi_j(D) u)$$
$$= \tfrac{i}{2}((\varphi_j(D) B_j^w(t, x, D) r_j^w(t, x, D) - r_j^w(t, x, D) B_j^w(t, x, D) \varphi_j(D)) u, u).$$

The last operator in R_j'' is a sum of three terms with a commutator as factor,

(4.4.9)
$$[\varphi_j(D), B_j^w(t, x, D)] r_j^w(t, x, D), \quad [B_j^w(t, x, D), r_j(t, x, D)] \varphi_j(D),$$
$$B_j^w(t, x, D)[\varphi_j(D), r_j^w(t, x, D)].$$

Here φ_j and r_j are uniformly bounded in $S(m_j^N, g_{1,0})$ for any N, so the Weyl symbol of $[\varphi_j(D), r_j^w(t, x, D)]$ is uniformly bounded in $S((1 + |\xi|)^{-1} m_j^N, g_{1,0})$ for any N, hence in $S(h_j m_j^N, g_j)$ too. This implies that the Weyl symbol of the last of the operators (4.4.9) is uniformly bounded in $S(m_j^N, g_j)$ and therefore also in $S(m_j^N, g_{\frac{1}{2}, \frac{1}{2}})$, for any N. To deal with the first two terms in (4.4.9) we observe that φ_j and r_j are in fact uniformly bounded in $S(m_j^N, G_j)$ where

$$G_j = |dx|^2 + |d\xi|^2/(2^j + |\xi|)^2, \quad G_j^\sigma = (2^j + |\xi|)^2 |dx|^2 + |d\xi|^2.$$

We have $g_j \leq 2^{-j} G_j^\sigma = h_j G_j^\sigma$, hence $g_j = g_j^\sigma \geq 2^j G_j$. Since B_j is uniformly bounded in $\dot{S}(h_j^{-\frac{1}{2}}, g_j)$ we conclude using [H1, Theorem 18.5.5] that the Weyl symbols of the commutators $[\varphi_j(D), B_j^w(t, x, D)]$ and $[B_j^w(t, x, D), r_j(t, x, D)]$ are uniformly bounded in $S(m_j^N, g_j)$ for any N, and this remains true for the first two operators in (4.4.9). As before this implies uniform boundedness in $S(m_j^N, g_{\frac{1}{2}, \frac{1}{2}})$ for any N. Summing up, we have proved that $R_j = (e_j^w(t, x, D) u, u)$ where e_j is uniformly bounded in $S(m_j, g_{\frac{1}{2}, \frac{1}{2}})$, so $\sum_0^\infty e_j \in S(1, g_{\frac{1}{2}, \frac{1}{2}})$ by (4.4.7). Thus the corresponding operator is L^2 continuous, and it follows in view of (4.4.3) that

$$\|u\|^2 + \sum_0^\infty h_j \|B_j^w(t, x, D) \varphi_j(D) u\|^2$$

$$\leq C_2 T \, \mathrm{Re}\left((f^w(t, x, D) - \partial/\partial t) u, \sum_0^\infty \varphi_j(D) B_j^w(t, x, D) \varphi_j(D) u\right) + C' T \|u\|^2,$$

when u is as in Theorem 4.4.1. Hence Theorem 4.4.1 follows with $C = 12 C_2$ when $C' T < \frac{1}{2}$, in addition to earlier requirements, if we prove that

(4.4.10)
$$2 \sum_0^\infty \varphi_j(D) B_j^w(t, x, D) \varphi_j(D) = B^w(t, x, D),$$

(4.4.11)
$$\|(1 + |D|^2)^{-\frac{1}{4}} B^w(t, x, D) u\|^2 \leq 12 \sum_0^\infty h_j \|B_j^w(t, x, D) \varphi_j(D) u\|^2,$$

where $B \in S^1_{\frac{1}{2},\frac{1}{2}} \cap \dot{S}^{\frac{1}{2}}_{\frac{1}{2},\frac{1}{2}}$. The Weyl symbol of $\varphi_j(D)B^w_j(t,x,D)\varphi_j(D)$ is uniformly bounded in $S(h_j^{-1}m_j^N, g_j)$ for any N, hence in $S(h_j^{-1}m_j^N, g_{\frac{1}{2},\frac{1}{2}})$. Now $h_j^{-1}m_j = 2^j m_j \le 1 + |\xi|$, so we have a uniform bound in $S((1+|\xi|)m_j, g_{\frac{1}{2},\frac{1}{2}})$, and it follows from (4.4.7) that (4.4.10) is valid with $B \in S^1_{\frac{1}{2},\frac{1}{2}}$. We must also prove that $\partial B/\partial x_k \in S^1_{\frac{1}{2},\frac{1}{2}}$ and that $\partial B/\partial \xi_k \in S^0_{\frac{1}{2},\frac{1}{2}}$, that is, that the Weyl symbol of $[B^w, D_k]$ is in $S^1_{\frac{1}{2},\frac{1}{2}}$ and that of $[B^w, x_k]$ is in $S^0_{\frac{1}{2},\frac{1}{2}}$. Since the symbol of $[B^w_j, D_k]$ is bounded in $S(h_j^{-1}, g_j)$ the first statement follows at once. The commutator with x_k has two terms where a factor $\varphi_j(D)$ is replaced by $(\partial_k\varphi_j)(D)$, which lowers the degree by one unit, and one where $B_j(t,x,\xi)$ is replaced by $\partial B_j(t,x,\xi)/\partial \xi_k$, which is uniformly bounded in $S(1, g_j)$. This completes the proof of the properties of B, for B is real valued since the sum in (4.4.10) defines a symmetric operator.

To prove (4.4.11) we first observe that since $1 + |\xi|^2 \ge (2h_j)^{-2}$ in $\mathrm{supp}\,\varphi_j$, we have

$$\|(1 + |D|^2)^{-\frac{1}{4}}v\|^2 \le \sum_0^\infty 2h_j\|\varphi_j(D)v\|^2, \quad v \in \mathcal{S}.$$

Only three supports of the functions φ_j can overlap, so we have

$$\varphi_j(D)B^w(t,x,D)u = \sum_{|k-j|\le 1} \varphi_j(D)\varphi_k(D)B^w_k(t,x,D)\varphi_k(D)u,$$

hence

$$\sum_0^\infty 2h_j\|\varphi_j(D)B^w(t,x,D)u\|^2 \le \sum_{|j-k|\le 1} 6h_j\|\varphi_j(D)B^w_k(t,x,D)\varphi_k(D)u\|^2$$

$$\le 12\sum_0^\infty h_k\|B^w_k(t,x,D)\varphi_k(D)u\|^2,$$

for $h_j \le 2h_k$ when $k \le j+1$.

REMARK. Theorem 4.4.1 remains valid, with smaller bounds for T, if to f one adds if_0 where f_0 is a continuous function of t with values in $S^0_{1,0}$ and f_0 is real valued. In fact, then we have

$$\mathrm{Re}(if^w_0(t,x,D), B^w(t,x,D)u) = \tfrac{i}{2}([B^w(t,x,D), f^w_0(t,x,D)]u, u),$$

and the Weyl symbol of the commutator is in $S^0_{\frac{1}{2},\frac{1}{2}}$ since the metrics $g_{1,0}$ and $g_{\frac{1}{2},\frac{1}{2}}$ are conformal and $B \in \dot{S}^{\frac{1}{2}}_{\frac{1}{2},\frac{1}{2}}$. Hence the commutator is L^2 continuous, so adding if_0 to f gives a new term in the right-hand side of (4.4.2) which is bounded by $C'T\|u\|^2$. For small T it can be cancelled against half of the left-hand side.

From (4.4.2) it follows at once that

$$\|(1+|D|^2)^{-\frac{1}{4}}B^w(t,x,D)u\| \le CT\|(1+|D|^2)^{\frac{1}{4}}(f^w(t,x,D) - \partial/\partial t)u\|,$$

$$\|u\| \le CT\|(1+|D|^2)^{\frac{1}{4}}(f^w(t,x,D) - \partial/\partial t)u\|.$$

This is an estimate of the desired form. The perturbation of f by f_0 which was possible in the estimate (4.4.2) could no longer have been done here. We also need the greater stability of (4.4.2) when proving the following local form which is an essential step toward the proof of solvability in Section 5. Let χ_0 a fixed function in $C_0^\infty((-1,1))$ such that $0 \le \chi_0 \le 1$ and $\chi_0 = 1$ in $[-\frac{1}{2}, \frac{1}{2}]$, and let φ_0 a function in $C_0^\infty([-1,1], S^0_{1,0}(T^*\mathbf{R}^n))$.

THEOREM 4.4.2. *Let $f(t,x,\xi)$ be a real valued continuous function of $t \in [-1,1]$ with values in $S_{1,0}^1(T^*\mathbf{R}^n)$ satisfying (2.4.2), the condition $(\bar{\Psi})$, let f_0 be real valued and a continuous function of $t \in [-1,1]$ with values in $S_{1,0}^1(T^*\mathbf{R}^n)$, and set $F = f + if_0$. If $T > 0$ is sufficiently small then*

$$(4.4.12) \quad \|\chi_0(t/T)\varphi_0^w(t,x,D)u\|^2 \le C_1 T^2 \|\varphi_0^w(t,x,D)(F^w(t,x,D) - \partial/\partial t)u\|_{(\frac{1}{2})}^2$$
$$+ C_2\|\chi_0'(t/T)u\|_{(\frac{1}{2})}^2 + C_3 T \|\chi_0(t/T)u\|^2, \quad u \in C_0^\infty(\mathbf{R}^{1+n}).$$

Here $\|v\|$ is the norm of v in $L^2(\mathbf{R}^{1+n})$, and $\|v\|_{(s)} = \|(1 + |D|^2)^{s/2}v\|$ when $s \in \mathbf{R}$.

PROOF. By Theorem 4.4.1 and a remark following its proof we have for small T

$$U = \|\chi_0(t/T)\varphi_0^w(t,x,D)u\|^2 + \|B^w(t,x,D)\chi_0(t/T)\varphi_0^w(t,x,D)u\|_{(-\frac{1}{2})}^2$$
$$\le CT \operatorname{Re}\left((F^w(t,x,D) - \partial/\partial t)\chi_0(t/T)\varphi_0^w(t,x,D)u, B^w(t,x,D)\chi_0(t/T)\varphi_0^w(t,x,D)u\right).$$

Here B depends on T but has bounds independent of T as function of t with values in $S_{\frac{1}{2},\frac{1}{2}}^1 \cap \dot{S}_{\frac{1}{2},\frac{1}{2}}^1$. We have

$$CT \operatorname{Re}\left(\chi_0(t/T)\varphi_0^w(t,x,D)(F^w(t,x,D) - \partial/\partial t)u, B^w(t,x,D)\chi_0(t/T)\varphi_0^w(t,x,D)u\right)$$
$$\le CT\|\varphi_0^w(t,x,D)(F^w(t,x,D) - \partial/\partial t)u\|_{(\frac{1}{2})}\|B^w(t,x,D)\chi_0(t/T)\varphi_0^w(t,x,D)u\|_{(-\frac{1}{2})},$$

which can be estimated by $\frac{1}{4}U$ plus the first term on the right-hand side of (4.4.12) with $C_1 = C^2$. What remains is to examine two commutator terms which appear when $\chi_0(t/T)\varphi_0^w(t,x,D)$ is moved to the left of $F^w(t,x,D) - \partial/\partial t$. Commuting $\chi_0(t/T)$ gives a term

$$-C \operatorname{Re}(\chi_0'(t/T)\varphi_0^w(t,x,D)u, B^w(t,x,D)\chi_0(t/T)\varphi_0^w(t,x,D)u)$$
$$\le C'\|\chi_0'(t/T)u\|_{(\frac{1}{2})}\|B^w(t,x,D)\chi_0(t/T)\varphi_0^w(t,x,D)u\|_{(-\frac{1}{2})}$$

which we estimate by $\frac{1}{4}U$ plus the second term on the right-hand side of (4.4.18) with $C_2 = C'^2$. The other commutator term involves

$$[F^w(t,x,D), \varphi_0^w(t,x,D)] = i\varphi_1^w(t,x,D) + R_1^w(t,x,D),$$

where $\varphi_1 = -\{f, \varphi_0\}$ (resp. R_1) is a continuous function of t with values in $S_{1,0}^0(T^*\mathbf{R}^n)$ (resp. $S_{1,0}^{-1}(T^*\mathbf{R}^n)$). Thus

$$CT \operatorname{Re}(\chi_0(t/T)R_1^w(t,x,D)u, B^w(t,x,D)\chi_0(t/T)\varphi_0^w(t,x,D)u)$$
$$\le C''T\|\chi_0(t/T)u\|_{(-\frac{1}{2})}\|B^w(t,x,D)\chi_0(t/T)\varphi_0^w(t,x,D)u\|_{(-\frac{1}{2})},$$

which can be estimated by $\frac{1}{4}U + (C''T)^2\|\chi_0(t/T)u\|_{(-\frac{1}{2})}^2$. We have

$$(4.4.13) \quad CT \operatorname{Re}(i\varphi_1^w(t,x,D)\chi_0(t/T)u, B^w(t,x,D)\chi_0(t/T)\varphi_0^w(t,x,D)u)$$
$$= \frac{1}{2}CT\big(i(\varphi_0^w(t,x,D)B^w(t,x,D)\varphi_1^w(t,x,D)$$
$$- \varphi_1^w(t,x,D)B^w(t,x,D)\varphi_0^w(t,x,D))\chi_0(t/T)u, \chi_0(t/T)u\big),$$

and the difference here is a sum of three terms with a commutator as a factor:

$$[\varphi_0^w(t,x,D), B^w(t,x,D)]\varphi_1^w(t,x,D), \quad [B^w(t,x,D), \varphi_1^w(t,x,D)]\varphi_0^w(t,x,D),$$
$$B^w(t,x,D)[\varphi_0^w(t,x,D), \varphi_1^w(t,x,D)].$$

The Weyl symbols of the first two are bounded with values in $S^0_{\frac{1}{2},\frac{1}{2}}$ since B is bounded in $\dot{S}^{\frac{1}{2}}_{\frac{1}{2},\frac{1}{2}}$ and so is that of the third since the symbol of the commutator there is bounded with values in $S^{-1}_{1,0}$. Hence (4.4.13) can be estimated by the last term in (4.4.12). Summing up, after cancellation of a term $\frac{3}{4}U$ we have obtained the estimate (4.4.12) with a factor $\frac{1}{4}$ in the left-hand side and some constants C_1, C_2, C_3 which we just have to multiply by 4 to obtain (4.4.12) as stated.

At the end of the proof we dropped the estimate of $B^w(t,x,D)\chi_0(t/T)\varphi_0^w(t,x,D)u$ for it will no longer be needed.

5. The sufficiency of condition (Ψ). Let P be a properly supported polyhomogeneous (see [H1, Def. 18.1.5], also called "classical") pseudodifferential operator of order m in an open set $\Omega \subset \mathbf{R}^n$, and let $p = p_1 + ip_2$ be the principal symbol. Let $x^0 \in \Omega$ and set $\Omega_\delta = \{x \in \Omega; |x - x^0| < \delta\}$. The following is the improvement by Lerner [L7] of the solvability theorem of Dencker [D1]:

THEOREM 5.1. *Assume that P is of principal type at x^0 in the sense that*

$$(5.1) \qquad \partial p(x^0, \xi)/\partial \xi \neq 0, \quad if\ 0 \neq \xi \in \mathbf{R}^n,\ p(x^0, \xi) = 0,$$

and that p satisfies condition (Ψ). *For every $s \in \mathbf{R}$ it follows that there is some $\delta > 0$ with $\Omega_\delta \Subset \Omega$ such that for every $f \in H_{(s)}(\mathbf{R}^n)$ there is some $u \in H_{(s+m-\frac{3}{2})}(\mathbf{R}^n)$ with $Pu = f$ in Ω_δ.*

A precise definition of condition (Ψ) is given in [H1, Definition 26.4.6]; see also [H1, Theorem 26.4.12]. In most of this section *we shall assume that $m = 1$ and that $s = 0$* but this restriction will be removed at the end. Then the statement follows from the Hahn-Banach theorem if we prove that for sufficiently small $\delta > 0$

$$(5.2) \qquad \|u\|_{(0)} \leq C\|P^*u\|_{(\frac{1}{2})}, \quad u \in C_0^\infty(\Omega_\delta).$$

Here $\|\cdot\|_{(s)}$ is the Sobolev norm in $\mathbf{R}^n$, $\|v\|_{(s)} = \|(1 + |D|^2)^{\frac{1}{2}s}v\|_{(0)}$ where $\|\cdot\|_{(0)}$ is the norm in $L^2(\mathbf{R}^n)$. It is sufficient to prove that

$$(5.3) \qquad \|u\|_{(0)} \leq C_1\|P^*u\|_{(\frac{1}{2})} + C_2\|u\|_{(-1)}, \quad u \in C_0^\infty(\Omega_\delta),$$

for if δ is sufficiently small the new term on the right is smaller than half of the left-hand side and can be cancelled. In fact, when $u \in C_0^\infty(\Omega_\delta)$ we have

$$\|u\|_{(-1)} \leq C_3\delta\|u\|_{(0)} \text{ if } n > 2, \quad \|u\|_{(-1)} \leq C_3\delta(\log(2/\delta))^{\frac{1}{2}}\|u\|_{(0)} \text{ if } n = 2,\ \delta < 1,$$

where $C_3 = (2n - 4)^{-\frac{1}{2}}$ when $n > 2$ and $C_3 = \sqrt{2}$ when $n = 2$. The proof of these well-known estimates is simple: If $n > 2$ it suffices to prove that $\int_{|x-x^0|<\delta} |v(x)|^2\,dx \leq C_3^2\delta^2 \int |v'(x)|^2\,dx$ when $v \in \mathcal{S}(\mathbf{R}^n)$. Introducing polar coordinates at x^0 we have

$$|v(r)|^2 \leq \left|\int_r^\infty v'(s)\,ds\right|^2 \leq \int_r^\infty |v'(s)|^2 s^{n-1}\,ds \int_r^\infty s^{1-n}\,ds, \quad \text{hence}$$

$$(n-2)\int_0^\delta |v(r)|^2 r^{n-1}\,dr \leq \iint_{r\leq\delta, r\leq s} r|v'(s)|^2 s^{n-1}\,dr\,ds = \tfrac{1}{2}\delta^2 \int s^{n-1}|v'(s)|^2\,ds.$$

When $n = 2$ we define $\chi(r) = 1$ when $r \leq 1$, $\chi(r) = 2 - r$ when $1 \leq r \leq 2$ and $\chi(r) = 0$ when $r > 2$. Then

$$\int |(\chi v)'|^2 r \, dr \leq 2 \int (|v'|^2 + |v|^2) r \, dr, \quad \text{hence } |v(r)|^2 \leq 2 \int_0^\infty (|v'(s)|^2 + |v(s)|^2) s \, ds \int_r^2 ds/s,$$

if $0 < r < 1$, and since $\int_0^\delta 2 \log(2/r) \, r \, dr < 2\delta^2 \log(2/\delta)$ the estimate follows then too.

It is sufficient to prove that for every $\xi^0 \in \mathbf{R}^n$ with $|\xi^0| = 1$ there is some $\varphi \in S_{1,0}^0$ which is non-characteristic at (x^0, ξ^0) such that

(5.4) $$\|\varphi^w(x, D)u\|_{(0)} \leq C_1 \|P^* u\|_{(\frac{1}{2})} + C_2 \|u\|_{(-1)}, \quad u \in C_0^\infty(\Omega_\delta),$$

for then there are finitely many $\varphi_1, \ldots, \varphi_\nu \in S_{1,0}^0$ for which (5.4) is valid such that $(\varphi_1^w(x, D), \ldots, \varphi_\nu^w(x, D))$ is an elliptic system at x^0, hence for sufficiently small δ

$$\|u\|_{(0)} \leq C_1' \sum_1^\nu \|\varphi_j^w(x, D)u\|_{(0)} + C_2' \|u\|_{(-1)} \leq C_1'' \|P^* u\|_{(\frac{1}{2})} + C_2'' \|u\|_{(-1)}, \ u \in C_0^\infty(\Omega_\delta).$$

It is even sufficient to have a weaker version of (5.4) such as

(5.5) $$\|\varphi^w(x, D)u\|_{(0)} \leq C_1 \|P^* u\|_{(\frac{1}{2})} + C_2(\delta) \|u\|_{(-1)} + C_3 \delta^{\frac{1}{2}} \|u\|_{(0)}, \quad u \in C_0^\infty(\Omega_\delta),$$

for the preceding summation argument then gives (5.3) for small δ with a constant C_2 depending on δ, but obviously the best constant in (5.3) cannot increase when $\delta \downarrow 0$. In the middle term we may replace $\|u\|_{(-1)}$ by $\|u\|_{(-\frac{1}{2})}$ for example, since $\|u\|_{(-\frac{1}{2})} \leq \|u\|_0^{\frac{1}{2}} \|u\|_{(-1)}^{\frac{1}{2}} \leq \varepsilon \|u\|_{(0)} + 4\|u\|_{(-1)}/\varepsilon$ for every $\varepsilon > 0$.

If $p(x^0, \xi^0) \neq 0$ and conesupp φ is in a sufficiently small conic neighborhood of (x^0, ξ^0) we have $\varphi^w(x, D) = \psi_1^w(x, D)P^* + \psi_2^w$ where $\psi_1 \in S_{1,0}^{-1}$ and $\psi_2 \in S_{1,0}^{-3}$, hence

(5.6) $$\|\varphi^w(x, D)u\|_{(s)} \leq C_1 \|P^* u\|_{(s-1)} + C_2 \|u\|_{(-1)}, \quad u \in C_0^\infty(\Omega_\delta), \ 0 \leq s \leq 2,$$

which is a much better estimate than (5.5). (By *conic* we mean invariance under multiplication of the ξ variable by positive real numbers.)

Now assume that $p(x^0, \xi^0) = 0$. We shall then conjugate P^* microlocally by means of Fourier integral operators to a form which is close to that studied in Theorem 4.4.2. This requires a basically well known result from symplectic geometry:

LEMMA 5.2. *Let $p(x, \xi)$ be a C^∞ function which is positively homogeneous of degree 1 in ξ in a conic neighborhood of $(x^0, \xi^0) \in T^*\mathbf{R}^n \setminus \{0\}$, and assume that $p(x^0, \xi^0) = 0$ and that $p_\xi'(x^0, \xi^0) \neq 0$. Then there exist C^∞ canonical coordinates (y, η) in a conic neighborhood Γ of (x^0, ξ^0), homogeneous of degree 0 and 1 respectively, such that $y_n(x, \xi) = 0$ when $x = x^0$ and*

(5.7) $$p(x, \xi) = q(x, \xi)(\eta_n - if(y, \eta)), \quad (x, \xi) \in \Gamma.$$

Here $q(x, \xi)$ is positively homogeneous in ξ of degree 0 and $q(x, \xi) \neq 0$ in Γ, and $f(y, \eta)$ is homogeneous of degree 0 in η, real valued and independent of η_n. We have $q \in C^\infty$ and $f \in C^\infty$, and $\eta_n = 0$, $f(y, \eta) = 0$ at (x^0, ξ^0).

PROOF. Replacing p by ip if necessary we may assume that $\operatorname{Re} p_\xi'(x^0, \xi^0) \neq 0$. Set $\eta_n(x, \xi) = \operatorname{Re} p(x, \xi) = p_1(x, \xi)$ and choose $y_n(x, \xi)$ satisfying the commutation relation $\{\eta_n, y_n\} = 1$, that is,

$$\langle \partial p_1(x, \xi)/\partial \xi, \partial y_n(x, \xi)/\partial x \rangle - \langle \partial y_n(x, \xi)/\partial \xi, \partial p_1(x, \xi)/\partial x \rangle = 1.$$

To do so we first choose j so that $\partial p_1(x,\xi)/\partial \xi_j \neq 0$ at (x^0, ξ^0). Then there is a unique local solution of this equation with $y_n = 0$ when $x_j = x_j^0$, for this is a non-characteristic Cauchy problem. Thus $y_n(x,\xi) = 0$ when $x = x^0$, and y_n is automatically homogeneous of degree 0 by the uniqueness. The Hamilton fields of y_n and η_n are linearly independent modulo the radial vector, for otherwise the Poisson bracket of y_n and η_n would vanish. By [H1, Theorem 21.1.9] we can now choose $y_1(x,\xi), \ldots, y_{n-1}(x,\xi)$ and $\eta_1(x,\xi), \ldots, \eta_{n-1}(x,\xi)$ homogeneous of degree 0 and 1 respectively to get a full system of canonical coordinates (y,η), $y(x^0, \xi^0) = 0$. With $\eta^0 = \eta(x^0, \xi^0)$ we have $\eta_n^0 = 0$, and regarding p as a function of η in a conic neighborhood of $(0, \eta^0)$, $p = \eta_n + ig(y,\eta)$, we obtain from Malgrange's preparation theorem as in the proof of [H1, Theorem 21.3.6] in a conic neighborhood of $(0, \eta^0)$

$$\eta_n = q(y,\eta)(\eta_n + ig(y,\eta)) + r_1(y,\eta) + ir_2(y,\eta)$$

where $q \neq 0$ and r_1, r_2 are real valued and independent of η_n. Thus $z_n = y_n$ and $\zeta_n = \eta_n - r_1(y,\eta)$ satisfy the commutation relation $\{\zeta_n, z_n\} = 1$, and we can again complete to a full set of canonical coordinates (z, ζ). With these new coordinates

$$q(y,\eta)p(x,\xi) = \zeta_n - if(z,\zeta)$$

where $f(z,\zeta) = r_2(y,\eta)$ is independent of ζ_n since $\{f, z_n\} = \{r_2, y_n\} = 0$. We have $z_n = 0$ when $x = x^0$ since this is true for y_n. If we now regard $q(y,\eta)$ as a function of (x,ξ) and rename the (z,ζ) coordinates (y,η), the lemma is proved.

REMARK. The condition $p'_\xi(x^0, \xi^0) \neq 0$ in the lemma is necessary for the conclusion to be valid, for

$$\{p(x,\xi), y_n(x,\xi)\} = q(x,\xi)\{\eta_n - if(y,\eta'), y_n\} = q(x,\xi) \neq 0 \quad \text{at } (x^0, \xi^0),$$

and if $\partial y_n/\partial \xi = 0$ when $x = x^0$, this implies that $\partial p(x,\xi)/\partial \xi \neq 0$ at (x^0, ξ^0).

For a pseudodifferential operator with principal symbol of the form suggested by Lemma 5.2 one can attain further simplifications of the lower orders terms using the following:

LEMMA 5.3. *Let P be a polyhomogeneous pseudodifferential operator in $\mathbf{R}^n$ of order 1 with principal symbol $p(x,\xi) = \xi_n + if(x,\xi)$ in a conic neighborhood Γ of $(0, \xi^0)$ where $\xi_n^0 = 0$, $f(0, \xi^0) = 0$, and assume that f is real valued and independent of ξ_n in Γ. Then one can find elliptic pseudodifferential operators A and B of order 0 such that the full symbol of BPA in a conic neighborhood of $(0, \xi^0)$ is equal to $\xi_n + if(x,\xi) + f_0(x,\xi) + r(x,\xi)$ where $f_0 \in C^\infty$ is real valued, homogeneous of degree 0 in ξ, and independent of ξ_n, and r is of order -1.*

PROOF. If p_0 is the term of order 0 in the symbol of P, then the Malgrange preparation theorem gives that in a conic neighborhood of $(0, \xi^0)$

$$p_0(x,\xi) = q_{-1}(x,\xi)(\xi_n + if(x,\xi)) + r_0(x,\xi)$$

where q_{-1} is homogeneous of degree -1 and r_0 is homogeneous of degree 0 and independent of ξ_n. If Q is a pseudodifferential operator with principal symbol q_{-1} and $A = \mathrm{Id} - Q$, $B = \mathrm{Id}$, it follows that the symbol of BPA is equal to

$$\xi_n + if(x,\xi') + p_0(x,\xi') - q_{-1}(x,\xi)(\xi_n + if(x,\xi')) = \xi_n + if(x,\xi') + r_0(x,\xi')$$

modulo terms of order -1, in a conical neighborhood of $(0, \xi^0)$. After this preliminary reduction it is sufficient to prove the lemma when $p_0(x,\xi)$ is independent of ξ_n. With

$a(x, \xi')$ and $b(x, \xi')$ homogeneous of degree 0 and $b(x, \xi') = 1/a(x, \xi')$ we choose A and B with principal symbols A and B in a conic neighborhood of $(0, \xi^0)$. Then $BPA = BAP + B[P, A]$ where $BA -$ Id is of order -2, so the principal symbol is equal to p and the term of order 0 is $-i\{p, a\}/a + p_0$. The imaginary part vanishes if a is real valued and $\partial a/\partial x_n = a \operatorname{Im} p_0$. We can solve this equation in a conic neighborhood of $(0, \xi_0)$ with the initial condition $a = 1$ when $x_n = 0$, which completes the proof.

PROOF OF THEOREM 5.1. Now we return to the proof of (5.6) for some φ of order 0 which is non-characteristic at (x^0, ξ^0) when $p(x^0, \xi^0) = 0$ but $p_\xi'(x^0, \xi^0) \neq 0$. Lemma 5.2 gives a canonical map $\chi : (x, \xi) \mapsto (y, \eta)$ from a conic neighborhood of $(x^0, \xi^0) \in T^*\mathbf{R}^n$ to a conic neighborhood of $(0, \eta^0) \in T^*\mathbf{R}^n$ (we may assume that $\eta^0 = (1, 0, \ldots, 0)$), such that locally by (5.7)

$$\overline{p(x, \xi)} = \overline{q(x, \xi)}(\eta_n + if(y, \eta')), \quad \eta' = (\eta_1, \ldots, \eta_{n-1}),$$

where f is real valued. Choose Fourier integral operators B and A of order 0 corresponding to χ and χ^{-1} with $WF'(B)$ and $WF'(A)$ close to $((0, \eta^0), (x^0, \xi^0))$ and $((x^0, \xi^0), (0, \eta^0))$ respectively, so that the principal symbol of AB is equal to $1/\overline{q(x, \xi)}$ in a conic neighborhood of (x^0, ξ^0). Since $\bar{p}$ is the principal symbol of P^* it follows that the principal symbol of the pseudodifferential operator BP^*A is equal to $\eta_n + if(y, \eta')$ in a conic neighborhood of $(0, \eta^0)$. Using Lemma 5.3 we can attain that in such a neighborhood the symbol of BP^*A is equal to $\eta_n + if(y, \eta') - f_0(y, \eta') + r(y, \eta)$ where f_0 is real valued and homogeneous of degree 0, and r is of order -1. Since P satisfies condition (Ψ) it follows that f does not change sign from $+$ to $-$ for increasing x_n. This property is not disturbed if we multiply by a non-negative cutoff function. Hence we can find $f(x, \xi') \in C_0^\infty([-1, 1], S_{1,0}^1(T^*\mathbf{R}^{n-1}))$ and $f_0(x, \xi') \in C_0^\infty([-1, 1], S_{1,0}^0(T^*\mathbf{R}^{n-1}))$, both real valued, and $r \in S_{1,0}^{-1}(T^*\mathbf{R}^{n-1})$ such that f *does not change sign from $+$ to $-$ for increasing x_n and the symbol of BP^*A/i is equal to* $-i\xi_n + f(x, \xi') + if_0(x, \xi') + r(x, \xi)$ *in a conic neighborhood* Γ *of* $(0, \eta^0)$. We can choose B_1 of order 0 corresponding to χ so that the pseudodifferential operator $AB_1 -$ Id is of order $-\infty$ in a conic neighborhood of (x^0, ξ^0) containing $\chi^{-1}\Gamma$, if Γ is chosen small enough.

By Theorem 4.4.2 with n replaced by $n - 1$, t replaced by x_n, and x replaced by x', but otherwise no change of the notation used there, we have with norms in $L^2(\mathbf{R}^n)$

$$(5.8) \quad \|\chi_0(x_n/T)\varphi_0^w(x, D')v\|^2 \leq C_1 T^2 \|(1 + |D'|^2)^{\frac{1}{4}}\varphi_0^w(x, D')(F^w(x, D') - \partial/\partial x_n)v\|^2$$
$$+ C_2 \|\chi'(x_n/T)(1 + |D'|^2)^{\frac{1}{4}}v\|^2 + C_3 T \|\chi_0(x_n/T)v\|_{(0)}^2, \quad v \in \mathcal{S}(\mathbf{R}^n),$$

if $\varphi_0 \in C_0^\infty([-1, 1], S_{1,0}^0(T^*\mathbf{R}^{n-1}))$. To be able to switch to pseudodifferential operators in $\mathbf{R}^n$ we choose φ_0 and a function $\psi_0 \in S_{1,0}^0(T^*\mathbf{R}^n)$ such that ψ_0 *is a function of the frequency variables* ξ *only*, and for some small positive number κ

$$(5.9) \qquad |\xi_n| \leq \kappa|\xi'|, \text{ if } \xi \in \operatorname{supp}\psi_0, \quad \tfrac{1}{2}\kappa|\xi'| \leq |\xi_n|, \text{ if } \xi \in \operatorname{supp} d\psi_0, \ |\xi| \geq 1,$$

$$(5.10) \qquad |x| \leq \kappa, \ |\xi_2| + \cdots + |\xi_{n-1}| \leq \kappa\xi_1, \ |\xi| \geq 1, \quad \text{if } (x, \xi) \in \operatorname{supp}\varphi_0.$$

Then it follows that $|\xi_2| + \cdots + |\xi_n| \leq \kappa(1 + \kappa)\xi_1$ when $\xi \in \operatorname{supp}\varphi_0 \cap \operatorname{supp}\psi_0$, hence $\operatorname{supp}\varphi_0 \cap \operatorname{supp}\psi_0 \subset \Gamma$ if κ is small enough. We can choose φ_0 and ψ_0 so that $\varphi_0\psi_0 = 1$ at infinity in a conic neighborhood Γ_0 of $(0, \eta^0)$.

To prove an estimate of the form (5.5) we shall apply (5.8) with v replaced by $\psi_0(D)v$ where $v = B_1 u$:

$(5.8)'$ $\quad \|\chi_0(x_n/T)\varphi_0^w(x, D')\psi_0(D)v\|^2$

$$\leq C_1 T^2 \|(1 + |D'|^2)^{\frac{1}{4}}\varphi_0^w(x, D')(F^w(x, D') - \partial/\partial x_n)\psi_0(D)v\|^2$$

$$+ C_2\|\chi'(x_n/T)(1 + |D'|^2)^{\frac{1}{4}}\psi_0(D)v\|^2 + C_3 T\|\chi_0(x_n/T)\psi_0(D)v\|_{(0)}^2, \quad v \in \mathcal{S}(\mathbf{R}^n),$$

In the first term in the right-hand side we shall commute the operator $\psi_0(D)$ to the left noting that

$$[F^w(x, D') - \partial/\partial x_n, \psi_0(D)] = [F^w(x, D'), \psi_0(D)] = \psi_1^w(x, D) + R_1^w(x, D),$$

where $\psi_1 = -i\{F, \psi_0\} = i\langle \partial F/\partial x, \partial\psi_0/\partial\xi\rangle$ and $R_1 \in S_{1,0}^{-1}(T^*\mathbf{R}^n)$. (This application of the calculus is legitimate by the first condition in (5.9).) Hence $\|R_1^w v\|_{(1)} \leq C\|v\|_{(0)}$. Since $|\xi| \leq (1 + 2/\kappa)|\xi_n|$ in the support of ψ_1, the principal symbol $\xi_n + if(x, \xi')$ of BP^*A is non-zero in $\mathrm{supp}\,\varphi_0 \cap \mathrm{supp}\,\psi_1$ if κ is small enough, so it follows as in the proof of (5.6) that $(1 + |D'|^2)^{\frac{1}{4}}\varphi_0^w(x, D')\psi_1^w(x, D) = \psi_2^w BP^*A + R_2^w$ where $\psi_2 \in S_{1,0}^{-\frac{1}{2}}$, $\mathrm{supp}\,\psi_2 \subset \Gamma$, and $R_2 \in S_{1,0}^{-\frac{1}{2}}$. Summing up,

(5.11) $\quad \|(1 + |D'|^2)^{\frac{1}{4}}\varphi_0^w(x, D')[F^w(x, D') - \partial/\partial x_n, \psi_0(D)]v\|$

$$\leq C\big(\|\psi_2^w(x, D)BP^*Av\| + \|v\|\big).$$

Commuting $\psi_0(D)$ to the left in the first term on the right-hand side of $(5.8)'$ leads to the pseudodifferential operator $(1+|D'|^2)^{\frac{1}{4}}\varphi_0^w(x, D')\psi_0(D) = \varphi^w(x, D)$ where $\varphi \in S_{1,0}^{\frac{1}{2}}(T^*\mathbf{R}^n)$ and $\varphi(x, \xi) - (1 + |\xi'|^2)^{\frac{1}{4}}\varphi_0(x, \xi')\psi_0(\xi) \in S_{1,0}^{-\frac{1}{2}}$, so φ is non-characteristic at $(0, \eta^0)$. (Here we have used the first part of (5.9) again.) We have

$$\varphi^w(x, D)(F^w(x, D') - \partial/\partial x_n) = \varphi^w(x, D)BP^*A/i + R_3^w(x, D), \quad R_3 \in S_{1,0}^{-\frac{1}{2}}(T^*\mathbf{R}^n),$$

for the symbol of $BP^*A/i - (F^w(x, D') - \partial/\partial x_n)$ is of order -1 in Γ. Hence

(5.12) $\qquad \|\varphi^w(x, D)(F^w(x, D') - \partial/\partial x_n)v\| \leq \|\varphi^w(x, D)BP^*Av\| + C\|v\|_{(0)}.$

Recalling that $v = B_1 u$ where $u \in C_0^\infty(\Omega_\delta)$ we have

$\|\varphi^w(x, D)BP^*Av\| = \|\varphi^w(x, D)BP^*AB_1 u\|$

$$\leq \|\varphi^w(x, D)BP^*u\| + \|\varphi^w(x, D)BP^*(AB_1 - \mathrm{Id})u\|.$$

Here $\varphi^w(x, D)BP^*(AB_1 - \mathrm{Id})$ is a smoothing operator since $AB_1 - \mathrm{Id}$ is of order $-\infty$ in $\chi^{-1}\mathrm{supp}\,\varphi$, so the last term can be estimated by $\|u\|_{(-1)}$ while the preceding one can be estimated by $\|P^*u\|_{(\frac{1}{2})}$. The term in (5.11) involving BP^*A can be handled in the same way, and since A, B, B_1 are $H_{(s)}$ continuous for every s, we have now proved that

(5.13) $\quad \|A\chi_0(x_n/T)\varphi_0^w(x, D)\psi_0(D)B_1 u\|^2$

$$\leq C_1'\|P^*u\|_{(\frac{1}{2})}^2 + C_2'\|\chi'(x_n/T)B_1 u\|_{(\frac{1}{2})}^2 + C_3'T\|u\|^2.$$

The Weyl symbol of the pseudodifferential operator $A\chi(x_n/T)\varphi^w(x, D)B_1$ in the left-hand side is in $S^0_{1,0}$ and non-characteristic at (x^0, ξ^0). For every $u \in H_{(-1)}$ with supp $u \subset \overline{\Omega_\delta}$ we have $|y_n| \le C\delta$ when $(y, \eta) \in WF(B_1 u)$ for the canonical transformation χ was chosen so that it maps $\{(0, \xi); \xi \in \mathbf{R}^n\}$ to the plane where $y_n = 0$. If $C\delta < T/3$ it follows that $\chi'_0(x_n/T)B_1 u \in C^\infty$. By the closed graph theorem this implies that there is an estimate

$$\|\chi'_0(x_n/T)B_1 u\|_{(\frac{1}{2})} \le C_T \|u\|_{(-1)}, \quad \text{when } u \in H_{(-1)}, \text{ supp } u \subset \overline{\Omega_\delta},$$

in particular when $u \in C^\infty_0(\Omega_\delta)$. Using this in (5.13) gives an estimate of the form (5.5) which completes the proof of Theorem 5.1 when $s = 0$ and $m = 1$.

To extend the conclusion to general s and m we choose for $\nu \in \mathbf{R}$ a properly supported pseudodifferential operator E_ν in Ω with total symbol $(1 + |\xi|^2)^{\nu/2}$. Then $E_{1-m-s}P^*E_s$ is of order 1 and the principal symbol $|\xi|^{1-m}\bar{p}(x, \xi)$ satisfies $(\bar{\Psi})$. Hence we can find $\delta_1 > 0$ such that

$$\|v\|_{(0)} \le \|E_{1-m-s}P^*E_s v\|_{(\frac{1}{2})}, \quad v \in C^\infty_0(\Omega_{\delta_1}).$$

Let $\chi_1 \in C^\infty_0(\Omega_{\delta_1})$ be equal to 1 in $\Omega_{\delta_1/2}$ and replace v by $\chi_1 E_{-s} v$. Then it follows that

$$\|v\|_{(-s)} \le C_1 \|P^*v\|_{(\frac{3}{2}-m-s)} + C_2 \|v\|_{(-s-1)}, \quad v \in C^\infty_0(\Omega_{\delta_1/3}).$$

We claim that for some C_0 and $\delta_2 > 0$

$$(5.14) \qquad \|v\|_{(-s)} \le C_0 \|P^*v\|_{(\frac{3}{2}-m-s)}, \quad v \in C^\infty_0(\Omega_{\delta_2}).$$

If this were not true we could find a sequence $v_j \in C^\infty_0(\Omega_{1/j})$ such that $\|v_j\|_{(-s)} = 1$ and $\|P^*v_j\|_{(\frac{3}{2}-m-s)} \to 0$. Hence $C_2 \|v_j\|_{(-s-1)} \ge \frac{1}{2}$ for large j, and since the sequence v_j is compact in $H_{(-s-1)}$ it has a strong limit $v \ne 0$ in $H_{(-s-1)}$ which is a weak limit in $H_{(-s)}$. We have $P^*v = 0$ and supp $v = \{x^0\}$, so $v = Q(D)\delta_{x^0}$ for some polynomial $Q \ne 0$. This means that $\bar{Q}(D)Pu(x^0) = 0$ for all $u \in C^\infty_0(\Omega)$, which implies that $\bar{q}(\xi)p(x^0, \xi) = 0$ when $\xi \in \mathbf{R}^n \setminus 0$, if q is the principal part of Q, of degree μ, for

$$\bar{q}(\xi)p(x, \xi)u(x) = \lim_{t \to +\infty} \left(\bar{Q}(D)P(u(x)e^{it\langle x, \xi \rangle}) \right) / \left(t^{\mu+m} e^{it\langle x, \xi \rangle} \right).$$

Since $p(x^0, \xi) \not\equiv 0$ we conclude that (5.14) is valid for small δ_2, which completes the proof of Theorem 5.1. (Note that the last part of the argument proves that assuming only that $p(x^0, \xi) \not\equiv 0$ one can find $u \in C^\infty_0$ with support in a given neighborhood of x^0 such that Pu has a given finite Taylor expansion at x^0.)

6. Some open problems on solvability. Theorem 5.1 immediately suggests several questions for the future study of solvability:

Q1. *Given P of order m with principal symbol satisfying (5.1) and condition (Ψ), and a number $\delta \in [0, \frac{3}{2}]$ does the equation $Pu = f$ have a local solution $u \in H_{(s+m-\delta)}(\mathbf{R}^n)$ at x^0 for every $f \in H_{(s)}(\mathbf{R}^n)$?*

For $\delta = 0$ the answer is yes if and only if P is elliptic, that is, $p(x^0, \xi) \ne 0$ when $\xi \in \mathbf{R}^n \setminus \{0\}$. If $0 < \delta < 1$ the answer is yes if and only if the adjoint of P is subelliptic in the sense that that for every $\xi \in \mathbf{R}^n \setminus \{0\}$ with $p(x^0, \xi) = 0$ some repeated Poisson bracket of $\mathrm{Re}\, p$ and $\mathrm{Im}\, p$ with at most $\delta/(1-\delta)$ factors is non-zero at (x^0, ξ).

For $\delta = 1$ the answer is yes if p satisfies the stronger condition (P), that is, p and $\bar{p}$ both satisfy condition (Ψ), but as proved by Lerner [L5] the answer is not always yes when p just satisfies condition (Ψ). (See also [H2].) However, his examples do not rule out the possibility that the answer could be positive for every $\delta > 1$, and they leave open the possibility of a positive answer when $\delta = 1$ if the principal symbol is real analytic. A number of cases where the answer is positive when $\delta = 1$ were listed in the survey paper [H2] but unfortunately there is an error in the statement of Theorem 8.4 and Corollary 8.5 there. A corrected version of Theorem 8.4 should read:

THEOREM 6.1. *Assume that $Q(t, x, \xi)$ is real valued, continuous in t, and that*

$$|D_x^\beta D_\xi^\alpha Q(t, x, \xi)| \le C_{\alpha,\beta} \lambda^{2-|\alpha|-|\beta|}, \quad t \in \mathbf{R}, \ x \in \mathbf{R}^n, \ \xi \in \mathbf{R}^n.$$

Assume also that $Q(s, x, \xi) > 0$ implies $Q(t, x, \eta) \ge 0$ when $s \le t$ and $x, \xi, \eta \in \mathbf{R}^n$. If δ and $1/\lambda$ are sufficiently small, it follows that

$$\sup \|u(t)\| \le 3 \int \|\partial u/\partial t - Q(t, x, D)u\| \, dt, \quad u \in C_0^\infty((-\delta, \delta) \times \mathbf{R}^n).$$

The proof began by defining

$$T(x) = \sup\{t \in [-\delta, \delta); Q(t, x, \xi) < 0 \text{ for some } \xi \text{ or } t = -\delta\},$$

followed by the statement that

(6.1) $Q(t, x, \xi) \le 0$ if $-\delta < t \le T(x), \quad Q(t, x, \xi) \ge 0$ if $T(x) \le t < \delta$.

This is indeed true under the sign condition made in Theorem 6.1, for if $Q(t, x, \xi) > 0$ we have $Q(s, x, \eta) \ge 0$ when $t \le s$, hence $T(x) \le t$. Conversely the existence of a function $T(x)$ such that (6.1) is valid implies the sign condition in Theorem 6.1, for if $Q(s, x, \xi) > 0$ and $Q(t, x, \eta) < 0$ then (6.1) implies that $s > T(x) > t$. The sign condition in Theorem 6.1 implies that $Q(t, x, \xi)$ has no sign change from $+$ to $-$ for increasing t, and that $Q(t, x, \xi)Q(t, x, \eta) \ge 0$ for all t, x, ξ, η, which is the hypothesis made in [H2], but not conversely. The hypothesis in [H2, Corollary 8.5] should be modified in the same way.

Q2. *Does it follow from the hypotheses of Theorem 5.1 that for every $f \in C^\infty$ there is some $u \in C^\infty$ such that $Pu = f$ in a neighborhood of x^0?*

This is true when p satisfies condition (P) but the only known proof requires a detailed study of the propagation of singularities.

Q3. *To what extent can the condition (5.1) be relaxed?*

As observed after the proof of Lemma 5.2 the condition (5.1) was very essential for the proof of Theorem 5.1. On the other hand, if p satisfies condition (P) then it follows from [H1, Theorems 26.11.1, 26.11.2] that (5.1) can be weakened to demanding only that if $\xi \in \mathbf{R}^n \setminus \{0\}$ and $p(x^0, \xi) = 0$ then there is a semi-bicharacteristic interval through (x^0, ξ) which does not remain over x^0 forever. However, the proof which again depends on the study of propagation of singularities only gives a solution as in **Q1** for every $\delta > 1$.

Q4. *Semiglobal existence theorems.*

The question is whether the point x^0 in Theorem 5.1 can be replaced by a compact set, if one allows P to have a range of finite codimension. The necessity of a corresponding version of condition (Ψ) is known (see [H1, Theorem 26.4.7 and Corollary 26.4.8]). When condition (P) is fulfilled sufficiency was proved in [H2, Theorem 26.11.1].

Q5. *Propagation of singularities.*

Many of the results for operators satisfying condition (P) mentioned above have been based on a study of propagation of singularities for solutions of the adjoint operator. (See [H1, Chapter 26].) Some of the results remain valid under condition (Ψ) but much remains to be explored.

References

[B]   J.-M. Bony, *Sur l'inégalité de Fefferman-Phong,*, Sém. Éq. Dér. Part., École Polytechnique 1998–1999, Exp. III, 14 pp..

[D1]   Nils Dencker, *The resolution of the Nirenberg-Treves conjecture*, Annals of Math., (to appear). Earlier manuscript versions February 4, 2004, 36 pp.; October 2, 2003, 36 pp.; September 20, 2003, 35 pp.; March 13, 2003, 53 pp..

[D2]   _____, *On the sufficiency of condition* (Ψ), Manuscripts May 22, 2001, 52 pp.; October 5, 2001, 61 pp.; April 24, 2002, 70 pp.; October 18, 2002, 55 pp..

[H1]   L. Hörmander, *The analysis of linear partial differential operators I–IV*, Springer Verlag, 1983–1985.

[H2]   _____, *On the solvability of pseudodifferential equations*, Structure of solutions of differential equations (M. Morimoto and T. Kawai, eds.), World Scientific, 1996, Taniguchi symposium June 1995.

[H3]   _____, *Notions of convexity*, Birkhäuser, 1994.

[L1]   N. Lerner, *Reading Dencker's paper on the Nirenberg-Treves conjecture*, Private communication April 16, 2002, 30 pp..

[L2]   _____, *Energy methods via coherent states and advanced pseudo-differential calculus*, Contemporary Math. **205** (1997), 177–201.

[L3]   _____, *More on the last version of Dencker's article with date October 3, 2003*, Private communication October 14, 2003, 5 pp..

[L4]   _____, *Comments on the last version of Dencker's article with date October 3, 2003*, Private communication October 22, 2003, 6 pp..

[L5]   _____, *Nonsolvability in L^2 for a first order operator satisfying condition* (ψ), Ann. of Math. **139** (1994), 363–393.

[L6]   _____, *Sufficiency of condition* (ψ) *for local solvability in two dimensions*, Ann. of Math. **128** (1988), 363–393.

[L7]   _____, *Cutting the loss of derivatives for solvability under condition* (Ψ), Manuscript June 21, 2005, 68 pp., available at http://www.perso.univ-rennes1.fr/nicolas.lerner.

LOWER BOUNDS FOR SUBELLIPTIC OPERATORS

LARS HÖRMANDER

0. Introduction. Let P be a pseudodifferential operator of order m in a C^∞ manifold X, which has a "classical" symbol with principal part p, defined on $T^*(X) \setminus 0$. If $0 < \delta < 1$ then P is called microsubelliptic with loss of δ derivatives at a point $\gamma \in T^*(X) \setminus 0$ if

$$(0.1) \qquad u \in \mathcal{D}'(X), \ Pu \in H^{\mathrm{loc}}_{(s)} \text{ at } \gamma \implies u \in H^{\mathrm{loc}}_{(s+m-\delta)} \text{ at } \gamma.$$

(Throughout we use the same standard notation as in [H].) Since $\delta < 1$ it follows easily that this condition only depends on the principal symbol of p in a neighborhood of γ. In fact it only depends on the derivatives of p at γ of order $\leq \delta/(1-\delta)$. If k is the largest integer $\leq \delta/(1-\delta)$ then it follows that P is microsubelliptic at γ with loss of $k/(k+1) \leq \delta$ derivatives, so the only relevant values of δ are of the form $k/(k+1)$ where k is a positive integer. (If $\delta < \frac{1}{2}$ then P is non-characteristic at γ so one can take $\delta = 0$.)

To simplify notation we shall assume from now on that P is of order 1 and that $X \subset \mathbf{R}^n$ which is no essential restriction in this local problem. If P is microsubelliptic at $\gamma = (x_0, \xi_0)$ with loss of $\delta = k/(k+1)$ derivatives and A is a pseudodifferential operator of order $1/(k+1)$ with $WF(A)$ contained in a sufficiently small conic neighborhood of γ, then there are constants C_1, C_2 such that with L^2 norms

$$(0.2) \qquad \|Au\|^2 \leq C_1 \|Pu\|^2 + C_2 \|u\|^2, \quad u \in C_0^\infty(K),$$

if $K \Subset X$ is a neighborhood of x_0. Conversely (0.2) implies that P is microsubelliptic at γ with loss of $k/(k+1)$ derivatives, if A is non-characteristic at γ.

The facts above are contained in Section 27.1 of [H], and Chapter XXVII of [H] gives complete proofs of the necessary and sufficient conditions for P to be microsubelliptic with loss of $k/(k+1)$ derivatives at γ, in terms of the derivatives of the principal symbol p at γ of order $\leq k$. However, there is no discussion in [H] of the quantitative aspect of (0.2), to determine the constants C_1 such that (0.2) is valid for some C_2 when $WF(A)$ is small. The purpose of this paper is to do so when P satisfies the condition (P) which in this situation means that microlocally the principal symbol can be reduced to the form $\xi_1 + iq(x, \xi')$ where $\xi' = (\xi_2, \ldots, \xi_n)$ and $q \geq 0$. The study of subellipticity is then far simpler than for operators which only satisfy condition $(\overline{\Psi})$, which required the major part of Chapter XXVII of [H] and is regrettably complicated. We hope to be able to return to this case also later on.

The plan of the paper is as follows. In Section 1 we prove (Theorem 1.3) that estimates such as (0.2) imply uniform estimates for a family of ordinary differential operators on $\mathbf{R}$ of the form $-d/dt + q(t)$ where $q \geq 0$ is a polynomial of degree k with leading term t^k. These localizations are not determined by the derivatives of p at γ of order $\leq k$ but constitute

© Springer International Publishing AG, part of Springer Nature 2018
L. Hörmander, *Unpublished Manuscripts*,
https://doi.org/10.1007/978-3-319-69850-2_24

a new invariant attached to the characteristic point γ. Section 2 is devoted to a detailed study of estimates for such families. One case where we have found the best constant is discussed further in Section 4 although this is not required for the proof of the main result Theorem 3.6, which is a converse of Theorem 1.3. Together these two theorems describe the infimum of the constants C_1 which can occur in (0.2).

The author is very much indebted to Professor Alberto Parmeggiani who has read and commented on an earlier version of this manuscript. His comments have led to the removal of numerous flaws.

1. Localization. In this section we shall deduce necessary conditions for (0.2) to be valid. Let $P = P(x, D)$ where $P \in S^1$ and $P(x, \xi) = \xi_1 + iq(x, \xi')$, $\xi = (\xi_1, \xi')$, when $|\xi|$ is large and $|\xi/|\xi| - \xi_0/|\xi_0|| + |x|$ is sufficiently small, $\xi_0 = (0, \xi_0') \neq 0$. Here q is positively homogeneous of degree 1, non-negative, and for some $k > 0$

$$(1.1) \qquad \partial_{x_1}^j q(0, \xi_0') = 0, \quad j < k; \quad \partial_{x_1}^k q(0, \xi_0')/k! = c > 0.$$

Then k is even, and since $q \geq 0$ it follows from Taylor's formula that for small x and ξ'

$$(1.1)' \qquad \begin{aligned} q(x_1, x', \xi_0' + \xi') &= cx_1^k + O(|x_1|^{k+1} + |x_1|^{k/2}(|x'| + |\xi'|) + |x'|^2 + |\xi'|^2) \\ \partial q(x_1, x', \xi_0' + \xi')/\partial \xi' &= O(|x_1|^{k/2} + |x'| + |\xi'|). \end{aligned}$$

Let $A = A(x, D)$ where $A \in S^{1/(k+1)}$ has principal symbol a.

PROPOSITION 1.1. *If* (0.2) *is valid and K is a neighborhood of* 0, *then*

$$(1.2) \qquad |a(0, \xi_0)|^2 \int_{\mathbf{R}} |U(t)|^2 \, dt \leq c^{2/(k+1)} C_1 \int_{\mathbf{R}} |U'(t) - t^k U(t)|^2 \, dt, \quad U \in C_0^\infty(\mathbf{R}).$$

We note that $|a(0, \xi_0)|^2/c^{2/(k+1)}$ is homogeneous of degree 0 in ξ_0. Inequalities such as (1.2) will be studied at great length in Section 2.

PROOF. Set $M_\varepsilon x = M_\varepsilon(x_1, x') = (\varepsilon x_1, \varepsilon^{(k+1)/2} x')$, $x \in \mathbf{R}^n$. To localize in the direction $(0, \xi_0)$ we observe that when $U \in \mathcal{S}(\mathbf{R}^n)$ then

$$(1.3) \qquad P(x, D)\left(e^{i\langle x, \xi_0\rangle/\varepsilon^{k+1}} U(M_\varepsilon^{-1} x)\right) = e^{i\langle x, \xi_0\rangle/\varepsilon^{k+1}} F_\varepsilon(M_\varepsilon^{-1} x),$$

$$(1.4) \qquad F_\varepsilon(x) = (2\pi)^{-n} \int e^{i\langle x, \xi\rangle} P(M_\varepsilon x, \xi_0/\varepsilon^{k+1} + M_\varepsilon^{-1}\xi)\widehat{U}(\xi) \, d\xi,$$

where $\widehat{U} \in \mathcal{S}(\mathbf{R}^n)$. When $|M_\varepsilon x|$, $\varepsilon^{(k+1)/2}|\xi|$ and ε are sufficiently small then

$$P(M_\varepsilon x, \xi_0/\varepsilon^{k+1} + M_\varepsilon^{-1}\xi) = \xi_1/\varepsilon + iq(M_\varepsilon x, \xi_0'/\varepsilon^{k+1} + \varepsilon^{-(k+1)/2}\xi')$$
$$= \varepsilon^{-1}(\xi_1 + i\varepsilon^{-k}q(M_\varepsilon x, \xi_0' + \varepsilon^{(k+1)/2}\xi')).$$

If $\varepsilon^{(k+1)/2}|\xi| > c > 0$ then

$$|P(M_\varepsilon x, \xi_0/\varepsilon^{k+1} + M_\varepsilon^{-1}\xi)| \leq C(\varepsilon^{-k-1} + \varepsilon^{-(k+1)/2}|\xi|) \leq C|\xi|^2(1/c^2 + 1/c),$$
$$|\partial P(M_\varepsilon x, \xi_0/\varepsilon^{k+1} + M_\varepsilon^{-1}\xi)/\partial \xi| \leq C|M_\varepsilon^{-1}| \leq C|\xi|/c,$$

since $P \in S^1$, and since $\widehat{U} \in \mathcal{S}$ the corresponding contribution to the integral $F_\varepsilon(x)$ is $O(\varepsilon^N)$ for any N. If $|M_\varepsilon x| > 2c > 0$ then $\varepsilon^{k+1}(|x_1|^{k+1} + |x'|^2) \geq c' = \min(c^{k+1}, c^2)$. If in addition $\varepsilon^{(k+1)/2}|\xi| < |\xi_0|/2$ it follows that

$$|P(M_\varepsilon x, \xi_0/\varepsilon^{k+1} + M_\varepsilon^{-1}\xi)| \leq C(|x_1|^{k+1} + |x'|^2),$$

$$|\partial P(M_\varepsilon x, \xi_0/\varepsilon^{k+1} + M_\varepsilon^{-1}\xi)/\partial\xi| \leq C(|x_1|^{(k+1)/2} + |x'|)$$

since $P \in S^1$, so we obtain a uniform bound for $F_\varepsilon(x)/(|x_1|^{k+1} + |x'|^2 + 1)$. When $\varepsilon^{(k+1)/2}|\xi| < c$, $|M_\varepsilon x| < 2c$, and c is small enough then it follows from $(1.1)'$ that

(1.5)
$$\varepsilon^{-k}q(M_\varepsilon x, \xi_0' + \varepsilon^{(k+1)/2}\xi') = cx_1^k + O(\varepsilon|x_1|^{k+1} + \varepsilon^{\frac{1}{2}}|x_1|^{k/2}(|\xi'| + |x'|) + \varepsilon(|\xi'|^2 + |x'|^2)),$$

$$|\varepsilon\partial P(M_\varepsilon x, \xi_0/\varepsilon^{k+1} + M_\varepsilon^{-1}\xi)/\partial\xi| = 1 + O(\sqrt{\varepsilon}x_1^{k/2} + \varepsilon(|x'| + |\xi'|)),$$

if ε is sufficiently small. Hence, uniformly on every compact set,

$$\varepsilon F_\varepsilon(x) \to D_1 U(x) + icx_1^k U(x)$$

when $\varepsilon \to 0$, and we have a uniform bound for $\varepsilon F_\varepsilon(x)/(|x_1|^{k+1} + |x'|^2 + 1)$ when $x \in \mathbf{R}^n$ and ε is small. If we multiply (1.4) by a monomial x^α and integrate by parts to obtain instead a derivative $(-D_\xi)^\alpha$ it follows that there is such a bound also for the product by x^α. In fact, in the term where $(-D_\xi)^\alpha$ acts on $\widehat{U}$ we obtain the same estimate as before since $(-D_\xi)^\alpha\widehat{U}(\xi) \in \mathcal{S}$. In the other terms where $\nu \geq 1$ of the derivatives act on $P(M_\varepsilon x, \xi_0/\varepsilon^{k+1} + M_\varepsilon^{-1}\xi)$ this gives a factor $\leq C\varepsilon^{-(k+1)\nu/2}(1 + |\xi_0/\varepsilon^{k+1} + M_\varepsilon^{-1}\xi|)^{1-\nu}$. When $2|M_\varepsilon^{-1}\xi| > |\xi_0/\varepsilon^{k+1}|$ then $2|\xi| > \varepsilon^{-(k+1)/2}|\xi_0|$ so this can be estimated by $C|\xi|^\nu$. When $2|M_\varepsilon^{-1}\xi| \leq |\xi_0/\varepsilon^{k+1}|$ we get the bound $O(\varepsilon^{(k+1)(\nu-1-\nu/2)})$ which is bounded if $\nu \geq 2$. A bound $O(1/\varepsilon)$ follows from the estimates of $\varepsilon\partial P(M_\varepsilon x, \xi_0/\varepsilon^{k+1} + M_\varepsilon^{-1}\xi)/\partial\xi$ above when $\nu = 1$, so in any case we have a bound for $x^\alpha\varepsilon F_\varepsilon(x)/(|x_1|^{k+1} + |x'|^2 + 1)$. Hence $\varepsilon F_\varepsilon(x)(1 + |x|)^N$ is bounded for arbitrary N, and by dominated convergence

$$\varepsilon^2\int_{\mathbf{R}^n} |F_\varepsilon(x)|^2\, dx \to \int_{\mathbf{R}^n} |D_1 U(x) + icx_1^k U(x)|^2\, dx, \quad \text{when } \varepsilon \to 0.$$

In (1.3), (1.4) we can replace $P(x, \xi)$ by $A(x, \xi)$ and $F_\varepsilon(x)$ by $G_\varepsilon(x)$. Since the limit of $\varepsilon A(M_\varepsilon x, \xi_0/\varepsilon^{k+1} + M_\varepsilon^{-1}\xi)$ when $\varepsilon \to 0$ is $a(0, \xi_0)$, it follows that

$$\varepsilon G_\varepsilon(x) \to a(0, \xi_0)U(x), \quad \text{when } \varepsilon \to 0,$$

uniformly on every compact set, which implies that $\underline{\lim}_{\varepsilon\to 0}\|\varepsilon G_\varepsilon\| \geq |a(0, \xi_0)|\|U\|$. If we apply (0.2) to $u_\varepsilon(x) = e^{i\langle x, \xi_0\rangle/\varepsilon^{k+1}}U(M_\varepsilon^{-1}x)$, it follows that $\|\varepsilon G_\varepsilon\|^2 \leq C_1\|\varepsilon F_\varepsilon\|^2 + C_2\|\varepsilon U\|^2$, and when $\varepsilon \to 0$ we obtain

$$|a(0, \xi_0)|^2\int_{\mathbf{R}^n} |U(x)|^2\, dx \leq C_1\int_{\mathbf{R}^n} |D_1 U(x) + icx_1^k U(x)|^2\, dx, \quad \widehat{U} \in C_0^\infty(\mathbf{R}^n).$$

Taking U as the product of a function of x_1 and one of x', and taking $t = c^{1/(k+1)}x_1$ as a new variable instead of x_1, we obtain (1.2).

So far we have followed the arguments of [H], p. IV:169 closely. However, we shall now see that the conclusion in Proposition 1.1 can be strengthened by a slight modification. First note that (1.3), (1.4) can be modified to

$$(1.3)' \qquad P(x,D)\big(e^{i\langle x,\eta\rangle/\varepsilon^{k+1}}U(M_\varepsilon^{-1}(x-y))\big) = e^{i\langle x,\eta\rangle/\varepsilon^{k+1}}F_\varepsilon(M_\varepsilon^{-1}(x-y)),$$

$$(1.4)' \qquad F_\varepsilon(x) = (2\pi)^{-n}\int e^{i\langle x,\xi\rangle}P(M_\varepsilon x + y, \eta/\varepsilon^{k+1} + M_\varepsilon^{-1}\xi)\widehat{U}(\xi)\,d\xi,$$

where $\eta = (0,\eta')$ is close to ξ_0 and y is close to 0. Suppose that we have sequences $\varepsilon_j \to 0$ in $\mathbf{R}$, $x(j) \to 0$ in $\mathbf{R}^n$ and $\xi'(j) \to \xi_0'$ in $\mathbf{R}^{n-1}$ such that with $e_1 = (1,0,\ldots,0) \in \mathbf{R}^n$

$$(1.6) \quad \varepsilon_j^{-k}q(\varepsilon_j t e_1 + x(j), \xi'(j)) = q_j(t) + O(\varepsilon_j)t^{k+1}, \quad q_j(t) = \sum_{\nu=0}^{k}\varepsilon_j^{\nu-k}\partial_{x_1}^\nu q(x(j),\xi'(j))t^\nu/\nu!,$$

has a limit $q_0(t)$ when $j \to \infty$; it is then a polynomial of degree k with leading term ct^k. This requires of course that $\partial_{x_1}^\nu q(x(j),\xi'(j)) = O(\varepsilon_j^{k-\nu})$ when $0 \le \nu \le k$. If only these bounds are assumed we can achieve convergence by passing to a subsequence. Using the positivity of q we obtain as in (1.5) when $\varepsilon_j^{(k+1)/2}|\xi|$, $|M_{\varepsilon_j}x|$ and ε_j are small enough

$$(1.5)' \quad \varepsilon_j^{-k}q(M_{\varepsilon_j}x + x(j), \varepsilon_j^{(k+1)/2}\xi' + \xi'(j))$$
$$= q_j(x_1) + O(\varepsilon_j|x_1|^{k+1} + \varepsilon_j^{\frac{1}{2}}|x_1|^{k/2}(|\xi'| + |x'|) + \varepsilon_j(|\xi'|^2 + |x'|^2)).$$

Making the small translations $x(j)$ and modifying the frequency ξ_0' to $\xi'(j)$ we obtain with obvious changes of the proof of Proposition 1.1 that (0.2) implies

$$|a(0,\xi_0)|^2\int_{\mathbf{R}^n}|U(x)|^2\,dx \le C_1\int_{\mathbf{R}^n}|D_1U(x) + iq_0(x_1)U(x)|^2\,dx, \quad U \in C_0^\infty(\mathbf{R}^n).$$

As before this implies with $t = c^{1/(k+1)}x_1$ and $\tilde{q}_0(t) = c^{-1/(k+1)}q(tc^{-1/(k+1)})$ that

$$(1.2)' \quad |a(0,\xi_0)|^2\int_{\mathbf{R}}|U(t)|^2\,dt \le c^{2/(k+1)}C_1\int_{\mathbf{R}}|U'(t) - \tilde{q}_0(t)U(t)|^2\,dt, \quad U \in C_0^\infty(\mathbf{R}).$$

The leading term of the polynomial $\tilde{q}_0(t)$ is t^k. We shall now determine which families of such limit polynomials that can occur and also discuss the invariance properties of the family which we have defined using a special local form of the principal symbol at the microsubelliptic point γ.

If $\varepsilon_j \to 0$, $x(j) \to 0$ and $\xi'(j) \to \xi_0'$ as in (1.6) and

$$(1.7) \qquad \sum_{\nu=0}^{k}\varepsilon_j^{\nu-k}\partial_{x_1}^\nu q(x(j),\xi'(j))t^\nu/\nu! \to q_0(t) \quad \text{when } j \to \infty,$$

then replacing ε_j by ε_j/ϱ where $\varrho > 0$ gives the limit

$$(1.8) \qquad \varrho^k q_0(t/\varrho) = (T_\varrho q_0)(t),$$

where T_ϱ is a one parameter group of linear transformations of polynomials of degree k in one variable preserving the leading term. The limit $T_0 q_0$ when $\varrho \to 0$ just gives the leading term and annihilates the others. Since $q_0(t)$ is the limit of $\varepsilon_j^{-k}q(x(j) + \varepsilon_j t, \xi'(j))$, replacing $x(j)$ by $x(j) + (\varepsilon_j\tau, 0,\ldots,0)$ will give the limit $q_0(t+\tau)$ so the set of limits is also translation invariant. An obvious diagonal procedure shows that it is closed, so we have proved the first part of the following proposition.

PROPOSITION 1.2. *Under the hypotheses in Proposition 1.1 the set of limits* (1.7) *which can be obtained as* $\varepsilon_j \to 0$, $x(j) \to 0$ *and* $\xi'(j) \to \xi'_0$ *is a closed subset of the set of polynomials* $q_0 \geq 0$ *with leading term* ct^k, *and it is translation invariant and invariant under the operators* T_ϱ *defined by* (1.8). *Examples show that all such sets can occur.*

PROOF. Let F be a set of polynomials $Q(t) \geq 0$ with leading term t^k and all coefficients ≤ 1 in absolute value. We shall construct an example $q(x_1, x_2)\xi_2$ such that the set of limits (1.7) with $\xi'_0 = 1$ is equal to the translation and T invariant closure of F. To do so we choose $\psi \in C_0^\infty((-2/5, 2/5))$ with $\psi \geq 0$ and $\psi(0) = 1$. Then the functions $\psi_j(x) = \psi(j^2(x - 1/j))/2^j$, $j = 1, 2, \ldots$, have disjoint supports $\subset (1/j - 2/(5j^2), 1/j + 2/(5j^2))$. Let Q_j, $j = 1, 2, \ldots$ be a sequence in F such that every element in F is a limit of some subsequence. We define $q(x_1, x_2) = T_{\psi_j(x_2)} Q_j(x_1)$ when $\psi_j(x_2) \neq 0$ and $q(x_1, x_2) = x_1^k$ when $\psi_j(x_2) = 0$ for every j. When $\psi_j(x_2) \neq 0$ we have

$$\partial_1^\nu q(0, x_2) = \psi_j(x_2)^{k-\nu} Q_j^{(\nu)}(0)$$

which is equal to $k!$ when $\nu = k$ while for $\nu < k$ all derivatives are bounded by 2^{-j} times a power of j so they tend to 0 when $j \to \infty$. Hence $q(x_1, x_2)$ is in C^∞ and is a nonnegative polynomial in x_1 with leading term x_1^k. When $\psi_j(x_2) \neq 0$ we have

$$\sum_{\nu=0}^k \varepsilon^{\nu-k} \partial_1^\nu q(x_1, x_2) t^\nu / \nu! = \varepsilon^{-k} (T_{\psi_j(x_2)} Q_j)(x_1 + \varepsilon t) = (T_{\psi_j(x_2)/\varepsilon} Q_j)(x_1/\varepsilon + t)$$

which as a polynomial in t is on a translation of an orbit of T generated by a point in F. When $\varepsilon = \psi_j(x_2)$ and $x_1 = 0$ it is equal to $Q_j(t)$. Hence the set of limits obtained when $x_1, x_2, \varepsilon \to 0$ is the closed translation and T invariant set generated by F, which completes the proof.

When $p(x, \xi) = \xi_1 + iq(x, \xi')$ at $\gamma = (0, \xi_0)$, $\xi_0 = (0, \xi'_0) \neq 0$, and (1.1) is fulfilled, $q \geq 0$, it follows from [H, Prop. 27.2.1] that if b is a smooth complex valued function at $\gamma \in T^*(X)$ then

$$(1.9) \qquad H_{\operatorname{Re} bp}^j \operatorname{Im}(bp) = \begin{cases} 0 & \text{when } j < k \\ k! c |b|^2 (\operatorname{Re} b)^{k-1} & \text{when } j = k \end{cases} \quad \text{at } \gamma,$$

so c is the maximum of the repeated Poisson brackets $H_{\operatorname{Re} bp}^k \operatorname{Im} bp / k!$ with $k + 1$ factors when $|b(\gamma)| = 1$. If k is even and for some b we have $H_{\operatorname{Re} bp}^j \operatorname{Im} bp = 0$ for $j < k$ but $H_{\operatorname{Re} bp}^k \operatorname{Im} bp \neq 0$ at $\gamma \in T^*(X)$ where p is the principal symbol, of order 1, of a psudodifferential operator in the manifold X, then P is microsubelliptic at γ with loss of $k/(k+1)$ derivatives and can be brought to the form in Proposition 1.1 by multiplication with an elliptic pseudodifferential operator and conjugation by a Fourier integral operator. (See Section 27.1 in [H].) In addition to the invariant description of the coefficient c in $(1.2)'$ given in (1.9), we shall now give an analogous description of the polynomial limits in (1.7).

Under the conditions of Proposition 1.1 we have $H_{\operatorname{Re} p} = \partial/\partial x_1$ so the polynomial q_0 in (1.7) is the limit of $\varepsilon^{-k} \operatorname{Im} p(\exp(\varepsilon t H_{\operatorname{Re} p})(x, \xi))$ when $\varepsilon = \varepsilon_j \to 0$, $x = x(j) \to 0$ and $\xi = (0, \xi'(j)) \to \xi_0 = (0, \xi'_0)$. Since $\operatorname{Im} p$ does not depend on ξ_1 nothing would change if we allow $\xi = \xi(j) \to \xi_0$ without insisting that the first coordinate is 0. This description is symplectically invariant but we must also prove that the same limit set is obtained if $p(x, \xi)$ is replaced by $b(x, \xi)p(x, \xi)$ where $b(0, \xi_0) = 1$. This can be done following the proof of

Proposition 27.2.1 in [H]. With $m_1 = 1$, $\mu_1 = k$ and $m_j = \mu_j = (k+1)/2$ when $1 < j \le n$ we have

$$(1.10) \qquad D_\xi^\alpha D_x^\beta p(x(j), \xi(j)) = O\big(\varepsilon_j^{k - \langle m, \beta \rangle - \langle \mu, \alpha \rangle}\big), \quad \alpha + \beta \ne 0.$$

This is trivial if $\alpha_1 \ne 0$ so we may assume that $\alpha_1 = 0$ and replace p by q. The existence of the limit (1.7) gives (1.10) when $\alpha' + \beta' = 0$ and in view of the positivity of q also that

$$D_\xi^\alpha D_x^\beta q(x(j), \xi'(j)) = O\big(\varepsilon_j^{k/2 - \beta_1}\big) = \varepsilon_j^{\frac{1}{2}} O\big(\varepsilon_j^{k - \beta_1 - (k+1)/2}\big)$$

if $|\alpha' + \beta'| = 1$, $\alpha_1 = 0$, which is better than (1.10) by a factor $\varepsilon_j^{\frac{1}{2}}$. When $|\alpha' + \beta'| \ge 2$ the trivial bound is better than (1.10) by a factor ε_j. Repeated Poisson brackets of $\nu + 1$ factors ξ_1 and $q(x, \xi')$ with bounds like (1.10) are $O\big(\varepsilon_j^{k - \nu}\big)$ and go to 0 when $j \to \infty$ if $\nu < k$. (Cf. [H, Lemma 27.1.10].) When $\nu = k$ they are bounded and the terms where derivatives with respect to x', ξ' occur are $O(\varepsilon_j^{\frac{1}{2}})$. If $b = b_1 + i b_2$ and $b = 1$ at $(0, \xi_0)$ it follows that

$$(1.11) \qquad \sum_{\nu \le k} \varepsilon_j^{\nu - k} t^\nu / \nu! \big(H_{b_1 \xi_1 - b_2 q}\big)^\nu (b_1 q + b_2 \xi_1)(x(j), \xi(j))$$

has bounded coefficients and those where derivatives with respect to x', ξ' occur go to 0 when $j \to \infty$. So do terms where some derivative falls on b_1 or b_2 for they involve fewer brackets of ξ_1 and $q(x, \xi')$. The limit of (1.11) is therefore the same as if $b = 1$ identically, so it is equal to the limit q_0 in (1.7).

If we replace $\xi'(j)$ by $\lambda \xi'(j)$ in (1.7) the limit becomes $\lambda q_0(t)$, and the leading term is λc, so the normalization which led from q_0 to $\tilde{q}_0$ in (1.2)' gives when applied to λq_0 the polynomial $(\lambda c)^{-1/(k+1)} \lambda q_0(t(c\lambda)^{-1/(k+1)}) = \lambda^{k/(k+1)} \tilde{q}_0(t\lambda^{-1/(k+1)})$, which is the polynomial $T_{\lambda^{1/(k+1)}} \tilde{q}_0$. Since the limit sets are T invariant it follows that after normalization they are the same at $(0, \lambda \xi_0)$ as at $(0, \xi_0)$, and we have proved:

THEOREM 1.3. *Let P be a first order pseudodifferential operator in the C^∞ manifold X satisfying condition (P), and assume that for some $\gamma \in T^*(X) \setminus 0$ and $b \in \mathbf{C}$*

$$(H_{\mathrm{Re}\, bp})^j \, \mathrm{Im}\, bp(\gamma) = 0 \quad when \ j < k, \qquad (H_{\mathrm{Re}\, bp})^k \, \mathrm{Im}\, bp(\gamma) \ne 0.$$

Then P is microsubelliptic at γ with loss of $k/(k+1)$ derivatives and there is an associated invariantly defined set $\mathcal{Q}(\gamma)$ of nonnegative polynomials on $\mathbf{R}$ with leading term $c(\gamma)x^k$, k even and

$$(1.12) \qquad c(\gamma) = \max_{|b|=1} (H_{\mathrm{Re}\, bp})^k \, \mathrm{Im}\, bp(\gamma)/k!,$$

such that $\mathcal{Q}(\gamma)$ is closed and invariant under translations and under the operators T_ϱ defined by (1.8). We have $\mathcal{Q}(\lambda\gamma) = \lambda\mathcal{Q}(\gamma)$ if $\lambda > 0$, and $\mathcal{Q}(\gamma)$ consists of all limits of $\varepsilon^{-k} \, \mathrm{Im}\, bp(\exp(\varepsilon t H_{\mathrm{Re}\, bp}(\gamma_\varepsilon)))$ when $\varepsilon \to 0$ and $\gamma_\varepsilon \to \gamma$, $|b| = 1$, and the maximum in (1.12) is attained. If (0.2) is valid for a pseudodifferential operator A of order $1/(k+1)$ with principal symbol a, then

$$(1.13) \qquad |a(\gamma)|^2 \int_{\mathbf{R}} |U(t)|^2 \, dt \le C_1 \int_{\mathbf{R}} |U'(t) - q(t)U(t)|^2 \, dt, \quad U \in C_0^\infty(\mathbf{R}), \ q \in \mathcal{Q}(\gamma).$$

In Section 2 we shall study bounds for ordinary differential operators of the form (1.13), and in Section 3 we shall prove that (0.2) is valid with C_1 replaced by $C_1(1 + \delta)$ and a constant $C_2(\delta)$ for every $\delta > 0$, provided that the necessary condition in Theorem 1.3 is fulfilled for all the characteristics over K.

2. The test estimates. Let p be a real valued polynomial in one variable with leading term cx^k where k is an *even* positive integer and $c > 0$. In this section we shall study L^2 estimates for the ordinary differential operator $d/dx - p(x)$ of the form

$$(2.1) \qquad \int_{\mathbf{R}} u(x)^2\, dx \le C \int_{\mathbf{R}} (u'(x) - p(x)u(x))^2\, dx, \quad u \in C_0^\infty(\mathbf{R});$$

the inequality follows automatically for all $u \in \mathcal{S}(\mathbf{R})$. If $f(x) = -u'(x) + p(x)u(x)$ then $f(x)e^{-P(x)} = -d(u(x)e^{-P(x)})/dx$ if $P' = p$, so

$$u(x) = \int_x^\infty e^{P(x)-P(y)} f(y)\, dy.$$

If $f \in C_0^\infty(\mathbf{R})$ then this defines a solution $u \in \mathcal{S}(\mathbf{R})$ of the equation $-u'(x) + p(x)u(x) = f(x)$, for $\exp P(x)$ decreases as $\exp(cx^{k+1}/(k+1))$ as $x \to -\infty$, and

$$\int_{\mathbf{R}} u(x)^2\, dx = \iint_{\mathbf{R}^2} (K^*K)(x,y) f(x) f(y)\, dx\, dy \le C \int_{\mathbf{R}} f(x)^2\, dx,$$

$$K(x,y) = \begin{cases} e^{P(x)-P(y)} & \text{if } x < y \\ 0 & \text{if } x \ge y \end{cases}, \quad (K^*K)(x,y) = \int_{z \le \min(x,y)} e^{2P(z)-P(x)-P(y)} dz,$$

so the best constant $C(p)$ in (2.1) is the norm of the integral operator with positive kernel K^*K. Note that when the extremum is attained then f does not change sign, so we may assume that f and u are nonnegative, and we have

$$(2.2) \qquad -u''(x) + (p(x)^2 + p'(x))u(x) = u(x)/C(p),$$

which shows that u and f are unique apart from a positive factor. Since $C(p)$ is a simple eigenvalue it follows that $C(p)$ is a C^∞ (in fact real analytic) function of p, and so are u and f if normalized so that $\int_{\mathbf{R}} u(x)^2\, dx = 1$ for example. From the equation

$$1/C(p) = \int_{\mathbf{R}} f(x)^2\, dx = \int_{\mathbf{R}} (-u'(x) + p(x)u(x))^2\, dx$$

it follows if we differentiate with respect to (the coefficients of) p that

$$-(dC(p))/C(p)^2 = 2 \int_{\mathbf{R}} f(x)u(x)dp(x)\, dx, \quad \text{that is,}$$

$$(2.3) \qquad dC(p) = -2C(p)^2 \int_{\mathbf{R}} f(x)u(x)dp(x)\, dx,$$

where $dp(x)$ is a polynomial of degree k.

Since $e^{P(x)-P(y)}$ is a logarithmically convex function of p and such functions form a convex cone, it follows that $\log C(p)$ is a convex function of p, and it is obvious that $C(p_1) \ge C(p_2)$ if $p_1 \le p_2$. Moreover, $\int u(x)^2\, dx \le C(p_1) \int (-u'(x) + p_2(x)u(x))^2\, dx$ if $p_1 \le p_2$ just in some interval containing the support of u. Replacing $u(x)$ by a translation $u(x + \xi)$ in (2.1) gives $C(p) = C(p(\cdot - \xi))$, and replacing $u(x)$ by $u(x\varrho)$ shows that $C(p) = \varrho^{-2}C(\varrho^{-1}p(\cdot/\varrho))$. Hence $C(\lambda p) = \varrho^{-2}C((\lambda/\varrho)p(\cdot/\varrho)) = \lambda^{-2/(k+1)}C(T_\varrho p)$ if $\varrho^{k+1} = \lambda$ and T_ϱ is defined by (1.8). The condition (1.13) can be written

$$|a(\gamma)|^2 \sup_{p \in \mathcal{Q}(\gamma)} C(p) \le C_1.$$

Since $\mathcal{Q}(\lambda\gamma) = \lambda\mathcal{Q}(\gamma)$ and $\mathcal{Q}(\gamma)$ is T invariant we have proved that the left-hand side is homogeneous of degree 0 in γ as expected.

As just observed, $\Phi(c_1, \ldots, c_k) = \log C(p)$, $p(t) = t^k + \sum_1^k c_j t^{k-j}$, is a convex function and so is $\log C(p)$ if $p(t) = \sum_0^k c_j t^{k-j}$ when $c_0 > 0$. Then we have $C(p) = c_0^{-2/(k+1)} C(\tilde{p})$ if $\tilde{p} = t^k + \sum_1^k x_j t^{k-j}$ where $x_j = c_j c_0^{j/(k+1)-1}$, so $\Psi(c_0, \ldots, c_k) = -(2/(k+1)) \log c_0 + \Phi(x)$ is also convex. This gives additional information on Φ, for

$$c_0^2 \partial^2 \Psi/\partial c_0^2 = 2/(k+1) + \sum_{j,l} \Phi_{jl}(x)(j/(k+1)-1)(l/(k+1)-1)x_j x_l$$

$$+ \sum_j \Phi_j(x)(j/(k+1)-2)(j/(k+1)-1)x_j,$$

$$c_0^2 \partial^2 \Psi/\partial c_0 \partial c_l = c_0^{l/(k+1)}\left(\Phi_l(x)(l/(k+1)-1) + \sum_j \Phi_{jl}(x)(j/(k+1)-1)x_j\right),$$

$$c_0^2 \partial^2 \Psi/\partial c_j \partial c_l = c_0^{(j+l)/(k+1)} \Phi_{jl}(x).$$

Here $j, l = 1, \ldots, k$ and $\Phi_j(x) = \partial\Phi(x)/\partial x_j$, $\Phi_{jl}(x) = \partial^2\Phi(x)/\partial x_j \partial x_l$. That Ψ is convex means that

$$t_0^2\left(2/(k+1) + \sum_{j,l} \Phi_{jl}(x)(j/(k+1)-1)(l/(k+1)-1)x_j x_l\right)$$

$$+ 2t_0 \sum_l t_l\left(\Phi_l(x)(l/(k+1)-1) + \sum_j \Phi_{jl}(x)(j/(k+1)-1)x_j\right) + \sum_{j,l} t_j t_l \Phi_{jl}(x) \geq 0$$

for all $t \in \mathbf{R}^{1+k}$. The coefficient of $\Phi_{jl}(x)$ is

$$t_0^2(j/(k+1)-1)(l/(k+1)-1)x_j x_l + 2(j/(k+1)-1)x_j t_0 t_l + t_j t_l = \tau_j \tau_l$$

where $\tau_j = t_j + (j/(k+1)-1)x_j t_0$. Hence the convexity of Ψ means that

$$t_0^2\left(2/(k+1) - \sum_j \Phi_j(x)(j/(k+1)-1)(j/(k+1))x_j\right)$$

$$+ 2t_0 \sum_j \tau_j \Phi_j(j/(k+1)-1) + \sum_{j,l} \Phi_{jl}(x)\tau_j \tau_l \geq 0.$$

With the notation $\gamma_j = j(k+1-j)$, $j = 1, \ldots, k$ the positivity of the coefficient of t_0^2 means that

$$\partial\Phi(e^{\gamma_1 \tau}x_1, \ldots, e^{\gamma_k \tau}x_k)/\partial\tau = \sum \gamma_j e^{\gamma_j \tau}x_j \Phi_j(e^{\gamma_1 \tau}x_1, \ldots, e^{\gamma_k \tau}x_k) \geq -2(k+1),$$

in other words, with $e^\tau = T$, that $C(t^k + c_1 t^{k-1}T^{\gamma_1} + \cdots + c_k T^{\gamma_k})T^{2(k+1)}$ is an increasing function of $T > 0$. If for given $c_1, \ldots, c_k$ we define $T = \max |c_j|^{1/\gamma_j}$ then there is a positive lower bound for $C(p)$ when $p = t^k + c_1 T^{-\gamma_1} t^{k-1} + \cdots + c_k T^{-\gamma_k}$. Hence there is a positive lower bound for $C(t^k + c_1 t^{k-1} + \cdots + c_k)T^{2(k+1)}$ if $T \geq 1$. The convexity of Ψ gives more information than that but we shall not pursue this matter further.

With the notation $\mu = 1/\sqrt{C(p)}$ the second order differential equation (2.2) is equivalent to the first order system

$$f(x) = -u'(x) + p(x)u(x), \quad f'(x) + p(x)f(x) = \mu^2 u(x), \quad \text{that is,}$$

(2.4) $$u'(x) = p(x)u(x) - f(x), \quad f'(x) = -p(x)f(x) + \mu^2 u(x),$$

so this system has apart from a constant factor a unique solution with $u, f \in \mathcal{S}(\mathbf{R})$ nonnegative, and $\|f\| = \mu\|u\|$. In fact, $f > 0$ as an eigenfunction of K^*K, hence $u = Kf > 0$ too. With the notation $\check{v}(x) = v(-x)$ it follows that

$$\check{f}'(x) = \check{p}(x)\check{f}(x) - \mu^2\check{u}(x), \quad \check{u}'(x) = -\check{p}(x)\check{u}(x) + \check{f}(x).$$

If we set $U(x) = \check{f}(x)/\mu$, $F(x) = \mu\check{u}(x)$ this means that

$$U'(x) = \check{p}(x)U(x) - F(x), \quad F'(x) = -\check{p}(x)F(x) + \mu^2 U(x),$$

and we can conclude that $\mu^2 = 1/C(\check{p})$, and that U and F are the solutions corresponding to $\check{p}$. Since $C(\check{p}) = C(p)$ the convexity gives that $C((p+\check{p})/2) \le C(p)$. If p is even then $U(x) = u(x)$, $F(x) = f(x)$, that is, $\check{f}(x) = \mu u(x)$, $\check{u}(x) = f(x)/\mu$ so $u(x)f(x)$ is even and $\mu u(0) = f(0) = -u'(0) + p(0)u(0)$, that is, $u'(0) = (p(0) - \mu)u(0)$.

With p not necessarily even we define

$$v_1(x) = 2u(x)f(x), \quad v_2(x) = \mu u(x)^2 - f(x)^2/\mu, \quad v_3(x) = \mu u(x)^2 + f(x)^2/\mu,$$

and obtain $v_1(x) = \sqrt{v_3(x)^2 - v_2(x)^2}$,

$$(2.5) \qquad v_1'(x) = 2\mu v_2(x), \quad v_2'(x) = 2p(x)v_3(x) - 2\mu v_1(x), \quad v_3'(x) = 2p(x)v_2(x).$$

The vector field in the right hand side is a linear combination of the tangential vector fields $v_2\partial/\partial v_1 - v_1\partial/\partial v_2$ and $v_3\partial/\partial v_2 + v_2\partial/\partial v_3$ of the cone where $v_1^2 + v_2^2 - v_3^2 = 0$. Conversely, if we have a solution of (2.5) with $v_3(0) > 0$ and $v_1 > 0$, then

$$u(x) = \left((v_3(x) + v_2(x))/(2\mu)\right)^{\frac{1}{2}}, \quad f(x) = \left((v_3(x) - v_2(x))\mu/2\right)^{\frac{1}{2}}$$

satisfies (2.4), for $v_1^2 + v_2^2 = v_3^2$ so u and f are positive C^∞ functions with $2uf = v_1$ and

$$2\mu u u' = \tfrac{1}{2}(v_3' + v_2') = p(v_3 + v_2) - \mu v_1 = 2p\mu u^2 - 2\mu u f,$$
$$2ff'/\mu = \tfrac{1}{2}(v_3' - v_2') = p(v_2 - v_3) + \mu v_1 = -2pf^2/\mu + 2\mu f u,$$

which implies (2.4). If p is even then $v_2(x) = \check{f}(x)^2/\mu - \mu\check{u}(x)^2$, $v_3(x) = \check{f}(x)^2/\mu + \mu\check{u}(x)^2$, so v_2 is odd and v_3 is even, $v_3(0) = v_1(0)$, $v_2(0) = 0$, which allows integration of (2.5) if μ is known. It is then convenient to set $W(x) = e^{2i\mu x}(v_1(x) + iv_2(x))$, which gives the equivalent equations

$$(2.5)' \qquad W'(x) = 2ip(x)v_3(x)e^{2i\mu x}, \quad v_3'(x) = 2p(x)\operatorname{Im}(W(x)e^{-2i\mu x}).$$

Assume now that $p(x) = x^k$ where k is even, and define $x_0 > 0$ so that $p(x_0) = \mu$. For small positive x we have $v_2(x) < 0$, and $d(v_1(x) - v_3(x))/dx = 2(\mu - p(x))v_2(x) < 0$ if $p(x) < \mu$ and $v_2(x) < 0$. Since $v_2(x) = 0$ would imply $v_1(x) = v_3(x)$ it follows that $v_2(x) < 0$ and $v_1(x) - v_3(x) < 0$ when $0 < x \le x_0$. When $x \ge x_0$ we have $v_2'(x) \ge 2\mu v_3(x) - 2\mu v_1(x) > 0$ so v_2 is increasing and $\to 0$ at infinity, hence $v_2(x) < 0$ when $x > 0$. When $0 < x \le x_0$ then

$$v_2''(x) = 2p'(x)v_3(x) + 2p(x)v_3'(x) - 2\mu v_1'(x) = 2p'(x)v_3(x) + 4(p(x)^2 - \mu^2)v_2(x) > 0,$$

so v_2 is convex in $[0, x_2]$ for some $x_2 > x_0$ and has a unique minimum point $x_1 \in (0, x_0)$. Thus v_2 is decreasing in $[0, x_1]$ and increasing in $[x_1, \infty)$, and v_1, v_3 are concave in $[0, x_1]$, convex in $[x_1, \infty)$ and decreasing in $[0, \infty)$. We have

$$v_1(0) = -2\mu \int_0^\infty v_2(x)\, dx = \mu \int_{-\infty}^\infty |v_2(x)|\, dx,$$

$$v_3(0) = -2 \int_0^\infty p(x)v_2(x)\, dx = \int_{-\infty}^\infty p(x)|v_2(x)|\, dx,$$

(2.6)

$$v_1(x) = 2\mu \int_x^\infty |v_2(y)|\, dy, \quad \int_0^\infty v_1(x)\, dx = 2\mu \int_0^\infty y|v_2(y)|\, dy,$$

$$\int_0^\infty v_3(x)\, dx \le \int_0^\infty (v_1(x) + |v_2(x)|)\, dx \le \int_0^\infty (1 + 2\mu x)|v_2(x)|\, dx.$$

The equations $(2.5)'$ with $p(x) = x^k$ and initial data $W(0) = 1, v_3(0) = 1$ can be integrated by setting

$$W(x) = \sum_{\nu=0}^\infty W^\nu(x), \quad v_3(x) = \sum_{\nu=0}^\infty v_3^\nu(x); \quad W^0 = 1, \ v_3^0 = 1, \ W^\nu(0) = v_3^\nu(0) = 0, \ \nu \ge 1,$$

$$dW^{\nu+1}(x)/dx = 2ip(x)e^{2i\mu x}v_3^\nu(x), \quad dv_3^{\nu+1}(x) = 2p(x)\operatorname{Im}(e^{-2i\mu x}W^\nu(x)), \ \nu \ge 0.$$

These equations imply that $d(|W_{\nu+1}(x)|^2 + v_3^{\nu+1}(x)^2)^{\frac{1}{2}}/dx \le (|W^\nu(x)|^2 + v_3^\nu(x)^2)^{\frac{1}{2}}$, hence by induction

$$(|W^\nu(x)|^2 + v_3^\nu(x)^2)^{\frac{1}{2}} \le \sqrt{2}(2z)^\nu/\nu!, \quad z = \int_0^x p(t)\, dt = x^{k+1}/(k+1), \ \nu \ge 0, x \ge 0,$$

so the expansion converges. Taking only the first two terms one obtains when z is small and k is large

$$v_1(x) = \cos(2\mu x) + 4\mu z x/(k+2) + O(z/(k+1)^2) + O(z^2),$$

$$v_2(x) = -\sin(2\mu x) + 2z + O(z/(k+1)^2) + O(z^2),$$

$$v_3(x) = 1 + 2z\cos(2\mu x) + 2z\sin(2\mu x)/(k+2) + O(z/(k+1)^2) + O(z^2).$$

The error terms are very large when $x > 1$ and will prevent v_1 from changing sign when $2\mu x = \pi/2$, which suggests that μ is close to $\pi/4$ when k is large. This will be proved after the following elementary lemma which takes care of the interval $(-1, 1)$ where $p(x) = x^k$ is small.

LEMMA 2.1. *If $u \in C^1([\alpha, \beta])$, $\alpha < \beta$, and $u(\beta) = 0$ then*

$$\int_\alpha^\beta u(x)^2\, dx \le (2(\beta - \alpha)/\pi)^2 \int_\alpha^\beta u'(x)^2\, dx,$$

with equality only if u is a multiple of $\sin(\pi(x - \beta)/(2(\beta - \alpha)))$.

PROOF. If we set $v(x) = u(\beta + x(\alpha - \beta))$ then $v \in C^1([0, 1])$, $v'(x) = (\alpha - \beta)u'(\beta + x(\alpha - \beta))$, $v(0) = 0$, and the inequality becomes

$$\int_0^1 v(x)^2\, dx \le (2/\pi)^2 \int_0^1 v'(x)^2\, dx.$$

The eigenfunctions with $v'' + \lambda v = 0$, $v(0) = 0$, and the free boundary condition $v'(1) = 0$ are $v(x) = \sin(\sqrt{\lambda}x)$ where $\sqrt{\lambda} = \pi(\frac{1}{2} + j)$ for some integer j, so the smallest eigenvalue is $(\pi/2)^2$.

PROPOSITION 2.2. *For $p(x) = x^k$ where k is an even positive integer the best constant $C(p)$ in (2.1) satisfies*

$$(2.7) \qquad C(p) = (4/\pi)^2 (1 + (\log k)/k + O(1/k)) = (4/\pi)^2 k^{1/k} + O(1/k).$$

PROOF. Let $u \in C_0^\infty(\mathbf{R})$ and set $f(x) = -u'(x) + x^k u(x)$. Assuming that $\int_{\mathbf{R}} f(x)^2\, dx = 1$ we shall estimate $\int_{\mathbf{R}} u(x)^2\, dx$. Since $f(x)^2 \geq x^{2k} u(x)^2 - x^k du(x)^2/dx$, we obtain

$$\int_a^\infty x^{2k} u(x)^2\, dx \leq \int_a^\infty f(x)^2\, dx, \quad \int_{-\infty}^{-b} (x^{2k} + kx^{k-1}) u(x)^2\, dx \leq \int_{-\infty}^{-b} f(x)^2\, dx + b^k u(-b)^2,$$

if $a > 1$ and $b > 1$. Choosing b so that $b^{k+1} = 2k$ we obtain

$$\int_a^\infty u(x)^2\, dx \leq a^{-2k} \int_a^\infty f(x)^2\, dx, \quad \int_{-\infty}^{-b} u(x)^2\, dx \leq 2b^{-2k} \int_{-\infty}^{-b} f(x)^2\, dx + 2b^{-k} u(-b)^2.$$

Next we shall estimate $u(a)$ and $u(-b)$ by using that

$$u(x) = \int_x^\infty \exp((x^{k+1} - t^{k+1})/(k+1)) f(t)\, dt.$$

If $0 < x < t = x + y$ then $y > 0$ and $(t^{k+1} - x^{k+1})/(k+1) \geq x^k y$, hence

$$\int_a^\infty e^{2(a^{k+1} - t^{k+1})/(k+1)} \leq \int_0^\infty e^{-2a^k y}\, dy = 1/(2a^k), \quad u(a)^2 \leq \int_a^\infty f(t)^2\, dt/(2a^k),$$

$$\int_{-b}^\infty e^{-2(b^{k+1} + t^{k+1})/(k+1)}\, dt \leq a + b + e^{-2(b^{k+1} + a^{k+1})/(k+1)}/(2a^k),$$

$$u(-b)^2 \leq (a + b + 1/(2a^k)) \int_{-b}^\infty f(t)^2\, dt,$$

$$\int_{-\infty}^{-b} u(x)^2\, dx \leq 2b^{-2k} \int_{-\infty}^{-b} f(t)^2\, dt + 2b^{-k}(a + b + 1/(2a^k)) \int_{-b}^\infty f(t)^2\, dt.$$

It remains to estimate u in $[-b, a]$, where we use that $|u(x)| \leq |u(a)| + V(x)$ where $V(a) = 0$ and $V'(x) = -|f(x)|$, for $u(x) = u(a) + \int_x^a \exp((x^{k+1} - t^{k+1})/(k+1)) f(t)\, dt$. Hence by Lemma 2.1

$$\int_{-b}^a V(x)^2\, dx \leq (2(a+b)/\pi)^2 \int_{-b}^a f(x)^2\, dx,$$

$$\left(\int_{-b}^a u(x)^2\, dx \right)^{\frac{1}{2}} \leq (a+b)\left((2/\pi)\left(\int_{-b}^a f(t)^2\, dt \right)^{\frac{1}{2}} + \left(\int_a^\infty f(t)^2\, dt \right)^{\frac{1}{2}}/\sqrt{2a^k} \right)$$

$$\int_{-b}^a u(x)^2\, dx \leq (a+b)^2 ((2/\pi)^2 + 1/(2a^k)) \int_{-b}^\infty f(x)^2\, dx.$$

If we define a by $a^k = k$ we have $a = 1 + (\log k)/k + O((\log k)/k)^2$ and $b = 1 + \log(2k)/k + O((\log k)/k)^2$, so it follows that

$$\int_{-\infty}^\infty u(x)^2\, dx \leq (4/\pi)^2 (1 + (\log k)/k + O(1/k)),$$

which is the upper bound in (2.7)

To prove the lower bound we set as suggested by the proof of Lemma 2.1

$$u(x) = \begin{cases} 0, & \text{if } x \geq 1, \\ \sin(\pi(x-1)/4)), & \text{if } -1 \leq x \leq 1, \\ -\exp((x^{k+1}+1)/(k+1)), & \text{if } x \leq -1. \end{cases}$$

Then u is Lipschitz continuous and $f(x) = -u'(x) + x^k u(x)$ has its support in $[-1, 1]$,

$$\int_{\mathbf{R}} f(x)^2 \, dx = \int_{-1}^{1} \left(\tfrac{\pi}{4} \cos(\pi(x-1)/4) - x^k \sin(\pi(x-1)/4) \right)^2 dx$$

$$= (\tfrac{\pi}{4})^2 - \tfrac{\pi}{4} \int_{-1}^{1} x^k \sin(\pi(x-1)/2) \, dx + \tfrac{1}{2} \int_{-1}^{1} x^{2k}(1 - \sin(\pi x/2)) \, dx$$

$$= (\tfrac{\pi}{4})^2 + \tfrac{\pi}{2} \int_{0}^{1} x^k \cos(\pi x/2) \, dx + \int_{0}^{1} x^{2k} \, dx < (\tfrac{\pi}{4})^2 + (\tfrac{\pi}{2})^2/((k+1)(k+2)) + 1/(2k+1).$$

We have $\int_{-1}^{1} u(x)^2 \, dx = 1$, and the integral from $-\infty$ to 0 is

$$\int_{-\infty}^{-1} e^{2(1+x^{k+1})/(k+1)} \, dx = (e^2/2)^{1/(k+1)}(k+1)^{-k/(k+1)} \int_{2/(k+1)}^{\infty} e^{-\xi} \xi^{-k/(k+1)} \, d\xi.$$

The integral from 1 to ∞ is < 1, and since $|e^{-\xi} - 1| \leq \xi$ the integral over $[2/(k+1), 1]$ differs at most by 1 from the integral of $\xi^{-k/(k+1)}$ which is

$$(k+1)(1 - (2/(k+1))^{1/(k+1)}) = (k+1)^{k/(k+1)}((k+1)^{1/(k+1)} - 2^{1/(k+1)}).$$

Altogether we obtain

$$1 + (\log(k+1))/(k+1) + O(1/k) \leq C(x^k)((\pi/4)^2 + O(1/k)),$$

which proves the lower bound in (2.7).

EXAMPLE. Numerical calculation gives $C(x^2) \approx 1.7789$ and $C(x^4) \approx 1.9998$, to be compared with the limit $(4/\pi)^2 \approx 1.6211$ for $C(x^k)$ when $k \to \infty$.

To study $C(p)$ when p has terms of lower order we need two lemmas.

LEMMA 2.3. *If p is a real valued polynomial with leading term x^k then*

(2.8) $$m(\{x; |p(x)| < a^k\}) \leq 4a2^{-1/k}, \quad a > 0.$$

If $p \geq 0$, thus k is even, then

(2.9) $$m(\{x; p(x) < a^k\}| \leq 4a2^{-2/k}, \quad a > 0.$$

PROOF. Set $E_a = \{x; |p(x)| < a^k\}$. If the zeros of p are replaced by their real parts then p does not increase on $\mathbf{R}$, so we may assume that all the zeros are real. The set E_a consists of at most k open intervals where $|p(x)| = a^k$ at the end points. Suppose that $[\alpha, \beta] \subset \complement E_a$, that $\alpha < \beta$, and that there are zeros of p both to the left of α and to the right of β, and write $p(x) = p_-(x)p_+(x)$ where $p_{\pm}(x) = \prod(x - x_j)$ with the product taken over the zeros of p to the right (left) of $[\alpha, \beta]$, and let $\tilde{p}(x) = p_-(x)p_+(x + \beta - \alpha)$. When $x < \alpha$ then $x < x + \beta - \alpha < \beta$, hence $|p_+(x + \beta - \alpha)| < |p_+(x)|$, so $|\tilde{p}(x)| < |p(x)| < a^k$ if $|p(x)| < a^k$. If $x > \alpha$ then $|p_-(x)| < |p_-(x + \beta - \alpha)|$, so $|\tilde{p}(x)| < |p(x + \beta - \alpha)| < a^k$, if $|p(x + \beta - \alpha)| < a^k$, hence $m(\{x; |\tilde{p}(x)| < a^k\}) \geq m(\{x; |p(x)| < a^k\})$. Repeating the argument we obtain a polynomial p_1 with leading term x^k and an interval of length $m(E_a)$ where $|p_1| < a^k$. By a translation we can make it equal to the interval $(-\nu, \nu)$ where $2\nu = m(E_a)$. Then $|p_1(\nu x)|/a^k \leq 1$ when $|x| < 1$, so it follows from Chebyshev's theorem that the leading coefficient ν^k/a^k is at most 2^{k-1}, thus $\nu \leq 2a2^{-1/k}$, that is, $m(E_a) = 2\nu \leq 4a2^{-1/k}$ which proves (2.8). To prove (2.9) we may again assume that all the zeros of p are real. Then $p(x) = q(x)^2$ where q is a real polynomial with leading term $x^{k/2}$. Since $p(x) < a^k$ is equivalent to $|q(x)| < a^{k/2}$ the estimate (2.9) follows from (2.8).

REMARK. If the zeros $x_1, \ldots, x_k$ are all real then $\log|p(x)| = \sum \log|x - x_j|$ so (2.8) states that if $E(x) = \log|x|$ then

$$m(\{x; E * d\mu(x) < \log a\}) \leq 4a2^{-1/k}$$

if $d\mu$ is the sum of Dirac measures with weight $1/k$ at the zeros $x_1, \ldots, x_k$. As a limiting case it follows that

$$m(\{x; E * d\mu(x) < \gamma\}) \leq 4e^\gamma, \quad \gamma \in \mathbf{R},$$

if $d\mu$ is any positive measure on $\mathbf{R}$ with compact support and total mass equal to 1.

We shall also use a small part of the following more detailed information about the level sets of p:

LEMMA 2.4. Let $p(x)$ be a nonnegative polynomial with leading term x^k and only real zeros $x_1 < x_2 < \cdots < x_\nu$, $\nu \leq k/2$. Then the equation $p(x) = a^k$ has for every $a > 0$ exactly one solution $X^+(a) > x_\nu$ and one solution $X^-(a) < x_1$, and $\pm X_\pm(a)$ is a convex function of a. The maximum M_j of $p(x)^{1/k}$ in (x_j, x_{j+1}) is taken at the zero ξ_j of p' there, and when $0 < a < M_j$ the equation $p(x) = a^k$ has precisely two solutions $X_j^-(a) < X_j^+(a)$ in the interval, $\mp X_j^\pm$ is convex, $X_j^\pm(a) - \xi_j \sim \pm\sqrt{2k(a/M_j - 1)p(\xi_j)/p''(\xi_j)}$ when $a \to M_j$ while the derivatives tend to $\mp\infty$. We have $X_j^-(a) \to x_j$, $X_j^+(a) \to x_{j+1}$ when $a \to 0$, and

$$(2.10) \qquad m(\{x; p(x) < a^k\}) = X^+(a) - X^-(a) - \sum_{a < M_j} (X_j^+(a) - X_j^-(a)).$$

The quotient $m(\{x; p(x) < a^k\})/a$ decreases to 2 when a increases from $\max M_j$ to ∞, and the maximum is attained at M_j for some $j = 1, \ldots, \nu - 1$.

PROOF. The equation $p(x) = a^k$ implies $p'(x)dx = ka^{k-1}da$, and x is locally a smooth function of a except at the zeros of p'. Another differentiation gives

$$p'(x)d^2x/da^2 + p''(x)(dx/da)^2 = k(k-1)a^{k-2} = k(k-1)p(x)/a^2,$$
$$k^{-1}a^2p'(x)^3p(x)^{-3}d^2x/da^2 = (k-1)(p'(x)/p(x))^2 - kp''(x)/p(x)$$
$$= -kd(p'(x)/p(x))/dx - (p'(x)/p(x))^2 = k\sum \mu_j/(x - x_j)^2 - \left(\sum \mu_j/(x - x_j)\right)^2$$

where μ_j is the multiplicity of x_j. This is positive by the Cauchy-Schwarz' inequality which proves the convexity. If $p(x) = a^k$ and $b = M_j - a$ and $x - \xi_j$ are small, then

$$a^k = M_j^k + p''(\xi_j)(x - \xi_j)^2/2 + O(x - \xi_j)^3,$$
$$M_j^k - a^k = kbM_j^{k-1} + O(b^2) = -kbp(\xi_j)/M_j + O(b^2),$$

which gives the stated behavior at M_j. If $p(x) = x^k + c_2 x^{k-2} + \ldots$ which can always be attained by a translation eliminating the term of degree $k-1$, then $\pm X^\pm(a) = a - c_2/(ka) + O(a^{-2})$ when $a \to \infty$, and $c_2 = -\sum \mu_j x_j^2 \leq 0$. The derivative is $1 + c_2 k^{-1} a^{-2} + O(a^{-3})$, and by the convexity it follows that $2a \leq X^+(a) - X^-(a) \leq 2a + O(a^{-1})$, and the convexity then gives that $(X^+(a) - X^-(a))/a$ is decreasing when $a > 0$, so $m(E_a)/a$ is decreasing when $a \geq \max M_j$ for then the sum in (2.10) disappears. By (2.10) $m(E_a)$ is a smooth strictly convex function in each interval containing no point M_j but at those points the left derivative is $+\infty$ and the right derivative is finite, so the convexity must break down. The maximum of $m(E_a)/a$ cannot be attained at a point where $m(E_a)$ is a strictly convex function of a which completes the proof.

We are now prepared to study $C(p)$ where p is an arbitrary nonnegative polynomial with leading term x^k. Besides the maximum of $C(p)$ for all such p we would like information on the maximum of $C(p_\varrho)$ for $\varrho > 0$, where as in (1.8)

$$p_\varrho(x) = T_\varrho p(x) = \varrho^k p(x/\varrho).$$

The operators T_ϱ are a one parameter group of linear transformations of the polynomials in question with the limit $p_0(x) = x^k$ when $\varrho \to 0$. If $p(x) = x^k + c_1 x^{k-1} + \cdots + c_k$ then $p(x - c_1/k) = x^k + c_2' x^{k-2} + \ldots$, $c_2' = c_2 - c_1^2(k-1)/(2k)$, and since $C(p)$ is translation invariant we can therefore assume that the coefficient of x^{k-1} in p is equal to 0. Then $p_\varrho(x) = x^k + c_2 \varrho^2 x^{k-2} + c_3 \varrho^3 x^{k-3} + O(\varrho^4)$, and it follows from (2.3) that

$$C(p_\varrho) = C(p_0) - 2C(p_0)^2 c_2 \varrho^2 \int_{\mathbf{R}} f(x)u(x)x^{k-2}\,dx + O(\varrho^4),$$

where $f = -u' + p_0 u \geq 0$, $u \geq 0$ and $\int u^2\,dx = 1 = C(p_0)\int f^2\,dx$ as in (2.3). (Recall that $f(x)u(x)$ is even.) If $x_1, \ldots, x_k$ are the zeros of $p(x)$ then $c_2 = \sum_{i<j} x_i x_j = -\frac{1}{2}\sum_1^k x_j^2$ since $\sum_1^k x_j = 0$. If all the zeros of p are real this means that $-2c_2$ is the variance of the zeros, and we conclude that $C(p_\varrho) > C(p_0)$ then when ϱ is small. Since c_2 is real we have in general $-2c_2 = \sum \mathrm{Re}\, x_j^2 = \sum(\mathrm{Re}\, x_j)^2 - \sum(\mathrm{Im}\, x_j)^2$ with summation over all zeros, repeated according to multiplicity, and this may be ≤ 0. It is then possible that $C(p_\varrho) \leq C(p_0)$ for every ϱ, which is true if $p_\varrho \geq p_0$ for every ϱ.

EXAMPLE. If $k = 4$ and $p(x) = x^4 + c_3 x + c_4$ then $p \geq 0$ if and only if $(c_4/3)^3 \geq (c_3/4)^4$, and with u and f as above, $k = 4$, we have by (2.3) for small (c_3, c_4) and $p_0(x) = x^4$

$$C(p) = C(p_0) - 2C(p_0)^2 \int_{\mathbf{R}} f(x)u(x)(c_3 x + c_4)\,dx + O(c_3^2 + c_4^2)$$

$$= C(p_0) - 2C(p_0)^2 c_4 \int_{\mathbf{R}} f(x)u(x)\,dx + O(c_4^{3/2}),$$

for $f(x)u(x)$ is an even function. Hence $C(p) < C(p_0)$ for small $c_4 > 0$, and we shall prove below (Corollary 2.6) that this is also true when c_4 is large. We have no proof but numerical evidence that this is always true; it suffices to verify this when $p(x) = x^4 + 4\varrho^3 x + 3\varrho^4$ where ϱ has known positive lower and upper bounds. If it is then $C(p) < C(p_0)$ for all nonnegative $p(x) = x^4 + c_2 x^2 + c_3 x + c_4$ with $c_2 \geq 0$. However, for polynomials $p \geq 0$ of degree $k \geq 6$ it is definitely not possible to decide if $\max_\varrho C(T_\varrho p) = C(p_0)$ from the sign of $C(T_\varrho p) - C(T_0 p) = C(T_\varrho p) - C(p_0)$ for small $\varrho > 0$, for let

$$p(x) = x^k + \sum_{j=0}^{k/2-1} c_{k-2j}\varepsilon^{j^2} x^{2j}, \quad p_0(x) = x^k,$$

where $c_{k-2j} = \pm 1$ when $0 < j < k/2$ and $c_k = 1$. Then $p \geq 0$ if ε is small, and if $\varrho = \varepsilon^\nu$, $0 < \nu < k/2$ then

$$T_\varrho p(x) = x^k + \sum_{j=0}^{k/2-1} c_{k-2j}\varepsilon^{j^2+\nu(k-2j)}x^{2j} = x^k + \sum_{j=0}^{k/2-1} c_{k-2j}\varepsilon^{(j-\nu)^2+\nu(k-\nu)}x^{2j}$$

hence $\quad C(T_\varrho p) = C(p_0) - 2C(p_0)^2 c_{k-2\nu}\varepsilon^{\nu(k-\nu)}\left(\int_{\mathbf{R}} f(x)u(x)x^{2\nu} + O(\varepsilon)\right).$

For small ε the sign of $C(T_\varrho p) - C(p_0)$ is equal to the sign of $-c_{k-2\nu}$. Taking $c_{2j} = (-1)^{j+1}$ when $2j < k$ gives an example where $C(T_\varrho p) < C(p_0)$ when $\varrho = \varepsilon^\nu$ and $k/2 - \nu$ is odd, $C(T_\varrho p) > C(p_0)$ when $\varrho = \varepsilon^\nu$ and $k/2 - \nu$ is even.

The maximum of $m(\{x; p(x) < a^k\})/a$ when $a > 0$ does not change if p is replaced by p_ϱ, and the minorant $(X^+(a) - X^-(a))/a$ when $a = \max M_j$, with the notation in Lemma 2.4, does not change either. It might be possible to relate these two quantities to the maximum of $C(p_\varrho)$ when $\varrho > 0$.

PROPOSITION 2.5. *If $\widehat{C}_k$ is the smallest constant such that (2.1) is valid with $C = \widehat{C}_k$ for all polynomials $p \geq 0$ with leading term x^k, then $\lim_{k\to\infty} \widehat{C}_k = (8/\pi)^2$. More precisely,*

$$(2.11) \qquad -2(\log k)/k + O(1/k) \leq \widehat{C}_k(\pi/8)^2 - 1 \leq 4(\log k)/k + O(1/k).$$

If $p_\varrho(x) = \varrho^k p(x/\varrho)$, $\varrho > 0$, then

$$(2.12) \quad \sup_{\varrho>0} C(p_\varrho) \leq \left((2/\pi)\sup_{a>0} a^{-1}m(\{x; p(x) < a^k\})\right)^2(1 + 4(\log k)/k + O(1/k)),$$

$$(2.13) \quad \sup_{\varrho>0} C(p_\varrho) \geq 4\sup_I(|I|/(\pi \sup_I p^{1/k}))^{2k/(k+1)}(1 - 2(\log k)/k + O(1/k)),$$

where I is an arbitrary interval and $|I|$ is the length of I.

PROOF. When $k = 2$ we know that the supremum $\widehat{C}_2$ is taken when $p(x) = x^k$ but when $k > 2$ it is not, so we assume $k \geq 4$ from now on. To prove an upper bound similar to that in (2.7) we let again a be a number slightly larger than 1 to be chosen later on, depending on k. We recall that if $E_a = \{x; p(x) < a^k\}$ then $m(E_a) \leq 4a2^{-2/k}$ by Lemma 2.3, and E_a consists of $\nu \leq k/2$ open intervals $I_1 = (a_1, b_1), \ldots, I_\nu = (a_\nu, b_\nu)$, for $p(x) = a^k$ at the end points. Then $q(x) = \min(p(x), a^k)$ is equal to $p(x)$ in E_a and $q(x) = a^k$ in the complement which consists of the closed intervals $J_0 = (-\infty, a_1], J_1 = [b_1, a_2], \ldots, J_\nu = [b_\nu, \infty)$. Since $q \leq p$ we have $\sup \|u\|/\| - u' + pu\| \leq \sup \|u\|/\| - u' + qu\|$, so it suffices to prove a bound for $\|u\|$ in terms of $\|f\|$ when $f = -u' + qu$ and $f \geq 0$, $u \geq 0$.

Let us first assume that $\mathrm{supp}\, f \subset E_a$. Set $\alpha_j = \sum_{i=1}^j (b_i - a_i)$, $i = 1, \ldots, \nu$, and define

$$U(x) = \begin{cases} u(x + a_1), & \text{if } 0 \leq x \leq \alpha_1 \\ u(x - \alpha_{j-1} + a_j), & \text{if } \alpha_{j-1} < x \leq \alpha_j, \ j = 2, \ldots, \nu, \end{cases}$$

and define F in $[0, \alpha_\nu]$ in the same way with u replaced by f. Then $F = -U' + qU \geq -U'$, except at the points α_j, and $U(\alpha_j + 0) = u(a_{j+1}) \geq u(b_j) = U(\alpha_j - 0)$ since $u(x) = c_j \exp(a^k x)$, $x \in J_j$, $c_\nu = 0$, so it follows that

$$U(x) \leq \int_x^{\alpha_\nu} F(t)\, dt, \quad 0 \leq x \leq \alpha_\nu.$$

We have $\alpha_\nu = m(E_a)$, which can be estimated by Lemma 2.3, and it follows that

$$\int_{E_a} u(x)^2\,dx = \int_0^{\alpha_\nu} U(x)^2\,dx \le (2m(E_a)/\pi)^2 \int_0^{\alpha_\nu} F(x)^2\,dx = (2m(E_a)/\pi)^2\|f\|^2,$$
$$U(x)^2 \le m(E_a)\|F\|^2 = m(E_a)\|f\|^2.$$

If $j < \nu$ then $\int_{J_j} u(x)^2\,dx \le u(a_{j+1})^2/(2a^k) \le m(E_a)\|f\|^2/(2a^k)$, and $u = 0$ in J_ν, so we have proved that

$$\|u\|^2 \le \big((2m(E_a)/\pi)^2 + \tfrac14 km(E_a)/a^k\big)\|f\|^2$$

if $\operatorname{supp} f \subset E_a$.

Now assume that $\operatorname{supp} f \subset \complement E_a$. We have

$$\int_{J_j} f(x)^2\,dx \ge a^{2k}\int_{J_j} u(x)^2\,dx + a^k u(b_j)^2 - a^k u(a_{j+1})^2, \quad 0 \le j \le \nu,$$

where $b_0 = -\infty$ and $a_{\nu+1} = +\infty$. Since $u' = pu \ge 0$ in I_j we have $u(a_j)^2 \le u(b_j)^2$ so adding we obtain $a^k u(b_j)^2 \le \|f\|^2$ for every j, hence

$$a^k \int_{E_a} u(x)^2\,dx \le m(E_a)\|f\|^2, \quad a^{2k}\sum\int_{J_j} u(x)^2\,dx \le \|f\|^2,$$
$$\|u\|^2 \le (m(E_a)a^{-k} + a^{-2k})\|f\|^2.$$

If $f = f_I + f_J$ where $\operatorname{supp} f_I \subset E_a$ and $\operatorname{supp} f_J \subset \complement E_a$ and u, u_I, u_J are the corresponding solutions then

$$\|u_I\|^2 \le (\tfrac14 km(E_a)a^{-k} + (2m(E_a)/\pi)^2)\|f_I\|^2, \quad \|u_J\|^2 \le (m(E_a)a^{-k} + a^{-2k})\|f_J\|^2,$$

and $\|u\| \le \|u_I\| + \|u_J\|$, $\|f\|^2 = \|f_I\|^2 + \|f_J\|^2$, so it follows that

$$\|u\|^2 \le \big((\tfrac14 k + 1)m(E_a)a^{-k} + (2m(E_a)/\pi)^2 + a^{-2k}\big)\|f\|^2$$
$$= a^2\big((2m(E_a)/(a\pi))^2 + (\tfrac14 k + 1)m(E_a)a^{-k-2} + a^{-2k-2}\big)\|f\|^2.$$

If we use that $m(E_a) \le 4a$ by Lemma 2.3 and define $a = k^{2/(k+1)}$, it follows that

$$(2.14) \qquad \|u\|^2 \le k^{4/(k+1)}\big((2m(E_a)/(a\pi))^2 + O(1/k)\big)\|f\|^2, \quad a = k^{2/(k+1)},$$

$$(2.15) \quad \widehat{C}_k \le a^2\big((8/\pi)^2 + (k+4)/k^2 + 1/k^4\big) = (8/\pi)^2(1 + 4(\log k)/k + O(1/k)).$$

With some numerical calculation it is easily seen that $\widehat{C}_k < 30$ for every k. The preceding proof gives a much better bound for a specific choice of p, and we shall discuss below how even better bounds can be obtained when the influence of the intervals J_j on the estimates in E_a is taken into account. However, we shall first establish some lower bounds for $C(p)$.

Let I be the interval $[-b, b]$ and let $M = \sup_I p^{1/k}$. As in the proof of the lower bound in (2.7) we set

$$u(x) = \begin{cases} 0, & \text{if } x \ge b, \\ \sin(\pi(x-b)/(4b)), & \text{if } -b \le x \le b, \\ -\exp(\int_x^{-b} p(t)\,dt), & \text{if } x \le -b. \end{cases}$$

Then u is Lipschitz continuous, and if $f(x) = -u'(x) + p(x)u(x)$ then supp $f \subset [-b, b]$ and

$$\int_{-b}^{b} f(x)^2 \, dx = \int_{-b}^{b} \big((\pi/4b) \cos(\pi(x - b)/(4b)) - p(x) \sin(\pi(x - b)/(4b)) \big)^2 \, dx$$

$$= \pi^2/(16b) - \pi/(4b) \int_{-b}^{b} p(x) \sin(\pi(x - b)/(2b)) \, dx + \tfrac{1}{2} \int_{-b}^{b} p(x)^2 (1 - \sin(\pi x/(2b))) \, dx$$

$$\leq \pi^2/(16b) + M^k + M^{2k}b, \qquad \int_{-b}^{b} u(x)^2 \, dx = b.$$

Hence

$$1 \leq C(p)(\pi^2/(16b^2) + M^k/b + M^{2k}).$$

If $p(x)$ is replaced by $p_\varrho(x) = \varrho^k p(x/\varrho)$ and b is replaced by ϱb then M is replaced by ϱM, so it follows that

$$1 \leq C(p_\varrho)\big(\pi^2/(16\varrho^2 b^2) + \varrho^{k-1} M^k/b + \varrho^{2k} M^{2k}\big).$$

To minimize the sum of the first two terms we shall choose ϱ so that $\pi^2/(8\varrho^3 b^2) = (k - 1)\varrho^{k-2} M^k/b$, that is, $\varrho^{k+1} = \pi^2/(8bM^k(k - 1))$, which gives

$$1 \leq C(p_\varrho)(\pi/4\varrho b)^2 (1 + 2/(k - 1) + 3/(k - 1)^2)$$

$$= C(p_\varrho)\tfrac{1}{4}(\pi M/2b)^{2k/(k+1)} (4(k - 1)/\pi)^{2/(k+1)} (1 + 2/(k - 1) + 3/(k - 1)^2).$$

Since $C(p)$ is invariant under translations of p we have therefore proved that

$$(2.16) \quad \sup_{\varrho > 0} C(p_\varrho) \geq 4 \sup_{I} (|I|/(\pi M_I))^{2k/(k+1)} (\pi/(4k - 4))^{2/(k+1)}/(1 + 2/(k-1) + 3/(k-1)^2),$$

where I is an arbitrary interval on $\mathbf{R}$ with length $|I| < \infty$ and $M_I = \sup_I p^{1/k}$. This implies (2.13). If we take $p(x) = 2^{1-k}(T_k(x) + 1)$, where $T_k(x)$ is now the kth Chebyshev polynomial, and $I = (-1, 1)$, $M_I^k = 2^{2-k}$, the lower bound in (2.11) follows. The quantities on the right-hand sides of (2.12) and (2.13) are invariant under T_ϱ, so (2.14) implies (2.12).

The supremum $\widehat{C}_k$ is attained:

COROLLARY 2.6. *If $p \geq 0$ is a polynomial of degree k and $p(x) - x^k$ is of degree $k - 2$ then $C(p) \leq CZ^{-4/(k-1)}$ if Z is the maximum of $|z|$ when $z \in \mathbf{C}$ and $p(z) = 0$. Hence the set of all such p with $C(p) \geq C(x^k)$ is compact; only they will play a role in our estimates.*

PROOF. If $k = 2$ then $p(x) = x^2 + Z^2$ and

$$\int |-u'(x) + p(x)u(x)|^2 \, dx = \int (|u'(x)|^2 + (p(x)^2 + p'(x))|u(x)|^2) \, dx$$

$$\geq \int |-u'(x) + x^2 u(x)|^2 \, dx + Z^4 \int |u(x)|^2 \, dx \geq (1/C(x^2) + Z^4) \int |u(x)|^2 \, dx,$$

so $C(p) \leq C(x^2)/(1 + Z^4 C(x^2))$ as stated. From now on we assume that $k \geq 4$.

If B is the maximum of $|\operatorname{Im} z|$ when $p(z) = 0$, and $B > 0$, then $p \geq B^2 p_1$ where $p_1 \geq 0$ is of degree $k - 2$ with leading coefficient 1. Hence $C(p) \leq B^{-4/(k-1)} C(T_\varrho p_1) \leq B^{-4/(k-1)} \widehat{C}_{k-2}$, where $\varrho = B^{2/(k-1)}$.

Next assume that p has only real zeros. Let $(\alpha - 2\gamma, \alpha + 2\gamma)$ be the largest interval where $p \neq 0$ but $p = 0$ at the end points. We can then write $p = p_+ p_-$ where the k_+ zeros of p_+ are $\geq \alpha + 2\gamma$ and the k_- zeros of p_- are $\leq \alpha - 2\gamma$. Both k_+ and k_- are even since all zeros have even multiplicity. Choose $\psi_+ \in C^\infty(\mathbf{R})$ equal to 1 in $[1, \infty)$ and 0 in $(-\infty, -1]$ so that $\psi_+^2 + \psi_-^2 = 1$ if $\psi_-(x) = \psi_+(-x)$. If $u \in \mathcal{S}(\mathbf{R})$ and $f = -u' + pu$ we set $u_\pm(x) = \psi_\pm((x - \alpha)/\gamma)u(x)$, $f_\pm(x) = \psi_\pm((x - \alpha)/\gamma)f(x)$. Since $-u'_\pm(x) + p(x)u_\pm(x) = f_\pm(x) - \psi'_\pm((x - \alpha)/\gamma)u(x)/\gamma$ and $p \geq p_\pm \gamma^{k_\mp}$ in the interval containing the support of $u_\pm$, it follows that

$$\int |u_\pm(x)|^2 \, dx \leq \widehat{C}_{k_\pm} \gamma^{-2k_\mp/(k_\pm + 1)} \int |f_\pm(x) - \psi'_\pm((x - \alpha)/\gamma)u(x)/\gamma|^2 \, dx.$$

Since $2k_\mp/(k_\pm + 1) \geq 4/(k - 1)$ adding gives if $\gamma > 1$

$$\int |u(x)|^2 \, dx \leq \max_{2 \leq j \leq k-2} \widehat{C}_j \gamma^{-4/(k-1)} \Big(2 \int |f(x)|^2 \, dx + C \int |u(x)|^2 \, dx/\gamma^2 \Big),$$

if $2(\psi'_+(x)^2 + \psi'_-(x)^2) \leq C$. For large γ it follows that $C(p) \leq C'\gamma^{-4/(k-1)}$. Since the mean value of the zeros is equal to 0 it follows that $Z \leq 2\gamma(k - 2)$ so $C(p)$ is bounded by a constant times $Z^{-4/(k-1)}$ for large Z, hence for all Z by the continuity.

If the zeros of p are not all real we let $\tilde{p}$ be the polynomial obtained when all zeros are replaced by their real parts. Then $\tilde{p} \leq p$, hence $C(p) \leq C(\tilde{p})$ which can be estimated by $C(\max |\operatorname{Re} z_j|)^{-4/(k-1)}$. At the beginning of the proof we also established the bound $C(\max |\operatorname{Im} z_j|)^{-4/(k-1)}$ which completes the proof.

REMARK. From the convexity discussion at the beginning of the section it follows that $C(p) \geq CZ^{-2(k+1)/k}$ for large $|Z|$.

EXAMPLE. All nonnegative polynomials with leading term x^4 are generated by translations and the scaling operators T_ϱ from the polynomials

$$p_{\alpha\beta}(x) = (x^2 - 1)^2 + (\alpha + \beta)(x^2 + 1) + 2(\alpha - \beta)x + \alpha\beta, \quad \alpha \geq 0, \ \beta \geq 0,$$

with the zeros $1 \pm i\sqrt{\alpha}$ and $-1 \pm i\sqrt{\beta}$. The sum of their squares is $2(2 - \alpha - \beta)$. When $\alpha + \beta \geq 2$ we have seen in a previous example that most likely $\max_\varrho C(T_\varrho p_{\alpha\beta}) = C(x^4)$. Since $\log C(T_\varrho p_{\alpha\beta})$ is a convex function of (α, β), the maximum with respect to ϱ is also convex so $\max_\varrho C(T_\varrho p_{\alpha\beta})$ would then be bounded by $\widehat{C}_4^{1-(\alpha+\beta)/2} C(x^4)^{(\alpha+\beta)/2}$ when $\alpha + \beta \leq 2$. In general $\max_\varrho C(T_\varrho p_{\alpha,\beta})$ attains its minimum for fixed $\alpha + \beta$ when $\alpha = \beta$. Numerical calculation gives $\widehat{C}_4 = C(p) \approx 2.4818$, $p(x) \approx (x^2 - 0.3739)^2$, and as already observed $C(x^4) \approx 1.9998$.

In the first part of the proof of (2.12) we ignored completely that the function U had a discontinuity at α_j where $U(\alpha_j - 0) = \kappa U(\alpha_j + 0)$ with $\kappa = \exp(-a^k|J_j|)$. If κ is very small this leads to improvements in view of the following variant of Lemma 2.1, which we only state for the case of one such discontinuity for the sake of simplicity.

LEMMA 2.1'. *Let* $\alpha_1 > 0$, $\alpha_2 > 0$ *and let* $v \in C^1([0, \alpha_1 + \alpha_2])$ *apart from jumps of* v *and* v' *at* α_1 *where* $v(\alpha_1 + 0) = \kappa v(\alpha_1 - 0)$, $0 < \kappa \leq 1$. *If* $v(0) = 0$ *and* $v' = f + (v(\alpha_1 + 0) - v(\alpha_1 - 0))\delta_{\alpha_1}$ *then*

$$(2.17) \qquad \int_0^{\alpha_1 + \alpha_2} v(x)^2 \, dx \leq \omega^{-2} \int_0^{\alpha_1 + \alpha_2} f(x)^2 \, dx,$$

where ω is the solution in $[\pi/(2(\alpha_1 + \alpha_2)), \pi/(2\max(\alpha_1, \alpha_2))]$ of the equation

$$(2.18) \qquad\qquad \arctan(\kappa^2 \tan(\omega\alpha_1)) + \omega\alpha_2 = \pi/2.$$

We have

$$(2.19) \qquad\qquad \omega\max(\alpha_1, \alpha_2) \geq \pi/2 - \arctan\kappa \geq \pi/4.$$

PROOF. If $K(x) = 1$ in $[0, \alpha_1]$ and $K(x) = \kappa$ in $(\alpha_1, \alpha_1 + \alpha_2]$ then $V(x) = v(x)/K(x)$ is a continuous function with derivative $F(x) = f(x)/K(x)$ in the sense of distribution theory so (2.17) means that $\int_0^{\alpha_1 + \alpha_2} V(x)^2 K(x)^2 \, dx \leq \omega^{-2} \int_0^{\alpha_1 + \alpha_2} V'(x)^2 K(x)^2 \, dx$ when $V(0) = 0$, and this implies when the constant is optimal that $V''(x) + \omega^2 V(x) = 0$ for $x \neq \alpha_1$ where V and $K^2 V'$ must be continuous. This gives apart from a constant factor that $v(x) = \sin(\omega x)$ when $0 \leq x < \alpha_1$ and that $v(x) = A\sin(\theta + \omega(x - \alpha_1))$ in $(\alpha_1, \alpha_1 + \alpha_2]$ where $\kappa\sin(\omega\alpha_1) = A\sin\theta$ and $\cos(\omega\alpha_1) = A\kappa\cos\theta$, which gives $\tan\theta = \kappa^2 \tan(\omega\alpha_1)$. The equation (2.18) is the free boundary condition $V'(\alpha_1 + \alpha_2) = 0$.

When $\kappa = 1$ then (2.18) means that $\omega(\alpha_1 + \alpha_2) = \pi/2$ and (2.17) is then equivalent to Lemma 2.1. When κ decreases it is clear that there is a unique solution $\omega(\kappa)$ of (2.18) with $\omega\max(\alpha_1, \alpha_2) < \pi/2$, and $\omega(\kappa)$ increases when $\kappa \downarrow 0$. For fixed α_1, α_2 we have when $\kappa \downarrow 0$

$$\omega(\kappa) = \begin{cases} (\pi/2 - \kappa^2 \tan(\pi\alpha_1/(2\alpha_2)) + O(\kappa^4))/\alpha_2, & \text{if } \alpha_1 < \alpha_2 \\ (\pi/2 - \kappa^2 \tan(\pi\alpha_2/(2\alpha_1)) + O(\kappa^4))/\alpha_1, & \text{if } \alpha_1 > \alpha_2 \\ (\pi/2 - \arctan\kappa)/\alpha, & \text{if } \alpha_1 = \alpha_2 = \alpha. \end{cases}$$

In the last case we have $\tan(\omega\alpha) = 1/\tan(\arctan\kappa) = 1/\kappa$ so the first term in (2.18) is $\arctan\kappa$ and (2.19) follows in this case. Assume now that $\alpha_1 < \alpha_2$. We wish to prove that $\omega(\kappa)\alpha_2 > \pi/2 - \arctan\kappa$, that is, that

$$\arctan(\kappa^2 \tan(\alpha_1\omega)) < \arctan\kappa \quad \text{if } \omega\alpha_2 = \pi/2 - \arctan\kappa.$$

This is true since $\arctan(\kappa^2 \tan(\alpha_2\omega)) = \arctan(\kappa^2/\tan(\arctan\kappa)) = \arctan\kappa$. If instead we assume that $\alpha_1 > \alpha_2$ and $\omega\alpha_1 = \pi/2 - \arctan\kappa$ then

$$\arctan(\kappa^2 \tan(\alpha_1\omega)) + \omega\alpha_2 = \arctan\kappa + \omega\alpha_2 < \pi/2,$$

which completes the proof of (2.19).

When κ is small the constant ω^{-2} in (2.17) is essentially $(2/\pi)^2 \max(\alpha_1^2, \alpha_2^2)$, to be compared with the constant $(2/\pi)^2(\alpha_1 + \alpha_2)^2$ from Lemma 2.1 obtained when the jump condition is disregarded. For small κ this means that there is little interaction between the intervals.

We shall now prove some estimates for the differential operator $-d/dx + p(x)$ which are essentially contained in [H, Section 27.3] but with rather different proofs. They are only required for estimating some error terms so the size of the constants is not as essential as in the results proved so far. Our aim is instead to make the proofs as short as possible.

PROPOSITION 2.7. *There is a constant $\tilde{C}$ such that for all polynomials $p \geq 0$ of degree k*

$$(2.20) \qquad \int_{\mathbf{R}} (p(x)u(x))^2 \, dx \leq \tilde{C}^k \int_{\mathbf{R}} (-u'(x) + p(x)u(x))^2 \, dx, \quad u \in \mathcal{S}(\mathbf{R}).$$

PROOF. If $k = 0$ then $\int_{\mathbf{R}} (-u'(x) + pu(x))^2 \, dx = \int (u'(x)^2 + p^2 u(x)^2) \, dx$ so (2.20) is valid then. We can therefore assume that $k \geq 2$ and that (2.20) is already proved when k is replaced by $k - 2$. Replacing x by ax shows that (2.20) is equivalent to

$$\int_{\mathbf{R}} (ap(ax)v(x))^2 \, dx \leq \widetilde{C}^k \int_{\mathbf{R}} (-v'(x) + ap(ax)v(x))^2 \, dx, \quad v \in \mathcal{S}(\mathbf{R}),$$

where $v(x) = u(ax)$, and we can choose a so that the leading term in $ap(ax)$ becomes x^k. We may therefore assume that $p(x) - x^k$ is of degree $< k$.

At first we assume that $u \in C_0^\infty(I)$, $I = [-T, T]$, where T will be chosen later. Let $B > T$. If $|z| < B$ for all the zeros $z \in \mathbf{C}$ of p then $p(x) \leq (B + T)^k$ when $|x| \leq T$, hence

$$(2.21) \qquad \int (p(x)u(x))^2 \, dx \leq (B + T)^{2k} \int u(x)^2 \, dx.$$

If p has a zero z_0 with $|z_0| \geq B$ then $p(x) = |1 - x/z_0|^2 p_1(x)$ where p_1 is a polynomial of degree $k - 2$ with leading coefficient $|z_0|^2$ and

$$(1 - T/B)^2 p_1(x) \leq p(x) \leq (1 + T/B)^2 p_1(x), \quad |x| < T.$$

If $P'(x) = p(x)$ and $P_1'(x) = (1 - T/B)^2 p_1(x)$, $P(0) = P_1(0) = 0$, and $f(x) = -u'(x) + p(x)u(x)$, then $-d(ue^{-P})/dx = fe^{-P}$ and

$$u(x) = \int_x^T e^{P(x) - P(y)} f(y) \, dy, \quad |u(x)| \leq U(x) = \int_x^T e^{P_1(x) - P_1(y)} |f(y)| \, dy, \quad |x| < T,$$

for $P_1(y) - P_1(x) \leq P(y) - P(x)$ when $-T < x < y < T$. Since p_1 is of degree $k - 2$ and $-U'(x) + (1 - T/B)^2 p_1(x)U(x) = |f(x)|$, it follows from the inductive hypothesis that

$$\int ((1 - T/B)^2 p_1(x)U(x))^2 \, dx \leq \widetilde{C}^{k-2} \int f(x)^2 \, dx, \quad \text{hence}$$

$$(2.22) \qquad \int (p(x)u(x))^2 \, dx \leq \widetilde{C}^{k-2}((B + T)/(B - T))^4 \int f(x)^2 \, dx.$$

From now on we just assume that $u \in C_0^\infty(\mathbf{R}^n)$. Choose $\varphi \in C_0^\infty([-1, 1])$ with $\int \varphi(t)^2 \, dt = 1$. If $\varphi(t) = \cos(\pi t/2)$, $|t| < 1$, then $\int_{-1}^1 \varphi(t)^2 \, dt = 1$ and $\int_{-1}^1 \varphi'(t)^2 \, dt = (\pi/2)^2 < 3$ so we can choose φ as a regularization of this function with $\int \varphi'(t)^2 \, dt < 3$. Set

$$u_t(x) = \varphi(x/T - t)u(x), \quad f_t(x) = \varphi(x/T - t)f(x), \quad F_t(x) = f_t(x) - T^{-1}\varphi'(x/T - t)u(x).$$

Then $-u_t'(x) + p(x)u_t(x) = F_t(x)$, and

$$\int_{\mathbf{R}} \|F_t\|^2 \, dt \leq 2\|f\|^2 + 6T^{-2}\|u\|^2,$$

$$\|pu_t\|^2 \leq (B + T)^{2k}\|u_t\|^2 + \widetilde{C}^{k-2}((B + T)/(B - T))^4\|F_t\|^2,$$

where the second inequality follows from (2.21) and (2.22) applied to a translation of u_t and p. Integration with respect to t gives

$$\|pu\|^2 \leq (B + T)^{2k}\|u\|^2 + \widetilde{C}^{k-2}\big((B + T)/(B - T)\big)^4(2\|f\|^2 + 6\|u\|^2).$$

By Proposition 2.5 $\|u\|^2 \leq \widehat{C}\|f\|^2$ where $\widehat{C} = \sup_j \widehat{C}_j < \infty$, and (2.20) follows if

$$\left((B+T)^{2k} + 6\widetilde{C}^{k-2}((B+T)/(B-T))^4\right)\widehat{C} + 2\widetilde{C}^{k-2}\left((B+T)/(B-T)\right)^4 \leq \widetilde{C}^k.$$

This is true if

$$\left((B+T)^4 + 6((B+T)/(B-T))^4\right)\widehat{C} + 2\left((B+T)/(B-T)\right)^4 \leq \widetilde{C}^2,$$

which implies that $\widetilde{C} > (B+T)^2$. This completes the proof of (2.20). We can for example take $T = 1$ and $B = 3$, $\widetilde{C}^2 = 352\widehat{C} + 32$ but this is far from optimal. For comparison we shall give in Section 4 the best possible estimate when $p(x) = x^k$.

We shall now take a first step toward a converse of Theorem 1.3.

PROPOSITION 2.8. *Let $\mathcal{Q} \subset C^{k+1}(\mathbf{R})$ and assume that*

(i) *if $q \in \mathcal{Q}$ then $q \geq 0$ and $q(\cdot + \tau) \in \mathcal{Q}$ for every $\tau \in \mathbf{R}$,*

(ii) *there is a constant C_0 such that*

$$(2.23) \qquad \sum_0^{k+1} |q^{(\nu)}(t)/\nu!| \leq C_0 \quad \text{when } t \in \mathbf{R}, \ q \in \mathcal{Q}.$$

Let $\mathcal{Q}_0$ be the set of all limits of $\varepsilon^{-k}q(\varepsilon t)$ when $q \in \mathcal{Q}$ and $\varepsilon \to 0$. If $q_0 \in \mathcal{Q}_0$ then q_0 is a polynomial ≥ 0 of degree $\leq k$, and $T_\varrho q_0 \in \mathcal{Q}_0$ when $\varrho > 0$, $T_\varrho q_0(t) = \varrho^k q_0(t/\varrho)$. If

$$(2.24) \qquad \int_{\mathbf{R}} |v(t)|^2 \, dt \leq A(\varepsilon) \int_{\mathbf{R}} |-v'(t) + \varepsilon^{-k}q(t\varepsilon)v(t)|^2 \, dt, \quad v \in \mathcal{S}(\mathbf{R}), \ q \in \mathcal{Q},$$

or equivalently, with $t\varepsilon$ as new variable,

$$(2.24)' \qquad \int_{\mathbf{R}} |v(t)|^2 \, dt \leq A(\varepsilon)\varepsilon^2 \int_{\mathbf{R}} |-v'(t) + \varepsilon^{-k-1}q(t)v(t)|^2 \, dt, \quad v \in \mathcal{S}(\mathbf{R}), \ q \in \mathcal{Q},$$

it follows if $A = \overline{\lim}_{\varepsilon \to 0} A(\varepsilon)$ that

$$(2.25) \qquad \int_{\mathbf{R}} |v(t)|^2 \, dt \leq A \int_{\mathbf{R}} |-v'(t) + q_0(t)v(t)|^2 \, dt, \quad v \in \mathcal{S}(\mathbf{R}), \ q_0 \in \mathcal{Q}_0.$$

Conversely, if (2.25) is valid, $A \geq 0$, and $\delta > 0$ then

$$(2.26) \qquad \int_{\mathbf{R}} |v(t)|^2 \, dt \leq (A+\delta) \int_{\mathbf{R}} |-v'(t) + \varepsilon^{-k}q(t\varepsilon)v(t)|^2 \, dt, \quad v \in \mathcal{S}(\mathbf{R}), \ q \in \mathcal{Q},$$

if $\varepsilon < \varepsilon(\delta)$. Thus $\overline{\lim}_{\varepsilon \to 0} A(\varepsilon) = A$ if $A(\varepsilon)$ and A are the best constants in (2.24) and (2.25). We have $A < \infty$ if and only if

$$(2.27) \qquad \sum_0^k |q^{(\nu)}(t)/\nu!| \geq C_1, \quad t \in \mathbf{R}, \ q \in \mathcal{Q},$$

for some $C_1 > 0$, and $A = 0$ if and only if for some $C_2 > 0$

$$(2.28) \qquad \sum_0^{k-1} |q^{(\nu)}(t)/\nu!| \geq C_2, \quad t \in \mathbf{R}, \ q \in \mathcal{Q}.$$

If (2.27) *is valid then*

$$(2.29) \quad \int_{\mathbf{R}} |\varepsilon^{-k} q(t\varepsilon) v(t)|^2 \, dt \leq 2\widetilde{C}^{k+2} \int_{\mathbf{R}} |-v'(t) + \varepsilon^{-k} q(t\varepsilon) v(t)|^2 \, dt, \quad v \in \mathcal{S}(\mathbf{R}), \ q \in \mathcal{Q},$$

where $\widetilde{C}$ is the constant in Proposition 2.7.

PROOF. If $q_j \in \mathcal{Q}$, $\varepsilon_j \to 0$, and $\varepsilon_j^{-k} q_j(\varepsilon_j t) \to q_0(t)$ when $j \to \infty$, it follows from Taylor's formula and (2.23) that $q_0 \in \mathcal{Q}_0$ is a polynomial of degree $\leq k$, and (2.25) follows from (2.24) when $v \in C_0^\infty(\mathbf{R})$, hence when $v \in \mathcal{S}(\mathbf{R})$. Replacing ε_j by ε_j/ϱ shows that $T_\varrho q_0 \in \mathcal{Q}_0$. If (2.27) is valid then

$$|q^{(k)}(0)/k!| \geq C_1 - \varepsilon \sum_0^{k-1} |q^{(\nu)}(0)\varepsilon^{\nu-k}/\nu!|, \quad 0 < \varepsilon < 1,$$

hence $|q_0^{(k)}(0)/k!| \geq C_1$ when $q_0 \in \mathcal{Q}_0$, and by Proposition 2.5 this implies (2.25) for $A = \widehat{C}_k/C_1^{2/(k+1)}$. On the other hand, if (2.27) is not valid we can choose a sequence $q_j \in \mathcal{Q}$ such that $q_j^{(\nu)}(0)/\nu! \to 0$ when $j \to \infty$ and $\nu \leq k$. If $\varepsilon_j \to 0$ sufficiently slowly then $q_j^{(\nu)}(0)\varepsilon_j^{\nu-k}/\nu! \to 0$ when $\nu \leq k$, so $0 \in \mathcal{Q}_0$ and (2.25) is not valid for any finite A. If (2.28) is valid then $\mathcal{Q}_0$ is empty and (2.25) is valid with $A = 0$. However, if (2.28) is not valid we can again find $q_j \in \mathcal{Q}$ and $\varepsilon_j \to 0$ such that $q_j^{(\nu)}(0)\varepsilon_j^{\nu-k}/\nu! \to 0$ when $j \to \infty$, now for $\nu < k$, so $\mathcal{Q}_0$ is not empty and (2.25) is not valid with $A = 0$.

Assume now that (2.25) is valid with a finite A. We shall first prove (2.26) for small $\varepsilon > 0$ when $v \in C_0^\infty(I)$ where $I = [-T, T]$ is a large finite interval. If (2.26) were not valid then we could find sequences $v_j \in C_0^\infty(I)$, $\varepsilon_j \to 0$, $q_j \in \mathcal{Q}$, such that

$$1 = \int_I |v_j(t)|^2 \, dt > (A + \delta) \int_I |-v_j'(t) + \varepsilon_j^{-k} q_j(t\varepsilon_j) v_j(t)|^2 \, dt.$$

By Taylor's formula and (2.23) we have

$$0 \leq \varepsilon_j^{-k} q_j(t\varepsilon_j) \leq \sum_0^k \varepsilon_j^{\nu-k} q_j^{(\nu)}(0) t^\nu/\nu! + C_0 \varepsilon_j |t|^{k+1},$$

hence $\quad \sum_0^k \varepsilon_j^{\nu-k} q_j^{(\nu)}(0) t^\nu/\nu! + C_0 \varepsilon_j (1 + t^{k+2}) = p_j(t) \geq 0,$

$$(A + \delta) \int_I |-v_j'(t) + p_j(t) v_j(t)|^2 \, dt \leq 1 + O(\varepsilon_j).$$

By (2.20) this implies a bound for $\int_I (|v_j'(t)|^2 + |p_j(t) v_j(t)|^2) \, dt$ so a subsequence of the sequence v_j has a uniform limit v with $\int_I v(t)^2 \, dt = 1$, which implies a bound for the coefficients of p_j. Passing to a subsequence we get a limit for p_j so $\varepsilon_j^{-k} q_j(t\varepsilon_j) = q_0(t) + r_j(t)$ where $q_0 \in \mathcal{Q}_0$ and $r_j \to 0$ uniformly in I when $j \to \infty$. Since

$$A \int_I |-v_j'(t) + q_0(t) v_j(t)|^2 \, dt$$

$$\leq (A + \delta/2) \int_I |-v_j'(t) + (q_0(t) + r_j(t)) v_j(t)|^2 \, dt + (A + 2A^2/\delta) \int_I |r_j(t) v_j(t)|^2 \, dt$$

it follows from (2.25) when j is large that

$$\int_I |v_j(t)|^2\, dt < (A+\delta)\int_I |-v_j'(t) + \varepsilon_j^{-k}q_j(t\varepsilon_j)v_j(t)|^2\, dt,$$

which is a contradiction proving (2.26) when $v \in C_0^\infty(I)$ and ε is sufficiently small.

To prove (2.29) for $v \in C_0^\infty(I)$ we observe as above that if $q \in \mathcal{Q}$ then

$$p(t) = \sum_0^k \varepsilon^{\nu-k}q^{(\nu)}(0)t^\nu/\nu! + C_0\varepsilon(1 + t^{k+2}) \geq 0,$$

so it follows from (2.20) that

$$\int_I |p(t)v(t)|^2\, dt \leq \widetilde{C}^{k+2}\int_I |-v'(t) + p(t)v(t)|^2\, dt.$$

Since

$$\|\varepsilon^{-k}q(\varepsilon\cdot)v\| \leq \|(\varepsilon^{-k}q(\varepsilon\cdot) - p)v\| + \|pv\| \leq \|(\varepsilon^{-k}q(\varepsilon\cdot) - p)v\|$$
$$+ \widetilde{C}^{(k+2)/2}\|v' - pv\| \leq (1 + \widetilde{C}^{(k+2)/2})\|(\varepsilon^{-k}q(\varepsilon\cdot) - p)v\| + \widetilde{C}^{(k+2)/2}\|-v' + \varepsilon^{-k}q(\varepsilon\cdot)v\|,$$

and $\|(\varepsilon^{-k}q(\varepsilon\cdot) - p)v\| \leq 2C_0\varepsilon(1 + T^{k+2})\|v\|$, it follows from the part of (2.26) already proved that $\|\varepsilon^{-k}q(\varepsilon\cdot)v\|^2 \leq 2\widetilde{C}^{(k+2)/2}\|-v' + \varepsilon^{-k}q(\varepsilon\cdot)v\|^2$ for small ε.

To prove (2.26) when $v \in \mathcal{S}(\mathbf{R})$ we choose $\chi \in C_0^\infty([-1,1])$ so that $\sum_{-\infty}^\infty \chi(t-\nu)^2 = 1$ and set $v_\nu(t) = \chi(t/T - \nu)v(t)$, $f_\nu(t) = \chi(t/T - \nu)f(t)$ where $f(t) = -v'(t) + \varepsilon^{-k}q(t\varepsilon)v(t)$. Then $-v_\nu'(t) + \varepsilon^{-k}q(t\varepsilon)v_\nu(t) = f_\nu(t) - \chi'(t/T - \nu)v_\nu(t)/T$, and by the translation invariance of $\mathcal{Q}$ it follows from the case of (2.26) already proved that for small ε, depending on T,

$$\int_{\mathbf{R}} |v_\nu(t)|^2\, dt \leq (A+\delta)\int_{\mathbf{R}} |f_\nu(t) - \chi'(t/T - \nu)v_\nu(t)/T|^2\, dt$$

$$\leq (A+\delta)\Big((1+\delta)\int_{\mathbf{R}} |f_\nu(t)|^2\, dt + \sup|\chi'|^2(1 + 1/\delta)/T^2\int_{\mathbf{R}} |v_\nu(t)|^2\, dt\Big).$$

Summing over ν gives if $K = \sup|\chi'|^2$

$$\big(1 - (A+\delta)(1 + 1/\delta)K/T^2\big)\int_{\mathbf{R}} |v(t)|^2\, dt \leq (A+\delta)(1+\delta)\int_{\mathbf{R}} |-v' + \varepsilon^{-k}q(t\varepsilon)v(t)|^2\, dt.$$

Choosing first a small δ and then a T so large that $(A+\delta)(1 + 1/\delta)K/T^2 < \delta$ we obtain (2.26) with a constant $(A+\delta)(1+\delta)/(1-\delta)$ arbitrarily close to A in the right-hand side when ε is small enough. In the same way (2.29) follows from the case where $v \in C_0^\infty(I)$ which has already been proved.

3. Local and global estimates. We shall begin by studying pseudodifferential operators $Q(x, D')$ in $x' = (x_2, \ldots, x_n)$ depending on x_1 as a parameter, assuming that $Q \in S^1(\mathbf{R}^n \times \mathbf{R}^{n-1})$, that is,

$$(3.1) \qquad |\partial_\xi^\alpha \partial_x^\beta Q(x, \xi')| \leq C_{\alpha\beta}(1 + |\xi'|)^{1-|\alpha|}, \quad x \in \mathbf{R}^n, \ \xi' \in \mathbf{R}^{n-1},$$

that $Q(x, \xi') \geq 0$ is positively homogeneous of degree 1 when $|\xi'| \geq 1$, and that

$$(3.2) \qquad |\xi'| \leq C\sum_0^k |\partial^j Q(x, \xi')/\partial x_1^j|/j!, \quad \text{if } |\xi'| \geq 1.$$

By $q(x, \xi')$ we shall denote the positively homogeneous principal symbol $|\xi'|Q(x, \xi'/|\xi'|)$ which is equal to $Q(x, \xi')$ when $|\xi'| \geq 1$. The following preliminary result is essentially contained in the proof of Proposition 27.3.1 in [H] but to prepare for an improvement we shall give a complete proof here.

PROPOSITION 3.1. *Under the preceding assumptions there is a constant B such that with L^2 norms in $L^2(\mathbf{R}^n)$*

$$(3.3) \quad \|D_1 u\|^2 + \|Q(x, D')u\|^2 + \|(1 + |D'|^2)^{1/(2k+2)}u\|^2$$
$$\leq B(\|D_1 u + iQ(x, D')u\|^2 + \|u\|^2), \quad u \in \mathcal{S}(\mathbf{R}^n).$$

PROOF. For the set $\mathcal{Q}$ of functions $\mathbf{R} \ni t \mapsto q(t + x_1, x', \xi')$ with $|\xi'| = 1$ the conditions (i) and (ii) of Proposition 2.8 are fulfilled, and (2.27) with $C_1 = 1/C$ follows from (3.2). For small $\varepsilon > 0$ and $|\xi'| = 1$ it follows from (2.26) and (2.29) that

$$\int_{\mathbf{R}} |v(t)|^2 \, dt \leq 2\widehat{C}_k C^{2/(k+1)} \int_{\mathbf{R}} |-v'(t) + \varepsilon^{-k} q(t\varepsilon, x', \xi')v(t)|^2 \, dt, \quad v \in \mathcal{S}(\mathbf{R}),$$

$$\int_{\mathbf{R}} |\varepsilon^{-k} q(t\varepsilon, x', \xi')v(t)|^2 \, dt \leq 2\widetilde{C}^{k+2} \int_{\mathbf{R}} |-v'(t) + \varepsilon^{-k} q(t\varepsilon, x', \xi')v(t)|^2 \, dt, \quad v \in \mathcal{S}(\mathbf{R}^n).$$

With $x_1 = \varepsilon t$ as a new variable this means that

$$(3.4) \quad |\eta'|^{2/(k+1)} \int_{\mathbf{R}} |v(x_1)|^2 \, dx_1 \leq 2\widehat{C}_k C^{2/(k+1)} \int_{\mathbf{R}} |-v'(x_1) + q(x_1, x', \eta')v(x_1)|^2 \, dx_1,$$

$$(3.5) \quad \int_{\mathbf{R}} |q(x_1, x', \eta')v(x_1)|^2 \, dx_1 \leq 2\widetilde{C}^{k+2} \int |-v'(x_1) + q(x_1, x', \eta')v(x_1)|^2 \, dx_1,$$

where $\eta' = \varepsilon^{-(k+1)}\xi'$, $|\xi'| = 1$ and ε is sufficiently small, that is, $|\eta'|$ is sufficiently large. A very important consequence of the assumption that $q \geq 0$ is that

$$(3.6) \quad |\partial q(x, \xi')/\partial \xi'|^2 |\xi'| + |\partial q(x, \xi')/\partial x|^2/|\xi'| \leq Cq(x, \xi'), \quad \xi' \neq 0,$$

where C only depends on the constants in (3.1) with $|\alpha + \beta| \leq 2$. This follows from Lemma 7.7.2 in [H]. We shall localize using the metric g of type $\frac{1}{2} + \varepsilon, \frac{1}{2} - \varepsilon$ defined by

$$(3.7) \quad g = |dx'|^2(1 + |\xi'|)^{1-2\varepsilon} + |d\xi'|^2(1 + |\xi'|)^{-1-2\varepsilon},$$

where $0 < \varepsilon < 1/(2k + 2)$. The dual metric with respect to the symplectic form is $g^\sigma = |dx'|^2(1 + |\xi'|)^{1+2\varepsilon} + |d\xi'|^2(1 + |\xi'|)^{-1+2\varepsilon} = (1 + |\xi'|)^{4\varepsilon}g$. As in [H], pp. IV:181, 182, it follows from Taylor's formula and (3.6) that

$$(3.8) \quad m(x, \xi') = (1 + |\xi'|^2)^\varepsilon + Q(x, \xi')$$

is uniformly g continuous and σ, g temperate. We have $Q \in S(m, g)$ and more precisely

$$(3.9) \quad |D_\xi^\alpha D_x^\beta Q(x, \xi')| \leq C'_{\alpha\beta}(1 + |\xi'|)^{2\varepsilon + |\beta|(\frac{1}{2}-\varepsilon)-|\alpha|(\frac{1}{2}+\varepsilon)}, \quad |\alpha| + |\beta| \geq 2,$$

$$(3.10) \quad |D_\xi^\alpha D_x^\beta Q(x, \xi')| \leq C'_{\alpha\beta} m(x, \xi')^{\frac{1}{2}}(1 + |\xi'|)^{\varepsilon + |\beta|(\frac{1}{2}-\varepsilon)-|\alpha|(\frac{1}{2}+\varepsilon)}, \quad |\alpha| + |\beta| = 1.$$

We choose a standard partition of unity $\{\chi_j\}$ for the g metric and $(x'_j, \xi'_j) \in \operatorname{supp} \chi_j$, thus $\chi_j \in C_0^\infty(\mathbf{R}^{2n-2})$ is real valued, $\sum \chi_j^2 = 1$, $\{\chi_j\}$ is uniformly bounded in $S(1, g)$ and there is a fixed bound for the number of overlapping supports. (Cf. [H], Section 18.4.) We can regard $\{\chi_j\}$ as a symbol in $S(1, g)$ with values in l^2, and

$$(3.11) \quad \chi_j(x', D')Q(x, D') = Q(x_1, x'_j, \xi'_j)\chi_j(x', D') + R_j(x, D'),$$

$$(3.12)$$
$$R_j(x, D') = (\chi_j(x', D')Q(x, D') - (\chi_j Q)(x, D')) + (\chi_j Q)(x, D') - Q(x_1, x'_j, \xi'_j)\chi_j(x', D')$$

By (3.9) and (3.10) the derivatives of Q although not Q itself have the bounds required for a symbol in $S((1 + |\xi'|)^\varepsilon m^{\frac{1}{2}}, g)$ so the symbol of the first difference on the right-hand side of (3.12) is in $S((1 + |\xi'|)^{-\varepsilon} m^{\frac{1}{2}}, g)$, and $\chi_j(x', \xi')(Q(x_1, x', \xi') - Q(x_1, x'_j, \xi'_j))$ is bounded in $S((1 + |\xi'|)^\varepsilon m^{\frac{1}{2}}, g)$, so $\{R_j\}$ is in this symbol space with values in l^2. If $f = (D_1 + iQ(x, D'))u$ then $\chi_j(x', D')f = (D_1 + iQ(x_1, x'_j, \xi'_j))\chi_j(x', D')u + iR_j(x, D')u$ and we obtain

$$(3.13) \quad \sum \|(D_1 + iQ(x_1, x'_j, \xi'_j))\chi_j(x', D')u\|^2 \leq 2((\chi(x', D')f, f) + (R(x, D')u, u)),$$

$$(3.14) \quad \chi(x', D') = \sum \chi_j(x', D')^*\chi_j(x', D'), \quad R(x, D') = \sum R_j(x, D')^* R_j(x, D').$$

The symbol of R is bounded in $S((1 + |\xi'|)^{2\varepsilon} m, g)$, and since $\sum \chi_j^2 = 1$ we have $\chi - 1 \in S((1 + |\xi'|)^{-2\varepsilon}, g)$. (There is a cancellation of first order terms in the expansion of the symbol so the exponent -2ε can be replaced by -4ε but that is not important now.)

Before combining (3.13) with (3.4) and (3.5) we note that (3.11) implies

$$(3.15) \quad (\chi(x', D')Q(x, D')u, Q(x, D')u)$$
$$\leq 2\sum \|Q(x_1, x'_j, \xi'_j)\chi_j(x', D')u\|^2 + 2(R(x, D')u, u).$$

Since $\|v\|_{(s)} = \|(1 + |D'|^2)^{s/2}v\|$, $v \in \mathcal{S}(\mathbf{R}^n)$, is a logarithmically convex function of s and $1 - \chi \in S((1 + |\xi'|)^{-2\varepsilon}, g)$, we have

$$\|v\|^2 \leq (\chi(x', D')v, v) + C'\|v\|^2_{\langle -\varepsilon \rangle} \leq (\chi(x', D')v, v) + \tfrac{1}{2}\|v\|^2 + C'\|v\|^2_{\langle -1 \rangle},$$

and it follows with $v = Q(x, D')u$ that

$$(3.16) \quad \|Q(x, D')u\|^2 \leq 2(\chi(x', D')Q(x, D')u, Q(x, D')u) + C''\|u\|^2.$$

Combination of (3.15) and (3.16) gives

$$\|Q(x, D')u\|^2 \leq 4\sum \|Q(x_1, x'_j, \xi'_j)\chi_j(x', D')u\|^2 + 4(R(x, D')u, u) + C''\|u\|^2,$$

and using (3.13) and (3.5) we obtain with summation only for large $|\xi'_j|$

$$\sum \|Q(x_1, x'_j, \xi'_j)\chi_j(x', D')u\|^2 \leq 4\widetilde{C}^{k+2}((\chi(x', D')f, f) + (R(x, D')u, u)),$$
$$(3.17) \quad \text{hence} \quad \|Q(x, D')u\|^2 \leq C'''(\|f\|^2 + (R(x, D')u, u) + \|u\|^2).$$

To estimate the error term $(R(x, D')u, u)$ we observe that since $m \in S(m, g)$ we have $\Phi \in S(1/m, g)$ if $\Phi = 1/m$ hence $\Phi(x, D')m(x, D') = \mathrm{Id} - T_1(x, D')$ where the symbol T_1 is in $S((1 + |\xi'|)^{-2\varepsilon}, g)$, and Id is the identity. This implies that

$$\mathrm{Id} = T_1(x, D')^N + (\mathrm{Id} + T_1(x, D') + \cdots + T_1(x, D')^{N-1})\Phi(x, D')((1 + |D'|^2)^\varepsilon + Q(x, D')).$$

for an arbitrary positive integer N. If we multiply from the left by $R(x, D')$, take N so large that the symbol of $R(x, D')T_1(x, D')^N$ is in $S(1, g)$, and recall that $R \in S((1 + |\xi'|^2)^\varepsilon m, g)$, it follows that

$$(3.18) \quad (R(x, D')u, u) \leq C^{(4)}(\|Q(x, D')u\| + \|u\|_{\langle 2\varepsilon \rangle})\|u\|_{\langle 2\varepsilon \rangle}.$$

If we use this estimate in (3.17) and that

$$C'''C^{(4)}\|Q(x,D')u\|\|u\|_{\langle 2\varepsilon\rangle} \leq \tfrac{1}{2}\|Q(x,D')u\|^2 + \tfrac{1}{2}(C'''C^{(4)})^2\|u\|_{\langle 2\varepsilon\rangle}^2,$$

it follows after cancellation of a term $\tfrac{1}{2}\|Q(x,D')u\|^2$ that

$$(3.19) \qquad \|Q(x,D')u\|^2 \leq 2C'''\|f\|^2 + C^{(5)}\|u\|_{\langle 2\varepsilon\rangle}^2.$$

Using (3.13) and (3.4) we obtain with summation only for large $|\xi'_j|$

$$\sum |\xi'_j|^{2/(k+1)}\|\chi_j(x',D')u\|^2 \leq 4\hat{C}_k C^{2/(k+1)}((\chi(x',D')f,f) + (R(x,D')u,u)),$$

$$(3.20) \qquad \text{hence} \quad \|u\|_{\langle 1/(k+1)\rangle}^2 \leq C^{(6)}(\|f\|^2 + (R(x,D')u,u) + \|u\|^2).$$

Estimating $(R(x,D')u,u)$ by (3.18) and then $\|Q(x,D')u\|$ by (3.19) gives

$$\|u\|_{\langle 1/(k+1)\rangle}^2 \leq C^{(7)}(\|f\|^2 + \|u\|_{\langle 2\varepsilon\rangle}^2).$$

Since $2\varepsilon < 1/(k+1)$ the logarithmic convexity of $\|u\|_{\langle s\rangle}$ implies that $C^{(7)}\|u\|_{\langle 2\varepsilon\rangle}^2$ can be estimated by $\tfrac{1}{2}\|u\|_{\langle 1/(k+1)\rangle}^2$ plus a constant times $\|u\|^2$, hence after a cancellation

$$(3.21) \qquad \|u\|_{\langle 1/(k+1)\rangle}^2 \leq 2C^{(7)}\|f\|^2 + C^{(8)}\|u\|^2.$$

(3.19) and (3.20) contain the desired estimate of the second and third terms on the left-hand side of (3.3), and the estimate of the first term is an immediate consequence. The proof is complete.

The constant B in (3.3) given by the proof is huge but the proof gives additional information on the error term $(R(x,D')u,u)$ which will be important for what follows. From (3.18) and (3.19) we obtain for any $\delta > 0$

$$(R(x,D')u,u) \leq \tfrac{1}{2}\delta\|Q(x,D')u\|^2 + C^{(4)}(\tfrac{1}{2}\delta^{-1}C^{(4)} + 1)\|u\|_{\langle 2\varepsilon\rangle}^2 \leq \delta C'''\|f\|^2 + C_\delta\|u\|_{\langle 2\varepsilon\rangle}^2.$$

By the logarithmic convexity in s of $\|\cdot\|_{\langle s\rangle}$ we can again estimate the last term by a small constant times $\|u\|_{\langle 1/(k+1)\rangle}^2$ and apply (3.21). This proves that for every $\delta > 0$ there is a (new) constant C_δ such that

$$(3.22) \qquad (R(x,D')u,u) \leq \delta\|D_1u + iQ(x,D')u\|^2 + C_\delta\|u\|^2, \quad u \in \mathcal{S}(\mathbf{R}^n).$$

We shall use this to give a precise estimate of $\|u\|_{\langle 1/(k+1)\rangle}^2$.

PROPOSITION 3.2. *Let the hypotheses of Proposition 3.1 be fulfilled, denote by $\mathcal{Q}_0$ the set of limits of $t \mapsto \varepsilon^{-k}q(\varepsilon t + x_1, x', \xi')$ when $\varepsilon \to 0$ and $|\xi'| = 1$, and assume that (2.25) is fulfilled, $A = 0$ if $\mathcal{Q}_0$ is empty. Then there exists for every $\delta > 0$ a constant C_δ such that*

$$(3.23) \quad \|(1 + |D'|^2)^{1/(2k+2)}u\|^2 \leq (A + \delta)\|D_1 + iQ(x,D')u\|^2 + C_\delta\|u\|^2, \quad u \in \mathcal{S}(\mathbf{R}^n).$$

PROOF. We keep the notation in the proof of Proposition 3.1. As in the proof of (3.4) it follows from (2.25) that for $v \in \mathcal{S}(\mathbf{R})$

$$(3.24) \quad (1 + |\eta'|^2)^{1/(k+1)} \int_{\mathbf{R}} |v(x_1)|^2 \, dx_1 \leq (A + \delta) \int_{\mathbf{R}} |-v'(x_1) + Q(x_1, x', \eta')v(x_1)|^2 \, dx_1,$$

when $|\eta'|$ is large enough, for $(1 + |\eta'|^2)/|\eta'|^2 \to 1$ when $|\eta'| \to \infty$. Thus

$$(3.25) \quad (1 + |\xi_j'|^2)^{1/(k+1)} \|\chi_j(x', D')u\|^2$$
$$\leq (A + \delta)\|(D_1 + iQ(x_1, x_j', \xi_j'))\chi_j(x', D')u\|^2, \quad u \in \mathcal{S}(\mathbf{R}^n),$$

if $|\xi_j'|$ is sufficiently large. The proof of (3.13) gives for arbitrary $\delta > 0$

$$(3.26) \quad \sum{}' \|(D_1 + iQ(x_1, x_j', \xi_j'))\chi_j(x', D')u\|^2$$
$$\leq (1 + \delta) \sum{}' \|\chi_j(x', D')f\|^2 + (1 + 1/\delta)(R(x, D')u, u).$$

Here $\sum'$ means summation only for $|\xi_j'| > X$ where X is some large number to be chosen later, and we recall that $f = D_1 u + iQ(x, D')u$, $u \in \mathcal{S}(\mathbf{R}^n)$. For the last term in (3.26) we have the estimate (3.22) but we must prove that the preceding sum is not much larger than $\|f\|^2$. When $|\xi_j'| > X$ then χ_j is bounded in $S(1, G)$ where

$$G = |dx'|^2(X + |\xi'|)^{1-2\varepsilon} + |d\xi'|^2(X + |\xi'|)^{-1-2\varepsilon}, \quad G^\sigma = (X + |\xi'|)^{4\varepsilon}G \geq X^{4\varepsilon}G.$$

Hence the symbol of $\sum'\chi_j(x', D')^*\chi_j(x', D')$ differs from $\sum'\chi_j(x', \xi')^2$ by a symbol in $S((X + |\xi'|)^{-2\varepsilon}, G)$ corresponding to an operator with norm $O(X^{-2\varepsilon})$. Since the nonnegative symbol $1 - \sum'\chi_j(x', \xi')^2$ is bounded in $S(1, G)$ it follows from the "sharp Gårding inequality" (Theorem 18.6.7 in [H]) that the corresponding operator has a lower bound $-C/X^{2\varepsilon}$. If X is large enough it follows from (3.26) that

$$(3.27) \quad \sum{}' \|(D_1 + iQ(x_1, x_j', \xi_j'))\chi_j(x', D')u\|^2$$
$$\leq (1 + 2\delta)\|f\|^2 + (1 + 1/\delta)(R(x, D')u, u),$$

and by (3.22) with δ replaced by δ^2 we have

$$(3.28) \quad (1 + 1/\delta)(R(x, D')u, u) \leq (\delta + \delta^2)\|f\|^2 + (1 + 1/\delta)C_{\delta^2}\|u\|^2.$$

Finally we shall prove that

$$(3.29) \quad ((1 + |D'|^2)^{1/(k+1)}u, u) \leq \sum{}'(1 + |\xi_j'|^2)^{1/(k+1)}\|\chi_j(x', D')u\|^2 + C\|u\|^2_{(1/(k+1)-\varepsilon)},$$

for some constant C. The main symbol of $\sum'\chi_j(x', D')^*(1+|\xi_j'|^2)^{1/(k+1)}\chi_j(x', D')$ is equal to $\sum'\chi_j(x', \xi')^2(1 + |\xi_j'|^2)^{1/(k+1)}$ apart from a remainder in $S((1 + |\xi'|^2)^{1/(k+1)-\varepsilon}, g)$, and

$$\sum{}'\chi_j(x', \xi')^2((1 + |\xi_j'|^2)^{1/(k+1)} - (1 + |\xi'|^2)^{1/(k+1)}) \in S((1 + |\xi'|)^{2/(k+1)+\varepsilon-\frac{1}{2}}, g)$$

for if $\chi_j(x', \xi') \neq 0$ then $|\xi' - \xi_j'| < C|\xi'|^{\frac{1}{2}+\varepsilon}$, so every term and therefore the sum $O(|\xi'|)^{2/(k+1)+\varepsilon-\frac{1}{2}}$. For the terms in the derivatives where the factor $(1 + |\xi'|^2)^{1/(k+1)}$ is differentiated we get even better bounds than required for a symbol with respect to g, and $(1 + |\xi'|^2)^{1/(k+1)}(1 - \sum'\chi_j(x', \xi')^2) \in S(1, g)$. (3.29) follows for $\varepsilon - \frac{1}{2} < -2\varepsilon$ since $\varepsilon < 1/(2k + 2) \leq 1/6$. If in (3.29) we use first (3.25), then (3.27) and (3.28) we obtain

$$\|u\|^2_{(1/(k+1))} \leq (A + \delta)(1 + 3\delta + \delta^2)\|D_1 u + iQ(x, D')u\|^2 + C_\delta'\|u\|^2_{(1/(k+1)-\varepsilon)}.$$

By the logarithmic convexity of $\| \cdot \|_{(s)}$ the last term can be estimated by $\delta \|u\|^2_{\langle 1/(k+1)\rangle}$ plus a constant times $\|u\|^2$, which gives the desired estimate (3.23) with the constant $(A + \delta)(1 + 3\delta + \delta^2)/(1 - \delta)$ instead of $(A + \delta)$. It converges to A when $\delta \to 0$ which completes the proof.

Let $q(x, \xi')$ be a nonnegative C^∞ function in $\mathbf{R}^n \times (\mathbf{R}^{n-1} \setminus 0)$ which is homogeneous of degree 1 in ξ', and assume as in (1.1) that

$$(3.30) \qquad \partial^j_{x_1} q(0, \xi'_0) = 0, \quad j < k; \quad \partial^k_{x_1} q(0, \xi'_0)/k! = c > 0,$$

where $\xi'_0 \in \mathbf{R}^{n-1}$ and $|\xi'_0| = 1$. Denote by $\mathcal{Q}_0(\xi'_0)$ the set of limits (1.7) when $\varepsilon_j \to 0$ and $x(j) \to 0, \xi'(j) \to \xi'_0$. Every $q_0 \in \mathcal{Q}_0(\xi'_0)$ is a polynomial in t of degree k with leading term ct^k.

LEMMA 3.3. *If (2.25) is valid when $q_0 \in \mathcal{Q}_0(\xi'_0)$ and $\delta > 0$ then there is a compact neighborhood V of $(0, \xi'_0)$ such that*

$$(3.31) \qquad \int_{\mathbf{R}} |v(t)|^2 \, dt \le (A + \delta) \int_{\mathbf{R}} |-v'(t) + q(t)v(t)|^2 \, dt, \quad v \in \mathcal{S}(\mathbf{R}), \ q \in \mathcal{Q}^V_0,$$

where $\mathcal{Q}^V_0$ is the set of limits of sequences (1.7) with $\varepsilon_j \to 0$ and $\lim (x(j), \xi'(j)) \in V$.

PROOF. If $q \in \mathcal{Q}^V_0$ then the leading coefficient is close to c if V is small. After a translation if necessary we can assume that the coefficient of t^{k-1} vanishes, and we can also assume that $C(q) \ge A$ since (3.31) is valid with $\delta = 0$ otherwise. By Corollary 2.6 the subset of $\mathcal{Q}^V_0$ with these two properties is compact and the intersection of them for all neighborhoods V of $(0, \xi'_0)$ is equal to $\mathcal{Q}_0(\xi'_0)$ so the distance to $\mathcal{Q}_0(\xi'_0)$ is arbitrarily small when V is small. Since $C(p)$ is a continuous function of p it follows that $C(q) \le A + \delta$ for all $q \in \mathcal{Q}^V_0$ if V is small enough.

If $\psi \in C^\infty_0(\mathbf{R})$, $0 \le \psi \le 1$, $\psi(t) = 1$ when $|t| \le 1$, and $\psi(t) = 0$ when $|t| \ge 2$ then

$$q_\varepsilon(x, \xi') = \psi(x_1/\varepsilon)q(x, \xi') + (1 - \psi(x_1/\varepsilon))|\xi'|$$

is nonnegative and equal to $q(x, \xi')$ when $|x_1| \le \varepsilon$, equal to $|\xi'|$ when $|x_1| \ge 2\varepsilon$ and $\ge |\xi'| \min(c\varepsilon^k/2, 1)$ when $\varepsilon \le |x_1| \le 2\varepsilon$ while $\partial^k_{x_1} q_\varepsilon(x, \xi')/k! \ge c|\xi'|/2$ when $|x_1| < \varepsilon$, if ε, $|x'|$ and $|\xi'/|\xi'| - \xi'_0|$ are small enough. If $\varphi(x', \xi')$ is a nonnegative C^∞ function in $\mathbf{R}^{n-1} \times (\mathbf{R}^{n-1} \setminus 0)$ with $0 \le \varphi \le 1$ which is homogeneous of degree 0 in ξ' and vanishes outside a small conic neighborhood of $(0, \xi'_0)$, $\varphi(0, \xi'_0) = 1$, then

$$q_1(x, \xi') = \varphi(x', \xi')q_\varepsilon(x, \xi') + (1 - \varphi(x', \xi'))|\xi'|$$
$$= \varphi(x', \xi')\psi(x_1/\varepsilon)q(x, \xi') + (1 - \varphi(x', \xi')\psi(x_1/\varepsilon))|\xi'|$$

has a positive lower bound for $|\xi'| = 1$ if $|x_1| > \varepsilon$ or $\varphi(x', \xi') \le \frac{1}{2}$ while $\partial^k_{x_1} q_1(x, \xi')/k! \ge c/4$ when $|\xi'| = 1$ if $|x_1| < \varepsilon$ and $\varphi(x', \xi') \ge \frac{1}{2}$. Hence (3.2) is fulfilled if the the principal symbol of Q is equal to q_1. If $\varphi(x', \xi') = 1$ in a neighborhood of $(0, \xi'_0)$ then $q_1(x, \xi') = q(x, \xi')$ in a neighborhood of $(0, \xi'_0)$.

If $\varepsilon^{-k}_j q_1(\varepsilon_j t + x_1(j), x'(j), \xi'(j))$ has a finite limit $q_0(t)$ and $\varepsilon_j \to 0$, $|\xi'(j)| = 1$, then $\varphi(x'(j), \xi'(j)) \to 1$ and $q_\varepsilon(x(j), \xi'(j)) \to 0$ so $|x_1| \le \varepsilon$ and $\varphi(x', \xi') = 1$ for a limit (x, ξ') of the sequence $(x(j), \xi'(j))$. Hence (x, ξ') is in as small a neighborhood V of $(0, \xi'_0)$ as we wish, and $\varepsilon^{-k}_j q(\varepsilon_j t + x_1(j), x'(j), \xi'(j)) \to q_0(t)$. Combining this with Lemma 3.3 and Proposition 3.2 we have proved:

PROPOSITION 3.4. *Let $q \in C^\infty(\mathbf{R}^n \times (\mathbf{R}^{n-1} \setminus 0))$ be a nonnegative function which is positively homogeneous of degree 1 with respect to ξ', and assume that that (3.30) is fulfilled for some ξ_0' with $|\xi_0'| = 1$. If $A \geq C(q_0)$ for every limit q_0 of $\varepsilon^{-k} q(\varepsilon t + x_1, x', \xi')$ when $\varepsilon \to 0$, $x \to 0$ and $\xi' \to \xi_0'$, then there is for every $\delta > 0$ another such function q_1 which is equal to q in a conic neighborhood of $(0, \xi_0')$ and a function $Q \in S^1(\mathbf{R}^n \times \mathbf{R}^{n-1})$ with principal symbol q_1 such that (3.23) is valid.*

We can now prove a converse of Theorem 1.3. Recall that the condition (1.13) means that $|a(\gamma)|^2 C(q) \leq C_1$, $q \in \mathcal{Q}(\gamma)$, and that $|a(\gamma)|^2 \max_{q \in \mathcal{Q}(\gamma)} C(q)$ is homogeneous in γ of degree 0.

PROPOSITION 3.5. *Let P, A, and γ satisfy the hypotheses of Theorem 1.3, and assume that*

$$(3.32) \qquad |a(\gamma)|^2 C(q) \leq C_1, \quad q \in \mathcal{Q}(\gamma),$$

with $\mathcal{Q}(\gamma)$ defined as in Theorem 1.3. Then there exists for every $\delta > 0$ a conic neighborhood Γ of γ such that

$$(3.33) \qquad \|A\Psi(x, D)u\|^2 \leq (1 + \delta) C_1 \|P\Psi(x, D)u\|^2 + C_{\delta, \Psi} \|u\|^2,$$

if $\Psi(x, D)$ is a pseudodifferential operator of order 0 with $WF(\Psi(x, D)) \subset \Gamma$.

PROOF. It follows from Corollary 27.1.9 in [H] that by conjugation with a Fourier integral operator the proof can be reduced to the case where $X \subset \mathbf{R}^n$, $\gamma = (0, \xi_0)$ with $\xi_0 = (0, \xi_0')$ and the principal symbol of P is equal to $\xi_1 + iQ(x, \xi')$ in a conic neighborhood of γ. By Propositions 3.2 and 3.4 we can choose Q so that

$$|a(\gamma)|^2 \|(1 + |D'|^2)^{1/(2k+2)} u\|^2 \leq C_1(1 + \delta) \|D_1 + iQ(x, D')u\|^2 + C_\delta \|u\|^2.$$

We replace u by $\Psi(x, D)u$ here and use that $(D_1 + iQ(x, D'))\Psi(x, D) - P\Psi(x, D)$ is of order 0 if Γ is sufficiently small, which gives

$$\|(D_1 + iQ(x, D'))\Psi(x, D)u\|^2 \leq (1 + \delta) \|P\Psi(x, D)u\|^2 + C_\delta' \|u\|^2.$$

We also have

$$\|A\Psi(x, D)u\|^2 \leq (1 + \delta) |a(\gamma)|^2 \|(1 + |D'|^2)^{1/(2k+2)} \Psi(x, D)u\|^2 + C_\delta'' \|u\|^2,$$

if Γ is sufficiently small, for

$$(A\Psi(x, D))^* A\Psi(x, D) - (1 + \delta) |a(\gamma)|^2 \Psi(x, D)^* (1 + |D'|^2)^{1/(k+1)} \Psi(x, D)$$

is then a pseudodifferential of order < 1 with principal symbol ≤ 0. Summing up we obtain

$$\|A\Psi(x, D)u\|^2 / (1 + \delta) \leq C_1(1 + \delta)^2 \|P\Psi(x, D)u\|^2 + C_\delta''' \|u\|^2,$$

which proves the statement.

Since the commutators $[\Psi(x, D), P]$ and $[\Psi(x, D), A]$ are of order 0 and $-k/(k + 1)$ respectively, it follows from (3.33) that with another C_δ

$$(3.34) \qquad \|\Psi(x, D)Au\|^2 \leq (1 + \delta) C_1 \|\Psi(x, D)Pu\|^2 + C_{\delta, \Psi} \|u\|^2,$$

Here is our final converse of Theorem 1.3:

THEOREM 3.6. *Let P be a first order "classical" pseudodifferential operator in the C^∞ manifold X satisfying condition (P). Let k be an even positive integer, and let K be a compact subset of X such that for every $\gamma \in T^*(X)|_K \setminus 0$ there is some $j \leq k$ and $b \in \mathbf{C}$, with $(H_{\mathrm{Re}\,bp})^j \operatorname{Im} bp(\gamma) \neq 0$, and define $\mathcal{Q}(\gamma)$ as in Theorem 1.3 if $(H_{\mathrm{Re}\,bp})^j \operatorname{Im} bp(\gamma) = 0$ for $j < k$. If A is a pseudodifferential operator of order $1/(k+1)$ with principal symbol a, and*

$$(3.35) \qquad |a(\gamma)|^2 C(q) \leq C_1, \quad q \in \mathcal{Q}(\gamma),$$

when $\mathcal{Q}(\gamma)$ is defined, it follows that for every $\delta > 0$

$$(3.36) \qquad \|Au\|^2 \leq C_1(1+\delta)\|Pu\|^2 + C_\delta\|u\|^2, \quad u \in C_0^\infty(K).$$

PROOF. For every γ such that $\mathcal{Q}(\gamma)$ is defined we have (3.34) when $\Psi(x,D)$ is of order 0 and $WF(\Psi(x,D))$ is contained in a sufficiently small neighborhood of γ. If $\mathcal{Q}(\gamma)$ is not defined then we have such an estimate even for A of higher order than $1/(k+1)$. If $X \subset \mathbf{R}^n$ we now choose finitely many real valued Ψ_j for which (3.34) is valid and $\sum \Psi_j(x,\xi)^2 = 1$ when x is in a neighborhood of K. Then $\sum \Psi_j(x,D)^* \Psi_j(x,D) - \mathrm{Id}$ is of order -2 in a neighborhood of K, and adding the estimates (3.34) with Ψ replaced by Ψ_j proves (3.36). A partition of unity in X is now sufficient to complete the proof in the same way.

The condition (3.35) implies that $|a(\gamma)|^2 C(x^k) \leq C_1$ when $\mathcal{Q}(\gamma) \neq \emptyset$, and (3.35) follows if $|a(\gamma)|^2 \widehat{C}_k \leq C_1$ then. The quotient $\widehat{C}_k / C(x^k)$ converges to 4 when $k \to \infty$ and might even be bounded by 4 for all k so the best constant C_1 can be deduced apart from a bounded factor without knowing the new invariants $\mathcal{Q}(\gamma)$. However, a precise result cannot be stated without them.

4. A precise commutator estimate. In this section we shall prove an analogue of Proposition 2.7 with an exact constant when $p(x) = x^k$. It plays no role in the proof of the main results of this paper but it proves that (2.20) cannot be valid for any constant independent of k and suggests that the exponential growth of the constant in (2.20) is far too large.

PROPOSITION 4.1. *If k is an even positive integer then*

$$(4.1) \qquad \int_{\mathbf{R}} (|u'(x)|^2 + |x^k u(x)|^2)\,dx \leq (1+k)\int_{\mathbf{R}} |u'(x) - x^k u(x)|^2\,dx, \quad u \in \mathcal{S}(\mathbf{R}),$$

with strict inequality except for the solutions in $\mathcal{S}(\mathbf{R})$ of the differential equation

$$(4.2) \qquad u''(x) = (x^{2k} + (k+1)x^{k-1})u(x).$$

PROOF. It is far from obvious that the equation (4.2) has a solution in $\mathcal{S}(\mathbf{R})$. To verify that it has a positive solution in $\mathcal{S}(\mathbf{R})$ and express it in terms of Kummer functions is the main part of the proof.

We shall now discuss the equation (4.2) and more generally the solution of

$$(4.3) \qquad u''(x) = (x^{2k} + \lambda x^{k-1})u(x),$$

which is rapidly decreasing at $-\infty$. We have to prove that when $\lambda = k + 1$ then the solution is rapidly decreasing also at $+\infty$ by comparing with the solution of the equation with $\lambda = -k - 1$ which is the reflection in the origin of a solution of (4.2) which decreases rapidly at $+\infty$. To study (4.3) when $x < 0$ we set $u(x) = e^{x^{k+1}/(k+1)}v(x)$ which reduces the equation to

$$(4.4) \qquad v''(x) + 2x^k v'(x) = (\lambda - k)x^{k-1}v(x).$$

We change variables by introducing $z = -2x^{k+1}/(k+1)$, noting that $-2x^k\, dx = dz$, $x\, dz = (k+1)z\, dx$, so that $xd/dx = (k+1)zd/dz$. Since $x^2(d/dx)^2 = (xd/dx)^2 - xd/dx$ the equation (4.4) becomes after multiplication by x^2 and division by $(k+1)^2z$, if $v(x) = w(z)$,

$$zw''(z) + \big(k/(k+1) - z\big)w'(z) = -(\lambda - k)w(z)/(2k+2).$$

This is Kummer's differential equation with the parameters $a = (k - \lambda)/(2k + 2)$ and $b = k/(k+1)$. (We owe the reference to Kummer's equation to the DSolve program of Mathematica.) When $x \to -\infty$ then $z \to +\infty$ and the solution $U(a,b,z)$ of Kummer's differential equation is then $O(z^{-a})$ while the solution $_1F_1(a,b,z)$ grows as $e^z z^{a-b}$ so $u(x)$ is rapidly decreasing when $x \to -\infty$ if and only if $w(z)$ is a multiple of $U(a,b,z)$. If we normalize so that $w(0) = 1$ then (see [MOS, p. 263] or [AS, p. 504])

$$(4.5) \quad w(z) = {}_1F_1(a,b,z) - z^{1-b}\,_1F_1(a+1-b,2-b,z)\Gamma(b)\Gamma(1+a-b)/(\Gamma(a)\Gamma(2-b))$$

when $z > 0$, and $-z^{1-b} = -z^{1/(k+1)} = (2/(k+1))^{1/(k+1)}x$. We have $_1F_1(\cdot,\cdot,0) = 1$,

$$\Gamma(b)\Gamma(1+a-b)/(\Gamma(a)\Gamma(2-b))$$
$$= \Gamma(k/(k+1))\Gamma((k+2-\lambda)/(2k+2))/\big(\Gamma((k+2)/(k+1))\Gamma((k-\lambda)/(2k+2))\big),$$

and

$$\Gamma((k+2-\lambda)/(2k+2))/\Gamma((k-\lambda)/(2k+2)) = \Gamma(1/(2k+2))/\Gamma(-1/(2k+2))$$
$$= -\Gamma((2k+3)/(2k+2))/\Gamma((2k+1)/(2k+2)),$$

if $\lambda = k + 1$, which is minus the value when $\lambda = -k - 1$. This proves that (4.2) has a solution in $\mathcal{S}(\mathbf{R})$ such that $u(0) = 1$ and $u'(0) = -c_k$ where

$$c_k = (2/(k+1))^{1/(k+1)}\frac{\Gamma(k/(k+1))}{\Gamma((k+2)/(k+1))}\frac{\Gamma((2k+3)/(2k+2))}{\Gamma((2k+1)/(2k+2))}$$
$$= (2/(k+1))^{1/(k+1)}\exp(\gamma/(k+1) + O(1/(k+1)^3)),$$

where γ is Euler's constant. Explicitly,

$$u(x) = e^{-z/2}\big(_1F_1(-1/(2k+2),k/(k+1),z) - c_k x\,_1F_1(1/(2k+2),(k+2)/(k+1),z)\big)$$

if $x \leq 0$ and $z = -2x^{k+1}/(k+1)$, and

$$u(x) = e^{-z/2}\big(_1F_1((2k+1)/(2k+2),k/(k+1),z) - c_k x\,_1F_1((2k+3)/(2k+2),(k+2)/(k+1),z)\big)$$

if $x > 0$ and $z = 2x^{k+1}/(k+1)$. We have $u(0) = 1$ and must prove that $u > 0$ on $\mathbf{R}$. If u has a zero $x_0 > 0$ but $u(x) > 0$ when $0 \leq x < x_0$ then $u'(x_0) \leq 0$, hence $u'(x_0) < 0$ since the Cauchy data of u at x_0 cannot both be equal to 0. Then it follows that $u(x) \leq u'(x_0)(x - x_0)$ when $x > x_0$ for (4.3) implies that u is a concave function when

$u < 0$. This is impossible since $u(x) \to 0$ when $x \to +\infty$, which proves that $u(x) > 0$ when $x \geq 0$. To prove that this is also true when $x < 0$ we first observe that Kummer's equation gives for every $\lambda \in [k, k+1]$ a solution u_λ of (4.3) with $u_\lambda(0) = 1$ which is rapidly decreasing at $-\infty$, and $u_k(x) = \exp(x^{k+1}/(k+1))$. If $X^{k+1} > k+1$ we have $x^{2k} + \lambda x^{k-1} > 0$ when $x \leq -X$, and if $u_\lambda(-X) > 0$ it follows as in the discussion above of u on $\mathbf{R}_+$ that $u_\lambda(x) > 0$ when $x \leq -X$. Hence the set of all $\lambda \in [k, k+1]$ such that $u_\lambda(x) > 0$ when $x \leq 0$ is open. It is also closed for if $u_\lambda(x) \geq 0$ when $x < 0$ it follows that $u_\lambda(x) > 0$ when $x < 0$, for the Cauchy data of u_λ cannot vanish at any point. Hence $u_\lambda(x) > 0$ when $x \leq 0$ if $\lambda \in [k, k+1]$, in particular when $\lambda = k+1$. This completes the proof that (4.2) has a positive solution $u \in \mathcal{S}(\mathbf{R})$.

If A is a constant so large that $p(x) = x^{2k} + (k+1)x^{k-1} + A > 0$ on $\mathbf{R}$, then U is an eigenfunction with eigenvalue A of the differential operator $u \mapsto -u'' + pu$ acting in L^2 with the Friedrichs domain where $u' \in L^2$, $x^k u \in L^2$ and $u'' \in L^2$. The infimum of the corresponding Dirichlet form

$$\int_{\mathbf{R}} (|u'(x)|^2 + p(x)|u(x)|^2)\, dx$$

when $\int_{\mathbf{R}} |u(x)|^2\, dx = 1$ is taken for the eigenfunction belonging to the lowest eigenvalue. Thus the eigenfunction is positive since replacing u by $|u|$ would otherwise decrease the Dirichlet form. Hence it is proportional to U, for otherwise it would be orthogonal, so the lowest eigenvalue is equal to A and (4.1) follows.

REMARK. The solutions of (4.2) which are not in $\mathcal{S}(\mathbf{R})$ grow as $|x|^{-k-\frac{1}{2}} \exp(-2x^{k+1}/(k+1))$ at $-\infty$ and as $x^{\frac{1}{2}} \exp(2x^{k+1}/(k+1))$ at $+\infty$.

There are several interesting equivalent versions of (4.1). Since

$$|u'(x) + x^k u(x)|^2 + |u'(x) - x^k u(x)|^2 = 2(|u'(x)|^2 + |x^k u(x)|^2),$$

it follows if we integrate and use (4.1) that

$$(4.6) \qquad \int_{\mathbf{R}} |u'(x) + x^k u(x)|^2\, dx \leq (1 + 2k) \int_{\mathbf{R}} |u'(x) - x^k u(x)|^2\, dx, \quad u \in \mathcal{S}(\mathbf{R}).$$

Conversely (4.6) implies (4.1). Replacing $u(x)$ by $u(-x)$ we also obtain

$$(4.7) \qquad (1 + 2k)^{-1} \int_{\mathbf{R}} |u'(x) - x^k u(x)|^2\, dx \leq \int_{\mathbf{R}} |u'(x) + x^k u(x)|^2\, dx, \quad u \in \mathcal{S}(\mathbf{R}).$$

Since

$$\int_{\mathbf{R}} |u'(x) - x^k u(x)|^2\, dx = \int_{\mathbf{R}} (|u'(x)|^2 + |x^k u(x)|^2)\, dx + \int_{\mathbf{R}} kx^{k-1}|u(x)|^2\, dx,$$

we can rewrite (4.1) in the form

$$(4.8) \qquad \int_{\mathbf{R}} |u'(x) - x^k u(x)|^2\, dx + \int_{\mathbf{R}} x^{k-1}|u(x)|^2\, dx \geq 0, \quad u \in \mathcal{S}(\mathbf{R}),$$

$$(4.9) \quad \text{or} \quad 0 \leq \int_{\mathbf{R}} (|u'(x)|^2 + |x^k u(x)|^2)\, dx + (1 + k) \int_{\mathbf{R}} x^{k-1}|u(x)|^2\, dx, \quad u \in \mathcal{S}(\mathbf{R}).$$

Replacing $u(x)$ by $u(\pm tx)$ in (4.9) we obtain after multiplication by t^k

$$(1+k)\left|\int_{\mathbf{R}} x^{k-1}|u(x)|^2\,dx\right| \leq t^{k+1}\int_{\mathbf{R}}|u'(x)|^2\,dx + t^{-k-1}\int_{\mathbf{R}}|x^k u(x)|^2\,dx$$

when $t > 0$, hence

$$(4.10)\qquad (1+k)\left|\int_{\mathbf{R}} x^{k-1}|u(x)|^2\,dx\right| \leq 2\left(\int_{\mathbf{R}}|u'(x)|^2\,dx\int_{\mathbf{R}}|x^k u(x)|^2\,dx\right)^{\frac{1}{2}}.$$

Since kx^{k-1} is the commutator $[d/dx, x^k]$ this motivates the title of the section.

It should be remembered that k has been an even integer so far. If k is a positive *odd* integer then we have instead of (4.1)

$$(4.11)\qquad \int_{\mathbf{R}}(|u'(x)|^2 + |x^k u(x)|^2)\,dx \leq \int_{\mathbf{R}}|u'(x) - x^k u(x)|^2\,dx,\quad u \in \mathcal{S}(\mathbf{R}),$$

for the difference between the two sides is $k\int_{\mathbf{R}} x^{k-1}|u(x)|^2\,dx > 0$ unless $u = 0$. The inequality cannot be improved by a constant < 1 in the right-hand side, for if $0 \neq u \in C_0^\infty(\mathbf{R})$ and u is replaced by $u(\cdot - \theta)$ then the two sides divided by θ^{2k} have the same limit when $\theta \to \infty$. Since

$$\int_{\mathbf{R}}(|u'(x)|^2 + |x^k u(x)|^2)\,dx = \int_{\mathbf{R}}|u'(x) + x^k u(x)|^2\,dx + k\int_{\mathbf{R}} x^{k-1}|u(x)|^2\,dx$$

we have $k\int_{\mathbf{R}} x^{k-1}|u(x)|^2\,dx \leq \int_{\mathbf{R}}(|u'(x)|^2 + |x^k u(x)|^2)\,dx$, with equality when $u(x) = e^{-x^{k+1}/(k+1)}$. As in the proof of (4.10) it follows when k is odd that

$$(4.12)\qquad k\int_{\mathbf{R}} x^{k-1}|u(x)|^2\,dx \leq 2\left(\int_{\mathbf{R}}|u'(x)|^2\,dx\int_{\mathbf{R}}|x^k u(x)|^2\,dx\right)^{\frac{1}{2}},\quad u \in \mathcal{S}(\mathbf{R}),$$

which is the uncertainty principle when $k = 1$. The analogy with (4.10) is obvious. When k is odd we have instead of (4.6)

$$(4.13)\qquad \int_{\mathbf{R}}|u'(x) + x^k u(x)|^2\,dx \leq \int_{\mathbf{R}}|u'(x) - x^k u(x)|^2\,dx,\quad u \in \mathcal{S}(\mathbf{R}),$$

and there is no inequality in the opposite direction since the left-hand side vanishes when $u(x) = \exp(-x^{k+1}/(k+1))$. This is the well known origin of condition (Ψ) for solvability of pseudodifferential equations.

A small modification of the proof of (4.1) also gives the best estimate of the form

$$(4.14)\qquad \int_{\mathbf{R}}|x|^{k-1}|u(x)|^2\,dx \leq C_k'\int_{\mathbf{R}}|u'(x) - x^k u(x)|^2\,dx,\quad u \in \mathcal{S}(\mathbf{R}).$$

As before the best constant is equal to $1/\mu$ if μ is a positive constant such that the differential equation

$$(4.15)\qquad (-d/dx - x^k)(d/dx - x^k)u = \mu|x|^{k-1}u(x)$$

has a positive solution which is rapidly decreasing at $\pm\infty$. The equation can be written $u''(x) = (x^{2k} + kx^{k-1} - \mu|x|^{k-1})u(x)$, that is,

$$u''(x) = (x^{2k} + (k - \mu)x^{k-1})u(x), \ x \geq 0; \ u''(x) = (x^{2k} + (k + \mu)x^{k-1})u(x), \ x \leq 0.$$

If u is a solution which is rapidly decreasing at $-\infty$ it follows from (4.5) with $\lambda = k + \mu$ that

$$u'(0)/u(0) = C_k\Gamma((2 - \mu)/(2k + 2))/\Gamma(-\mu/(2k + 2)),$$
$$C_k = (2/(k + 1))^{1/(k+1)}\Gamma(k/(k + 1))/\Gamma((k + 2)/(k + 1)),$$

and for the solution u which is rapidly decreasing at $+\infty$ we have by (4.5) with $\lambda = -k + \mu$

$$u'(0)/u(0) = -C_k\Gamma((2k + 2 - \mu)/(2k + 2))/\Gamma(1 - (\mu + 2)/(2k + 2)).$$

There is a solution decreasing rapidly at $\pm\infty$ precisely when these logarithmic derivatives are equal, that is,

$$(\mu/(2-\mu))\Gamma(1+(2-\mu)/(2k+2))/\Gamma(1-\mu/(2k+2)) = \Gamma(1-\mu/(2k+2))/\Gamma(1-(\mu+2)/(2k+2)).$$

With the notation $\kappa = 1/(2k + 2)$ this equation can be written

$$\log(\mu/(2 - \mu)) = 2\log\Gamma(1 - \kappa\mu) - \log\Gamma(1 - \kappa(\mu - 2)) - \log\Gamma(1 - \kappa(\mu + 2)).$$

The right-hand side is negative since $\log\Gamma$ is a convex function so it follows that $\mu < 1$. Since

$$\log\Gamma(1 + x) = -\gamma x + \sum_{\nu=2}^{\infty}(-x)^\nu S_\nu/\nu, \quad S_\nu = \sum_{1}^{\infty} n^{-\nu}, \quad S_2 = \pi^2/6,$$

the equation for μ can be written

$$\log(\mu/(2 - \mu)) = -\sum_{\nu=2}^{\infty} S_\nu\kappa^\nu((\mu - 2)^\nu + (\mu + 2)^\nu - 2\mu^\nu)/\nu$$

where the terms in the sum are positive. This proves that μ increases to 1 when $\kappa \downarrow 0$ and that $\mu = 1 - 2S_2\kappa^2 - 4S_3\kappa^3 - 10S_4\kappa^4 + O(\kappa^5)$ when $\kappa \to 0$. Combined with a numerical calculation this gives for the best constant C'_k in (4.14) that $C'_2 \approx 1.13881$, $C'_4 \approx 1.04041$, $C'_6 \approx 1.01919$, $C'_8 \approx 1.01121$, $C'_{10} \approx 1.00735$, that $C'_k \downarrow 1$ when $k \to \infty$, and that

$$(4.16) \qquad C'_k = 1 + \tfrac{1}{2}S_2(k + 1)^{-2} + \tfrac{1}{2}S_3(k + 1)^{-3} + (\tfrac{5}{8}S_4 + \tfrac{1}{4}S_2^2)(k + 1)^{-4} + O(k^{-5}).$$

The constant is of course larger than in (4.8) where some cancellation occurs between the contributions for $x \lessgtr 0$, but for large k the difference is small.

References

[AS]   M. Abramowitz and I. A. Stegun, *Handbook of Mathematical Functions with Formulas, Graphs and Mathematical Tables*, Dover Publications, New York, 1970.

[H]   L. Hörmander, *The Analysis of Linear Partial Differential Operators I–IV*, Springer Verlag, 1983–85, Grundl. d. Math. Wiss. Band 256, 257, 274, 275.

[MOS] W. Magnus, F. Oberhettinger and R.P. Soni, *Formulas and Theorems for the Special Functions of Mathematical Physics*, Springer Verlag, 1966, Grundl. d. Math. Wiss. Band 52.

APPROXIMATION OF SOLUTIONS OF CONSTANT COEFFICIENT BOUNDARY PROBLEMS AND OF ENTIRE FUNCTIONS

LARS HÖRMANDER

1. Introduction. If Ω is an open convex subset of $\mathbf{R}^n$ and $P(D)$, $D = -i\partial/\partial x$, is a differential operator with constant coefficients, then sums of exponential solutions of the equation $P(D)u = 0$ are dense in the set of all solutions in $C^\infty(\Omega)$. (Recall that u is called an exponential solution if $u(x) = v(x)e^{i\langle x, \zeta \rangle}$ where $\zeta \in \mathbf{C}^n$ and v is a polynomial.) The proof is easy: If $\nu \in \mathcal{E}'(\Omega)$ is orthogonal to all exponential solutions with $v = 1$ then $\hat{\nu}(\zeta) = 0$ when $P(-\zeta) = 0$; if ν is orthogonal to all exponential solutions then $\hat{\nu}(\zeta) = P(-\zeta)f(\zeta)$ where f is entire analytic. By Paley-Wiener's theorem it follows that $f = \hat{\mu}$ where $\mu \in \mathcal{E}'(\Omega)$, hence $\nu(u) = (P(-D)\mu)(u) = \mu(P(D)u) = 0$ if $u \in C^\infty(\Omega)$ and $P(D)u = 0$. The approximation theorem is thus a consequence of the Hahn-Banach theorem. By [1, Theorem 7.6.14] it remains valid for arbitrary systems $P(D)$ with constant coefficients but the proof is less elementary then.

In this paper we shall study an analogous density property for boundary problems.

DEFINITION 1.1. Let $P(D)$ be a differential operator with constant coefficients in $\mathbf{R}^n$, of order m with principal part P_m, let the hyperplane $\mathbf{R}^{n-1} = \{x; x \in \mathbf{R}^n, x_n = 0\}$ be non-characteristic with respect to $P(D)$, and let $Q_j(D)$, $j = 1, \ldots, J$, be differential operators with constant coefficients of transversal order $< m$. If $\Omega \subset \mathbf{R}^n$ is an open set, we denote by $C^\infty_{P,\vec{Q}}(\Omega)$ the set of all $u \in C^\infty(\Omega)$ such that

$$(1.1) \qquad P(D)u = 0 \quad \text{in } \Omega; \quad Q_j(D)u = 0 \quad \text{in } \omega = \Omega \cap \mathbf{R}^{n-1}, \ j = 1, \ldots, J.$$

We denote by $E_{P,\vec{Q}}$ the linear subspace of $C^\infty(\mathbf{R}^n)$ generated by *exponential solutions*, that is, solutions of the form

$$(1.2) \qquad u(x) = \sum v_k(x)e^{i\langle x', \zeta' \rangle + i\lambda_k x_n}, \quad x = (x', x_n),$$

where v_k are polynomials and λ_k are the zeros of $P(\zeta', \lambda)$, $\zeta' = (\zeta_1, \ldots, \zeta_{n-1}) \in \mathbf{C}^{n-1}$.

The original purpose of this paper was thus to examine when $E_{P,\vec{Q}}$ (restricted to Ω) is dense in $C^\infty_{P,\vec{Q}}(\Omega)$, but the main interest may be the problems on entire functions to which this leads. We shall prove (Corollary 4.4) that $E_{P,\vec{Q}}$ is dense in $C^\infty_{P,\vec{Q}}(\mathbf{R}^n)$, so an equivalent problem is to decide if $C^\infty_{P,\vec{Q}}(\mathbf{R}^n)$ is dense in $C^\infty_{P,\vec{Q}}(\Omega)$, that is, if an analogue of Runge's theorem holds. We assume of course that $J \geq 1$ and that $\omega \neq \emptyset$, for otherwise (1.1) is not a boundary problem. We shall also assume throughout that Ω is convex. In general convexity does not guarantee density:

© Springer International Publishing AG, part of Springer Nature 2018
L. Hörmander, *Unpublished Manuscripts*,
https://doi.org/10.1007/978-3-319-69850-2_25

EXAMPLE 1.1. For the Cauchy boundary conditions, $Q_j(D) = D_n^{j-1}$, $j = 1, \ldots, m$, all exponential solutions are equal to 0 since they are analytic and vanish of infinite order when $x_n = 0$. Vanishing Cauchy data in ω implies by Holmgren's uniqueness theorem that $u = 0$ in the intersection of all half spaces with characteristic boundary containing ω, and in no larger set since for every closed half space with characteristic boundary one can find a solution of the homogeneous equation with support equal to the half space. (Cf. [3, Theorem 8.6.8].) If we introduce the supporting functions

$$H_\Omega(\xi) = \sup_{x \in \Omega} \langle x, \xi \rangle, \ \xi \in \mathbf{R}^n; \quad h_\omega(\xi') = \sup_{x' \in \omega} \langle x', \xi' \rangle = \inf_{\xi_n} H_\Omega(\xi', \xi_n), \ \xi' \in \mathbf{R}^{n-1},$$

this means that $E_{P,\vec{Q}} = \{0\}$ is dense in $C_{P,\vec{Q}}^\infty(\Omega)$ if and only if

$$(1.3) \qquad\qquad H_\Omega(\xi) = h_\omega(\xi') \quad \text{if } \xi = (\xi', \xi_n) \in \mathbf{R}^n, \ P_m(\xi) = 0.$$

This condition is of course empty if $P(D)$ is elliptic. If K is a convex compact subset of Ω it follows from (1.3) that there is a convex compact subset k of ω such that

$$(1.3)' \qquad\qquad H_K(\xi) \leq h_k(\xi') \quad \text{if } \xi = (\xi', \xi_n) \in \mathbf{R}^n, \ P_m(\xi) = 0.$$

In fact, for every $\xi' \in \mathbf{R}^{n-1}$ with $|\xi'| = 1$ and $\xi_n \in \mathbf{R}$ with $P_m(\xi', \xi_n) = 0$, it follows from (1.3) that $H_K(\xi', \xi_n) < H_\Omega(\xi', \xi_n) = h_\omega(\xi')$. Hence $H_K(\xi', \xi_n) < \langle x', \xi' \rangle$ for some $x' \in \omega$. This remains true in a conic neighborhood of (ξ', ξ_n). We can cover the real characteristics (ξ', ξ_n) with $|\xi'| = 1$ by a finite number of such neighborhoods, and then $(1.3)'$ is valid for the convex hull k of the corresponding points $x' \in \omega$.

EXAMPLE 1.2. If $m = 1$ then (1.3) guarantees that $E_{P,\vec{Q}}$ is dense in $C_{P,\vec{Q}}^\infty(\Omega)$. In fact, if $u \in C_{P,\vec{Q}}^\infty(\Omega)$ and $v_j = Q_j(D)u$, then $v_j = 0$ in ω and $P(D)v_j = Q_j(D)P(D)u = 0$ in Ω, hence $v_j = 0$ in Ω since (1.3) guarantees uniqueness for the Cauchy problem for $P(D)$ in Ω. Thus (1.1) is equivalent to

$$P(D)u = 0 \text{ and } Q_j(D)u = 0, \ j = 1, \ldots, J, \text{ in } \Omega,$$

and $E_{P,\vec{Q}}$ is dense in $C_{P,\vec{Q}}^\infty(\Omega)$ by [1, Theorem 7.6.14].

EXAMPLE 1.3. In Example 1.2 it would have been sufficient to consider the two examples $P(D) = D_n$ and $P(D) = D_{n-1} + iD_n$. We shall now examine their powers, first $P(D) = D_n^m$. Then (1.3) means that $\Omega \subset \omega \times \mathbf{R}$, and $P(D)u = 0$ means that

$$u(x) = \sum_{k < m} u_k(x')(ix_n)^k/k!.$$

We can write $Q_j(D) = \sum_{k < m} Q_{j,k}(D')D_n^k/k!$. The boundary conditions $Q_j(D)u = 0$ when $x_n = 0$ become

$$\sum_{k < m} Q_{j,k}(D')u_k(x')/k! = 0 \quad \text{in } \omega.$$

Thus we can approximate $\{u_k\}$ by sums of exponential solutions of this system in ω, so (1.3) implies that $E_{P,\vec{Q}}$ is dense in $C_{P,\vec{Q}}^\infty(\Omega)$.

If $P(D) = (D_{n-1} + iD_n)^m$ then $P(D)u = 0$ in Ω means that there exist unique functions u_k in Ω, $0 \le k < m$, such that $(D_{n-1} + iD_n)u_k = 0$ and

$$u(x) = \sum_{k<m} u_k(x)(ix_n)^k/k!.$$

This is well known, and the proof by induction is easy: If u has this representation then

$$(D_{n-1} + iD_n)u = i \sum_{k<m-1} u_{k+1}(x)(ix_n)^k/k!,$$

and since $(D_{n-1} + iD_n)^{m-1}((D_{n-1} + iD_n)u) = 0$, it follows from the case with m replaced by $m - 1$ that this is true with uniquely defined $u_1, \ldots, u_{m-1}$. We can then choose u_0 uniquely so that the decomposition is valid. When $x_n = 0$ then $Q_j(D)u$ is equal to

$$\sum_{k+l<m} Q_{j,k+l}(D')D_n^k u_l(x)/(k!l!).$$

This is an analytic function of $x_{n-1} + ix_n$, and since (1.3) implies uniqueness for the Cauchy problem for $D_{n-1} + iD_n$ in Ω, it follows that in Ω

$$(D_{n-1} + iD_n)u_k = 0, \ k = 0, \ldots, m-1; \ \sum_{k+l<m} Q_{j,k+l}(D')D_n^k u_l(x)/(k!l!) = 0, \ j = 1, \ldots, J.$$

Solutions of this system can be approximated by sums of exponential solutions so (1.3) implies that $E_{P,\bar{Q}}$ is dense in $C^\infty_{P,\bar{Q}}$. — It is easy to extend the conclusions in this example to $P(D) = p(D_n)$ or $P(D) = p(D_{n-1} + iD_n)$ where p is an arbitrary polynomial.

EXAMPLE 1.4. For the Laplace operator with Dirichlet or Neumann boundary conditions we can extend u by reflection from Ω to $\Omega \cup \check{\Omega}$ where $\check{\Omega}$ is the reflection in the boundary plane $x_n = 0$. This set is starshaped with respect to any point in ω so the complement is connected. Hence the extension of u can be approximated by harmonic polynomials h in the whole space, and replacing h by $(h - \check{h})/2$ or $(h + \check{h})/2$ we can respect the boundary conditions. Thus we can indeed approximate in Ω by exponential solutions.

EXAMPLE 1.5. If $P(D) = D_1^2 - D_2^2$ is the wave operator, then every solution of $P(D)u = 0$ in $\Omega \subset \mathbf{R}^2$ can be uniquely extended to the smallest rectangle $\tilde{\Omega} \supset \Omega$ with sides of slope ± 1. A sequence in $E_{P,\bar{Q}}$ converging to u in $C^\infty(\Omega)$ must also converge in $C^\infty(\tilde{\Omega})$. If $u \in C^\infty_{P,\bar{Q}}(\Omega)$ is in the closure of $E_{P,\bar{Q}}$, then the extension $\tilde{u}$ to $\tilde{\Omega}$ must satisfy the boundary conditions when $x_1 \in I = \{x_1; (x_1, 0) \in \tilde{\Omega}\}$. We have $\tilde{u}(x_1, x_2) = f_+(x_1 + x_2) - f_-(x_1 - x_2)$ where $f_\pm$ are uniquely determined up to an additive constant in $I_\pm = \{x_1 \pm x_2; (x_1, x_2) \in \tilde{\Omega}\} = \{x_1 \pm x_2; (x_1, x_2) \in \Omega\}$; $I_+ \cap I_- = I$. A boundary condition $Q_j(D)\tilde{u} = 0$ means that

$$Q_j(D_1, D_1)f_+(x_1) = Q_j(D_1, -D_1)f_-(x_1), \quad x_1 \in I.$$

We cannot have $Q_j(D_1, \pm D_1) = 0$ for both signs, for Q_j would then be divisible by $D_1^2 - D_2^2$ which is impossible since the degree in D_2 is less than two. Hence $I = \omega$ if $E_{P,\bar{Q}}$ is dense in $C^\infty_{P,\bar{Q}}(\Omega)$, which means that Ω must have a supporting line with slope $+1$ or -1 at each end point of ω. The sufficiency of this condition when $J = 1$ and $Q_1(D_1, D_1)Q_1(D_1, -D_1) \not\equiv 0$

will be established in Corollary 3.8. That this condition cannot be omitted is seen from the example $P(D) = D_1^2 - D_2^2$, $Q(D) = D_1 - D_2$. For an exponential solution u we have $v = (D_1 - D_2)u = 0$ when $x_2 = 0$ and $(D_1 + D_2)v = 0$ in $\mathbf{R}^2$, hence $v = 0$ in $\mathbf{R}^2$. On the other hand, if $\psi \in C^\infty(\mathbf{R})$ vanishes in ω and $\psi' \neq 0$ in $\mathbf{R} \setminus \bar{\omega}$, and $u(x) = \psi(x_1 - x_2)$, then $P(D)u = 0$ in $\mathbf{R}^2$ and $Q(D)u = 0$ in ω. Since $(D_1 - D_2)u = 2\psi'(x_1 - x_2)/i$ it follows that $E_{P,Q}$ cannot be dense in $C_{P,Q}^\infty$ unless $x_1 - x_2 \in \omega$ when $x \in \Omega$.

This example shows that (1.3) is not always a necessary condition for every $\vec{Q}$.

EXAMPLE 1.6. If $P(D)$ is a second order differential operator invariant under the reflection $i : (x', x_n) \mapsto (x', -x_n)$, and Ω is also invariant under reflection, $Q(D) = D_n^j$ where $j = 0$ or $j = 1$, then $E_{P,Q}$ is dense in $C_{P,Q}^\infty(\Omega)$ if and only if (1.3) is valid. In fact, if $u \in C_{P,Q}^\infty(\Omega)$ then $\check{u} = u \circ i \in C_{P,Q}^\infty(\Omega)$, and $\check{u}$ has the same Cauchy data as $(-1)^{j+1}u$, hence $\check{u} = (-1)^{j+1}u$ in Ω if (1.3) is valid. If $u_k \in E_{P,Q}$ is a sequence of linear combinations of exponential solutions of $P(D)u_k = 0$ converging to u in $C^\infty(\Omega)$, then $v_k = (u_k + (-1)^{j+1}\check{u}_k)/2 \in E_{P,Q}$ converges to $(u + (-1)^{j+1}\check{u})/2 = u$ in $C^\infty(\Omega)$ so u is in the closure of $E_{P,Q}$. On the other hand, if $E_{P,Q}$ is dense in $C_{P,Q}^\infty$ then $\check{u} = (-1)^{j+1}u$ in Ω for every $u \in C_{P,Q}^\infty$, since this is true for every $u \in E_{P,Q}$. Hence every solution $u \in C^\infty(\Omega)$ of the equation $P(D)u = 0$ with vanishing Cauchy data in ω must vanish in Ω, for

$$u_\pm(x) = \begin{cases} u(x), & \text{if } \pm x_n \geq 0, \\ 0, & \text{if } \mp x_n \geq 0, \end{cases} \quad x \in \Omega,$$

is in $C_{P,Q}^\infty(\Omega)$, thus equal to 0 in Ω. This implies that (1.3) must be fulfilled.

Written in the weak form the equation $P(D)u = 0$ in Ω means that

$$\langle u, P(-D)\varphi \rangle = 0, \quad \varphi \in C_0^\infty(\Omega).$$

The boundary conditions in (1.1) mean that

$$\langle Q_j(D)u, \varphi_j \otimes \delta_0 \rangle = \langle u, Q_j(-D)(\varphi_j \otimes \delta_0) \rangle = 0, \quad \varphi_j \in C_0^\infty(\omega).$$

Thus $\langle u, \nu \rangle = 0$ if $\nu = P(-D)\varphi + \sum_1^J Q_j(-D)(\varphi_j \otimes \delta_0)$, so (1.1) means that $\langle u, \nu \rangle = 0$ if

$$(1.4) \qquad \hat{\nu}(\zeta) = P(-\zeta)\hat{\varphi}(\zeta) + \sum_1^J Q_j(-\zeta)\hat{\varphi}_j(\zeta'), \quad \varphi \in C_0^\infty(\Omega), \ \varphi_j \in C_0^\infty(\omega).$$

If $u \in C_{P,\vec{Q}}^\infty(\Omega)$ this is true for $\varphi \in \mathcal{E}'(\Omega)$ and $\varphi_j \in \mathcal{E}'(\omega)$. Thus $\langle u, \nu \rangle = 0$ if $u \in C_{P,\vec{Q}}^\infty(\Omega)$, $\nu \in \mathcal{E}'(\Omega)$ and

$$(1.5) \qquad \hat{\nu}(-\zeta) = P(\zeta)\Phi(\zeta) + \sum_1^J Q_j(\zeta)a_j(\zeta'), \quad \zeta \in \mathbf{C}^n,$$

where Φ and a_j are entire functions in $\mathbf{C}^n$ and $\mathbf{C}^{n-1}$ respectively, and

$$(1.6) \qquad |a_j(\zeta')| \leq C(1 + |\zeta'|)^N e^{h(-\operatorname{Im}\zeta')}, \quad \zeta' \in \mathbf{C}^{n-1},$$

where $h(\eta') = \sup_{x' \in k} \langle x', \eta' \rangle$, $\eta' \in \mathbf{R}^{n-1}$, is the supporting function of a convex compact set $k \subset \omega$. In fact, by the Paley-Wiener theorem (1.6) implies that $a_j(-\zeta') = \hat{\varphi}_j(\zeta')$ for some $\varphi_j \in \mathcal{E}'(k)$. We have

$$|\hat{\nu}(\zeta)| \leq C'(1 + |\zeta|)^{N'} e^{H(\operatorname{Im}\zeta)}, \quad \zeta \in \mathbf{C}^n,$$

where H is the supporting function of some compact set $K \subset \Omega$. Thus

$$|P(-\zeta)\Phi(-\zeta)| \le C'(1+|\zeta|)^{N'}e^{H(\mathrm{Im}\,\zeta)} + C''(1+|\zeta|)^{N''}e^{h(\mathrm{Im}\,\zeta')}, \quad \zeta \in \mathbf{C}^n,$$

which implies that there is such a bound for $\Phi(-\zeta)$, with different constants. (See e.g. [3, Theorem 7.3.2].) Hence $\check{\Phi} = \hat{\varphi}$ where $\varphi \in \mathcal{E}'(\Omega)$ is supported by the convex hull of K and $k \times \{0\}$, so the statement follows from (1.4).

The bound (1.6) implies that

$$(1.7) \quad |a_j(\xi')| \le C(1+|\xi'|)^N, \quad \varlimsup_{t \to +\infty} t^{-1}\log|a_j(\xi'+it\eta')| \le h(-\eta'), \quad \xi', \eta' \in \mathbf{R}^{n-1}.$$

Conversely, (1.7) implies (1.6) with another constant. This follows from the Phragmén-Lindelöf principle. If $\xi', \eta' \in \mathbf{R}^{n-1}$ and $\langle \xi', \eta' \rangle = 0$ then $1+|\xi'+z\eta'|^2 = 1+|\xi'|^2+|z|^2|\eta'|^2 \ge \frac{1}{2}|\sqrt{1+|\xi'|^2} - iz|\eta'||^2$, if $z \in \mathbf{C}$ and $\mathrm{Im}\,z \ge 0$. The function

$$A(z) = a_j(\xi'+z\eta')(\sqrt{1+|\xi'|^2} - iz|\eta'|)^{-N}e^{izh(-\eta')}$$

is analytic when $\mathrm{Im}\,z \ge 0$, and bounded by $2^{N/2}C$ when $z \in \mathbf{R}$, by the first part of (1.7). By the second part the same bound is valid when $\mathrm{Im}\,z \ge 0$. which proves (1.6) with C replaced by 2^NC.

After recalling a well known division algorithm in Section 2 we prove in Section 3 that a distribution $\nu \in \mathcal{E}'$ is orthogonal to $E_{P,\vec{Q}}$ if and only if $\hat{\nu}$ has a decomposition of the form (1.5). However, we only obtain a much weaker bound than (1.6) if $\nu \in \mathcal{E}'(\Omega)$. To conclude that the restriction of $E_{P,\vec{Q}}$ to Ω is dense in $C^\infty_{P,\vec{Q}}(\Omega)$ we need to approximate the entire functions of exponential type which occur by functions satisfying (1.6). A partial solution of this problem will occupy the rest of the paper.

Section 4 is an improved exposition of an approximation method suggested in [2]. It proves that for a given convex $\omega \subset \mathbf{R}^{n-1}$ and a given $P(D)$ there always exist convex Ω with $\Omega \cap \mathbf{R}^{n-1} = \omega$ such that $E_{P,\vec{Q}}$ is dense in $C^\infty_{P,\vec{Q}}(\Omega)$ for every $\vec{Q}$, but we can only prove the sufficiency of (1.3) in a few cases. In Section 5 we use a different approach to prove that $E_{P,Q}$ is dense in $C^\infty_{P,Q}(\Omega)$ for every convex Ω if $P(D)$ is elliptic and $J = 1$. Moreover, we prove that real analytic elements in $C^\infty_{P,Q}(\Omega)$ can always be approximated, and we reduce the approximation problem in the non-analytic case to a problem on plurisubharmonic functions.

In Section 6 we prove a general approximation theorem for entire functions with given bounds for the indicator function which was used in Section 5. We also study another more subtle approximation problem for entire functions which is closely related to the questions raised in Section 5.

2. A division algorithm. If F is an entire analytic function in $\mathbf{C}^n$, then it is well known (cf. e.g. Kiselman [6]) that there is a unique decomposition

$$(2.1) \qquad\qquad F(\zeta) = P(\zeta)\Phi(\zeta) + \Psi(\zeta), \quad \zeta \in \mathbf{C}^n,$$

where Φ and Ψ are entire analytic and $\Psi(\zeta) = \sum_0^{m-1} \Psi_l(\zeta')\zeta_n^l$ is a polynomial of degree $< m$ with respect to ζ_n. To prove uniqueness we observe that if for a given ζ we choose an open bounded set $A_\zeta \subset \mathbf{C}$ with piecewise C^1 boundary such that $\tau P(\zeta',\zeta_n+\tau) \ne 0$ when $\tau \notin A_\zeta$, then

$$(2.2) \qquad\qquad \Phi(\zeta) = \frac{1}{2\pi i}\int_{\partial A_\zeta} \frac{F(\zeta',\zeta_n+\tau)}{P(\zeta',\zeta_n+\tau)}\frac{d\tau}{\tau},$$

since $\Psi(\zeta', \zeta_n + \tau)/(\tau P(\zeta', \zeta_n + \tau))$ is analytic in the complement of A_ζ and $O(1/\tau^2)$ at ∞. The right-hand side is independent of the choice of A_ζ which can be made locally constant, so (2.2) defines an entire function Φ. An entire function Ψ is then given by (2.1),

$$(2.3) \quad \Psi(\zeta) = \frac{1}{2\pi i} \int_{\partial A_\zeta} \frac{F(\zeta', \zeta_n + \tau)\, d\tau}{\tau} - P(\zeta', \zeta_n)\Phi(\zeta)$$

$$= \frac{1}{2\pi i} \int_{\partial A_\zeta} \frac{P(\zeta', \zeta_n + \tau) - P(\zeta', \zeta_n)}{\tau P(\zeta', \zeta_n + \tau)} F(\zeta', \zeta_n + \tau)\, d\tau$$

$$= \frac{1}{2\pi i} \int_{\zeta_n + \partial A_\zeta} \frac{P(\zeta', \tau) - P(\zeta', \zeta_n)}{(\tau - \zeta_n)P(\zeta', \tau)} F(\zeta', \tau)\, d\tau.$$

Here $(P(\zeta', \tau) - P(\zeta', \zeta_n))/(\tau - \zeta_n)$ is a polynomial in ζ', ζ_n and τ of degree $m - 1$. If $B_{\zeta'}$ is a bounded open set with piecewise C^1 boundary such that $P(\zeta', \tau) \neq 0$ when $\tau \notin B_{\zeta'}$ we can therefore integrate over $\partial B_{\zeta'}$ instead, and the same contour can also be used for a neighborhood of ζ'. We can choose for $B_{\zeta'}$ the union of the discs of radius r with center at $\{\lambda; P(\zeta', \lambda) = 0\}$. The length of $\partial B_{\zeta'}$ is then $\leq 2\pi rm$, and $|P(\zeta', \tau)| \geq |P_m(0, 1)| r^m$ on $\partial B_{\zeta'}$. Hence we have proved, with $A_\zeta = \{\tau \in \mathbf{C}; |\tau| < r\} \cup (B_{\zeta'} - \zeta_n)$:

PROPOSITION 2.1. *Every entire analytic function F in $\mathbf{C}^n$ has a unique decomposition (2.1) where Φ and Ψ are entire and Ψ is a polynomial of degree $< m$ with respect to ζ_n. If $0 < r \leq 1$ we have*

$$(2.4) \qquad |\Psi(\zeta)| \leq C(1 + |\zeta|)^{m-1} r^{1-m} \sup_{|\tau| \leq r, P(\zeta', \lambda) = 0} |F(\zeta', \lambda + \tau)|, \quad \zeta \in \mathbf{C}^n,$$

$$(2.5) \quad |\Phi(\zeta)| \leq C r^{-m} \Big(\sup_{|\tau| \leq r} |F(\zeta', \zeta_n + \tau)| + \sup_{|\tau| \leq r, P(\zeta', \lambda) = 0} |F(\zeta', \lambda + \tau)| \Big), \quad \zeta \in \mathbf{C}^n.$$

If F is of exponential type, it follows from (2.4) and (2.5) that Φ and Ψ are also of exponential type, but even if F is bounded in $\mathbf{R}^n$ they may grow exponentially in some real direction unless $P_m(\xi', \zeta_n) = 0$, $\xi' \in \mathbf{R}^{n-1} \setminus \{0\}$, implies $\mathrm{Im}\, \zeta_n = 0$, that is, when the principal part is hyperbolic.[19]

For later reference we make some observations on suprema such as those in (2.4).

LEMMA 2.2. *If $h(\zeta)$ is a continuous plurisubharmonic function in $\mathbf{C}^n$, then*

$$(2.6) \qquad \tilde{h}(\zeta') = \sup_{|\tau| \leq 1, P(\zeta', \zeta_n) = 0} h(\zeta', \zeta_n + \tau)$$

is also continuous and plurisubharmonic.

PROOF. If $R(\zeta')$ is the product of the discriminants of the irreducible factors of P then the zeros $\lambda_k(\zeta')$ of $\zeta_n \mapsto P(\zeta', \zeta_n)$ are locally analytic when $R(\zeta') \neq 0$. Hence

$$\tilde{h}(\zeta') = \sup_{|\tau| \leq 1} \max_k h(\zeta', \lambda_k(\zeta') + \tau)$$

is plurisubharmonic then, and $\tilde{h}$ is continuous everywhere. Thus $\tilde{h}(\zeta') + \varepsilon \log |R(\zeta')|$ is plurisubharmonic in $\mathbf{C}^{n-1}$ when $\varepsilon > 0$, which proves that the continuous distribution limit $\tilde{h}$ when $\varepsilon \to 0$ is plurisubharmonic.

We shall also encounter more general constructions than (2.6):

---

[19]Kiselman [6] circumvented this difficulty by studying the analogous approximation problem for holomorphic functions instead. This problem has properties similar to the hyperbolic case just mentioned, for the Cauchy problem is then correctly posed. Here we shall discuss the approximation problem in the real domain.

LEMMA 2.3. *Let $F(\zeta', w_1, \ldots, w_m)$ be a real valued Lipschitz continuous function in $\mathbf{C}^{n-1+m}$ which is symmetric in $w_1, \ldots, w_m$. If $F_P(\zeta') = F(\zeta', \lambda_1(\zeta'), \ldots, \lambda_m(\zeta'))$ where $\lambda_1(\zeta'), \ldots, \lambda_m(\zeta')$ are the zeros of $P(\zeta', w)$, then*

$$(2.7) \qquad |F_P(\zeta' + z') - F_P(\zeta')| \leq C|z'|^{1/m}(1 + |\zeta'| + |z'|)^{(m-1)/m}, \quad z', \zeta' \in \mathbf{C}^{n-1},$$

$$(2.8) \qquad |F_P(\zeta') - F_{P_m}(\zeta')| \leq C(1 + |\zeta'|)^{(m-1)/m}, \quad \zeta' \in \mathbf{C}^{n-1}.$$

PROOF. In the open set

$$Z_{\zeta', r} = \bigcup_{j=1}^{m} \{w \in \mathbf{C}; |w - \lambda_j(\zeta')| < r\}, \quad \zeta' \in \mathbf{C}^{n-1}, \ r > 0,$$

the diameter of every component is $\leq mr$. On the boundary we have $|P(\zeta', w)| \geq |P_m(0,1)|r^m$ since $|w - \lambda_j(\zeta')| \geq r$ for $j = 1, \ldots, m$, hence

$$|P(\zeta' + z', w)| \geq |P_m(0,1)|r^m - C\varrho(1 + |\zeta'| + r + \varrho)^{m-1}, \quad |z'| \leq \varrho,$$

for $|w| \leq C'(1 + |\zeta'| + r)$. When $r = R\varrho^{1/m}(1 + |\zeta'| + \varrho)^{(m-1)/m} < R(1 + |\zeta'| + \varrho)$ the right-hand side is positive if $|P_m(0,1)|R^m > C(1+R)^{m-1}$, so there are as many zeros $\lambda_j(\zeta' + z')$ as $\lambda_j(\zeta')$ in each component of $Z_{\zeta', r}$ when $|z'| \leq \varrho$. Hence $|\lambda_j(\zeta' + z') - \lambda_j(\zeta')| \leq mr$ with a suitable labelling, which proves (2.7) by the Lipschitz continuity.

To prove (2.8) we homogenize P to $\zeta_0^m P(\zeta/\zeta_0)$ which is equal to $P(\zeta)$ when $\zeta_0 = 1$ and $P_m(\zeta)$ when $\zeta_0 = 0$. If we apply (2.7) to this polynomial with ζ' replaced by $(0, \zeta')$ and z' replaced by $(1, 0)$ then (2.8) follows.

When F_P is not plurisubharmonic we can use the Hölder continuity to get some information on plurisubharmonic minorants:

LEMMA 2.4. *Let φ be an upper semicontinuous function in $\mathbf{C}^N$ such that $|\mathrm{Im}\,\zeta| \leq C_0\varphi(\zeta)$, $\zeta \in \mathbf{C}^N$, for some constant C_0. Then the supremum φ_0 of all plurisubharmonic functions $\psi \leq \varphi$ is plurisubharmonic, φ_0 is positively homogeneous of degree one if φ is, and φ_0 is then Hölder continuous of order $\mu \in (0,1)$ if φ is.*

PROOF. Since the upper semicontinuous regularization of φ_0 is plurisubharmonic and $\leq \varphi$, it is equal to φ_0 so φ_0 is plurisubharmonic. If $\psi(\zeta) \leq \varphi(\zeta)$, $\zeta \in \mathbf{C}^N$, then $t^{-1}\psi(t\zeta) \leq t^{-1}\varphi(t\zeta) = \varphi(\zeta)$, $t > 0$, if φ is positively homogeneous of degree one, so φ_0 is also positively homogeneous of degree one. What remains is to verify that φ_0 then inherits Hölder continuity from φ.

If φ is Hölder continuous of order μ then

$$(2.9) \qquad |\varphi(\zeta + w) - \varphi(\zeta)| \leq C(|\zeta|^{1-\mu}|w|^\mu + |w|), \quad \zeta, w \in \mathbf{C}^N.$$

It suffices to verify this when $|\zeta| = 1$, and then it follows from the Hölder continuity if $|w| < \frac{1}{2}$ and it is trivial if $|w| \geq \frac{1}{2} = \frac{1}{2}|\zeta|$. We can choose a plurisubharmonic function ϱ in $\mathbf{C}^N$ such that

$$(2.10) \qquad |\mathrm{Im}\,\zeta| - a_2(1 + |\zeta|)^\kappa \leq \varrho(\zeta) \leq |\mathrm{Im}\,\zeta| - a_1|\zeta|^\kappa, \quad \zeta \in \mathbf{C}^N,$$

where $\kappa = 1 - \mu$ and a_1, a_2 are positive constants. If $N = 1$ we define such a function ϱ_1 by

$$\varrho_1(z) = \mathrm{Im}\,z - \mathrm{Re}(1 + z/i)^\kappa = \varrho_1(\bar{z}), \quad \mathrm{Im}\,z \geq 0,$$

where the argument of z is taken in $[0, \pi]$. Then $\varrho_1(x + iy) = |y| - \mathrm{Re}(1 + |y| + x/i)^\kappa$ is continuous and

$$\partial \varrho_1(x + i0)/\partial y = 1 - \kappa \, \mathrm{Re}(1 - ix)^{\kappa-1} \geq 1 - \kappa = \mu > 0$$

which proves that $\Delta \varrho_1$ is a positive measure on $\mathbf{R}$ so that ϱ_1 is subharmonic. We have

$$|z|^\kappa \cos(\kappa\pi/2) \leq \mathrm{Re}(1 + z/i)^\kappa \leq (1 + |z|)^\kappa, \quad \mathrm{Im}\, z \geq 0,$$

which proves (2.10) when $N = 1$. For higher dimensions we just have to take

$$\varrho(\zeta) = c \int_{S^{N-1}} \varrho_1(\langle \zeta, \xi \rangle) \, dS(\xi)$$

where $c \int_{S^{N-1}} |\xi_1| dS(\xi) = 1$ and S^{N-1} is the unit sphere in $\mathbf{R}^{N-1}$.

If $\psi \leq \varphi$ then (2.9) gives

$$\psi(\zeta + w) \leq \varphi(\zeta + w) \leq \varphi(\zeta) + C(|\zeta|^\kappa |w|^\mu + |w|), \quad \zeta, w \in \mathbf{C}^N.$$

Adding $Ca_1^{-1}|w|^\mu \varrho(\zeta) - C|w|$ to both sides we obtain by (2.10) since $|\, \mathrm{Im}\, \zeta| \leq C_0 \varphi(\zeta)$

$$\psi(\zeta + w) + Ca_1^{-1}|w|^\mu \varrho(\zeta) - C|w| \leq \varphi(\zeta) + Ca_1^{-1}|w|^\mu |\, \mathrm{Im}\, \zeta| \leq \varphi(\zeta)(1 + C'|w|^\mu).$$

If ψ is plurisubharmonic it follows from the definition of φ_0 that

$$\psi(\zeta + w) + Ca_1^{-1}|w|^\mu \varrho(\zeta) - C|w| \leq \varphi_0(\zeta)(1 + C'|w|^\mu), \quad \zeta, w \in \mathbf{C}^N,$$

and taking the supremum with respect to ψ we can replace $\psi(\zeta + w)$ by $\varphi_0(\zeta + w)$. Hence

$$\varphi_0(\zeta + w) - \varphi_0(\zeta) \leq C''(|w|^\mu + |w|) \quad \text{if } |\zeta| = 1,$$

which proves that φ_0 is Hölder continuous of order μ.

3. The annihilator of the exponential solutions. In this section we shall determine the conditions which $\nu \in \mathcal{E}'$ must satisfy if ν is orthogonal to $E_{P,\bar{Q}}$. It is obvious that the Cauchy data of an exponential solution (1.2) are of the form $f_j(x')e^{i\langle x', \zeta' \rangle}$ where f_j is a polynomial. Conversely, we have

LEMMA 3.1. *For arbitrary given polynomials* $w_j(x')$, $j = 0, \ldots, m - 1$, *and* $\zeta' \in \mathbf{C}^{n-1}$ *the solution of the Cauchy problem*

$$(3.1) \qquad P(D)u = 0, \quad D_n^j u(x)|_{x_n=0} = w_j(x')e^{i\langle x', \zeta' \rangle}, \; j = 0, \ldots, m - 1,$$

is of the form

$$u(x) = \sum_{j=1}^m v_j(x)e^{i\langle x', \zeta' \rangle + i\lambda_j x_n}$$

where v_j *are polynomials and* λ_j *are the zeros of* $P(\zeta', \lambda)$; *thus* u *is an exponential solution of* (1.1) *if* $Q_j(D)u = 0$, $j = 1, \ldots, J$, *when* $x_n = 0$.

PROOF. (Cf. Kiselman [6, Lemma 4.1].) When w_j are constants, the solution is given by

$$(3.2) \qquad u(x) = e^{i\langle x', \zeta' \rangle} \frac{1}{2\pi i} \int_{\partial B_{\zeta'}} \frac{[P(\zeta', \tau) \sum_0^{m-1} w_k/\tau^{k+1}]}{P(\zeta', \tau)} e^{i\tau x_n} \, d\tau$$

where $B_{\zeta'}$ is the union of the discs of radius 1 centered at a zero of $P(\zeta', \cdot)$ and $[\ ]$ means dropping negative powers of τ in the argument. The solution with Cauchy data $w_j x'^\beta e^{i\langle x', \zeta' \rangle}$, $j = 0, \ldots, m - 1$, is obtained by applying $D_{\zeta'}^\beta$ in the right-hand side. The resulting contour integrals can only have powers of $P(\zeta', \tau)$ in the denominator so the claim follows by residue calculus.

LEMMA 3.2. *Let $\nu \in \mathcal{E}'(\mathbf{R}^n)$ and write using Proposition 2.1*

$$(3.3) \qquad \hat{\nu}(-\zeta) = P(\zeta)\Phi(\zeta) + \Psi(\zeta)$$

where Φ and Ψ are entire and $\Psi(\zeta) = \sum_0^{m-1} \Psi_l(\zeta')\zeta_n^l$ is a polynomial of degree $< m$ with respect to ζ_n. If $w_0, \ldots, w_{m-1}$ are polynomials in x' and u is the solution of the Cauchy problem (3.1) given by Lemma 3.1 then

$$(3.4) \qquad \langle u, \nu \rangle = \sum_0^{m-1} w_l(D')\Psi_l(\zeta').$$

PROOF. Assume at first that $w_0, \ldots, w_{m-1}$ are constants. By (3.2) we have

$$\langle u, \nu \rangle = \frac{1}{2\pi i} \int_{\partial B_{\zeta'}} \frac{\hat{\nu}(-\zeta', -\tau)}{P(\zeta', \tau)} [P(\zeta', \tau) \sum_0^{m-1} w_k \tau^{-k-1}] \, d\tau$$

$$= \frac{1}{2\pi i} \int_{\partial B_{\zeta'}} \sum_0^{m-1} \frac{\Psi_l(\zeta')\tau^l}{P(\zeta', \tau)} [P(\zeta', \tau) \sum_0^{m-1} w_k \tau^{-k-1}] \, d\tau = \sum_0^{m-1} w_l \Psi_l(\zeta').$$

Replacing w_j by $w_j x'^\beta$ means replacing u by $D_{\zeta'}^\beta u$, and the right-hand side becomes $\sum_0^{m-1} w_l D_{\zeta'}^\beta \Psi_l(\zeta')$. This completes the proof.

The solution u of (3.1) is in $E_{P,\vec{Q}}$ if and only if

$$(3.5) \qquad Q_j(D)e^{i\langle x', \zeta' \rangle} \sum_0^{m-1} w_l(x')(ix_n)^l/l! = 0 \quad \text{when } x_n = 0, \ j = 1, \ldots, J.$$

If we write $Q_j(\zeta) = \sum_0^{m-1} Q_{jl}(\zeta')\zeta_n^l$ the boundary conditions (3.5) become

$$0 = \sum_{l=0}^{m-1} Q_{jl}(D_{x'})(w_l(x')e^{i\langle x', \zeta' \rangle}) = \sum_{l=0}^{m-1} w_l(D_{\zeta'})(Q_{jl}(\zeta')e^{i\langle x', \zeta' \rangle}), \quad j = 1, \ldots, J,$$

and since linear combinations of exponentials are dense this means that

$$(3.6) \qquad \sum_{l=0}^{m-1} w_l(D_{\zeta'})(Q_{jl}(\zeta')a_j(\zeta')) = 0, \quad j = 1, \ldots, J,$$

for arbitrary entire functions a_j. Thus ν is orthogonal to $E_{P,\vec{Q}}$ if and only if (3.6) implies $\sum_0^{m-1} w_l(D')\Psi_l(\zeta') = 0$. Equivalently, the system of equations

$$\Psi_l(\cdot) = \sum_{j=1}^{J} a_j(\cdot)Q_{jl}(\cdot), \quad l = 0, \ldots, m-1,$$

can be solved in terms of formal power series a_j at ζ'. (See [1, p. 224].) This implies first local solvability and then, by Cartan's Theorem B, global solvability with entire analytic functions a_j, so we have proved:

PROPOSITION 3.3. *A distribution $\nu \in \mathcal{E}'(\mathbf{R}^n)$ is orthogonal to $E_{P,\bar{Q}}$ if and only if*

$$(3.7) \qquad \hat{\nu}(-\zeta) = P(\zeta)\Phi(\zeta) + \sum_1^J a_j(\zeta')Q_j(\zeta), \quad \zeta \in \mathbf{C}^n,$$

where Φ and a_j are entire functions in $\mathbf{C}^n$ and $\mathbf{C}^{n-1}$ respectively.

Using cohomology with bounds as in [1, Section 7.6] we can obtain bounds for a_j. If H is the supporting function of ch supp ν then

$$|\hat{\nu}(-\zeta)| \le C(1 + |\zeta|)^N e^{H(-\operatorname{Im}\zeta)}, \quad \zeta \in \mathbf{C}^n,$$

so it follows from (2.4) and (2.5) that

$$(3.8) \quad |\Psi_k(\zeta')| \le C'(1 + |\zeta'|)^{N+m-1} e^{\tilde{H}(-\zeta')}, \quad \tilde{H}(\zeta') = \max_{P(-\zeta', -\zeta_n)=0} H(\operatorname{Im}\zeta', \operatorname{Im}\zeta_n)$$

$$(3.9) \qquad |\Phi(\zeta)| \le C'(1 + |\zeta|)^N (e^{H(-\operatorname{Im}\zeta)} + e^{\tilde{H}(-\zeta')}).$$

Hence

$$\int |\Psi_k(\zeta')|^2 e^{-2\tilde{H}(-\zeta')}(1 + |\zeta'|^2)^{-(N+m-1+n)}\, d\lambda(\zeta') < \infty$$

where $d\lambda$ is the Lebesgue measure in $\mathbf{C}^{n-1}$, and it follows from [1, Theorem 7.6.11] that a_j can be chosen analytic with

$$\int |a_j(\zeta')|^2 e^{-2\tilde{H}(-\zeta')}(1 + |\zeta'|^2)^{-N'}\, d\lambda(\zeta') < \infty$$

for some other N'. In a ball with center ζ' and radius $(1 + |\zeta'|)^{1-m}$ it follows from (2.7) that $\tilde{H}$ can only oscillate a bounded amount, so the mean value of $|a_j|$ over such a ball is $\le C'' e^{\tilde{H}(-\zeta')}(1 + |\zeta'|)^{N''}$ where $N'' = N' + (n-1)(m-1)$. We have therefore proved:

PROPOSITION 3.4. *The space $C^\infty_{P,\bar{Q}}(\Omega)$ of solutions in $C^\infty(\Omega)$ of the boundary problem* (1.1) *is the orthogonal space of the set of all $\nu \in \mathcal{E}'(\Omega)$ such that* (3.7) *is valid for some entire functions Φ and a_j in $\mathbf{C}^n$ and $\mathbf{C}^{n-1}$ respectively satisfying* (1.6) *where h is the supporting function of some compact convex set $k \subset \omega$. On the other hand, ν is orthogonal to $E_{P,\bar{Q}}$ if and only if* (3.7) *is valid with Φ and a_j entire and for some C and N*

$$(3.10) \qquad |a_j(\zeta')| \le C(1 + |\zeta'|)^N e^{\tilde{H}(-\zeta')}, \quad \zeta' \in \mathbf{C}^{n-1}.$$

Here $\tilde{H}(\zeta') = \max_{P(-\zeta', -\zeta_n)=0} H(\operatorname{Im}\zeta', \operatorname{Im}\zeta_n)$ where H is the supporting function of ch supp ν, *and Φ has the bound* (3.9).

The problem is thus to bridge the gap between the conditions (1.6) and (3.10). A standard regularization procedure using the invariance of $E_{P,\bar{Q}}$ under translation in the tangential x' variables can be used to give some improvement of (3.10):

LEMMA 3.5. *There exist even functions $\psi \in C_0^\infty([-1,1])$ such that $\psi \ge 0$, $\hat{\psi}(0) = 1$,*

$$(3.11) \qquad |\hat{\psi}(\zeta)| \le C \exp(|\operatorname{Im}\zeta| - |\zeta|/(\log|\zeta|)^2), \quad \zeta \in \mathbf{C}, \ |\zeta| > e,$$

and $\lim_{t\to+\infty} t^{-1} \log|\hat\psi(t\zeta)| = |\operatorname{Im}\zeta|$ *uniformly on compact sets where* $\operatorname{Im}\zeta \neq 0$.

PROOF. This follows from a standard construction of functions in Denjoy-Carleman classes as in Section 1.3 of [3]. We first note that the Fourier-Laplace transform of the characteristic function χ of $[-1,1]$ is $2E(\zeta)$ where $E(\zeta) = (\sin\zeta)/\zeta$. We have $E(0) = 1$, $|E(\zeta)| \le e^{|\operatorname{Im}\zeta|}$, $|\zeta E(\zeta)| \le e^{|\operatorname{Im}\zeta|}$ and $|\zeta E(\zeta)|^2 = (\sin\operatorname{Re}\zeta)^2 + (\sinh\operatorname{Im}\zeta)^2 \neq 0$ when $\operatorname{Im}\zeta \neq 0$. If $a_j = j^{-1}(\log j)^{-2}$ when $j \ge 4$ then

$$A = \sum_4^\infty a_j \le \int_3^\infty dt/(t(\log t)^2) = \int_{\log 3}^\infty ds/s^2 = 1/\log 3 < 1.$$

The convolution of the functions $\chi(x/a_j)/(2a_j)$, $j = 4,5,\dots$ converges to a function ψ in $C_0^\infty([-A,A])$ with $\hat\psi(\zeta) = \prod_4^\infty E(a_j\zeta)$, we have $\hat\psi(\zeta) \neq 0$ when $\operatorname{Im}\zeta \neq 0$, and

$$|\hat\psi(\zeta)| \prod_{|a_j\zeta| \ge 1} |a_j\zeta| \le e^{A|\operatorname{Im}\zeta|}, \quad \zeta \in \mathbf{C}.$$

If $T(\log T)^2 = |\zeta|$ then

$$\sum_{|a_j\zeta|\ge 1} (\log a_j + \log|\zeta|) \ge \int_4^T (\log|\zeta| - \log t - 2\log\log t)\,dt = [t(\log|\zeta| - \log t - 2\log\log t)]_4^T$$

$$+ \int_4^T (1 + 2/\log t)\,dt \ge T - 4 - 4(\log|\zeta| - \log 4 - 2\log\log 4) \ge T - 4\log|\zeta|.$$

When $|\zeta| > e$ we have $T < |\zeta|$, hence $T(\log|\zeta|)^2 > T(\log T)^2 = |\zeta|$ and conclude that $|\hat\psi(\zeta)| \exp(|\zeta|/(\log|\zeta|)^2 - 4\log|\zeta|) \le e^{A|\operatorname{Im}\zeta|}$ then. When $\operatorname{Im}\zeta \neq 0$ the function $\log|\hat\psi(\zeta)|$ is harmonic and $\le A|\operatorname{Im}\zeta|$,

$$t^{-1}\log|\hat\psi(it)| = \sum_4^\infty t^{-1}\log|E(ia_jt)| = \sum_4^\infty t^{-1}\log(\sinh(a_jt)/(a_jt)) \to A, \quad t \to +\infty,$$

for the terms are positive and bounded by the limit a_j. Hence it follows from Harnack's inequality that $t^{-1}\log|\hat\psi(t\zeta)| \to A|\operatorname{Im}\zeta|$ uniformly on every compact set where $\operatorname{Im}\zeta \neq 0$. If we replace $\psi(x)$ by $A\psi(Ax)$ then $\hat\psi(\zeta)$ is replaced by $\hat\psi(\zeta/A)$ and the lemma is proved.

REMARK. In (3.11) we could replace the square of $\log|\zeta|$ by any power greater than 1. However, what we shall only use that (3.11) implies for any $\kappa \in (0,1)$ that

(3.11)′ $$|\hat\psi(\zeta)| \le C_\kappa \exp(|\operatorname{Im}\zeta| - |\zeta|^\kappa), \quad \zeta \in \mathbf{C}.$$

If $\Upsilon(x) = \prod_1^N \psi(x_j)$, $x \in \mathbf{R}^N$, then $\operatorname{supp}\Upsilon = \{x \in \mathbf{R}^N, |x_j| \le 1, j = 1,\dots,N\}$ is the unit cube, $\hat\Upsilon(\zeta) = \prod_1^N \hat\psi(\zeta_j)$, $\zeta \in \mathbf{C}^N$, $\hat\Upsilon(0) = 1$,

(3.11)″ $$|\hat\Upsilon(\zeta)| \le C_\kappa \exp\Big(\sum_1^N |\operatorname{Im}\zeta_j| - |\zeta|^\kappa\Big), \quad \zeta \in \mathbf{C}^N,$$

if $0 < \kappa < 1$, and $t^{-1}\log|\hat\Upsilon(t\zeta)| \to \sum_1^N |\operatorname{Im}\zeta_j|$ when $t \to +\infty$ if $\prod_1^N |\operatorname{Im}\zeta_j| \neq 0$.

Let $\Upsilon \in C_0^\infty(\mathbf{R}^{n-1})$ be an even function with support in the unit cube $B \subset \mathbf{R}^{n-1}$ such that $\Upsilon \geq 0$, $\hat{\Upsilon}(0) = 1$, and $(3.11)''$ is valid with $N = n - 1$. Set $\Upsilon_\varepsilon(x') = \varepsilon^{1-n}\Upsilon(x'/\varepsilon)$. If $\nu \in \mathcal{E}'(\Omega)$ then $\nu_\varepsilon = \nu * (\Upsilon_\varepsilon(x') \otimes \delta(x_n)) \in \mathcal{E}'(\Omega)$ and $\nu_\varepsilon \to \nu$ when $\varepsilon \to 0$, so ν is orthogonal to $C_{P,\vec{Q}}^\infty(\Omega)$ if ν_ε is orthogonal to $C_{P,\vec{Q}}^\infty(\Omega)$ for small $\varepsilon > 0$. If ν is orthogonal to $E_{P,\vec{Q}}$ then $\hat{\nu}$ has the decomposition (3.7), and multiplication by $\hat{\Upsilon}(\varepsilon\zeta')$ gives

$$\hat{\nu}_\varepsilon(-\zeta) = P(\zeta)(\hat{\Upsilon}(\varepsilon\zeta')\Phi(\zeta)) + \sum_1^J (\hat{\Upsilon}(\varepsilon\zeta')a_j(\zeta'))Q_j(\zeta).$$

With the notation in (3.10) we have

$$\sum_1^{n-1} |\operatorname{Im}\varepsilon\zeta_j| + \tilde{H}(-\zeta') = \max_{P(-\zeta',-\zeta_n)=0} H_\varepsilon(\operatorname{Im}\zeta',\operatorname{Im}\zeta_n)$$

where $H_\varepsilon(\eta) = \sum_1^{n-1}|\varepsilon\eta_j| + H(\eta)$ is the supporting function of ν_ε by the theorem of supports. By Lemma 2.3 we have

$$\max_{P(-\zeta',-\zeta_n)=0} H_\varepsilon(\operatorname{Im}\zeta',\operatorname{Im}\zeta_n) \leq \max_{P_m(\zeta',\zeta_n)=0} H_\varepsilon(\operatorname{Im}\zeta',\operatorname{Im}\zeta_n) + C(1+|\zeta'|)^{(m-1)/m}$$

so it follows from (3.10) and $(3.11)''$ that

$$|\hat{\Upsilon}(\varepsilon\zeta')a_j(\zeta')| \leq C_\varepsilon \exp(\tilde{H}_\varepsilon(\zeta')), \quad H_\varepsilon(\zeta') = \max_{P_m(\zeta',\zeta_n)=0} H_\varepsilon(\operatorname{Im}\zeta',\operatorname{Im}\zeta_n).$$

Replacing ν_ε by ν and using the Hahn-Banach theorem we have proved:

PROPOSITION 3.6. *The restriction of $E_{P,\vec{Q}}$ to Ω is dense in $C_{P,\vec{Q}}^\infty(\Omega)$ if $C_{P,\vec{Q}}^\infty$ is orthogonal to every $\nu \in \mathcal{E}'(\Omega)$ such that for $0 < \kappa < 1$*

$$(3.12) \qquad |\hat{\nu}(\zeta)| \leq C_\kappa(1+|\zeta_n|)^N \exp(H(\operatorname{Im}\zeta) - |\zeta'|^\kappa), \quad \zeta \in \mathbf{C}^n,$$

where H is the supporting function of $\operatorname{ch}\operatorname{supp}\nu$, *and (3.7) is valid with entire Φ and a_j such that when $\zeta' \in \mathbf{C}^{n-1}$*

$$(3.13) \qquad |a_j(\zeta')| \leq C_\kappa \exp(\tilde{H}(-\operatorname{Im}\zeta') - |\zeta'|^\kappa); \quad \tilde{H}(\zeta') = \max_{P_m(\zeta',\zeta_n)=0} H(\operatorname{Im}\zeta',\operatorname{Im}\zeta_n).$$

COROLLARY 3.7. *If the principal part P_m is hyperbolic and (1.3) is fulfilled, then the restriction of $E_{P,\vec{Q}}$ to Ω is dense in $C_{P,\vec{Q}}^\infty(\Omega)$.*

PROOF. Let $\nu \in \mathcal{E}'(\Omega)$ satisfy (3.12) and assume that (3.13) is valid for the decomposition (3.7). If $K \Subset \Omega$ is the convex hull of $\operatorname{supp}\nu$ there is by $(1.3)'$ a convex compact set $k \subset \omega$ such that $H_K(\eta',\eta_n) \leq h_k(\eta')$ when $\eta = (\eta',\eta_n) \in \mathbf{R}^n$ and $P_m(\eta',\eta_n) = 0$. Since P_m is hyperbolic it follows that $\tilde{H}(\xi') = 0$ and that $\tilde{H}(i\eta') \leq h_k(\eta')$ when $\xi',\eta' \in \mathbf{R}^{n-1}$, so (1.7) is fulfilled with $h = h_k$. Hence ν is orthogonal to $C_{P,\vec{Q}}^\infty(\Omega)$ which proves the corollary.

A slight change of the proof shows that (1.3) is not always a necessary condition for $E_{P,\vec{Q}}$ to be dense in $C_{P,\vec{Q}}^\infty$ as suggested in Examle 1.5. We shall assume for a moment that $J = 1$ and write Q instead of $\vec{Q}$.

COROLLARY 3.8. *If the principal part P_m is hyperbolic and P and Q have no common factor then the restriction of $E_{P,Q}$ to Ω is dense in $C^\infty_{\bar P,Q}(\Omega)$ if*

$$(3.14) \qquad \min_{P_m(\xi',\xi_n)=0} H_\Omega(\xi',\xi_n) = h_\omega(\xi'), \quad \xi' \in \mathbf{R}^{n-1}.$$

Note that (1.3) requires that (3.14) is true with maximum instead of minimum in the left-hand side. In Section 5 we shall justify the hypothesis that P and Q are relatively prime.

PROOF. Let $\nu \in \mathcal{E}'(\Omega)$ satisfy (3.12) and assume that

$$\hat\nu(-\zeta) = P(\zeta)\Phi(\zeta) + a(\zeta')Q(\zeta)$$

with entire Φ and a. Then

$$|a(\zeta')||Q(\zeta',\zeta_n)| \le C_\kappa \exp(H_K(-\operatorname{Im}\zeta', -\operatorname{Im}\zeta_n) - |\zeta'|^\kappa), \quad \text{when } P(\zeta',\zeta_n) = 0,$$

if $0 < \kappa < 1$. Since P and Q are relatively prime there are polynomials R_0 and R_1 such that

$$P(\zeta)R_0(\zeta) + Q(\zeta)R_1(\zeta) = q(\zeta') \not\equiv 0.$$

Hence $|q(\zeta')| \le C'|Q(\zeta',\zeta_n)|(1+|\zeta'|)^{N'}$ when $P(\zeta',\zeta_n) = 0$, and it follows from Lemma 2.3 that

$$|a(\zeta')q(\zeta')| \le C'' \exp(h(-\zeta'))), \quad h(\eta') = \min_{P_m(\eta',\eta_n)=0} H_K(\eta',\eta_n).$$

From (3.14) it follows, as in the proof that (1.3) implies (1.3)$'$, that $h(\eta') \le h_k(\eta')$ for some convex compact set $k \subset \omega$. Hence $a(-\zeta')$ is the Fourier-Laplace transform of a distribution $\in \mathcal{E}'(\omega)$ and the corollary follows from Proposition 3.6.

When P_m is not hyperbolic then the right-hand side of (3.13) will increase exponentially in some directions in $\mathbf{R}^{n-1}$. The simple regularization procedure used in the proof of Proposition 3.6 is no longer applicable then and other methods will be introduced in the following sections. We end this section with two examples which illustrate why we had to use powerful tools to prove Propositions 3.3 and 3.4 in full generality.

EXAMPLE 3.1. Let $P(D)$ be the Laplace operator Δ in $\mathbf{R}^2$. Then $\nu \in \mathcal{E}'$ is orthogonal to the exponential solutions of the Dirichlet problem if and only if

$$\hat\nu(-\zeta) = (\zeta_1^2 + \zeta_2^2)\Phi(\zeta) + a(\zeta_1)$$

with analytic Φ and a. This follows from (3.7) with $Q_1 = 1$. The decomposition just means that $\hat\nu(-\zeta_1, i\zeta_1) = \hat\nu(-\zeta_1, -i\zeta_1)$ so the necessity follows at once by using the exponential solutions $u(x) = e^{ix_1\zeta_1}(e^{x_2\zeta_1} - e^{-x_2\zeta_1})$ with $\zeta_1 \ne 0$. The sufficiency is also elementary. However, if we take instead the boundary operator $Q_1(D) = D_1^2$ then all first order polynomials u also become exponential solutions, which gives $a(0) = a'(0) = 0$. Thus $a(\zeta_1) = \zeta_1^2 b(\zeta_1)$ with b analytic, that is,

$$\hat\nu(-\zeta) = (\zeta_1^2 + \zeta_2^2)\Phi(\zeta) + \zeta_1^2 b(\zeta_1)$$

as stated in Proposition 3.3. This shows that it would not have been enough to study only generic points ζ' in the proof of Proposition 3.3.

EXAMPLE 3.2. Let $P(D)$ be the Laplacian in $\mathbf{R}^3$. Again it is elementary that $\nu \in \mathcal{E}'$ is orthogonal to the exponential solutions of the Dirichlet problem if and only if

$$\hat{\nu}(-\zeta) = \langle \zeta, \zeta \rangle \Phi(\zeta) + a(\zeta')$$

with analytic Φ and a. However, if instead we take the two boundary operators $Q_j(D) = D_j$, $j = 1, 2$, then all constants are also exponential solutions which means that $a(0) = 0$. Thus $a(\zeta') = \zeta_1 a_1(\zeta') + \zeta_2 a_2(\zeta')$ where a_j is entire analytic but not uniquely determined. Choosing them with essentially the same bounds as for a already requires some sophisticated tools.

4. Gaussian regularization. In [2] a similar difficulty in approximating solutions of convolution equations was solved by means of a regularization procedure which is also applicable to the boundary problems studied here under restrictive conditions on Ω and P but no hypotheses on the boundary conditions. In [2, p. 314] we wrote: "This remark can be used to extend the approximation theorems of Kiselman [6] to functions of real variables. We hope to return to this question at some other time." A preliminary manuscript on this approach was written in 1967. It has been resting for 40 years but will be elaborated here.

If we multiply an entire function of exponential type in $\mathbf{C}^{n-1}$ by a Gaussian such as $\exp(-t\langle \zeta', \zeta' \rangle)$ with $t > 0$, then it becomes rapidly decreasing in a conic neighborhood of $\mathbf{R}^{n-1}$, but this will cause problems when $\mathrm{Re}\langle \zeta', \zeta' \rangle < 0$. We therefore choose $\chi \in C^\infty(\mathbf{R})$ such that $\chi(s) = 0$ when $s < 0$ and $\chi(s) = 1$ when $s > 1$, and shall use the cutoff function $\tilde{\chi}(\zeta') = \chi(\mathrm{Re}\langle \zeta', \zeta' \rangle)$ which vanishes then. In the support of $\bar{\partial}\chi(\zeta')$ we have $0 \leq |\mathrm{Re}\,\zeta'|^2 - |\mathrm{Im}\,\zeta'|^2 \leq 1$, hence $|\mathrm{Im}\,\zeta'| \leq |\mathrm{Re}\,\zeta'| \leq |\mathrm{Im}\,\zeta'| + 1$. Starting from the decomposition (3.7) of the Fourier-Laplace transform of a distribution $\nu \in \mathcal{E}'(\Omega)$ orthogonal to the exponential solutions of (1.1) we set

$$(4.1) \qquad a_j^t(\zeta') = (\tilde{\chi}(\zeta')\exp(-t\langle \zeta', \zeta' \rangle) + (1 - \tilde{\chi}(\zeta')))a_j(\zeta') - v_j^t(\zeta'),$$

$$(4.2) \qquad \Phi^t(\zeta) = (\tilde{\chi}(\zeta')\exp(-t\langle \zeta', \zeta' \rangle) + (1 - \tilde{\chi}(\zeta')))\Phi(\zeta) - w^t(\zeta),$$

where v_j^t, w^t shall be chosen so that

$$(4.3) \qquad \bar{\partial}v_j^t(\zeta') = a_j(\zeta')(\exp(-t\langle \zeta', \zeta' \rangle) - 1)\bar{\partial}\tilde{\chi}(\zeta') = r_j^t(\zeta'),$$

$$(4.4) \qquad \bar{\partial}w^t(\zeta) = \Phi(\zeta)(\exp(-t\langle \zeta', \zeta' \rangle) - 1)\bar{\partial}\tilde{\chi}(\zeta') = R^t(\zeta).$$

Then we have

$$\hat{\nu}(-\zeta) - P(\zeta)\Phi^t(\zeta) - \sum_1^J a_j^t(\zeta')Q_j(\zeta)$$

$$= \hat{\nu}(-\zeta)(1 - \exp(-t\langle \zeta', \zeta' \rangle))\tilde{\chi}(\zeta') + P(\zeta)w^t(\zeta) + \sum_1^J Q_j(\zeta)v_j^t(\zeta').$$

If we can find Φ^t with inverse Fourier transform supported by a fixed compact subset of $-\Omega$ and a_j^t with inverse Fourier transform supported by a fixed compact subset of $-\omega$ so that w^t and v_j^t converge to 0 when $t \to 0$ in a suitable topology, we shall obtain an approximation theorem.

When $\mathrm{Re}\langle \zeta', \zeta' \rangle \leq 0$ we have $a_j^t(\zeta') = a_j(\zeta') - v_j^t(\zeta')$, and since no cancellation can be expected between the two terms, the method proposed gives no improvement at all of the estimate (3.10) in this set. Ignoring for a moment the lower order terms of P, this means that for every convex compact subset K of Ω we need to be able to find a convex compact subset k of ω such that for the corresponding support functions

$$(4.5) \qquad H_K(\mathrm{Im}\,\zeta', \mathrm{Im}\,\zeta_n) \leq h_k(\mathrm{Im}\,\zeta') \quad \text{when } \mathrm{Re}\langle \zeta', \zeta' \rangle \leq 0, \; P_m(\zeta', \zeta_n) = 0.$$

LEMMA 4.1. *The inequality (4.5) is equivalent to*

$$(4.6) \qquad H_K(\xi', \mathrm{Re}\, w) \le h_k(\xi') \quad \text{if } \xi', \eta' \in \mathbf{R}^{n-1}, \ P_m(\xi' + i\eta', w) = 0 \text{ and } |\eta'| = |\xi'|,$$

which is the special case of (4.5) when $\mathrm{Re}\langle \zeta', \zeta' \rangle = 0$.

PROOF. With ξ', η', w as in (4.6) and $\zeta' = i(\xi' + i\eta')$, $\zeta_n = iw$, we have $\mathrm{Re}\langle \zeta', \zeta' \rangle = 0$ and $P_m(\zeta', \zeta_n) = 0$, which makes (4.6) the special case of (4.5) when $\mathrm{Re}\langle \zeta', \zeta' \rangle = 0$. For arbitrary $\xi', \eta' \in \mathbf{R}^{n-1} \setminus \{0\}$ the condition $\mathrm{Re}\langle \xi' + z\eta', \xi' + z\eta' \rangle \le 0$ can be written

$$(\mathrm{Re}\, z + \langle \eta', \xi' \rangle / |\eta'|^2)^2 - |\mathrm{Im}\, z|^2 + (|\eta'|^2 |\xi'|^2 - \langle \eta', \xi' \rangle^2)/|\eta'|^4 \le 0$$

which means the interior of two hyperbolic arcs containing the points at infinity on the imaginary z axis. Since $h_k(\mathrm{Im}(\xi' + z\eta')) = |\mathrm{Im}\, z| h_k(\eta' \operatorname{sgn} \mathrm{Im}\, z)$ is harmonic in each of these components and each of them lies between their orthogonal asymptotes, it follows from the Phragmén-Lindelöf theorem applied to the subharmonic function $z \mapsto \max_{P_m(\xi' + z\eta', w) = 0} H_K(\mathrm{Im}(\xi' + z\eta'), \mathrm{Im}\, w)$ that (4.5) is valid when $\zeta' = \xi' + z\eta'$ and $\mathrm{Re}\langle \zeta', \zeta' \rangle \le 0$ if this is true in the special case where $\mathrm{Re}\langle \zeta', \zeta' \rangle = 0$.

LEMMA 4.2. *In order that for every convex compact subset K of Ω there shall exists a convex compact subset k of ω such that (4.5), (4.6) are valid, it is necessary that*

$$(4.7) \qquad H_\Omega(\xi', \mathrm{Re}\, w) = h_\omega(\xi') \quad \text{if } \xi', \eta' \in \mathbf{R}^{n-1}, \ P_m(\xi' + i\eta', w) = 0, \text{ and } |\eta'| \le |\xi'|,$$

and sufficient that this is true when $|\eta'| = |\xi'|$.

PROOF. Since $h_k \le h_\omega$ and H_Ω is the supremum of H_K with respect to compact convex subsets K of Ω, it is clear that (4.7) is necessary. Assume now that (4.7) is fulfilled when $|\eta'| = |\xi'|$. Given a convex compact subset K of Ω we fix ξ', η', w in (4.6) with $|\xi'| = |\eta'| = 1$ and $P_m(\xi' + i\eta', w) = 0$. Then it follows from (4.7) that $H_K(\xi', \mathrm{Re}\, w) < H_\Omega(\xi', \mathrm{Re}\, w) \le h_\omega(\xi')$, so we can find a point $x' \in \omega$ such that $H_K(\xi', \mathrm{Re}\, w) < \langle x', \xi' \rangle$. For reasons of continuity this inequality remains valid in a neighborhood of ξ', η', w. By the Borel-Lebesgue theorem we conclude that (4.6) is valid with k equal to the convex hull of a finite number of points in ω. The proof is complete.

We shall now fill in the details of the approach outlined above.

THEOREM 4.3. *If (4.7) is valid then $E_{P,\vec{Q}}$ is dense in $C^\infty_{P,\vec{Q}}(\Omega)$.*

Condition (4.7) is void when $\Omega = \mathbf{R}^n$ so we obtain, as mentioned in the introduction:

COROLLARY 4.4. *$E_{P,\vec{Q}}$ is dense in $C^\infty_{P,\vec{Q}}(\mathbf{R}^n)$.*

PROOF OF THEOREM 4.3. By Proposition 3.6 it suffices to prove that $\nu \in \mathcal{E}'(\Omega)$ is orthogonal to $C^\infty_{P,\vec{Q}}(\Omega)$ if (3.12) and (3.13) are valid for the decomposition (3.7) of $\hat{\nu}$, where $H = H_K$ is the supporting function of the convex hull K of $\operatorname{supp}\nu$. Let k be a convex compact subset of ω such that (4.5) and (4.6) are valid. When $\mathrm{Re}\langle \zeta', \zeta' \rangle \le 1$ then $|\mathrm{Re}\, \zeta'| \le 1 + |\mathrm{Im}\, \zeta'|$ so ζ' is at distance ≤ 1 from the set where (4.5) is applicable and it follows from (2.7) in Lemma 2.3 that

$$\tilde{H}(\zeta') \le h_k(\mathrm{Im}\, \zeta') + C_1 (1 + |\zeta'|)^{(m-1)/m}, \quad \mathrm{Re}\langle \zeta', \zeta' \rangle \le 1.$$

Here $\tilde{H}$ is defined by (3.13). When $\mathrm{Re}\langle \zeta', \zeta' \rangle \ge 0$ we have $|\exp(-t\langle \zeta', \zeta' \rangle) - 1| \le t|\zeta'|^2$; with r^t_j defined by (4.3) we therefore obtain

$$|r^t_j(\zeta')| \le C_2 t \exp(h_k(-\mathrm{Im}\, \zeta')).$$

By Lemma A.2 we can now choose a solution v_j^t of (4.3) such that

$$(4.8) \qquad |v_j^t(\zeta')| \le C_3 t(1 + |\zeta'|)^{n+\frac{3}{2}} \exp(h_k(-\operatorname{Im}\zeta')).$$

Hence $a_j^t(\zeta') = \hat{\varphi}_j^t(-\zeta')$ where $\varphi_j^t \in \mathcal{E}'(K)$, for (1.7) is fulfilled.
   By (3.7), (3.12) and (3.13) we have for $0 < \kappa < 1$

$$\Phi(\zeta) \le C_\kappa (1 + |\zeta_n|)^N \exp(H_{K_1}(-\operatorname{Im}\zeta) - |\zeta'|^\kappa), \quad \zeta \in \mathbf{C}^n,$$

if $\operatorname{Re}\langle\zeta',\zeta'\rangle \le 1$, hence in $\operatorname{supp}\bar{\partial}\tilde{\chi}$. Here K_1 is the convex hull of K and $k \times \{0\}$. Repeating the arguments concerning v_j^t above, now in $\mathbf{C}^n$, we obtain that there is a solution w^t of (4.4) such that

$$|w^t(\zeta)| \le C_4 t(1 + |\zeta_n|)^{N+1}(1 + |\zeta'|)^{n+\frac{5}{2}} \exp(H_{K_1}(-\operatorname{Im}\zeta)), \quad \zeta \in \mathbf{C}^n.$$

Hence there is a distribution $\varphi^t \in \mathcal{E}'(K_1)$ such that $\Phi^t(\zeta) = \hat{\varphi}^t(-\zeta)$. By (4.8) and (4.9)

$$\left|\hat{\nu}(-\zeta) - \left(P(\zeta)\Phi^t(\zeta) + \sum_1^J a_j^t(\zeta')Q_j(\zeta)\right)\right|$$

$$= \left|\hat{\nu}(-\zeta)(1 - \exp(-t\langle\zeta',\zeta'\rangle))\tilde{\chi}(\zeta') + P(\zeta)w^t(\zeta) + \sum_1^J Q_j(\zeta)v_j^t(\zeta')\right|$$

$$\le C_5 t(1 + |\zeta|)^{N'} \exp(H_{K_1}(-\operatorname{Im}\zeta)),$$

so $P(-D)\varphi^t + \sum_1^J Q_j(-D)(\varphi_j^t \otimes \delta_0) \in \mathcal{E}'(K_1)$ and converges to ν when $t \to 0$. Hence the restriction of $E_{P,\vec{Q}}$ to Ω is dense in $C_{P,\vec{Q}}^\infty(\Omega)$.

There are many sets Ω which satisfy (4.7):

PROPOSITION 4.5. *For every open convex set* $\omega \subset \mathbf{R}^{n-1}$ *there exist open convex sets* $\Omega \subset \mathbf{R}^n$ *satisfying* (4.7) *such that* $\omega = \Omega \cap \mathbf{R}^{n-1}$.

PROOF. Assuming as we may that the origin is an interior point of ω we can let Ω be the convex hull of $\omega \subset \{x \in \mathbf{R}^n; x_n = 0\}$ and $\{(0, x_n) \in \mathbf{R}^n; |x_n| < \delta\}$ with a sufficiently small $\delta > 0$. In fact, then we have

$$H_\Omega(\xi', \xi_n) = \max(h_\omega(\xi'), \delta|\xi_n|),$$

and the condition (4.7) reduces to

$$\delta|\operatorname{Re}w| \le h_\omega(\xi') \quad \text{if } \xi', \eta' \in \mathbf{R}^{n-1}, \ P_m(\xi' + i\eta', w) = 0 \text{ and } |\eta'| \le |\xi'|.$$

Since $|\xi'| \le C_1 h_\omega(\xi')$ and $|w| \le C_2|\xi'|$, this is true when $\delta C_1 C_2 \le 1$.

The proof of Theorem 4.3 has essentially followed an old manuscript referred to in [2]. However, the result can be improved with no major change of the arguments if $\langle\zeta',\zeta'\rangle$ is replaced by $A(\zeta')^\mu$ where $A(\zeta')$ is an arbitrary quadratic form such that A is never ≤ 0 in $\mathbf{R}^{n-1} \setminus \{0\}$ and $\mu \in (1/2, 1)$ is close to $1/2$. This is of course only useful where $|\arg A(\zeta')| < \pi/(2\mu)$, so nothing is improved when $A(\zeta') < 0$, but we can obtain the following improvement of Theorem 4.3:

THEOREM 4.6. *If $A(\zeta')$ is a quadratic form in $\zeta' \in \mathbf{C}^{n-1}$ such that*

$$(4.10) \qquad A(\xi') \leq 0,\ \xi' \in \mathbf{R}^{n-1} \implies \xi' = 0,$$

then $A(\zeta')$ is non-singular, and $E_{P,\bar{Q}}$ is dense in $C^\infty_{P,\bar{Q}}(\Omega)$ if

$$(4.11)\quad H_\Omega(\xi', \operatorname{Re} w) = h_\omega(\xi')\ \text{when}\ \xi', \eta' \in \mathbf{R}^{n-1},\ P_m(\xi' + i\eta', w) = 0,\ A(\xi' + i\eta') \geq 0.$$

PROOF. To prove that A is non-singular assume the contrary, that is, that $\langle t, \partial/\partial\zeta'\rangle A(\zeta') \equiv 0$ for some $t \in \mathbf{C}^{n-1} \setminus \{0\}$. Then $A(\zeta') = 0$ when $\zeta' \in \mathbf{C}t$, so $\mathbf{R}^{n-1} \cap \mathbf{C}t = \{0\}$ by (4.10). Thus $\operatorname{Re} t$ and $\operatorname{Im} t$ are linearly independent, and by a linear change of coordinates preserving $\mathbf{R}^{n-1}$ we may assume that $t = (1, i, 0, \ldots, 0)$. Then $(\partial/\partial\xi_1 + i\partial/\partial\xi_2)A(\xi') = 0$ in $\mathbf{R}^{n-1}$, so $A(\xi')$ is for fixed $(\xi_3, \ldots, \xi_{n-1}) \neq 0$ a quadratic polynomial in $\xi_1 + i\xi_2$ which has no zeros by (4.10). Hence A is independent of ξ_1 and ξ_2 which contradicts (4.10).

Since A is non-singular we can choose complex coordinates $w = (w_1, \ldots, w_{n-1})$ such that $A(z) = \sum_1^{n-1} w_j^2$, thus $\operatorname{Re} A(z) = |w'|^2 - |w''|^2$ and $\operatorname{Im} A(z) = 2\langle w', w''\rangle$ if $w' = \operatorname{Re} w$ and $w'' = \operatorname{Im} w$. The quadratic forms $\operatorname{Re} A$ and $\operatorname{Im} A$ in $\mathbf{C}^{n-1} = \mathbf{R}^{n-1} \oplus \mathbf{R}^{n-1}$ are therefore nondegenerate, the zero sets intersect transversally and $\{z \in \mathbf{C}^{n-1} \setminus \{0\}; A(z) > 0\}$ is outside the origin a conic $2n-3$ dimensional manifold bounded by $\{z \in \mathbf{C}^{n-1} \setminus \{0\}; A(z) = 0\}$. It is fibered by the analytic hypersurfaces $\{z \in \mathbf{C}^{n-1}; A(z) = a\}$ where $a > 0$, and $\arg(-A(z))$ is uniquely defined with values in $(-\pi, \pi)$ in the complement of the closure.

As in the proof of Lemma 4.2 it follows from (4.11) that for every convex compact subset K of Ω there exists a convex compact subset k of ω such that

$$H_K(\operatorname{Re}\zeta', \operatorname{Re} w) \leq h_k(\operatorname{Re}\zeta')\ \text{when}\ P_m(\zeta', w) = 0,\ A(\zeta') \geq 0.$$

We have $|\operatorname{Im}\zeta'| \leq C|\operatorname{Re}\zeta'|$ when $A(\zeta') \geq 0$, for if $\operatorname{Re}\zeta' = 0$ and $\eta' = \operatorname{Im}\zeta' \neq 0$ then $A(\eta') = -A(\zeta') \leq 0$ which contradicts (4.10). Hence it follows for reasons of continuity that for every $\delta > 0$ there is a conic neighborhood V_δ of $\{\zeta' \in \mathbf{C}^{n-1} \setminus \{0\}; A(\zeta') \geq 0\}$ such that

$$H_K(\operatorname{Re}\zeta', \operatorname{Re} w) \leq h_k(\operatorname{Re}\zeta') + \delta|\operatorname{Re}\zeta'|,\quad \text{if}\ \zeta' \in V_\delta,\ P_m(\zeta', w) = 0.$$

We can choose δ so that $k + \{x' \in \mathbf{R}^{n-1}; |x'| \leq \delta\}$ is a compact subset of ω. Replacing k by this set we have

$$H_K(\operatorname{Re}\zeta', \operatorname{Re} w) \leq h_k(\operatorname{Re}\zeta')\ \text{if}\ P_m(\zeta', w) = 0,\ \text{and}\ \zeta' \in V_\delta,$$

that is,

$$(4.12) \qquad H_K(\operatorname{Im}\zeta) \leq h_k(\operatorname{Im}\zeta'),\quad \text{if}\ P_m(\zeta) = 0\ \text{and}\ \zeta' \in iV_\delta.$$

When $\zeta' \notin iV_\delta$ we have $\arg A(\zeta') \in (-\pi, \pi)$, and by the homogeneity it follows that $\arg A(\zeta') \in [-\alpha, \alpha]$ then, for some $\alpha \in (\pi/2, \pi)$. Choose $\mu \in (1/2, \pi/(2\alpha)) \subset (1/2, 1)$ and $\chi \in C^\infty(\mathbf{R})$ so that $\chi(s) = 0$ when $s < 1/2$ and $\chi(s) = 1$ when $s > 1$. Set $\tilde\chi(\zeta') = \chi(\operatorname{Re} A(\zeta')^\mu)$, defined as 0 when $A(\zeta') < 0$. This is a C^∞ function since $\tilde\chi(\zeta') = 0$ when $\operatorname{Re} A(\zeta')^\mu < 1/2$, which is a neighborhood of the set where $A(\zeta') \leq 0$. We have

$$(4.13) \qquad \operatorname{Re} A(\zeta')^\mu \geq \cos(\alpha\mu)|A(\zeta')|^\mu \geq c|\zeta'|^{2\mu},\quad \zeta' \notin iV_\delta,$$

where $c > 0$, for $A(\zeta') \neq 0$ when $\zeta' \notin iV_\delta$. In the support of $\bar\partial\tilde\chi$ we have $\zeta' \neq 0$ and $1/2 \leq \operatorname{Re} A(\zeta')^\mu \leq 1$, which implies that $\zeta' \in iV_\delta$ if $c|\zeta'|^{2\mu} > 1$.

Let $\nu \in \mathcal{E}'(\Omega)$ be orthogonal to $E_{P,\bar{Q}}$ and assume that (3.12), (3.13) are valid for the decomposition (3.7) of $\hat{\nu}$. Let K be the convex hull of $\operatorname{supp} \nu$. If we replace (4.1) and (4.2) by

$$(4.14) \qquad a_j^t(\zeta') = (\tilde{\chi}(\zeta') \exp(-tA(\zeta')^\mu) + (1 - \tilde{\chi}(\zeta')))a_j(\zeta') - v_j^t(\zeta'),$$

$$(4.15) \qquad \Phi^t(\zeta) = (\tilde{\chi}(\zeta') \exp(-tA(\zeta')^\mu) + (1 - \tilde{\chi}(\zeta')))\Phi(\zeta) - w^t(\zeta),$$

where v_j^t and w^t are chosen so that

$$(4.16) \qquad \bar{\partial} v_j^t(\zeta') = a_j(\zeta')(\exp(-tA(\zeta')^\mu) - 1)\bar{\partial}\tilde{\chi}(\zeta') = r_j^t(\zeta'),$$

$$(4.17) \qquad \bar{\partial} w^t(\zeta) = \Phi(\zeta)(\exp(-tA(\zeta')^\mu) - 1)\bar{\partial}\tilde{\chi}(\zeta') = R^t(\zeta).$$

then we have

$$\hat{\nu}(-\zeta) - P(\zeta)\Phi^t(\zeta) - \sum_1^J a_j^t(\zeta')Q_j(\zeta)$$

$$= \hat{\nu}(-\zeta)(1 - \exp(-tA(\zeta')^\mu))\tilde{\chi}(\zeta') + P(\zeta)w^t(\zeta) + \sum_1^J Q_j(\zeta)v_j^t(\zeta').$$

From (4.12) and (2.7) in Lemma 2.3 it follows that

$$\tilde{H}(\zeta') = \max_{P(-\zeta', -\zeta_n)=0} H_K(\operatorname{Im}\zeta', \operatorname{Im}\zeta_n)$$

$$\leq \max_{P_m(\zeta', \zeta_n)=0} H_K(\operatorname{Im}\zeta', \operatorname{Im}\zeta_n) + C_1(1 + |\zeta'|)^{(m-1)/m} \leq h_k(\operatorname{Im}\zeta') + C_1(1 + |\zeta'|)^{(m-1)/m},$$

if $\zeta' \in iV_\delta$. Since $\operatorname{supp} \bar{\partial}\tilde{\chi} \setminus (iV_\delta)$ is bounded it follows that

$$\tilde{H}(\zeta') \leq h_k(\operatorname{Im}\zeta') + C_2(1 + |\zeta'|)^{(m-1)/m}, \quad \text{if } \zeta' \in \operatorname{supp} \bar{\partial}\tilde{\chi}.$$

From this point on we can repeat the proof of Theorem 4.3; v_j^t and w^t can be chosen satisfying (4.16) and (4.17) so that the estimates (4.8) and (4.9) hold, and the proof is completed as before.

When $n = 2$ the condition (4.11) is easy to analyze, for $\zeta' = \zeta_1$ then, and apart from an irrelevant positive factor $A(\zeta') = e^{2i\theta}\zeta_1^2$ where $|\theta| < \pi/2$. The condition (4.11) becomes

$$H_\Omega(\xi_1, \operatorname{Re} w) = h_\omega(\xi_1) \text{ when } \xi_1, \eta_1 \in \mathbf{R}, \ P_m(\xi_1 + i\eta_1, w) = 0, \ e^{i\theta}(\xi_1 + i\eta_1) = t \in \mathbf{R}.$$

With $W = e^{i\theta}w$, this means that $P_m(t, W) = 0$, $\xi_1 = t\cos\theta$, $\eta_1 = -t\sin\theta$, so

$$H_\Omega(t, \operatorname{Re} W + \tan\omega \operatorname{Im} W) = h_\omega(t) \quad \text{when } P_m(t, W) = 0, \ t \in \mathbf{R}.$$

If $\omega = (a, b) \subset \mathbf{R}$ and $\lambda_1, \ldots, \lambda_m$ are the zeros of $P_m(1, W)$, it follows from Theorem 4.6 that the restriction of $E_{P,\bar{Q}}$ to Ω is dense in $C_{P,\bar{Q}}^\infty$ if for some $c \in \mathbf{R}$

$$(4.18) \qquad a < x_1 + (\operatorname{Re}\lambda_j + c\operatorname{Im}\lambda_j)x_2 < b, \quad x \in \Omega, \ j = 1, \ldots, m.$$

(Theorem 4.3 required this to be true for $c = 1$ and $c = -1$, which shows that Theorem 4.6 is much stronger.) If $P(D)$ is not elliptic and λ_-, λ_+ are the smallest and largest real zero λ_j then condition (1.3) means that

$$a < x_1 + \lambda_{\pm} x_2 < b, \quad x \in \Omega,$$

which implies (4.18) if

(4.19) $$\lambda_- \leq \operatorname{Re} \lambda_j + c \operatorname{Im} \lambda_j \leq \lambda_+, \quad j = 1, \ldots, m,$$

that is, all λ_j lie in a strip intersecting $\mathbf{R}$ in $[\lambda_-, \lambda_+]$. When $\operatorname{Im} \lambda_j \neq 0$ this means that c shall belong to an interval with center $\frac{1}{2}(\lambda_- + \lambda_+ - 2 \operatorname{Re} \lambda_j)/\operatorname{Im} \lambda_j$ and length $(\lambda_+ - \lambda_-)/|\operatorname{Im} \lambda_j|$. This means that (4.19) is valid for some c if and only if

(4.20) $\quad |(\lambda_- + \lambda_+ - 2 \operatorname{Re} \lambda_j)/\operatorname{Im} \lambda_j - (\lambda_- + \lambda_+ - 2 \operatorname{Re} \lambda_k)/\operatorname{Im} \lambda_k|$

$$\leq (\lambda_+ - \lambda_-)(1/|\operatorname{Im} \lambda_j| + 1/|\operatorname{Im} \lambda_k|),$$

when $\operatorname{Im} \lambda_j \neq 0$ and $\operatorname{Im} \lambda_k \neq 0$. Thus there are many operators $P(D)$ for which (1.3) implies that $E_{P,\bar{Q}}$ is dense in $C^{\infty}_{P,\bar{Q}}(\Omega)$. When $m \leq 3$ then the condition (4.20) is void, so we have:

THEOREM 4.7. *For a second order non-elliptic operator $P(D)$ in $\mathbf{R}^2$ the condition (1.3) implies that $E_{P,\bar{Q}}$ is dense in $C^{\infty}_{P,\bar{Q}}(\Omega)$.*

When $n > 2$ there are so many admissible choices of the quadratic form A in Theorem 4.6 that a geometric analysis of the consequences seems extremely difficult. The elliptic case will be discussed in Section 5.

5. The case of a single boundary operator. In this section we shall assume that there is only one boundary operator $Q(D)$ as in Corollary 3.8, but the principal part $p(\zeta) = P_m(\zeta)$ may be an arbitrary homogeneous polynomial of degree m in $\mathbf{C}^n$ with $P_m(0,1) \neq 0$. As in the preceding sections we assume throughout that $\Omega \subset \mathbf{R}^n$ is a convex open set with $\omega = \Omega \cap \mathbf{R}^{n-1} \neq \emptyset$. It is convenient and no restriction to assume that $0 \in \omega$.

If $P(\zeta)$ and $Q(\zeta)$ have a non-trivial polynomial factor $A(\zeta)$ in common, then the principal part of A does not vanish at $(0,1)$, and if $u \in C^{\infty}_{P,Q}(\Omega)$ then $v = A(D)u \in C^{\infty}_{P/A,Q/A}(\Omega)$. Conversely, if $v \in C^{\infty}_{P/A,Q/A}(\Omega)$ we can choose $u \in C^{\infty}(\Omega)$ with $P(D)u = v$, and then $u \in C^{\infty}_{P,Q}(\Omega)$. If the restriction of $C^{\infty}_{P,Q}(\mathbf{R}^n)$ to Ω is dense in $C^{\infty}_{P,Q}(\Omega)$ it follows that the restriction of $C^{\infty}_{P/A,Q/A}(\mathbf{R}^n)$ to Ω is dense in $C^{\infty}_{P/A,Q/A}(\Omega)$. Conversely, if this is true and $u \in C^{\infty}_{P,Q}(\Omega)$ we can find a sequence $v_j \in C^{\infty}_{P/A,Q/A}(\mathbf{R}^n)$ such that $v_j \to v = A(D)u$ in $C^{\infty}(\Omega)$. We have $v_j = A(D)u_j$ for some $u_j \in C^{\infty}_{P,Q}(\mathbf{R}^n)$, and $A(D)(u - u_j) = v - v_j \to 0$ in $C^{\infty}(\Omega)$. Hence we can find $w_j \in C^{\infty}(\Omega)$ such that $A(D)w_j = v - v_j = A(D)(u - u_j)$ and $w_j \to 0$ in $C^{\infty}(\Omega)$. Then $A(D)(u - u_j - w_j) = 0$ in Ω so there is a sequence $h_j \in C^{\infty}(\mathbf{R}^n)$ with $A(D)h_j = 0$ such that $u - u_j - w_j - h_j \to 0$ in $C^{\infty}(\Omega)$. Since $C^{\infty}_{P,Q}(\mathbf{R}^n) \ni u_j + h_j \to u$ in $C^{\infty}(\Omega)$ this proves that $C^{\infty}_{P,Q}(\mathbf{R}^n)$ is dense in $C^{\infty}_{P,Q}(\Omega)$. Without any serious restriction we can therefore assume that $P(D)$ and $Q(D)$ are relatively prime. (This argument remains valid with no change if we have $J > 1$ boundary operataors.)

If $\nu \in \mathcal{E}'(\Omega)$ is orthogonal to $E_{P,Q}$ and (3.12) is valid then $\hat{\nu}$ has a decomposition

(5.1) $$\hat{\nu}(-\zeta) = \Phi(\zeta)P(\zeta) + a(\zeta')Q(\zeta)$$

with entire functions $\Phi(\zeta)$ and $a(\zeta')$ such that when $0 < \kappa < 1$

$$(5.2) \qquad |a(\zeta')| \leq C'_\kappa \exp\left(\min_{P(\zeta',\zeta_n)=0} H(-\operatorname{Im}\zeta', -\operatorname{Im}\zeta_n) - |\zeta'|^\kappa\right), \quad \zeta' \in \mathbf{C}^{n-1},$$

$$(5.3) \qquad |a(\zeta')| \leq C'_\kappa \exp\left(\min_{P_m(\zeta',\zeta_n)=0} H(-\operatorname{Im}\zeta', -\operatorname{Im}\zeta_n) - |\zeta'|^\kappa\right), \quad \zeta' \in \mathbf{C}^{n-1}.$$

The proof is a repetition of the first part of the proof of Corollary 3.8. The following lemma will show that this estimate is optimal. To prepare for the proof of Theorem 5.2 we state it for supporting functions of compact sets in $\mathbf{C}^n$ and not only for sets in $\mathbf{R}^n$.

LEMMA 5.1. *Let M be a compact convex subset of $\mathbf{C}^n$ and let $H_M(\zeta) = \sup_{z \in M} \operatorname{Im}\langle z, \zeta\rangle$ be the supporting function. If a is an entire function in $\mathbf{C}^{n-1}$ such that*

$$(5.4) \qquad |a(\zeta')| \leq e^{\varphi(\zeta')}, \quad \varphi(\zeta') = \min_{P(\zeta',\zeta_n)=0} H_M(\zeta', \zeta_n), \quad \zeta' \in \mathbf{C}^{n-1},$$

then one can find an entire analytic function Φ_1 in $\mathbf{C}^n$ such that

$$(5.5) \qquad |\Phi_1(\zeta)P(\zeta) + a(\zeta')| \leq C(1 + |\zeta|)^{m^2+m+n+2} e^{H_M(\zeta)}, \quad \zeta \in \mathbf{C}^n,$$

where C does not depend on a.

PROOF. Choose $\chi \in C^\infty(\mathbf{C}^n)$ so that $\chi(\zeta) = 1$ when the distance from ζ to the zeros Z_0 of P is $\leq \frac{1}{2}(1 + |\zeta|)^{1-m}$, but $\chi(\zeta) = 0$ when the distance is $\geq (1 + |\zeta|)^{1-m}$, and $|\chi'(\zeta)|(1 + |\zeta|)^{1-m}$ is bounded, $0 \leq \chi \leq 1$. Set

$$N(\zeta) = \chi(\zeta)a(\zeta') - P(\zeta)v(\zeta), \quad \text{where} \quad \bar{\partial}v(\zeta) = a(\zeta')P(\zeta)^{-1}\bar{\partial}\chi(\zeta) = r(\zeta),$$

which makes N entire. When $\zeta \in \operatorname{supp}\chi$ we have $\varphi(\zeta') \leq H_M(\zeta) + C$ for some constant C, for we can choose $\tilde{\zeta} \in Z_0$ with $|\tilde{\zeta} - \zeta| \leq (1 + |\zeta|)^{1-m}$, hence $\varphi(\zeta') \leq \varphi(\tilde{\zeta}') + C'$ by Lemma 2.3, and $\varphi(\tilde{\zeta}') \leq H_M(\tilde{\zeta}) \leq H_M(\zeta) + C'$ since H_M is Lipschitz continuous. Since $(1 + |\zeta|)^{-m(m-1)}P(\zeta)^{-1}$ is bounded in $\operatorname{supp}\bar{\partial}\chi$ it follows that

$$|r(\zeta)| \leq C(1 + |\zeta|)^{m^2-1} e^{H_M(\zeta)},$$

and there is a bound for $\chi(\zeta)a(\zeta')e^{-H_M(\zeta)}$. By Lemma A.2 we can choose v so that

$$|v(\zeta)| \leq C'(1 + |\zeta|)^{m^2+n+2} e^{H_M(\zeta)}, \quad \zeta \in \mathbf{C}^n.$$

We have $N(\zeta) = \Phi_1(\zeta)P(\zeta) + a(\zeta')$ where $\Phi_1(\zeta) = (N(\zeta) - a(\zeta'))/P(\zeta)$ is also entire since $\Phi_1 = -v$ in a neighborhood of Z_0, and (5.5) follows from our estimates of $v(\zeta)$ and $\chi(\zeta)a(\zeta')$.

Lemma 5.1 shows that (5.3) is the best estimate of a that one can hope for, and it is also essential in the proof of the following theorem.

THEOREM 5.2. *The closure of $E_{P,Q}$ in $C^\infty(\Omega)$ contains all real analytic elements in $C^\infty_{P,Q}(\Omega)$. Thus $E_{P,Q}$ is dense in $C^\infty_{P,Q}(\Omega)$ if $P(D)$ is elliptic.*

PROOF. As already observed we may assume that $0 \in \Omega$, hence $0 \in \omega$, and that $P(\zeta)$ and $Q(\zeta)$ have no non-trivial factor in common. We must prove that if $u \in C^\infty_{P,Q}(\Omega)$ is real analytic, then u is orthogonal to every $\nu \in \mathcal{E}'(\Omega)$ which is orthogonal to $E_{P,Q}$. We may then assume that an estimate of the form (3.12) is valid for ν itself, which implies that we have a decomposition (5.1) with entire Φ and a satisfying (5.2), (5.3) with $H(\xi) = \sup_{x \in K} \langle x, \xi \rangle$, $\xi \in \mathbf{R}^n$, for some compact convex set $K \subset \Omega$. We can choose $\delta > 0$ so small that u is analytic in a neighborhood of the convex compact set $K_\delta = K + \{z \in \mathbf{C}^n; |z| \leq \delta\}$. The supporting function of K_δ as in Lemma 5.1 is $H_\delta(\zeta) = H(\mathrm{Im}\,\zeta) + \delta|\zeta|$. From (5.2) it follows, as will be proved in Section 6, that there is a sequence of entire functions a_j in $\mathbf{C}^{n-1}$ such that for $\zeta' \in \mathbf{C}^{n-1}$

$$(5.6) \qquad |a(\zeta') - a_j(\zeta')| \leq j^{-1} \exp \big(\min_{P(\zeta', \zeta_n) = 0} H_\delta(-\zeta', -\zeta_n) \big), \quad |a_j(\zeta')| \leq C_j e^{\delta|\zeta'|}.$$

If we apply Lemma 5.1 to $a - a_j$, with $M = -K_\delta$, we obtain a sequence of entire functions Φ_j such that

$$|\Phi_j(\zeta)P(\zeta) + (a(\zeta') - a_j(\zeta'))| \leq Cj^{-1}(1 + |\zeta|)^N e^{H_\delta(-\zeta)}, \quad \zeta \in \mathbf{C}^n,$$

where $N = m^2 + m + n + 2$. Thus

$$\Phi_j(\zeta)P(\zeta) + (a(\zeta') - a_j(\zeta')) = \hat{\nu}_j(-\zeta),$$

where ν_j is an analytic functional carried by K_δ, and $\nu_j \to 0$ as $j \to \infty$ so $\langle \nu_j, u \rangle \to 0$ as $j \to \infty$. We have

$$\hat{\nu}(\zeta) - Q(-\zeta)\hat{\nu}_j(\zeta) = (\Phi(-\zeta) - Q(-\zeta)\Phi_j(-\zeta))P(-\zeta) + Q(-\zeta)a_j(-\zeta').$$

Here $\Phi(-\zeta) - Q(-\zeta)\Phi_j(-\zeta)$ is the Fourier-Laplace transform of an analytic functional F_j carried by K_δ and $a_j(-\zeta')$ is the Fourier-Laplace transform of an analytic functional A_j in $\mathbf{C}^{n-1}$ carried by $\{z' \in \mathbf{C}^{n-1}; |z'| \leq \delta\}$ by the second part of (5.6). Since u is analytic in a neighborhood of these sets we obtain

$$\langle \nu, u \rangle - \langle \nu_j, Q(D)u \rangle = \langle F_j, P(D)u \rangle + \langle A_j \otimes \delta(x_n), Q(D)u \rangle = 0.$$

When $j \to \infty$ it follows that $\langle \nu, u \rangle = 0$ as claimed.

The preceding proof used only a fairly elementary approximation theorem for entire analytic functions. The remainder of this section will be devoted to determining the precise approximation theorems for entire functions which would be required to decide when $E_{P,Q}$ is dense in $C^\infty_{P,Q}(\Omega)$. They will then be studied in Section 6.

THEOREM 5.3. *Let Ω be a bounded open convex neighborhood of $0 \in \mathbf{R}^n$ and let $\omega = \{x \in \Omega; x_n = 0\}$. If $P(\zeta)$ and $Q(\zeta)$ have no non-trivial factor in common then the following conditions are equivalent:*

(i) *$E_{P,Q}$ (restricted to Ω) is dense in $C^\infty_{P,Q}(\Omega)$.*
(ii) *$E_{P,Q}$ (restricted to $r\Omega$) is dense in $C^\infty_{P,Q}(r\Omega)$ for every $r > 0$.*
(iii) *If $\varphi_0(\zeta') = \min_{P_m(\zeta', \zeta_n) = 0} H_\Omega(-\mathrm{Im}\,\zeta', -\mathrm{Im}\,\zeta_n)$, $\zeta' \in \mathbf{C}^{n-1}$, and $0 < r < R$, then for every entire analytic function a in $\mathbf{C}^{n-1}$ such that*

$$(5.7) \qquad |a(\zeta')| \leq (1 + |\zeta'|)^{-1} e^{\varphi_0(r\zeta')}, \quad \zeta' \in \mathbf{C}^{n-1},$$

there is a sequence $A_j \in C^\infty_0(-R\omega)$ such that

$$(5.8) \qquad \sup_{\zeta'} |\hat{A}_j(\zeta') - a(\zeta')| e^{-\varphi_0(R\zeta')} \to 0 \quad \text{when } j \to \infty.$$

PROOF. It is obvious that (ii) $\Longrightarrow$ (i) so we must prove that (iii) $\Longrightarrow$ (ii) and (i) $\Longrightarrow$ (iii).

(iii) $\Longrightarrow$ (ii). We must prove that if $\nu \in \mathcal{E}'(r\Omega)$ is orthogonal to $E_{P,Q}$ and satisfies (3.12) then ν is orthogonal to $C^\infty_{P,Q}(r\Omega)$. By (3.12)

$$|\hat{\nu}(\zeta)| \le C_\kappa (1 + |\zeta|)^N \exp(r_0 H_\Omega(\operatorname{Im} \zeta) - |\zeta'|^\kappa), \quad \zeta \in \mathbf{C}^n, \ 0 < \kappa < 1,$$

for some $r_0 < r$. In the decomposition (5.1) of ν we therefore have by (5.3)

$$|a(\zeta')| \le C(1 + |\zeta'|)^{-1} e^{r_0 \varphi_0(\zeta')}, \quad \zeta' \in \mathbf{C}^{n-1}.$$

If $r_0 < R < r$ it follows from (iii) that there is a sequence $A_j \in C^\infty_0(-R\omega)$ such that

$$\sup |\hat{A}_j(\zeta') - a(\zeta')| e^{-R\varphi_0(\zeta')} \to 0 \text{ when } j \to \infty.$$

We now regularize again by convolution with $\Upsilon_\varepsilon(x') = \varepsilon^{1-n} \Upsilon(x'/\varepsilon)$ in the x' variables as in the proof of Proposition 3.6. If $R < R_1 < r$ and ε is sufficiently small then

$$\sup |\widehat{\Upsilon}(\varepsilon \zeta')(\hat{A}_j(\zeta') - a(\zeta'))| \exp \Big(- \min_{P(\zeta', \zeta_n) = 0} R_1 H_\Omega(-\operatorname{Im} \zeta', -\operatorname{Im} \zeta_n) \Big) \to 0 \quad \text{when } j \to \infty,$$

by Lemma 2.3. By Lemma 5.1 it follows that there is a sequence of entire functions Φ_j and distributions $\nu_j \in \mathcal{E}'(R_1 \bar{\Omega})$ converging to 0 such that

$$\hat{\nu}_j(-\zeta) = \Phi_j(\zeta) P(\zeta) + \widehat{\Upsilon}(\varepsilon \zeta')(\hat{A}_j(\zeta') - a(\zeta')) Q(\zeta).$$

This implies that

$$\hat{\nu}_j(-\zeta) + \widehat{\Upsilon}(\varepsilon \zeta') \hat{\nu}(-\zeta) = (\Phi_j(\zeta) + \widehat{\Upsilon}(\varepsilon \zeta') \Phi(\zeta)) P(\zeta) + \widehat{\Upsilon}(\varepsilon \zeta') \hat{A}_j(\zeta') Q(\zeta).$$

The support of $\nu_j + (\Upsilon_\varepsilon \otimes \delta(x_n)) * \nu$ is contained in $R_1 \bar{\Omega}$ and the support of $\Upsilon_\varepsilon * A_j$ is contained in $-R_1 \omega$. Hence $\nu_j + (\Upsilon_\varepsilon \otimes \delta(x_n)) * \nu$ is orthogonal to $C^\infty_{P,Q}(r\Omega)$. When $j \to \infty$ then $\nu_j \to 0$ so we may conclude that $(\Upsilon_\varepsilon \otimes \delta(x_n)) * \nu$ is also orthogonal which implies that ν is orthogonal if we let $\varepsilon \to 0$ afterwards.

(i) $\Longrightarrow$ (iii). Let a be an entire function in $\mathbf{C}^{n-1}$ satisfying (5.7) for some $r < 1$, and let $r < r_1 < 1$. If $\varepsilon > 0$ is sufficiently small we have by Lemma 2.3

$$|\widehat{\Upsilon}(\varepsilon \zeta') a(\zeta')| \le C_{\kappa,\varepsilon} (1 + |\zeta'|)^{-1} \exp(\varphi_0(r\zeta') + C\varepsilon|\operatorname{Im} \zeta'| - |\zeta'|^\kappa)$$
$$\le C'_\varepsilon (1 + |\zeta'|)^{-1} \exp \Big(\min_{P(\zeta', \zeta_n) = 0} r_1 H_\Omega(-\operatorname{Im} \zeta', -\operatorname{Im} \zeta_n) \Big).$$

By Lemma 5.1 with $K = -r_1 \bar{\Omega}$ it follows that we can find an entire function Φ_ε and $\nu_\varepsilon \in \mathcal{E}'(r_1 \bar{\Omega})$ such that

$$\hat{\nu}_\varepsilon(-\zeta) = \Phi_\varepsilon(\zeta) P(\zeta) + \widehat{\Upsilon}(\varepsilon \zeta') a(\zeta') Q(\zeta).$$

If $E_{P,Q}$ is dense in $C^\infty_{P,Q}(\Omega)$ then ν_ε must be in the closed hull of the distributions

$$(5.9) \qquad P(-D)\phi + Q(-D)\chi \otimes \delta_0, \quad \phi \in C^\infty_0(\Omega), \ \chi \in C^\infty_0(\omega),$$

for every locally convex topology in $\mathcal{E}'(\Omega)$ such that the dual space is $C^\infty(\Omega)$, for by the Hahn-Banach theorem there would otherwise exist some $u \in C^\infty(\Omega)$ with $P(D)u = 0$ in Ω and $Q(D)u = 0$ in ω but $\langle \nu_\varepsilon, u \rangle \neq 0$. Since $C^\infty(\Omega)$ is a Montel space, this is true for the strong dual.

If $\sigma > 0$ then the set of all $u \in C^\infty(\Omega)$ such that

$$|D^\alpha u(x)| \leq (|\alpha|!)^\sigma, \quad x \in \Omega, \ \forall \alpha,$$

is bounded in $C^\infty(\Omega)$, so the set of distributions $\nu \in \mathcal{E}'(\Omega)$ such that

$$|\langle \nu, u \rangle| \leq \sup_{x \in \Omega, \alpha} |D^\alpha u(x)|(|\alpha|!)^{-\sigma}, \quad u \in C^\infty(\Omega),$$

is a neighborhood of the origin in the strong dual of $C^\infty(\Omega)$. This implies that

(5.10) $$|\hat{\nu}(\zeta)| \leq \exp(H_\Omega(\operatorname{Im}\zeta) + \sigma|\zeta|^{1/\sigma}), \quad \zeta \in \mathbf{C}^n,$$

for $s^k/k!^\sigma = ((s^{1/\sigma})^k/k!)^\sigma \leq \exp(\sigma s^{1/\sigma})$ if $s \geq 0$. Thus the set of all $\nu \in \mathcal{E}'(\Omega)$ satisfying (5.10) is a neighborhood of the origin in the strong dual.

In particular the distribution $\nu_\varepsilon \in \mathcal{E}'(r_1\bar{\Omega})$ is a limit of distributions of the form (5.9) with respect to the seminorm corresponding to (5.10). Thus we can find $A_{\varepsilon,j} \in C_0^\infty(-\omega)$ such that

$$|\hat{A}_{\varepsilon,j}(\zeta') - \widehat{\Upsilon}(\varepsilon\zeta')a(\zeta')| \leq j^{-1}\exp(H_\Omega(-\operatorname{Im}\zeta) + \sigma|\zeta|^{1/\sigma}), \quad \zeta' \in \mathbf{C}^{n-1}, \ P(\zeta) = 0,$$

hence, again by Lemma 2.3,

$$|\hat{A}_{\varepsilon,j}(\zeta') - \widehat{\Upsilon}(\varepsilon\zeta')a(\zeta')| \leq j^{-1}\exp\left(\varphi_0(\zeta') + C_0|\zeta'|^{1-1/m}\right), \quad \zeta' \in \mathbf{C}^{n-1},$$

if $\sigma < m/(m-1)$. The extra term in the exponential can be removed by multiplication with another factor $\widehat{\Upsilon}(\varepsilon\zeta')$, which gives

$$|\widehat{\Upsilon}(\varepsilon\zeta')\hat{A}_{\varepsilon,j}(\zeta') - \widehat{\Upsilon}(\varepsilon\zeta')^2 a(\zeta')| \leq C_\varepsilon j^{-1}\exp(\varphi_0(\zeta') + C\varepsilon|\operatorname{Im}\zeta'|), \quad \zeta' \in \mathbf{C}^{n-1}.$$

If $R > 1$ and ε is sufficiently small, then $\hat{\chi}(\varepsilon\zeta')\hat{A}_{\varepsilon,j}(\zeta')$ is the Fourier-Laplace transform of a function in $C_0^\infty(-R\omega)$, and

$$|\widehat{\Upsilon}(\varepsilon\zeta')\hat{A}_{\varepsilon,j}(\zeta') - a(\zeta')| \leq C_\varepsilon j^{-1}\exp(\varphi_0(\zeta') + C\varepsilon|\operatorname{Im}\zeta'|) + |\widehat{\Upsilon}(\varepsilon\zeta')^2 - 1||a(\zeta')|.$$

Since $|\widehat{\Upsilon}(\varepsilon\zeta')^2 - 1| \leq C\varepsilon|\zeta'|e^{2\varepsilon|\operatorname{Im}\zeta'|}$ it follows from (5.7) that the second term can be estimated by $C\varepsilon e^{(r+\varepsilon c)\varphi_0(\zeta')}$. If we choose $\varepsilon = \varepsilon_j$ converging to 0 so slowly that $C_{\varepsilon_j}^2 < j$, then $r + \varepsilon_j c < R$ for large j and $C_{\varepsilon_j}j^{-1} < j^{-\frac{1}{2}} \to 0$, so we have proved (iii) if $r < 1 < R$. Since φ_0 is homogeneous, replacing ζ' by $t\zeta'$ in (5.7) and (5.8) proves (iii) when $tr < 1 < tR$, which completes the proof.

6. Approximation of entire functions of exponential type. For an entire function F in $\mathbf{C}^N$ such that

$$|F(z)| \leq C\exp(M|z|), \quad z \in \mathbf{C}^N,$$

for some constants C and M, the *indicator function* i_F is defined by

$$i_F(z) = \varlimsup_{\tilde{z} \to z, t \to +\infty} t^{-1}\log|F(t\tilde{z})|, \quad z \in \mathbf{C}^N.$$

It is a plurisubharmonic function which is positively homogeneous of degree one.

DEFINITION 6.1. If φ is a plurisubharmonic function in $\mathbf{C}^N$ which is positively homogeneous of degree one we shall denote by $\mathcal{A}_\varphi$ the set of entire functions of exponential type such that $i_F \leq \varphi$.

If $\tilde{\varphi}$ is a continuous function in $\mathbf{C}^N$ which is positively homogeneous of degree 1 and $\tilde{\varphi}(z) > \varphi(z)$ when $0 \neq z \in \mathbf{C}^N$ then

$$(6.1) \qquad\qquad \sup |F(z)| e^{-\tilde{\varphi}(z)} < \infty,$$

if $F \in \mathcal{A}_\varphi$. Conversely, if F is an entire function such that (6.1) is valid then $i_F \leq \tilde{\varphi}$, so $F \in \mathcal{A}_\varphi$ if and only if (6.1) is valid for all such majorants $\tilde{\varphi}$ of φ. We may replace the condition (6.1) by

$$(6.1)' \qquad\qquad \int |F(z)|^2 e^{-2\tilde{\varphi}(z)} \, d\lambda(z) < \infty,$$

for (6.1) implies that

$$\int |F(z)|^2 e^{-2(\tilde{\varphi}(z)+\delta|z|)} \, d\lambda(z) < \infty$$

for every $\delta > 0$, and (6.1)' implies that

$$|F(z)| \leq C \exp\Big(\sup_{|w| \leq 1} \tilde{\varphi}(z+w) \Big) \leq C_\delta e^{\tilde{\varphi}(z)+\delta|z|}$$

for every $\delta > 0$, since $F(z)$ is equal to the mean value of F over a ball with center at z. We shall also need a notation for spaces of entire functions satisfying conditions such as (6.1)':

DEFINITION 6.2. If φ is a continuous real valued function in $\mathbf{C}^N$ we shall denote by $\mathcal{H}_\varphi$ the set of entire functions F in $\mathbf{C}^N$ such that

$$(6.2) \qquad\qquad \|F\|_\varphi = \left(\int |F(z)|^2 e^{-2\varphi(z)} \, d\lambda(z) \right)^{\frac{1}{2}} < \infty.$$

It is clear that $\mathcal{H}_\varphi$ is a Hilbert space with the norm (6.2).

LEMMA 6.3. Let φ be plurisubharmonic and positively homogeneous of degree 1 in $\mathbf{C}^N$. Then there exists a decreasing sequence of plurisubharmonic functions φ_ν in $\mathbf{C}^N$ such that φ_ν is positively homogeneous of degree 1, $\varphi_\nu \in C^\infty$ and $\varphi_\nu > \varphi_{\nu+1} > \cdots \to \varphi$ in $\mathbf{C}^N \setminus \{0\}$, and the Levi form $\mathcal{L}_{\varphi_\nu}(z;w) = \sum_{j,k=1}^n \partial^2 \varphi_\nu(z)/\partial z_j \partial \bar{z}_k w_j \bar{w}_k$ is strictly positive definite when $z \neq 0$. Thus $\mathcal{A}_\varphi = \cap \mathcal{H}_{\varphi_\nu}$ is a Fréchet space with the topology defined by the seminorms $\| \cdot \|_{\varphi_\nu}$.

PROOF. (Cf. [5, Appendix B].) To regularize φ we choose a non-negative $\chi \in C_0^\infty([-1,1])$ such that

$$\int \chi(\|A\|^2) \, d\lambda(A) = 1, \quad A = (A_{jk})_{j,k=1}^N, \quad \|A\|^2 = \sum_{j,k=1}^N |A_{jk}|^2,$$

where A is a matrix with complex coefficients and $d\lambda(A)$ is the Lebesgue measure in $\mathbf{C}^{N^2} = \mathbf{R}^{2N^2}$. Since

$$\int G(Az)\chi(\nu\|A\|^2) \, \nu^{N^2} d\lambda(A)$$

is a rotation symmetric smooth average of G in a ball of radius $|z|/\sqrt{\nu}$, it follows that

$$\varphi_\nu(z) = \int \varphi(z + Az)\chi(\nu\|A\|^2)\,\nu^{N^2}\,d\lambda(A) + |z|/\nu$$

decreases to $\varphi(z)$ when $\nu \to \infty$ and is a C^∞ homogeneous plurisubharmonic function of z with positive definite Levi form when $z \neq 0$.

By the indicator theorem (see e.g. Sigurdsson [7, Theorem 1.4.1]) there exist entire functions F with $i_F = \varphi$. It is exceptional that $i_F \neq \varphi$ when $F \in \mathcal{A}_\varphi$:

PROPOSITION 6.4. *Let $\mathcal{F}$ be a Fréchet space $\subset C^\infty(\mathbf{C}^N)$ of entire functions of exponential type in $\mathbf{C}^N$. Then $\varphi = \sup_{F \in \mathcal{F}} i_F$ is plurisubharmonic and positively homogeneous of degree 1, and $i_F = \varphi$ for all $F \in \mathcal{F}$ outside a set of first category.*

PROOF. For $k = 1, 2, \ldots$ the set $\mathcal{F}_k$ of all $F \in \mathcal{F}$ with

$$|F(z)| \leq ke^{k|z|}, \quad z \in \mathbf{C}^N,$$

is closed, convex and balanced in $\mathcal{F}$. Since $\cup \mathcal{F}_k = \mathcal{F}$ by hypothesis it follows that $\mathcal{F}_k$ is a neighborhood of the origin in $\mathcal{F}$ when k is large enough. This implies that $i_F(z) \leq k|z|$, $z \in \mathbf{C}^N$, for every $F \in \mathcal{F}$, so it follows that the upper semicontinuous regularization $\tilde{\varphi}$ of φ is plurisubharmonic and positively homogeneous of degree 1; we have $i_F \leq \tilde{\varphi}$ for all $F \in \mathcal{F}$. Fix $z_0 \in \mathbf{C}^N \setminus \{0\}$. If $F \in \mathcal{F}$ and $i_F(z_0) < \tilde{\varphi}(z_0)$ we can choose a_0 so that $i_F(z_0) < a_0 < \tilde{\varphi}(z_0)$. Then

$$\log|F(tz)| \leq ta_0, \quad \text{when } t > T, \ |z - z_0| < 1/T,$$

if T is large enough. Given T the set E_T of $f \in \mathcal{F}$ for which this is true with F replaced by f is closed, and if E_T has interior points for some T then

$$|g(tz)| \leq 2e^{ta_0}, \quad t > T, \ |z - z_0| < 1/T,$$

for all $g \in \mathcal{F}$ in a neighborhood of 0. But this implies $i_g(z) \leq a_0$ for all $g \in \mathcal{F}$ when $|z - z_0| < 1/T$, hence $\tilde{\varphi}(z_0) \leq a_0$, which is a contradiction proving that E_T has no interior points.

If S is a countable set in $\mathbf{C}^N \setminus \{0\}$ it follows that the set of $F \in \mathcal{F}$ such that $i_F(z) < \tilde{\varphi}(z)$ for some $z \in S$ is of the first category. Now one can choose S so that $i_F = \tilde{\varphi}$ if $i_F(z) = \tilde{\varphi}(z)$ when $z \in S$ (see Sigurdsson [7, Lemma 1.4.6]), so this proves that $i_F = \tilde{\varphi}$ for all $F \in \mathcal{F}$ outside a set of first category. In particular $\varphi = \tilde{\varphi}$ which completes the proof.

It is clear that $\mathcal{A}_\varphi$ increases with φ. We shall prove:

THEOREM 6.5. *If ψ and φ are positively homogeneous of degree 1 and plurisubharmonic in $\mathbf{C}^N$, and $\psi \leq \varphi$, then $\mathcal{A}_\psi$ is dense in $\mathcal{A}_\varphi$.*

We postpone the proof to give a corollary.

COROLLARY 6.6. *If ψ and φ are positively homogeneous of degree 1 and plurisubharmonic in $\mathbf{C}^N$ and $\mathcal{A}_\psi \cap \mathcal{A}_\varphi \neq \{0\}$, then $\mathcal{A}_\psi \cap \mathcal{A}_\varphi = \mathcal{A}_V$ where V is the supremum of all plurisubharmonic functions $\leq \min(\psi, \varphi)$ in $\mathbf{C}^n$, which is also a positively homogeneous plurisubharmonic function.*

PROOF. If $F \in \mathcal{A}_\psi \cap \mathcal{A}_\varphi$ and $F \not\equiv 0$ then $i_F \not\equiv -\infty$ and $i_F \leq \min(\psi, \varphi)$. The supremum V of all plurisubharmonic functions $v \leq \min(\psi, \varphi)$ is plurisubharmonic by [4, Theorem 3.4.4] and a remark on page 230 there, and it is obvious that V is also positively homogeneous of degree 1. Since $i_F \leq V$ we have $F \in \mathcal{A}_V \subset \mathcal{A}_\psi \cap \mathcal{A}_\varphi$, for $V \leq \psi$ and $V \leq \varphi$, so $\mathcal{A}_\psi \cap \mathcal{A}_\varphi \subset \mathcal{A}_V \subset \mathcal{A}_\psi \cap \mathcal{A}_\varphi$.

EXAMPLE 6.1. If $N = 1$ then ψ and φ are convex functions and so is V, so V is the largest convex minorant of ψ and φ; it is the supporting function of the convex compact set which is the intersection of those with supporting functions ψ and φ. Convex functions are plurisubharmonic also when $N > 1$ but not conversely. If ψ and φ are convex and the intersection of the corresponding convex compact sets is empty, we can achieve by a translation and unitary transformation that they are located so that $\psi(z) < 0$ in a conic neighborhood of the negative x_1 axis while $\varphi(z) < 0$ in a conic neighborhood of the positive x_1 axis. Then $V < 0$ in a conic neighborhood of the whole axis, so the subharmonic function $\mathbf{C} \ni w \mapsto V(w + z_1, z_2, \ldots, z_n)$ is $< -c|w|$ at infinity on the real axis, hence $\equiv -\infty$, so $V \equiv -\infty$. If $\mathcal{A}_\psi \cap \mathcal{A}_\varphi \neq \{0\}$ it follows that there exists a largest homogeneous convex function $V_c \leq \min(\psi, \varphi)$, and $V_c \leq V$, but in general there is inequality. An example is given by

$$\varphi(z) = |z_1|/4 + |z_2|, \quad \psi(z) = |z_1| + |z_2|/4, \quad z \in \mathbf{C}^2.$$

The largest convex minorant is $V_c(z) = (|z_1| + |z_2|)/4$, but $V(z) \geq \sqrt{|z_1 z_2|} > V_c(z)$ if $|z_1| = |z_2|$. (It is easy to see that $V(z) = v(|z_1|, |z_2|)$ and that the homogeneity and plurisubharmonicity mean that $\log v(e^s, 1)$ is a convex function of s, which gives

$$V(z) = \begin{cases} \sqrt{|z_1 z_2|}, & \text{if } |z_2/4| \leq |z_1| \leq 4|z_2|, \\ |z_1| + |z_2|/4, & \text{if } |z_2| \geq 4|z_1|, \\ |z_1|/4 + |z_2| & \text{if } |z_1| \geq 4|z_2|. \end{cases}$$

Thus it is not obvious how to find the largest minorant V in Corollary 6.6 even in simple examples.)

To prove Theorem 6.5 we need a preliminary lemma.

LEMMA 6.7. *Let φ and ψ be positively homogeneous of degree 1 and plurisubharmonic in $\mathbf{C}^N$, and assume that φ and ψ are in C^∞, $\psi \leq \varphi$, and that the Levi forms $\mathcal{L}_\varphi$ and $\mathcal{L}_\psi$ are strictly positive definite in $\mathbf{C}^N \setminus \{0\}$. Then $\mathcal{H}_\psi$ is dense in $\mathcal{H}_\varphi$.*

PROOF. By hypothesis we have $\varphi(z) - \psi(z) \leq c|z|$ for some positive constant c. Let $\chi \in C^\infty(\mathbf{R})$ be a decreasing function which is equal to 1 on $(-\infty, 0]$ and vanishes on $[1, \infty)$, and set with a constant $A > 1$

$$\varphi_A(z) = \psi(z)(1 - \chi_A(z)) + \varphi(z)\chi_A(z), \quad \chi_A(z) = \chi((\log|z|)/A).$$

Then $\varphi_A(z) \leq \varphi(z)$ with equality when $|z| < 1$, and $\varphi_A(z) \leq \psi(z) + ce^A$. We shall now prove that φ_A is strictly plurisubharmonic if A is sufficiently large. The Levi form of φ_A is given by

$$\mathcal{L}_{\varphi_A}(z; w) = (1 - \chi_A(z))\mathcal{L}_\psi(z; w) + \chi_A(z)\mathcal{L}_\varphi(z; w) + (\varphi(z) - \psi(z))\mathcal{L}_{\chi_A}(z; w)$$
$$+ 2\operatorname{Re}\langle \partial\chi_A(z)/\partial z, w\rangle\langle \partial(\varphi(z) - \psi(z))/\partial\bar{z}, \bar{w}\rangle.$$

With a positive constant c_0 independent of the choice of A, the sum of the first two terms is $\geq c_0|w|^2/|z|$. Direct computation shows that $|\chi_A'(z)| \leq C/(|z|A)$ and $|\chi_A''(z)| \leq C/(|z|^2 A)$, where C is independent of A and z. Hence the last two terms in the Levi form can then be estimated by $C'|w|^2/(|z|A)$. If A is sufficiently large it follows that the Levi form of φ_A is bounded below by $c_0|w|^2/(2|z|)$.

Choose a cutoff function $\gamma \in C_0^\infty(\mathbf{C}^N)$ such that $\gamma(z) = 1$ when $|z| < \frac{1}{2}$ and $\gamma(z) = 0$ when $|z| > 1$. Given $F \in \mathcal{H}_\varphi$ we set

$$F_R(z) = \gamma_R(z)F(z) - v_R(z), \quad \gamma_R(z) = \gamma(z/R),$$

where v_R is a solution of the equation $\bar{\partial} v_R = F \bar{\partial} \gamma_R = g_R$. Since $\varphi(z) = R\varphi_A(z/R)$ in the support of g_R and the Levi form of $R\varphi_A(z/R)$ is bounded below by $c_0|w|^2/(2|z|)$, it follows from [1, Lemma 4.4.1] that we can choose v_R so that

$$\int |v_R(z)|^2 e^{-2R\varphi_A(z/R)}\, d\lambda(z) \leq 4c_0^{-1} \int |z||g_R(z)|^2 e^{-2\varphi(z)}\, d\lambda(z)$$

$$\leq CR^{-1} \int_{|z|>R/2} |F(z)|^2 e^{-2\varphi(z)}\, d\lambda(z) \to 0, \text{ when } R \to \infty.$$

Since $F_R(z) - F(z) = (\gamma_R(z) - 1)F(z) - v_R(z)$ and $R\varphi_A(z/R) \leq \varphi(z)$, it follows that $\|F_R - F\|_\varphi \to 0$ when $R \to \infty$, and since

$$\int_{|z|>R} |F_R(z)|^2 e^{-2\psi(z)}\, d\lambda(z) \leq C\exp(2ce^A R),$$

this completes the proof of the lemma.

PROOF OF THEOREM 6.5. Choose a sequence $\varphi_\nu \downarrow \varphi$ using Lemma 6.3 and similarly a sequence $\psi_\nu \downarrow \psi$, making sure that $\psi_\nu < \varphi_\nu$ in $\mathbf{C}^N \setminus \{0\}$ for every ν. We have to prove that if $F \in \mathcal{A}_\varphi$ we can for every ν and $\varepsilon > 0$ find $G \in \mathcal{A}_\psi$ such that

$$\int |F(z) - G(z)|^2 e^{-2\varphi_\nu(z)}\, d\lambda(z) < \varepsilon^2.$$

We shall in fact prove that this is possible for every $F \in \mathcal{H}_{\varphi_\nu}$, that is, when

$$\int |F(z)|^2 e^{-2\varphi_\nu(z)}\, d\lambda(z) < \infty.$$

Using Lemma 6.7 we first choose an entire function G_ν such that

$$\int |G_\nu|^2 e^{-2\psi_\nu}\, d\lambda < \infty, \quad \int |F - G_\nu|^2 e^{-2\varphi_\nu}\, d\lambda < (\varepsilon/2)^2.$$

We can then successively choose $G_{\nu+1}, G_{\nu+2}, \ldots$ so that for $\mu \geq \nu$

$$\int |G_{\mu+1}|^2 e^{-2\psi_{\mu+1}}\, d\lambda < \infty, \quad \int |G_{\mu+1} - G_\mu|^2 e^{-2\psi_\mu}\, d\lambda < (\varepsilon 2^{\nu-\mu-2})^2.$$

This implies that when $\kappa \geq \mu \geq \nu$

$$\left(\int |G_{\kappa+1} - G_\kappa|^2 e^{-2\psi_\mu}\, d\lambda\right)^{\frac{1}{2}} < \varepsilon 2^{\nu-\kappa-2}.$$

Hence there is an entire function G such that $G_\kappa \to G$ locally uniformly and

$$\int |G - G_\mu|^2 e^{-2\psi_\mu}\, d\lambda \leq (\varepsilon 2^{\nu-\mu-1})^2, \quad \mu \geq \nu.$$

This proves that $G \in \mathcal{A}_\psi$. Taking $\mu = \nu$ we obtain

$$\int |F - G|^2 e^{-2\varphi_\nu}\, d\lambda < \varepsilon^2,$$

which completes the proof.

The strong regularity hypotheses can be removed in Lemma 6.7 at least when φ and ψ are continuous and positive:

PROPOSITION 6.8. *Let φ and ψ be positively homogeneous of degree 1 and continuous and plurisubharmonic in $\mathbf{C}^N$, and assume that $0 < \psi(z) \leq \varphi(z)$ when $0 \neq z \in \mathbf{C}^N$. Then $\mathcal{H}_\psi$ is dense in $\mathcal{H}_\varphi$.*

PROOF. If $F \in \mathcal{H}_\varphi$ then

$$\lim_{t \to 1-0} \int |F(z) - F(tz)|^2 e^{-2\varphi(z)} \, d\lambda(z) = 0.$$

This is obvious for the integration over a compact set. Now

$$\int_{|z|>R} |F(tz)|^2 e^{-2\varphi(z)} \, d\lambda(z) = t^{-2N} \int_{|z|>tR} |F(z)|^2 e^{-2\varphi(z)/t} \, d\lambda(z)$$

$$\leq t^{-2N} \int_{|z|>tR} |F(z)|^2 e^{-2\varphi(z)} \, d\lambda(z) \to 0 \quad \text{when } R \to \infty$$

uniformly in t when $\frac{1}{2} \leq t \leq 1$, which proves the statement. Since $\varphi > 0$ outside the origin we can choose φ_t strictly plurisubharmonic with $\varphi(t\cdot) \leq \varphi_t \leq \varphi$ and correspondingly ψ_t with $\psi(t\cdot) \leq \psi_t \leq \psi$ and $\psi_t \leq \varphi_t$. For the function $F_t = F(t\cdot)$ above, with t close to 1, we have $\int |F_t(z)|^2 e^{-2\varphi_t(z)} \, d\lambda(z) < \infty$, so we can find an entire approximation G_t such that

$$\int |F_t(z) - G_t(z)|^2 e^{-2\varphi_t(z)} \, d\lambda(z)$$

is as small as we please while $\int |G_t(z)|^2 e^{-2\psi_t(z)} \, d\lambda(z) < \infty$. This proves the proposition.

With $\varphi(\zeta')$ equal to the largest plurisubharmonic function $\leq H(-\operatorname{Im} \zeta', -\operatorname{Im} \zeta_n)$ when $P_m(\zeta) = 0$, and $\psi = 0$, Theorem 6.5 proves the existence of the approximating sequence in (5.6) and completes the proof of Theorem 5.2. In what follows we shall analyse the approximation condition (iii) in Theorem 5.3 which is a much more difficult problem. Note that in condition (iii) there we can replace φ_0 by the largest plurisubharmonic minorant. By Lemma 2.4 it is also Hölder continuous of order $1/m$. However, we shall first make some remarks on the plurisubharmonic functions which can occur in this context.

PROPOSITION 6.9. *If φ is plurisubharmonic in $\mathbf{C}^N$, positively homogeneous of degree 1, and Hölder continuous, and if $\varphi(z) \geq c|\operatorname{Im} z|$, $z \in \mathbf{C}^N$, for some $c > 0$, then there exist entire analytic functions F in $\mathbf{C}^N$ such that $i_F = \varphi$ and*

$$|F(z)| \leq C_\kappa \exp(\varphi(z) - |z|^\kappa), \quad z \in \mathbf{C}^N,$$

for every $\kappa \in (0,1)$.

PROOF. If φ is Hölder continuous of order $\mu \in (0,1)$ then

$$|\varphi(z+w) - \varphi(z)| \leq C(1+|z|)^{1-\mu}, \quad |w| \leq 1.$$

Hence Theorem 1.4.1 in Sigurdsson [7] proves that there are entire analytic functions f with $i_f = \varphi$ and

$$\int |f(z)|^2 (1+|z|^2)^{-1-3N} \exp(-2\varphi(z) - 4C|z|^{1-\mu}) \, d\lambda(z) < \infty.$$

Since $f(z)$ is equal to the average of f over the ball with radius 1 and center z it follows that

$$|f(z)| \leq C_1 (1 + |z|)^{1+3N} \exp(\varphi(z) + 4C|z|^{1-\mu}), \quad z \in \mathbf{C}^\nu.$$

To improve the behavior at infinity we choose as in the remark following Lemma 3.5 a function $\Upsilon \in C_0^\infty(\mathbf{R}^N)$ with $\operatorname{supp} \Upsilon = \{x \in \mathbf{R}^n; |x_j| \leq 1, j = 1, \dots, N\}$ such that $\widehat{\Upsilon}(0) = 1$ and

(6.3)
$$|\widehat{\Upsilon}(z)| \leq C_\kappa \exp\Big(\sum_1^N |\operatorname{Im} z_j| - |z|^\kappa\Big), \quad z \in \mathbf{C}^N,$$

when $0 < \kappa < 1$, and $\lim_{t\to+\infty} t^{-1}\log|\widehat{\Upsilon}(tz)| = \sum_1^N |\operatorname{Im} z_j|$ when $\prod_1^N |\operatorname{Im} z_j| \neq 0$. If we set $f_\delta(z) = f((1-\delta)z)\widehat{\Upsilon}(\delta cz/N)$ where $0 < \delta < 1$, then $|f_\delta(z)|\exp(|z|^\kappa - \varphi(z))$ is bounded for every $\kappa \in (0,1)$. When $z \in \mathbf{C}^N$ and $\prod_1^N |\operatorname{Im} z_j| \neq 0$ we have

$$i_{f_\delta}(z) = (1-\delta)\varphi(z) + \delta c/N \sum_1^N |\operatorname{Im} z_j| \geq (1-\delta)\varphi(z).$$

If we apply Proposition 6.4 to the Fréchet space $\mathcal{F}$ of all entire functions F with the seminorms $\sup |F(z)|\exp(|z|^\kappa - \varphi(z)) < \infty$, when $\kappa \in (0,1)$ is rational, it follows that $i_F = \varphi$ for all $F \in \mathcal{F}$ outside a set of the first category.

EXAMPLE 6.2. The condition $\varphi(z) \geq c|\operatorname{Im} z|$ in Corollary 6.9 cannot be omitted, for if $\varphi(z) = |\operatorname{Im}\sqrt{\langle z,z\rangle}|$, $z \in \mathbf{C}^N$, and $N \geq 2$, then φ is Hölder continuous of order $\frac{1}{2}$ by Lemma 2.3, and $F \equiv 0$ if

$$|F(z)| \leq C_k(1+|z|)^{-k}e^{\varphi(z)}, \quad z \in \mathbf{C}^N,$$

for every integer k. In fact, if $0 \neq z_0$ and $\langle z_0, z_0\rangle = 0$ then $F(wz_0) \to 0$ as $\mathbf{C} \ni w \to \infty$ so $F(wz_0) = 0$. For every α and k we have

$$|F^{(\alpha)}(z)| \leq C_{\alpha k}(1+|z|)^{-k}e^{\varphi(z)}, \quad z \in \mathbf{C}^N,$$

for $|F^{(\alpha)}(z)| \leq |\alpha|! r^{-|\alpha|}\sup_{|Z-z|\leq r}|F(Z)|$ and $\varphi(Z) \leq \varphi(z) + C|z|^{\frac{1}{2}}r^{\frac{1}{2}} \leq \varphi(z) + C$ when $|Z - z| \leq r \leq |z|$ and $r/|z| \leq 1$. Hence $F^{(\alpha)}(wz_0) = 0$ for every α so $F = 0$.

REMARK. The proof of Theorem 1.4.1 in Sigurdsson [7] shows that given $z_0 \in \mathbf{C}^N$ with $\operatorname{Im} z_0 \neq 0$ we can choose f in the proof of Proposition 6.9 so that

$$\varlimsup_{t\to+\infty} t^{-1}\log|f(tz_0)| = \varphi(z_0),$$

which implies that

$$\varlimsup_{t\to+\infty} t^{-1}\log|f_\delta(tz_0)| \geq (1-\delta)\varphi(z_0),$$

if no coordinate of $\operatorname{Im} z_0$ vanishes. By the category argument used to prove Proposition 6.4 it follows that if φ satisfies the hypotheses of Proposition 6.9 and $\operatorname{Im} z_0 \neq 0$ then

$$\varlimsup_{t\to+\infty} t^{-1}\log|F(tz_0)| = \varphi(z_0)$$

for all $F \in \mathcal{H}_\varphi$ outside a set of the first category.

LEMMA 6.10. *Let φ be plurisubharmonic in $\mathbf{C}^N$ and positively homogeneous of degree one. If φ is Hölder continuous of order $\mu \in (0,1)$ and $F \in \mathcal{H}_\varphi$ then*

$$(6.4) \qquad |F(z)| \leq C\|F\|_\varphi e^{\varphi(z)} (1+|z|)^{N(1-\mu)/\mu} \quad z \in \mathbf{C}^N.$$

PROOF. In the ball with radius $r = (1+|z|)^{(\mu-1)/\mu}$ and center at z we have $\varphi \leq \varphi(z) + C$ by (2.9), hence the average of $|F|$ over this ball can be estimated by $Cr^{-N}e^{\varphi(z)}\|F\|_\varphi$ with another constant C, which proves (6.4) with yet another constant.

We shall now give a reformulation of the condition (iii) of Theorem 5.3.

PROPOSITION 6.11. *Let φ and ψ be plurisubharmonic functions in $\mathbf{C}^N$ which are positively homogeneous of degree one such that φ is Hölder continuous and $c|\operatorname{Im} z| \leq \psi(z) \leq \varphi(z)$ when $z \in \mathbf{C}^N$, for some $c > 0$. Then the following conditions are equivalent:*

(i) $\mathcal{H}_\psi$ *is dense in $\mathcal{H}_\varphi$.*
(ii) *If $0 < r < R$ and F is an entire function in $\mathbf{C}^N$ such that $|F(z)|e^{-r\varphi(z)} \to 0$ when $z \to \infty$, then there is a sequence of entire functions F_j such that $|F_j(z)|e^{-R\psi(z)}$ is bounded for every j and*

$$\sup |F(z) - F_j(z)|e^{-R\varphi(z)} \to 0, \quad j \to \infty.$$

(iii) *The same conclusion is valid for all entire functions F such that $|F(z)|e^{|z|^\kappa - r\varphi(z)}$ is bounded for every $\kappa \in (0,1)$.*
(iv) *φ is almost everywhere equal to the supremum S of the plurisubharmonic functions $s \leq \varphi$ such that $s - \psi$ is bounded above.*

PROOF. By the homogeneity of φ and ψ we may assume that $r < 1 < R$ when proving that (i) $\implies$ (ii). Choose $\Upsilon \in C_0^\infty(\mathbf{R}^N)$ with $\int \Upsilon(x)\,dx = 1$ and support in the unit cube, as in the proof of Proposition 6.9, so that (6.3) is valid when $0 < \kappa < 1$. If $|F(z)|e^{-r\varphi(z)}$ is bounded as in condition (ii), it follows then that $z \mapsto \widehat{\Upsilon}(\varepsilon z)F(z)$ is in $\mathcal{H}_\varphi$ if $r + \varepsilon N/c \leq 1$. By condition (i) we can therefore choose a sequence $F_{\varepsilon,j} \in \mathcal{H}_\psi$ such that $\|\widehat{\Upsilon}(\varepsilon\cdot)F - F_{\varepsilon,j}\|_\varphi \to 0$ when $j \to \infty$. By Lemma 6.10 it follows that

$$r_{\varepsilon,j} = \sup |\widehat{\Upsilon}(\varepsilon z)F(z) - F_{\varepsilon,j}(z)|e^{-\varphi(z)}(1+|z|)^{-N(1-\mu)/\mu} \to 0, \quad j \to \infty.$$

Hence

$$\sup |\widehat{\Upsilon}(\varepsilon z)^2 F(z) - \widehat{\Upsilon}(\varepsilon z)F_{\varepsilon,j}(z)|e^{-R\varphi(z)} \leq r_{\varepsilon,j}\sup e^{(1-R)\varphi(z)}|\widehat{\Upsilon}(\varepsilon z)|(1+|z|)^{N(1-\mu)/\mu} \to 0,$$

when $j \to \infty$ if $\varepsilon N/c \leq R - 1$. Since $|\widehat{\Upsilon}(\varepsilon z)^2 - 1||F(z)|e^{-R\varphi(z)} \to 0$ uniformly on every compact set when $\varepsilon \to 0$ and $|\widehat{\Upsilon}(\varepsilon z)|^2 e^{(r-R)\varphi(z)}$ is uniformly bounded when $\varepsilon \to 0$, we have proved (ii) with $F_j(z) = \widehat{\Upsilon}(\varepsilon_j z)F_{\varepsilon_j,j}(z)$ if $\varepsilon_j \to 0$ sufficiently slowly when $j \to \infty$.

That (ii) $\implies$ (iii) is obvious. To prove that (iii) implies (iv) we choose an entire function F with $i_F = \varphi$ and $|F| \leq e^\varphi$, as in Proposition 6.9. If $R > 1$ there is then by condition (iii) a sequence of entire functions F_j such that $|F_j(z)|e^{-R\psi(z)}$ is bounded for every fixed j and

$$\sup |F(z) - F_j(z)|e^{-R\varphi(z)} \to 0, \quad \text{when } j \to \infty.$$

Thus $|F_j(z)| \leq 2e^{R\varphi(z)}$ for large j, so

$$s_j(z) = R^{-1}\log |F_j(z)/2| \leq \varphi(z),$$

and $s_j(z) - \psi(z)$ is bounded above. When $j \to \infty$ it follows that $S(z) \geq \lim_{j \to \infty} s_j(z) = R^{-1} \log |F(z)/2|$. Hence $\log |F/2| \leq S \leq \varphi$ and $t^{-1} \log |F(tz)/2| \leq S(z) \leq \varphi(z)$. The upper limit of the left-hand side as $t \to +\infty$ is equal to $i_F(z) = \varphi(z)$ almost everywhere which proves (iv).

To prove that (iv) $\Longrightarrow$ (i) finally we take an increasing sequence of plurisubharmonic functions ψ_j with $\psi \leq \psi_j \leq \psi + C_j$ so that $\psi_j \uparrow \varphi$ almost everywhere. Let $F \in \mathcal{H}_\varphi$, $\|F\|_\varphi \leq 1$. Choose $\gamma \in C_0^\infty(\mathbf{C}^N)$ so that $0 \leq \gamma \leq 1$, $\gamma(z) = 1$ when $|z| < 1$ and $\gamma(z) = 0$ when $|z| > 2$, and set

$$G_k(z) = \gamma_k(z) F(z) - v_k(z), \quad \gamma_k(z) = \gamma(z/k),$$

where v_k is a solution of the equation

$$\bar{\partial} v_k(z) = r_k(z) = F(z) \bar{\partial} \gamma_k(z)$$

which makes G_k entire. We have

$$\int |r_k(z)|^2 e^{-2\psi_j(z)} \, d\lambda(z) \leq Ck^{-2} \int_{k < |z| < 2k} |F(z)|^2 e^{-2\psi_j(z)} \, d\lambda(z)$$

$$\to Ck^{-2} \int_{k < |z| < 2k} |F(z)|^2 e^{-2\varphi(z)} \, d\lambda(z), \quad j \to \infty.$$

Hence we can choose j_k so that for large k

$$\int |r_k(z)|^2 e^{-2\psi_j(z)} \, d\lambda(z) \leq k^{-2}, \quad j \geq j_k.$$

Since ψ_j is plurisubharmonic, it follows from [1, Theorem 4.4.2] that we can choose v_k so that

$$\int |v_k(z)|^2 e^{-2\psi_j(z)} (1 + |z|^2)^{-2} \, d\lambda(z) \leq k^{-2}, \quad j \geq j_k.$$

(It suffices to make such a choice when $j = j_k$.) Hence

$$\int |G_k(z)|^2 e^{-2\psi(z)} (1 + |z|^2)^{-2} \, d\lambda(z) < \infty,$$

$$\int |F(z) - G_k(z)|^2 e^{-2\varphi(z)} (1 + |z|^2)^{-2} \, d\lambda(z)$$

$$\leq 2 \int_{|z| > k} |F(z)|^2 e^{-2\varphi(z)} (1 + |z|^2)^{-2} \, d\lambda(z) + 2k^{-2} \to 0, \quad k \to \infty,$$

and it only remains to remove the factor $(1 + |z|^2)^{-2}$ by another regularization. If $R > 1 + \varepsilon N/c$ we obtain

$$\int |\widehat{\Upsilon}(\varepsilon z) G_k(z)|^2 e^{-2R\psi(z)} \, d\lambda(z) < \infty,$$

$$\int |\widehat{\Upsilon}(\varepsilon z) F(z) - \widehat{\Upsilon}(\varepsilon z) G_k(z)|^2 e^{-2R\varphi(z)} \, d\lambda(z) \to 0, \quad k \to \infty,$$

$$\int |\widehat{\Upsilon}(\varepsilon z) F(z) - F(z)|^2 e^{-2R\varphi(z)} \, d\lambda(z) \to 0, \quad \varepsilon \to 0.$$

If $F \in \mathcal{H}_\varphi$ have therefore found that F is in the closure of $\mathcal{H}_{R\psi}$ in $\mathcal{H}_{R\varphi}$, for every $R > 1$. Changing scales we conclude that if $F \in \mathcal{H}_{\varphi/R}$ then F is in the closure of $\mathcal{H}_\psi$ in $\mathcal{H}_\varphi$. Since $F \in \mathcal{H}_\varphi$ implies that $F_R = F(\cdot/R) \in \mathcal{H}_{\varphi/R}$ and that $F_R \to F$ in $\mathcal{H}_\varphi$ when $R \downarrow 1$, this completes the proof.

PROPOSITION 6.12. *Let ψ and φ be plurisubharmonic functions in $\mathbf{C}^N$ such that $c|\operatorname{Im} z| \leq \psi(z) \leq \varphi(z)$, $z \in \mathbf{C}^N$ for some $c > 0$, and assume that both are positively homogeneous of degree 1 and Hölder continuous. If $\mathcal{H}_\psi$ is dense in $\mathcal{H}_\varphi$ then*

$$(6.5) \qquad \varphi(wx) = \psi(wx) \quad \text{when } x \in \mathbf{R}^N, \ \varphi(x) = \varphi(-x) = 0, \ w \in \mathbf{C}.$$

PROOF. It suffices to prove this when $x = (1, 0, \ldots, 0)$. If $f \in \mathcal{H}_\psi$ it follows from Lemma 6.10 that

$$(6.6) \qquad |f(z_1, 0)| \leq C\|f\|_\psi e^{\psi(z_1, 0)}(1 + |z_1|)^{N(1-\mu)/\mu}, \quad z_1 \in \mathbf{C}.$$

Since $z_1 \to \psi(z_1, 0)$ is convex and vanishes on $\mathbf{R}$, we have

$$(6.7) \qquad \psi(z_1, 0) = \pm\psi(\pm i, 0)\operatorname{Im} z_1, \quad \text{when } \pm \operatorname{Im} z_1 \geq 0.$$

We can replace ψ by φ in (6.6) and (6.7), so it follows from the Phragmén-Lindelöf theorem that with another constant C

$$|f(z_1, 0)| \leq C\|f\|_\varphi e^{\psi(z_1, 0)}(1 + |z_1|)^{N(1-\mu)/\mu}, \quad z_1 \in \mathbf{C},$$

if $f \in \mathcal{H}_\psi \subset \mathcal{H}_\varphi$. Here we have used that since $\log(1 + |x|) \leq (\log(1 + x^2) + \log 2)/2 = \operatorname{Re}\log(x \pm i) + \log\sqrt{2}$ when $x \in \mathbf{R}$, the Poisson integral of $\log(1+|x|)$ is $\leq \log(1+|w|) + \log\sqrt{2}$ at $w \in \mathbf{C}$. If $\mathcal{H}_\psi$ is dense in $\mathcal{H}_\varphi$ the inequality remains true for all $f \in \mathcal{H}_\varphi$. By the remark following Example 6.2 we can choose $f \in \mathcal{H}_\varphi$ so that $\overline{\lim}_{t\to+\infty} t^{-1}\log|f(tz_1, 0)| = \varphi(z_1, 0)$, if $\operatorname{Im} z_1 \neq 0$, so we obtain $\varphi(z_1, 0) \leq \psi(z_1, 0) \leq \varphi(z_1, 0)$, which proves the proposition.

The condition (6.5) is of course fulfilled for every $w \in \mathbf{C}$ if it is valid when $\operatorname{Re} w = 0$.

For arbitrary plurisubharmonic functions in $\mathbf{C}^N$ which are positively homogeneous of degree 1 it is obvious that

(i) If $\psi \leq \varphi \leq \chi$ and $\mathcal{H}_\psi$ is dense in $\mathcal{H}_\varphi$ then $\mathcal{H}_\psi$ is dense in $\mathcal{H}_\chi$ if $\mathcal{H}_\varphi$ is dense in $\mathcal{H}_\chi$.

(ii) If $\psi \leq \varphi \leq \chi$ and $\mathcal{H}_\psi$ is dense in $\mathcal{H}_\chi$ then $\mathcal{H}_\varphi$ is dense in $\mathcal{H}_\chi$.

Assuming in addition Hölder continuity and a lower bound $c|\operatorname{Im} z|$ which allows application of Proposition 6.11 we have

(iii) If $\psi_j \leq \varphi_j$ and $\mathcal{H}_{\psi_j}$ is dense in $\mathcal{H}_{\varphi_j}$ for $j = 1, \ldots, k$, then $\mathcal{H}_\psi$ is dense in $\mathcal{H}_\varphi$ if $\psi = \max \psi_j$ and $\varphi = \max \varphi_j$, or $\psi = \sum \lambda_j \psi_j$, $\varphi = \sum \lambda_j \varphi_j$ where $\lambda_j > 0$.

(iv) Same conclusion with $\psi = h(\psi_1, \ldots, \psi_k)$ and $\varphi = h(\varphi_1, \ldots, \varphi_k)$ where h is the supporting function of a convex compact subset of the first open octant in $\mathbf{R}^k$.

(v) If the sequences ψ_j and φ_j are infinite but bounded above, then we can take $\psi = \overline{\lim}\, \psi_j$ and $\varphi = \overline{\lim}\, \varphi_j$.

When ψ and φ are convex the general necessary condition in Proposition 6.12 is also sufficient:

PROPOSITION 6.13. *Let ψ and φ be convex functions in $\mathbf{C}^N$ which are positively homogeneous of degree 1. If $c|\operatorname{Im} z| \leq \psi(z) \leq \varphi(z)$ when $z \in \mathbf{C}^N$ for some $c > 0$ then $\mathcal{H}_\psi$ is dense in $\mathcal{H}_\varphi$ if and only if $\psi(iy) = \varphi(iy)$ when $y \in \mathbf{R}^N$ and $\varphi(y) = \varphi(-y) = 0$.*

Before the proof of Proposition 6.13 we shall list some generalities concerning $\mathcal{H}_\psi$ when ψ is an arbitrary convex function which is positively homogeneous of degree one. This means that ψ is the supporting function of a convex compact set $K_\psi \subset \mathbf{C}^N$,

$$\psi(z) = \sup_{\zeta \in K_\psi} \operatorname{Im}\langle z, \zeta\rangle, \ z \in \mathbf{C}^N; \quad K_\psi = \{\zeta \in \mathbf{C}^N; \operatorname{Im}\langle z, \zeta\rangle \leq \psi(z), \ z \in \mathbf{C}^N\}.$$

Since ψ is subadditive we have $\psi(z) + \psi(-z) \geq 0$ and

$$L_\psi = \{z \in \mathbf{C}^N; \psi(z) + \psi(-z) = 0\}$$

is a linear space over $\mathbf{R}$ where ψ is linear. If $\theta \in \mathbf{C}^N$ and $\psi_\theta(z) = \psi(z) - \mathrm{Im}\langle z, \theta \rangle$ then $L_\psi = L_{\psi_\theta}$ and

$$\|F\|_{\mathcal{H}_\psi}^2 = \int |F(z)|^2 e^{-2\psi(z)}\, d\lambda(z) = \int |F(z)e^{i\langle z,\theta\rangle}|^2 e^{-2\psi_\theta(z)}\, d\lambda(z),$$

so $F \mapsto Fe^{i\langle z,\theta\rangle}$ is an isomorphism $\mathcal{H}_\psi \to \mathcal{H}_{\psi_\theta}$. (If $\psi \leq \varphi$ then $\mathcal{H}_\psi$ is dense in $\mathcal{H}_\varphi$ if and only if $\mathcal{H}_{\psi_\theta}$ is dense in $\mathcal{H}_{\varphi_\theta}$.) Since $K_{\psi_\theta} = K_\psi - \{\theta\}$ we have $0 \in K_{\psi_\theta}$ if we choose $\theta \in K_\psi$ so $\psi_\theta \geq 0$ then. If we choose θ in the interior of K_ψ in its affine hull then $\psi_\theta(z) = 0$ implies that $\mathrm{Im}\langle z, \zeta \rangle = 0$ when $\zeta \in K_{\psi_\theta}$ so $\psi_\theta(-z) = 0$ too, and $z \in L_{\psi_\theta}$. Replacing ψ by ψ_θ we can therefore always assume that $\psi(z) \geq 0$ with equality only when $z \in L_\psi$. Then

$$\psi(z + w) \leq \psi(z) + \psi(w) = \psi(w) \leq \psi(-z) + \psi(z + w) = \psi(z + w), \quad z \in L_\psi,$$

that is, ψ is a convex function ≥ 0 in $\mathbf{C}^N/L_\psi$ which is strictly positive except in L_ψ.

If L_ψ contains a complex line, then ψ is constant in that direction so $\mathcal{H}_\psi = \{0\}$ since a holomorphic function in $L^2(\mathbf{C})$ is identically 0. If $\mathcal{H}_\psi \neq \{0\}$ it follows that L_ψ is totally real which we assume from now on. If T is an invertible complex linear map in $\mathbf{C}^N$, then composition with T gives an isomorphism of $\mathcal{H}_\psi$ on $\mathcal{H}_{\psi_T}$ where $\psi_T = \psi \circ T$, and $L_{\psi_T} = T^{-1}L_\psi$. We can choose T so that $T^{-1}L_\psi = \{z = (z', z'') \in \mathbf{C}^n; \mathrm{Im}\, z' = 0, z'' = 0\}$, where $z' = (z_1, \ldots, z_\nu)$ and $z'' = (z_{\nu+1}, \ldots, z_N)$ for some $\nu \leq N$. After a change of variables we may therefore also assume that ψ is a convex function of $(\mathrm{Im}\, z', z'')$ which is strictly positive when $(\mathrm{Im}\, z', z'') \neq 0$, which means that

$$(6.8) \qquad c_1\psi(z) \leq |\mathrm{Im}\, z'| + |z''| \leq c_2\psi(z), \quad z \in \mathbf{C}^N,$$

for some positive constants c_1 and c_2 which implies the hypothesis on ψ in Proposition 6.13. To prove this proposition we also need the following elementary lemma:

LEMMA 6.14. *Let ψ be a continuous plurisubharmonic function in $\mathbf{C}^N$ which is positively homogeneous of degree 1. If $|\mathrm{Im}\, z_1| \leq C\psi(z)$, $z \in \mathbf{C}^N$, then*

$$(6.9) \qquad \varphi(z) = \max\left(\psi(z), \mathrm{Im}(az_1)\right), \quad z \in \mathbf{C}^N,$$

is plurisubharmonic if $a \in \mathbf{C}$, and if $a \in \mathbf{C}\backslash\mathbf{R}$ then φ is the supremum of plurisubharmonic functions s such that $s - \psi$ is bounded above.

PROOF. Let $a \in \mathbf{C} \setminus \mathbf{R}$ and set

$$(6.10)$$
$$\Phi_\varepsilon(z) = \begin{cases} \max(\psi(z), \mathrm{Im}(az_1)/(1+\varepsilon) - \varepsilon\,\mathrm{Re}((-iaz_1)^\gamma)), & \text{when } \mathrm{Im}(az_1) > \psi(z), \\ \psi(z), & \text{when } \mathrm{Im}(az_1) \leq \psi(z), \end{cases}$$

where ε is a small positive number and $\gamma > 1$ will be chosen in a moment. Since $az_1 = \mathrm{Re}(az_1) + i\,\mathrm{Im}(az_1)$ we have $|a|^2\,\mathrm{Im}\, z_1 = -\mathrm{Re}(az_1)\,\mathrm{Im}\, a + \mathrm{Im}(az_1)\,\mathrm{Re}\, a$ which implies

$$|\mathrm{Re}(az_1)||\mathrm{Im}\, a| \leq |a|^2|\mathrm{Im}\, z_1| + |\mathrm{Re}\, a||\mathrm{Im}(az_1)|.$$

By hypothesis $|\operatorname{Im} z_1| \leq C\psi(z)$, $z \in \mathbf{C}^N$, so it follows in the first case of (6.10) that $\operatorname{Im}(az_1) > \psi(z) \geq |\operatorname{Im} z_1|/C$, hence

$$|\operatorname{Re}(az_1)||\operatorname{Im} a| \leq (C|a|^2 + |\operatorname{Re} a|)\operatorname{Im}(az_1).$$

If we now choose $\gamma > 1$ so that $\tan(\pi/2\gamma) > (C|a|^2 + |\operatorname{Re} a|)/|\operatorname{Im} a|$, it follows that $\operatorname{Re}((-iaz_1)^\gamma) \geq c|z_1|^\gamma$ for some $c > 0$. Hence

$$\varphi(z)/(1+\varepsilon) - c\varepsilon|z_1|^\gamma \leq \Phi_\varepsilon(z) \leq \varphi(z), \quad \Phi_\varepsilon(z) \leq \psi(z) + C_\varepsilon, \quad C_\varepsilon = \max(|a||w| - c\varepsilon|w|^\gamma).$$

We have $\Phi_\varepsilon(z) = \psi(z)$ when $\operatorname{Im}(az_1) < \psi(z)(1+\varepsilon)$ so Φ_ε is plurisubharmonic except possibly where $\psi(z)(1+\varepsilon) \leq \operatorname{Im}(az_1) \leq \psi(z)$, hence $\psi(z) = \operatorname{Im}(az_1) = 0$. Since $\Phi_\varepsilon = 0$ then and $\Phi_\varepsilon \geq 0$ everywhere, the mean value property is obvious at such points so Φ_ε is plurisubharmonic. Since $\Phi_\varepsilon \to \varphi$ when $\varepsilon \to 0$, the lemma is proved.

PROOF OF PROPOSITION 6.13. Replacing $\psi(z)$ by $\psi(z) - \operatorname{Im}\langle z, \theta\rangle$ for a suitable θ we may assume that (6.8) is valid so that K_ψ is a neighborhood of the origin in the subspace of $\mathbf{C}^N$ defined by $\operatorname{Im} z' = 0$ where $z' = (z_1, \ldots, z_\nu)$. The linear space L_φ is a subset of $L_\psi = \{(x', 0), x' \in \mathbf{R}^\nu\}$, and after a linear change of the z' variables we may assume that $L_\varphi = \{x \in \mathbf{R}^N; x_j = 0, j > \mu\}$ for some $\mu \leq \nu$. The linear hull of K_φ is then defined by $\operatorname{Im} z_j = 0$ when $j = 1, \ldots, \mu$. If $\mu < \nu$ we can choose the x_j coordinate axes for $\mu < j \leq \nu$ so that their unit vectors e_j are in K_φ. We can then apply Lemma 6.15 with z_1 replaced by z_j and $a = i$ to conclude using (iv) in Proposition 6.11 that $\mathcal{H}_\psi$ is dense in $\mathcal{H}_{\psi_1}$ if $\psi_1(z) = \max(\psi(z), \operatorname{Im} iz_j, j = \mu + 1, \ldots, \nu)$. Since K_{ψ_1} is the convex hull of K_ψ and the vectors ie_j, $j = \mu + 1, \ldots, \nu$, in K_φ, its linear hull is equal to that of K_φ. To complete the proof it suffices to prove that $\mathcal{H}_{\psi_1}$ is dense in $\mathcal{H}_\varphi$. By another translation as in the discussion preceding the proof we may then assume that $K_{\psi_1} \subset K_\varphi$ are both open neighborhoods of the origin in $\{\zeta \in \mathbf{C}^N; \operatorname{Im} \zeta_j = 0, j \leq \mu\}$. Then we have

$$(6.11) \qquad c_1\varphi(z) \leq \sum_1^\mu |\operatorname{Im} z_j| + \sum_{\mu+1}^N |z_j| \leq c_2\psi_1(z), \quad z \in \mathbf{C}^N,$$

with positive constants c_1, c_2, so we can apply Lemma 6.14 with z_1 replaced by any complex linear combination of z_j with $j > \mu$ and conclude using Proposition 6.11 again that $\mathcal{H}_{\psi_1}$ is dense in $\mathcal{H}_{\psi_2}$ if $\psi_2(z) = \max(\psi_1(z), \operatorname{Im}\langle z, \zeta\rangle; \zeta \in K_\varphi, \zeta_j = 0$ if $j \leq \mu)$. Note that K_{ψ_2} is the convex hull of K_{ψ_1} and $\{\zeta \in K_\varphi; \Pi\zeta = 0\}$ where Π is the projection $\mathbf{C}^N \ni \zeta \mapsto (\zeta_1, \ldots, \zeta_\mu) \in \mathbf{C}^\mu$. We can argue in the same way after moving an arbitrary point in the relative interior of K_{ψ_1} to the origin, which proves that $\mathcal{H}_\psi$ is dense in $\mathcal{H}_\chi$ if

$$(6.12) \qquad K_\chi = \{\zeta \in K_\varphi; \Pi\zeta \in \Pi K_{\psi_1}\}.$$

Now $(\zeta_1, \ldots, \zeta_\mu) \in \Pi K_\varphi$ means that

$$\sum_1^N \operatorname{Im} z_j \zeta_j \leq \varphi(z), \quad z \in \mathbf{C}^N,$$

for some $(\zeta_{\mu+1}, \ldots, \zeta_N)$, and it follows from the Hahn-Banach theorem that this is equivalent to

$$\sum_1^\mu \operatorname{Im} z_j \zeta_j \leq \varphi(z, 0), \quad z \in \mathbf{C}^\mu.$$

Since $\varphi(z,0)$ and $\psi_1(z,0)$ are convex functions of $\operatorname{Im} z$ when $z \in \mathbf{C}^\mu$ which are equal by hypothesis, we can replace φ by ψ_1 here so $\Pi K_\varphi = \Pi K_{\psi_1}$ which completes the proof.

REMARK. We have actually proved more: If $\psi \leq \varphi$ are arbitrary convex functions in $\mathbf{C}^N$ which are positively homogeneous of degree 1 then $\mathcal{H}_\psi \neq \{0\}$ if and only if L_ψ is totally real, and $\mathcal{H}_\psi$ is then dense in $\mathcal{H}_\varphi$ if and only if $\varphi = \psi$ in $\mathbf{CL}_\psi$. In general the largest convex positively homogeneous function χ with $\psi \leq \chi \leq \varphi$ such that $\mathcal{H}_\psi$ is dense in $\mathcal{H}_\varphi$ is the largest convex function $\leq \varphi$ such that $\chi = \psi$ in $\mathbf{CL}_\psi$. In fact, since $L_\varphi \subset L_\chi$ it follows from Proposition 6.12 that $\chi = \psi$ in $\mathbf{CL}_\varphi$ if $\mathcal{H}_\psi$ is dense in $\mathcal{H}_\chi$. If χ is the largest convex positively homogeneous function with $\psi \leq \chi \leq \varphi$ such that $\chi = \psi$ in $\mathbf{CL}_\varphi$ it follows from the proof of Theorem 6.13 that $L_\chi = L_\varphi$, hence that $\mathcal{H}_\psi$ is dense in $\mathcal{H}_\chi$.

Proposition 6.13 remains valid if the convexity of ψ is relaxed to plurisubharmonicity provided that $\varphi(y) = \varphi(-y) = 0$, $y \in \mathbf{R}^N$, implies $y = 0$, for then we can replace $\psi(z)$ by $c|\operatorname{Im} z|$.

The proof of Lemma 6.14 is closely related to the Gaussian regularization in Section 4 which can be adapted to proving more generally:

PROPOSITION 6.15. *Let φ and ψ be as in Proposition 6.11, and assume that $\psi(z) = \varphi(z)$ when $A(z) < 0$, $z \in \mathbf{C}^N$, where A is a quadratic form such that*

(6.13) $$\{x \in \mathbf{R}^N; A(x) \leq 0\} = \{0\}.$$

Then $\mathcal{H}_\psi$ is dense in $\mathcal{H}_\varphi$.

PROOF. The condition (6.13) is the same as (4.10) with $n = N + 1$, so we have already proved that A is non-singular and that $\{z \in \mathbf{C}^N \setminus \{0\}; A(z) < 0\}$ is a conic $2N - 1$ dimensional manifold with boundary where $|z| \leq C\psi(z)$ since $\operatorname{Im} z \neq 0$. Hence we can for every $\delta > 0$ find an open conic neighborhood V_δ where $\varphi(z) \leq (1 + \delta)\psi(z)$. Outside this neighborhood we have $\arg A(z) \in [-\alpha, \alpha]$ for some $\alpha \in (\pi/2, \pi)$, and we choose $\mu \in (1/2, \pi/2\alpha) \subset (1/2, 1)$. Then

$$\psi_{\delta,\varepsilon}(z) = \max\left(\varphi(z) - \varepsilon \operatorname{Re} A(z)^\mu, (1 + \delta)\psi(z)\right),$$

defined as $(1 + \delta)\psi(z)$ when $A(z) < 0$, is plurisubharmonic when $z \neq 0$ and equal to $(1 + \delta)\psi$ in V_δ. Since $\psi_{\delta,\varepsilon} \geq 0$ with equality at the origin we have plurisubharmonicity in $\mathbf{C}^N$, and $\psi_{\delta,\varepsilon}/(1 + \delta) \leq \varphi$. We have a bound $\psi_{\delta,\varepsilon}/(1 + \delta) \leq \psi + C_{\delta,\varepsilon}$ when $\varepsilon > 0$, where $C_{\delta,\varepsilon}$ is a constant. The limit when $\varepsilon \to 0$ is $\max(\varphi/(1 + \delta), \psi)$ which converges to φ when $\delta \to 0$ and proves the proposition.

EXAMPLE 6.3. Let ω be a bounded open convex neighborhood of $0 \in \mathbf{R}^N = \mathbf{R}^{n-1}$, and let $\Omega = \{(tx', \pm(1 - t)) \in \mathbf{R}^n; 0 < t \leq 1, x' \in \omega\}$ be the interior of the convex hull of ω and $(0, \pm 1)$, thus $H_\Omega(\xi', \xi_n) = \max(h_\omega(\xi'), |\xi_n|)$, and let $P(D) = D_n^2 - B(D')$ where B is a quadratic form in $\mathbf{C}^N = \mathbf{C}^{n-1}$. Then we know from Example 1.6 that $E_{P,1}$ is dense in $C_{P,1}^\infty(\Omega)$ if and only if $\sqrt{B(\xi')} \leq h_\omega(\xi')$ when $\xi' \in \mathbf{R}^N$ and $B(\xi') \geq 0$. The function φ_0 in Theorem 5.3 becomes $\max(h_\omega(-\operatorname{Im}\zeta'), |\operatorname{Im}\sqrt{B(\zeta')}|)$, so Proposition 6.11 shows that if $\psi(z) = h_\omega(-\operatorname{Im} z)$ and $\varphi(z) = \max(\psi(z), |\operatorname{Im}\sqrt{B(z)}|)$ then $\mathcal{H}_\psi$ is dense in $\mathcal{H}_\varphi$ if and only if $\sqrt{B(\xi')} \leq \psi(\xi')$ when $\xi' \in \mathbf{R}^N$ and $B(\xi') \geq 0$. If $\psi(z) = h(\operatorname{Im} z)$ where h is a convex positively homogeneous function in $\mathbf{R}^N$ with $h(x) \geq c|x|$ for some $c > 0$ and $\varphi(z) = \max(\psi(z), |\operatorname{Im}\sqrt{B(z)}|)$ we conclude that $\mathcal{H}_\psi$ is dense in $\mathcal{H}_\varphi$ if and only if $\sqrt{B(x)} \leq h(x)$ when $x \in \mathbf{R}^N$ and $B(x) \geq 0$. When B is positive semidefinite this implies that $|\operatorname{Im}\sqrt{B(z)}| \leq h(\operatorname{Im} z)$, $z \in \mathbf{C}^N$, so $\varphi = \psi$ and the result is trivial. If B is negative definite the condition is empty and the result follows from Proposition 6.15 with $A = B$.

In Proposition 6.13 the necessary condition given by Proposition 6.12 concerned only a linear set. However, in general the set where Proposition 6.12 is applicable can be quite arbitrary:

PROPOSITION 6.16. *For every closed conic set $E \subset \mathbf{R}^N$ with $E = -E$ there exists a plurisubharmonic function $\varphi \geq 0$ in $\mathbf{C}^N$ which is positively homogeneous of degree one and Hölder continuous of order $\frac{1}{2}$, such that $E = \{z \in \mathbf{C}^N; \varphi(z) = 0\}$ and $h(y) = \varphi(iy)$ is convex, $y \in \mathbf{R}^N$. If $\psi(z) = h(\operatorname{Im} z)$ then $\mathcal{H}_\psi$ is dense in $\mathcal{H}_{\max(\varphi,\psi)}$.*

PROOF. If $E = \{0\}$ we can simply take $\varphi(z) = |z|$, and if E is contained in the subspace of $\mathbf{R}^N$ defined by $x_{\nu+1} = \cdots = x_N = 0$ we can take $\varphi(z) = \varphi_0(z') + |z''|$ where $z' = (z_1, \ldots, z_\nu)$, $z'' = (z_{\nu+1}, \ldots, z_N)$, and φ_0 has the required properties in $\mathbf{C}^\nu$. Hence it is no restriction to assume that $\mathbf{R}^N$ is the linear hull of E, and even that E contains all N coordinate axes.

We start from the familiar plurisubharmonic function

$$u(z) = |\operatorname{Im}\sqrt{\langle z, z\rangle}|, \quad z \in \mathbf{C}^N.$$

Since $(\operatorname{Im} w)^2 = \frac{1}{2}(|w|^2 - \operatorname{Re}(w^2))$, $w \in \mathbf{C}$, we obtain with $w = \sqrt{\langle z, z\rangle}$

$$u(z)^2 = \frac{1}{2}(|\langle z, z\rangle| - \operatorname{Re}\langle z, z\rangle), \quad z \in \mathbf{C}^N.$$

Since $\mathbf{C} \ni w \mapsto \frac{1}{2}(|w| - \operatorname{Re} w)$ has Lipschitz constant 1, it follows that $u(z + Z)^2 \leq u(z)^2 + 2|z||Z| + |Z|^2$, hence $u(z + Z) \leq u(z) + \sqrt{2|z||Z|} + |Z|$ and

(6.14) $\qquad |u(Z) - u(z)| \leq \sqrt{2\max(|z|,|Z|)}|Z - z|^{\frac{1}{2}} + |Z - z|, \quad Z, z \in \mathbf{C}^N.$

If $\omega \in S^{N-1}$, the unit sphere in $\mathbf{R}^N$, and $0 \leq a \leq 1$, then

$$\varphi_{\omega,a}(z) = |\operatorname{Im}\sqrt{a\langle z, z\rangle - \langle z, \omega\rangle^2}|$$

is also plurisubharmonic and satisfies (6.14). It suffices to prove this when $\omega = (1, 0, \ldots, 0)$, and then it is clear since

$$\varphi_{\omega,a}(z) = u(i\sqrt{1 - a}\,z_1, \sqrt{a}\,z_2, \ldots, \sqrt{a}\,z_n).$$

When $x \in \mathbf{R}^N$ and $|x| = 1$ we have $\varphi_{\omega,a}(x) = 0$ if $\langle x, \omega\rangle^2 \leq a$ but $\varphi_{\omega,a}(x) = \sqrt{\langle x, \omega\rangle^2 - a}$ if $\langle x, \omega\rangle^2 > a$. Hence

$$\varphi(z) = \max_{\omega \in S^{N-1}} \varphi_{\omega,a(\omega)}(z), \quad a(\omega) = \max_{x \in E, |x|=1} \langle \omega, x\rangle^2,$$

is plurisubharmonic, positively homogeneous of degree 1, and Hölder continuous of order $\frac{1}{2}$. When $x \in E \cap S^{N-1}$ then $\langle x, \omega\rangle^2 \leq a(\omega)$, $\omega \in S^{N-1}$, so $\varphi(x) = 0$, but $\varphi(\omega) \geq \sqrt{1 - a(\omega)} > 0$ when $\omega \in S^{N-1} \setminus E$. Since

$$1 - a(\omega) = \min_{x \in E, |x|=1}(1 - \langle \omega, x\rangle^2) = \min_{x \in E, |x|=1}|\omega - x\langle \omega, x\rangle|^2 = \min_{x \in E}|\omega - x|^2,$$

it follows that $\sqrt{1 - a(\omega)}$ is the distance from ω to E. By the homogeneity it follows that $\varphi(\omega)$ is bounded below by the distance from ω to E for every $\omega \in \mathbf{R}^N$.

We shall now determine all the zeros of φ in $\mathbf{C}^N$. That $\varphi(z) = 0$ means precisely that

$$a(\omega)\langle z, z\rangle - \langle z, \omega\rangle^2 \geq 0, \quad \omega \in S^{N-1}.$$

Hence

$$a(\omega)\langle \operatorname{Re} z, \operatorname{Im} z\rangle = \langle \operatorname{Re} z, \omega\rangle\langle \operatorname{Im} z, \omega\rangle, \quad \omega \in S^{N-1}.$$

If $\operatorname{Re} z$ and $\operatorname{Im} z$ are not linearly dependent and the angle between them is denoted by 2α, $0 < \alpha < \frac{1}{4}\pi$, then the maximum of the right-hand side is $|\operatorname{Re} z||\operatorname{Im} z|\cos^2 \alpha$, and since $\cos^2 \alpha > \cos^2 \alpha - \sin^2 \alpha = \cos(2\alpha)$, the left-hand side is smaller. Hence $\varphi(z) = 0$ implies $z \in \mathbf{C}\mathbf{R}^N$, that is, $z = wx$ where $w \in \mathbf{C}$ and $x \in \mathbf{R}^N$, $|x| = 1$. Then we have

$$\varphi(wx) = \max\left(|\operatorname{Im} w| \max \sqrt{a(\omega) - \langle x, \omega\rangle^2}, |\operatorname{Re} w| \max \sqrt{\langle x, \omega\rangle^2 - a(\omega)}\right),$$

where the maxima of the square roots are taken for $\omega \in S^{N-1}$ such that the square root is real. We have

$$\max_{|\omega|=1}(a(\omega) - \langle x, \omega\rangle^2) = \max_{y \in E, |y|=1} \max_{|\omega|=1}(\langle y, \omega\rangle^2 - \langle x, \omega\rangle^2), \quad |x| = 1.$$

Since $\langle y, \omega\rangle^2 - \langle x, \omega\rangle^2 = \langle y - x, \omega\rangle\langle y + x, \omega\rangle$ and $y - x$ is orthogonal to $y + x$, the maximum with respect to ω is

$$|y - x||y + x|/2 = \sqrt{2 - 2\langle x, y\rangle}\sqrt{2 + 2\langle x, y\rangle}/2 = \sqrt{1 - \langle x, y\rangle^2},$$

which is the distance from x to the line generated by y, so we have by the homogeneity

$$h(x) = \varphi(ix) = \max_{y \in E, |y|=1} |x - y\langle y, x\rangle|^{\frac{1}{2}}|x|^{\frac{1}{2}}, \quad x \in \mathbf{R}^N.$$

This is a positively homogeneous function of x of degree 1 which is strictly positive when $x \neq 0$. To prove that h is convex we must prove that

$$A = \{x \in \mathbf{R}^N; |x - y\langle x, y\rangle|^2|x|^2 \leq 1, y \in E, |y| = 1\},$$

is convex. Taking for y the unit vectors on the coordinate axes we obtain for $k = 1, \ldots, N$

$$\sum_{j \neq k} |x_j|^2|x|^2 \leq 1, \quad x \in A,$$

and adding we obtain $|x|^4 \leq N/(N-1) \leq 2$ if $N \geq 2$. (The case $N = 1$ is trivial.) Hence the convexity of A will follow if we prove that for every $y \in \mathbf{R}^N$ with $|y| = 1$ the set

$$A_0 = \{x \in \mathbf{R}^N; |x - y\langle x, y\rangle|^2|x|^2 \leq 1, |x|^4 \leq 2\}$$

is convex. It is sufficient to do so when y is the unit vector on the x_1 axis, that is, using the rotation invariance in $x_2, \ldots, x_N$, to prove the convexity of

$$\{x \in \mathbf{R}^2; x_2^2(x_1^2 + x_2^2) \leq 1, x_1^2 + x_2^2 \leq \sqrt{2}\}.$$

In polar coordinates (r, θ) this is defined by $r \leq \min(|\sin \theta|^{-\frac{1}{2}}, 2^{\frac{1}{4}})$. For a curve $r = r(\theta)$ the curvature counted positive toward the origin is equal to

$$(r(\theta)^2 + 2r'(\theta)^2 - r''(\theta)r(\theta))(r(\theta)^2 + r'(\theta)^2)^{-\frac{3}{2}}.$$

When $r(\theta) = (\sin \theta)^{-\frac{1}{2}}$ the first parenthesis multiplied by $(\sin \theta)^3$ becomes $\frac{1}{2}\sin^2 \theta - \frac{1}{4}\cos^2 \theta = \frac{3}{4}\sin^2 \theta - \frac{1}{4} > 0$ when $\sin^2 \theta > \frac{1}{3}$, hence when $\sin^2(\theta) \geq \frac{1}{2}$. This proves the convexity of A_0 and of h.

By Example 6.3 $\mathcal{H}_\psi$ is dense in $\mathcal{H}_{\max(\varphi_{\omega,a},\psi)}$ if $\varphi_{\omega,a}(iy) \leq h(y)$ when $a\langle y, y\rangle - \langle y, \omega\rangle^2 \geq 0$. Since $h(y)$ is the supremum of $\varphi_{\omega,a(\omega)}(iy)$ when $a(\omega)\langle y, y\rangle \geq \langle y, \omega\rangle^2$, this is true when $a = a(\omega)$. Hence it follows from Proposition 6.11 that $\max(\varphi_{\omega,a(\omega)}, \psi)$ is the supremum of all plurisubharmonic functions $s \leq \max(\varphi_{\omega,a(\omega)}, \psi)$ such that $s - \psi$ is bounded above, so $\max(\varphi, \psi)$ is the supremum of all plurisubharmonic functions $s \leq \max(\varphi, \psi)$ such that $s - \psi$ is bounded above. The proof is complete, again by Proposition 6.11.

It is tempting to conjecture that Proposition 6.13 remains valid when ψ and φ are just plurisubharmonic, Hölder continuous and positively homogeneous of degree one and not necessarily convex. However, Proposition 6.16 suggests that new obstacles may then occur so we content ourselves with raising this as an interesting question in the theory of entire functions of exponential type.

Appendix. In this paper existence theorems for the $\bar{\partial}$ operator with pointwise estimates have been used repeatedly. To avoid repetitions we shall here recall how they can be obtained from the basic existence theorems in L^2 spaces with weights given in [1].

LEMMA A.1. *If $u \in L^1(B_r)$, $B_r = \{z \in \mathbb{C}^n; |z| < r\}$, and $\bar{\partial}u$ is bounded in B_r, then u is continuous in B_r and*

$$(A.1) \qquad |u(0)| \leq 2r \sup_{B_r} |\bar{\partial}u| + \int_{B_r} |u|\, d\lambda \Big/ \int_{B_r} d\lambda.$$

PROOF. If c_{2n} is the area of the unit sphere $S^{2n-1} \subset \mathbb{R}^{2n}$ and $n > 1$, then

$$\Delta|z|^{2-2n}/((2-2n)c_{2n}) = 2\sum_1^n \frac{\partial^2}{\partial z_j \partial \bar{z}_j}|z|^{2(1-n)}/((1-n)c_{2n}) = 2/c_{2n}\sum_1^n \frac{\partial}{\partial \bar{z}_j}(\bar{z}_j|z|^{-2n}) = \delta_0.$$

The last equality is also true when $n = 1$. If $u \in C^1(\bar{B}_r)$ and $\chi(z) = 1 - |z/r|^{2n}$ this gives

$$u(0) = -2/c_{2n}\int_{B_r}\sum_1^n \bar{z}_j|z|^{-2n}\partial(\chi(z)u(z))/\partial\bar{z}_j\, d\lambda(z)$$

$$= -2/c_{2n}\int_{B_r}\sum_1^n \bar{z}_j|z|^{-2n}\chi(z)\partial u(z)/\partial\bar{z}_j\, d\lambda(z) + 2n/c_{2n}\int_{B_r} r^{-2n}u(z)\, d\lambda(z).$$

Since $c_{2n}/2n = C_{2n}$ is the volume of the unit ball, the last term can be estimated by the second term in (A.1). With polar coordinates we have $2/c_{2n}\int_{B_r}|z|^{1-2n}\, d\lambda(z) = 2\int_0^r d\varrho = 2r$, which completes the proof of (A.1) when $u \in C^1(\bar{B}_r)$. If u only satisfies the hypotheses in the lemma we just apply this result and the preceding identity to a sequence of regularizations in balls with radii $\uparrow r$.

LEMMA A.2. *Let f be a $\bar{\partial}$ closed $(0,1)$ form in $\mathbf{C}^n$ such that*

(A.2)
$$|f(z)| \leq Ce^{\varphi(z)}, \quad z \in \mathbf{C}^n,$$

where φ is a plurisubharmonic function. Then the equation $\bar{\partial}u = f$ has a solution u such that

(A.3)
$$|u(z)| \leq 3Cr^{-n}(1 + r + |z|)^{n+\frac{5}{2}} \exp(\sup_{|w|<r} \varphi(z+w)), \quad z \in \mathbf{C}^n, \ r > 0.$$

PROOF. By (A.2) we have

$$\int |f(z)|^2 (1+|z|)^{-2n-1} e^{-2\varphi(z)} \, d\lambda(z) \leq C^2 \int (1+|z|)^{-2n-1} \, d\lambda(z)$$

$$\leq C^2 c_{2n} \int_0^\infty (1+r)^{-2n-1} r^{2n-1} \, dr = C^2 C_{2n}$$

for with $s = r/(1+r)$ as a new variable the integral becomes $\int_0^1 s^{2n-1} \, ds = 1/2n$. From [1, Theorem 4.4.2] it follows that the equation $\bar{\partial}u = f$ has a solution such that

$$\int |u(z)|^2 (1+|z|)^{-2n-5} e^{-2\varphi(z)} \, d\lambda(z) \leq C^2 C_{2n}.$$

By Lemma A.1 it follows that

$$|u(z)| \leq C(2r + r^{-n}(1+r+|z|)^{n+\frac{5}{2}}) \exp(\sup_{|w|<r} \varphi(z+w)),$$

where we have used the Cauchy-Schwarz inequality to estimate the mean value of $|u|$. This implies (A.3) and completes the proof.

Since $\log(1 + |z|)$ is plurisubharmonic we can apply Lemma A.2 with $\varphi(z)$ replaced by $\varphi(z) + N \log(1 + |z|)$ for any $N > 0$. This means that if in the right-hand side of (A.2) we have another factor $(1 + |z|)^N$, then we obtain (A.3) with the exponent of $(1 + r + |z|)$ increased by N.

References

[1] L. Hörmander, *An introduction to complex analysis in several variables.*, North Holland Publ. Co., Amsterdam, 1990, Third extended edition.

[2] ———, *Convolution equations in convex domains*, Inv. Math. 4 (1968), 306–317.

[3] ———, *The analysis of linear partial differential operators I, II.* Grundlehren d.Math.Wiss. 256, 257, Springer Verlag, Berlin, Heidelberg, New York, Tokyo, 1983, Russian translation MIR 1986. Second revised printing 1990.

[4] ———, *Notions of convexity*, Birkhäuser, Boston, 1994.

[5] L. Hörmander and R. Sigurdsson, *Growth properties of plurisubharmonic functions related to Fourier-Laplace transforms*, J. Geom. Anal. 8 (1998), 251–311.

[6] C. O. Kiselman, *Existence and approximation theorems for solutions of complex analogues of boundary problems*, Ark. för Mat. 6 (1965), 193–207.

[7] R. Sigurdsson, *Growth properties of analytic and plurisubharmonic functions of finite order*, Math. Scand. 59 (1986), 235–304.

Autobiography, and
Looking Forward from ICM 1962

LARS HÖRMANDER

Reprinted, with kind permission from world Scientific Publishing, from the book *Fields Medallists' Lectures,* Sir M. Atiyah, and D. Iagolnitzer Eds., in World Scientific Series in 20th Century Mathematics vol. 5, pp. 82-103 1st Ed.1997; 2nd Ed. published as vol. 9 of the same series, pp. 91-111, 2003. © World Scientific Publishing.

© World Scientific Publishing 2003
L. Hörmander, Fields Medallists' Lectures, pp. 91-111
https://doi.org/10.1142/9789812564856_0005

AUTOBIOGRAPHY OF LARS HÖRMANDER

I was born on January 24, 1931, in a small fishing village on the southern coast of Sweden where my father was a teacher. After elementary school there and "realskola" in a nearby town which could be reached daily by train I went to Lund to attend "gymnasium", as my older brothers and sisters had done before me. I was more fortunate than they, for the principal was just starting an experiment which meant that three years were decreased to two with only three hours daily in school. This meant that I could mainly work on my own, with much greater freedom than the universities in Sweden offer today, and that suited me very well. I was also lucky to get an excellent and enthusiastic mathematics teacher who was a docent at the University of Lund. He encouraged me to start reading mathematics at the university level, and it was natural to follow his advice and go on to study mathematics at the University of Lund when I finished "gymnasium" in 1948.

In 1950 I got a masters degree and started as a graduate student. Marcel Riesz was my advisor, as he had been for my "gymnasium" mathematics teacher. Riesz was close to his retirement in 1952, and his lectures which I had actually attended since 1948 were not devoted to partial differential equations where he had recently made major contributions but rather to his earlier interests in classical function theory and harmonic analysis. My first mathematical attempts were therefore in that area. Although they did not amount to much this turned out to be an excellent preparation for working in the theory of partial differential equations. That became natural when Marcel Riesz retired and left for the United States while the two new professors Lars Gårding and Åke Pleijel appointed in Lund were both working on partial differential equations.

After a year's absence for military service 1953–1954, spent largely in defense research which gave ample opportunity to read mathematics, I finished my thesis in 1955 on the theory of linear partial differential operators. It was to a large extent inspired by the thesis of B. Malgrange which was announced in 1954, combined with techniques developed for hyperbolic differential operators by J. Leray and L. Gårding. Soon after that I was ready for my first visit to the United States, where in 1956 I spent the Winter and Spring quarters at the University of Chicago, the summer at the Universities of Kansas and Minnesota, and the fall in New York at what is now called the Courant Institute. (At the time R. Courant was still the director and it was called the Institute of Mathematical Sciences.) In Chicago there was no activity at all in my field, but the Zygmund seminar, conducted in his absence by E. M. Stein and G. Weiss, gave a useful addition to my background in harmonic analysis. At the other places I visited there was much to learn in my proper field.

At the end of this stay I was appointed to a full professorship at the University of Stockholm (called Stockholms Högskola then), which I had applied for before leaving for the United States. I took up my duties there in January 1957 and remained as professor until 1964. However, already during the academic year 1960–61 I was back in the United States as a member of the Institute for Advanced Study. During the summers of 1960 and 1961 I lectured at Stanford University and wrote a major part of my first book on partial differential equations. It was published by Springer Verlag in the Grundlehren Series in 1963 after the manuscript had been completed and polished back in Stockholm during the academic year 1961–62.

The 1962 International Congress of Mathematicians was held in Stockholm. In view of the small number of professors in Sweden at the time it was inevitable that I should be rather heavily involved in the preparations but it came as a complete surprise to me when I was informed that I would receive one of the Fields medals at the congress.

Some time after the two summers at Stanford I received an offer of a part time appointment as professor at Stanford University. I had declared that I did not want to leave Sweden, so the idea was that I should spend the Spring and Summer quarters at Stanford but remain in Stockholm most of the academic year there, from September through March. A corresponding partial leave of absence was granted by the ministry of education in Sweden and the arrangement became effective in 1963. However, I had barely arrived at Stanford when I received an offer to come to the Institute for Advanced Study as permanent member and professor. Although I had previously been determined not to leave Sweden, the opportunity to do research full time in a mathematically very active environment was hard to resist. After an attempt to create a research professorship for me in Sweden had failed, I finally decided in the fall of 1963 to accept the offer from the Institute and resign from the universities of Stockholm and Stanford to take up the new position in Princeton in the fall of 1964.

At that time the focus of interest in Princeton was definitely not in analysis which was felt both as a challenge and as a great opportunity to broaden my mathematical outlook. However, it turned out that I found it hard to stand the demands on excellence which inevitably accompany the privilage of being an Institute professor. After two years of very hard work I felt that my results were not up to the level which could be expected. Doubting that I would be able to stand a lifetime of this pressure I started to toy with the idea of returning to Sweden when a regular professorship became vacant. An opportunity arose in 1967, and I decided to take it and return as professor in Lund from the fall term 1968. After the decision had been taken I felt much more relaxed, and my best work at the Institute was done during the remaining year.

So in 1968 I had completed a full circle and was back in Lund where I had started as an undergraduate in 1948. I have remained there since then, with interruptions for some visits mainly to the United States. During the Fall term of 1970 I was visiting professor at the Courant Institute in New York, during the Spring term

1971 I was a member of the Institute for Advanced Study, and during the Summer quarter 1971 I was back at Stanford University, where I also lectured during the Summer quarters of 1977 and 1982. During the academic year 1977–1978 I was again a member of the Institute for Advanced Study which had a special year in microlocal analysis then, and during the Winter quarter 1990 I was visiting professor at the University of California in San Diego.

After five years devoted to writing a four volume monograph on linear partial differential operators I spent the academic years 1984–1986 as director of the Mittag–Leffler Institute in Stockholm. I had only accepted a two year appointment with a leave of absence from Lund since I suspected that the many administrative duties there would not agree very well with me. The hunch was right, and since 1986 I have been in Lund where I became professor emeritus in January 1996.

In my contribution to this volume I have summed up my work in partial differential equations and microlocal analysis which can be considered as the continuation of the work which gave me a Fields medal. Another major interest has been the theory of functions of several complex variables and its applications to the theory of partial differential equations. Lecture notes on this subject written at Stanford during the summer of 1964 became a book published in 1966, and extended editions were published in 1973 and 1990. A book on convexity theory published in 1994 is also to a large extent devoted to this field.

LOOKING FORWARD FROM ICM 1962

by

LARS HÖRMANDER

Department of Mathematics, University of Lund

1. Introduction

When I received the invitation to contribute to a volume entitled *Fields Medallists' Lectures*, intended to be similar to *Nobel Lectures in Physics*, my first reaction was quite negative. The Nobel prizes and the Fields medals are so very different in character; while Nobel prizes are supposed to be given for work of already recognized importance, often the work of a lifetime, the Fields medals are given "in recognition of work already done, and *as an encouragement for further achievement on the part of the recipients*". However, since this may imply an obligation to account for the expected further achievements, I have decided to contribute to this volume a brief survey of the later development of the work for which I assume that I received a Fields medal in 1962. It will not be a complete survey even of the development of these topics in the theory of linear partial differential equations since 1962, for I shall concentrate on my own work and only mention work by others interacting with it.

At the ICM in 1962 I gave a half hour lecture [H1] with the following table of contents:

1. Notations
2. Equations without solutions
3. Existence theorems
4. Hypoelliptic operators
5. Holmgren's uniqueness theorem
6. Carleman estimates
7. Uniqueness of the Cauchy problem
8. Unique continuation of singularities

The paper [H1] was written just after the manuscript of my first book [H2] had been completed, and all the topics 2–8 are presented in detail there. When discussing the later development I shall often compare it with the new version [H3] published in four volumes about 20 years later.

2. Pseudodifferential Operators and The Wave Front Set

In [H2] the main tool for proving *a priori* estimates for differential operators was the Fourier transformation. When the coefficients were variable their arguments were first "frozen" at a point, sometimes after a preliminary integration by parts as in [H2, Chap. VIII]. For the success of the procedure it was of course necessary that the error committed in freezing the coefficients was in some sense small compared to the quantities to be estimated. This worked well also for the proof of uniqueness theorems originally proved by Calderón [Ca] using singular integral operators; in fact, it was possible to reduce his regularity assumptions. Singular integral operators were not included in [H2], for they appeared to have many drawbacks: they required an artificial reduction to operators of order 0, only principal symbols were handled, and the role of the Fourier transformation was so suppressed that the calculations involved seemed artificial. These objections were removed by the introduction of pseudodifferential operators by Kohn and Nirenberg [KN]. They considered operators of the form

$$a(x, D)u(x) = (2\pi)^{-n} \int a(x, \xi)e^{i\langle x,\xi\rangle}\hat{u}(\xi)d\xi,$$

$$\hat{u}(\xi) = \int e^{-i\langle x,\xi\rangle}u(x)dx, \quad u \in C_0^\infty(\mathbf{R}^n),$$

(2.1)

where a is asymptotically a sum of terms which are homogeneous of integer order, which means that this algebra of operators is essentially generated by singular integral operators, differential operators and standard potential operators. However, the important point was that the calculus formulas for the symbols a were essentially the same as the familiar formulas for differential operators, not only on a principal symbol level. It was then easy and natural to introduce more general symbols which were useful in the further development of the theory of linear differential operators. The properties of parametrices of hypoelliptic operators with constant coefficients led me to introduce in [H4] the symbol class $S_{\varrho,\delta}^m$ of C^∞ functions in $\mathbf{R}^n \times \mathbf{R}^n$ such that

$$\left|\partial_x^\beta \partial_\xi^\alpha a(x, \xi)\right| \leq C_{\alpha,\beta}(1 + |\xi|)^{m-\varrho|\alpha|+\delta|\beta|}.$$

(2.2)

When $0 \leq \delta < \varrho \leq 1$ operators with such symbols give a calculus with very good properties, and if $\delta \geq 1 - \varrho$, hence $\varrho > \frac{1}{2}$, it is invariant under changes of variables so it can be transplanted to smooth manifolds. (Such operators had already been used in [H5] to construct left parametrices of hypoelliptic differential operators, but I had not understood that an algebra of operators could be obtained in this way.)

Pseudodifferential operators were used at once to localize the study of singularities of solutions of partial differential equations. However, it took a few years before this was codified in the notion of wave front set: if $u \in \mathcal{D}'(\mathbf{R}^n)$ then the wave front set $WF(u) \subset T^*(\mathbf{R}^n)$ was defined in [H6] as the intersection of the characteristic sets of all pseudodifferential operators A such that $Au \in C^\infty(\mathbf{R}^n)$. By standard elliptic regularity theory it follows that the projection of $WF(u)$ in $\mathbf{R}^n$ is equal to the singular support of u, and the definition guarantees that $WF(Au) \subset WF(u)$ if A is a pseudodifferential operator. The wave front set describes both the location of the singularities and the directions of the frequencies which cause the singularities.

A similar resolution of analytic singularities was given independently by Sato [Sa] but is technically more difficult to explain. For the original definitions we refer to [SKK], and various alternative definitions can be found in [H3, Chap. VIII, IX].

The proof of estimates is often reduced to positivity of an operator in L^2. The classical Gårding inequality states that if $a(x, D)$ is a differential operator with $a \in S_{1,0}^m$, $m > 0$, and $a(x, \xi) \geq c(1 + |\xi|)^m$ for some $m > 0$ and $c > 0$, then

$$\text{Re } (a(x, D)u, u) \geq - C(u, u), \quad u \in C_0^\infty(\mathbf{R}^n), \tag{2.3}$$

where $(\cdot, \cdot)$ is the scalar product in L^2. This remains true for pseudodifferential operators with symbol in $S_{\varrho, \delta}^m$, where $0 \leq \delta < \varrho \leq 1$. The proof is quite trivial in the pseudodifferential framework since $a(x, D) + a(x, D)^* = b(x, D)^* b(x, D) + c(x, D)$ for some $b \in S_{\varrho, \delta}^{m/2}$ and $c \in S_{\varrho, \delta}^0$. A stronger result, called the "sharp Gårding inequality" was proved in [H7]; an extended version given in [H3, Theorem 18.6.7] states that if $a \in S_{\varrho, \delta}^{\varrho - \delta}$ and $a \geq 0$ then

$$\text{Re } (a(x, D)u, u) \geq - C(u, u), \quad u \in C_0^\infty(\mathbf{R}^n). \tag{2.4}$$

An important refinement of the right-hand side was given by Melin [Mel] (see also [H3, Theorem 22.3.3]), and Fefferman–Phong [FP] (see also [H3, Theorem 18.6.8]) proved that (2.4) remains valid when $a \in S_{\varrho, \delta}^{2(\varrho - \delta)}$ and $a \geq 0$. This is often a very significant technical improvement.

The study of solvability of (pseudo)differential equations led Beals–Fefferman [BF1] to introduce much more general symbol classes than the classes $S_{\varrho, \delta}^m$ above, which could be tailored to the equation being studied. A further extension was made in [H8] (see [H3, Chap. XVIII]) using a variant of the definition (2.1) proposed already by Weyl [W] in his work on quantum mechnanics. It is the symplectic invariance of the approach of Weyl which allows a greater generality. Even more general symbols have been studied by Bony and Lerner [BL].

3. Fourier Intergral Operators

Since pseudodifferential operators cannot increase the wave front set, a (pseudo)differential operator cannot have a pseudodifferential fundamental solution (or parametrix) unless it is hypoelliptic. Parametrices of a different kind were constructed already in 1957 by Lax [La] for certain hyperbolic differential operators, as linear combinations of what is now called Fourier integral operators. In the simplest case these are of the form (2.1) with a modified exponent,

$$Au(x) = (2\pi)^{-n} \int a(x, \eta) e^{i\varphi(x, \eta)} \hat{u}(\eta) \, d\eta,$$

$$\hat{u}(\eta) = \int e^{-i\langle y, \eta \rangle} u(y) \, dy, \quad u \in C_0^\infty(\mathbf{R}^n), \tag{3.1}$$

where $\varphi(x, \eta)$ is positively homogeneous of degree 1 in η and $\det \partial^2 \varphi / \partial x \partial \eta \neq 0$. (A reference to this construction in [La] was given in [H2, p. 230] but it could not be studied with the techniques used in [H2].) With an operator of the form (3.1) there is associated a canonical transformation χ with the generating function φ,

$$\chi : (\partial\varphi(x,\eta)/\partial\eta, \eta) \mapsto (x, \partial\varphi(x,\eta)/\partial x) ; \qquad (3.2)$$

it has the important property that

$$WF(Au) \subset \chi WF(u) , \quad u \in \mathcal{D}' . \qquad (3.3)$$

A systematic study of such operators was initiated in [H9] in connection with a study of spectral asymptotics (see Section 9 below), and a thorough, more general and global theory was presented in [H10]. Fourier integral operators permit simplifications of (pseudo)differential operators, for as observed by Egorov [Eg1] conjugation by an invertible Fourier integral operator changes the principal symbol by composition with the canonical transformation. This was exploited in [DH], a sequel to [H10] written jointly with J. J. Duistermaat. A somewhat different exposition of these matters is given in [H3, Chap. XXV]. Some early papers in the area have been collected and provided with an introduction and bibliography by Brüning and Guillemin [BG].

4. Hypoellipticity

A (pseudo)differential operator P in an open set $X \subset \mathbf{R}^n$ (or a manifold X) is called hypoelliptic if

$$\text{sing supp } u = \text{sing supp } Pu , \quad u \in \mathcal{D}'(X) . \qquad (4.1)$$

Hypoelliptic differential operators with constant coefficients in $\mathbf{R}^n$ were characterised in my thesis (see [H2, Chap. IV]). The simplest class of hypoelliptic differential operators with variable coefficients is the class of elliptic operators: A differential operator P of order m is elliptic if the principal symbol $p(x,\xi)$, which is homogeneous of degree m, never vanishes when $\xi \in \mathbf{R}^n \setminus \{0\}$. If $H^{\text{loc}}_{(m)} = \{u; D^\alpha u \in L^2_{\text{loc}}, |\alpha| \leq m\}$, then P is elliptic if and only if $Pu \in H^{\text{loc}}_{(0)}$ implies $u \in H^{\text{loc}}_{(0)}$. The definition of the Sobolev spaces $H^{\text{loc}}_{(s)}$ can be extended to all $s \in \mathbf{R}$ in a unique way such that for all elliptic pseudodifferential operators of order $m \in \mathbf{R}$ we have $Pu \in H^{\text{loc}}_{(0)} \Longleftrightarrow u \in H^{\text{loc}}_{(m)}$, and then we have more generally $Pu \in H^{\text{loc}}_{(s)} \Longleftrightarrow u \in H^{\text{loc}}_{(s+m)}$.

Elliptic operators (with smooth coefficients) have been understood for a very long time, for they can easily be treated as mild perturbations of elliptic operators with constant coefficients. A class of hypoelliptic differential operators which can be studied similarly starting from hypoelliptic differential operators with constant coefficients was investigated by B. Malgrange and myself; the results were included in [H2, Chap. VII]. At the time they seemed quite general, but hypoellptic operators not covered by these results were soon found by Treves [T1]. In [H5] his ideas were developed as a very primitive and incomplete version of pseudodifferential operator theory already mentioned. In the more mature form of pseudodifferential operators in [H4] it was a simple consequence of the calculus, that if P is a pseudodifferential operator with symbol $p \in S^m_{\varrho,\delta}$, where $0 \leq \delta < \varrho \leq 1$, and if

$$|p(x,\xi)^{-1}\partial^\beta_x \partial^\alpha_\xi p(x,\xi)| \leq C_{\alpha,\beta}|\xi|^{-\varrho|\alpha|+\delta|\beta|}, \quad |\xi| > C,$$

$$|p(x,\xi)^{-1}| \leq C|\xi|^{m'}, \qquad\qquad |\xi| > C,$$

then P is hypoelliptic. (In [H4, Section 4] there is actually a more general result for systems.) When $1 - \varrho \leq \delta$ this gives a class of hypoelliptic operators which is invariant under changes of variables. However, that requires $\varrho > \frac{1}{2}$, and it is easy to see that second order differential operators satisfying this condition must in fact be elliptic. This led to the detailed study in [H11] of second order hypoelliptic operators.

A classical model equation, the Kolmogorov equation, was known in the theory of Brownian motion but was more familiar to probabilists than to experts in partial differential equations. The Kolmogorov operator is

$$P = (\partial/\partial x)^2 + x\partial/\partial y - \partial/\partial t$$

in $\mathbf{R}^3$ with coordinates (x, y, t). Kolmogorov [Ko] constructed an explicit fundamental solution, singular only on the diagonal, which implies hypoellipticity. Freezing the coefficients would give an operator with constant coefficients acting only along a two dimensional subspace, so the operator is obviously not covered by the results mentioned so far. However, the vector fields $\partial/\partial x$ and $x\partial/\partial y - \partial/\partial t$ do not satisfy the Frobenius integrability condition so the operator does not act only along submanifolds. The importance of this fact was established in [H11] where it was proved more generally that if

$$P = \sum_1^r X_j^2 + X_0 + c, \tag{4.2}$$

where $X_0, \ldots, X_r$ are smooth real vector fields in a manifold M of dimension n, and if among the operators $X_j, [X_j, X_k], [X_j, [X_k, X_l]], \ldots$ obtained by taking repeated commutators it is possible to find a basis for the tangent space at any point in M, the P is hypoelliptic. The condition on the vector fields is necessary in the sense that if the rank is smaller than n in an open subset then the operator only acts in less than n local coordinates, if they are suitably chosen, so it cannot be hypoelliptic. However, the condition can be relaxed at smaller subsets. (See e.g. [OR], [KS], [BM].) Simplified proofs of these results (with slightly less precise regularity) due to J. J. Kohn can be found in [H3, Section 22.2]. There is now a very extensive literature on operators of the form (4.2) (see e.g. [RS], [HM], [NSW], [SC]), partly motivated by the importance in probability theory.

If P is of the form (4.2), then the principal symbol of P is ≤ 0. Many of the results concerning operators of the form (4.2) have been extended to general pseudodifferential operators such that the principal symbol (microlocally) takes its values in an angle $\subset \mathbf{C}$ with opening $< \pi$. We refer to [H3, Chap. XXII] for such results and references to their origins.

A pseudodifferential operator of order m and type $1, 0$ is called *subelliptic* with loss of δ derivaties, if $0 < \delta < 1$ and

$$u \in \mathcal{D}'(\Omega), \quad Pu \in H^{loc}_{(s)} \Longrightarrow u \in H^{loc}_{(s+m-\delta)}. \qquad (4.3)$$

For $\delta = 0$ this condition would be equivalent to ellipticity, and the assumption $\delta < 1$ implies that (4.3) depends only on the principal symbol $p(x, \xi)$ of P, which we assume to be homogeneous. In [H7] it was proved that $\delta \geq \frac{1}{2}$ when (4.3) is valid and P is not elliptic, and that (4.3) is valid for $\delta = \frac{1}{2}$ if and only if

$$\{\operatorname{Re} p(x, \xi), \operatorname{Im} p(x, \xi)\} > 0, \quad \text{when } p(x, \xi) = 0. \qquad (4.4)$$

(The term subellipticity was introduced in [H3] just for the case $\delta = \frac{1}{2}$. There are pseudodifferential operators with this property but no differential operators since the Poisson bracket is then an odd function of ξ.) In [H7] a very implicit condition for subellipticity with loss of $\frac{1}{2}$ derivatives was also given for systems. The condition was made somewhat more explicit in [H12] but there are no really satisfactory results on subellipticity for systems with loss of δ derivatives even for $\delta = \frac{1}{2}$, and very little is known for $\delta \in (\frac{1}{2}, 1)$. However, for the scalar case Egorov [Eg2] found necessary and sufficient conditions for subellipticity with loss of δ derivatives; the proof of sufficiency was completed in [H13]. The results prove that the best δ is always of the form $k/(k+1)$ where k is a positive integer, and the conditions replacing (4.4) then involve the Poisson brackets of $\operatorname{Re} p$ and $\operatorname{Im} p$ of order $\leq k$, at the characteristics. A slight modification of the presentation in [H13] is given in [H3, Chap. XXVII], but it is still very complicated technically. Another approach which also covers systems operating on scalars has been given by Nourrigat [No] (see also the book [HN] by Helffer and Nourrigat), but it is also far from simple so the study of subelliptic operators may not yet be in a final form.

The hypoelliptic operators discussed here are all microlocally hypoelliptic in the sense that (4.1) can be strengthened to

$$WF(u) = WF(Pu), \quad u \in \mathcal{D}'(X). \qquad (4.1)'$$

This remains valid in an open conic subset of the cotangent bundle if the conditions above are only satisfied there.

5. Solvability and Propagation of Singularities

The discovery by Lewy [Lew] that the equation

$$\partial u/\partial x_1 + i\partial u/\partial x_2 + 2i(x_1 + ix_2)\partial u/\partial x_3 = f$$

for most $f \in C^\infty(\mathbf{R}^3)$ (in the sense of category) has no solution in any open subset of $\mathbf{R}^3$ led to a systematic study of local solvability of differential equations in [H14],

[H15] and [H2, Chap. VI, VIII]. It was proved there that if $p(x, \xi)$ is the principal symbol of a differential operator P of order m in $\mathbf{R}^n$ then the equation $Pu = f$ has no distribution solution in any neighborhood of x^0 for most $f \in C^\infty$ unless

$$p(x, \xi) = 0 \Longrightarrow \{\operatorname{Re} p(x, \xi), \operatorname{Im} p(x, \xi)\} = 0\,, \tag{5.1}$$

when x is in a neighborhood of x^0. On the other hand, if

$$\{\operatorname{Re} p(x, \xi), \operatorname{Im} p(x, \xi)\} = \operatorname{Re}(p(x, \xi)a(x, \xi))\,, \tag{5.2}$$

in a neighborhood of x^0 for some polynomial a in ξ with C^1 coefficients, and

$$p'_\xi(x, \xi) \neq 0 \quad \text{for} \ \xi \in \mathbf{R}^n \setminus \{0\}\,, \tag{5.3}$$

it was proved that the equation $Pu = f$ has a solution $u \in H^{\text{loc}}_{(s+m-1)}$ in a neighborhood of x^0 for every $f \in H^{\text{loc}}_{(s)}$. There is of course a substantial gap between the conditions (5.1) and (5.2). For first order differential equations it was filled to a large extent by Nirenberg–Treves [NT1]. The study of boundary problems for elliptic operators led to the extension of the solvability problem from differential to pseudodifferential operators in [H7]. There it was proved that the results of [H14], [H15] remain valid for pseudodifferential operators if (5.1), (5.2) are modified to

$$p(x, \xi) = 0 \Longrightarrow \{\operatorname{Re} p(x, \xi), \operatorname{Im} p(x, \xi)\} \leq 0\,, \tag{5.1$'$}$$

$$\{\operatorname{Re} p(x, \xi), \operatorname{Im} p(x, \xi)\} \leq \operatorname{Re}(p(x, \xi)a(x, \xi))\,. \tag{5.2$'$}$$

Thus (5.1)$'$ forbids $\operatorname{Im} p$ to change sign from $-$ to $+$ at a simple zero in the forward direction on a bicharacteristic of $\operatorname{Re} p$. Nirenberg and Treves [NT2] proved that such sign changes must not occur at any zeros of finite order either, and later work based on an idea of Moyer [Mo] has shown that local solvability implies that there are no such sign changes at all. (See [H3, Theorem 26.4.7].) This is called the condition (Ψ). For differential operators we have $p(x, -\xi) = (-1)^m p(x, \xi)$ and it follows that (Ψ) implies that there are no sign changes at all, which is called condition (P). It was also proved in [NT2] that local solvability follows from condition (P) and (5.3) provided that p is real analytic, a condition which was later removed by Beals and Fefferman [BF2]. However, it is still not known whether there is local solvability under condition (Ψ); there are many positive and negative results by Lerner [Ler1], [Ler2], [Ler3] and others. For these matters we refer to a recent survey paper [H16] and to [H3, Chap. XXVI].

The existence theorems just mentioned, first established in [H17], are not only local. They are valid for arbitrary compact sets which do not trap any complete bicharacteristics for the operator (see [H3, Theorem 26.11.1]), and they also give C^∞ solutions for C^∞ data. The key to such semiglobal results is the study of propagation of singularities of solutions of pseudodifferential equations. In [H1] there was just a very primitive result of this kind, Theorem 8.1, stating that if

the principal symbol p of a differential operator P is real and satisfies (5.3), then
a distribution u such that $Pu \in C^\infty$ is in C^∞ in a neighborhood of a point x^0
if there is a C^2 hypersurface through x^0 which has positive curvature at x^0 with
respect to tangential bicharacteristics such that $u \in C^\infty$ outside the surface. A
first step toward more precise results was taken by Grushin [Gr] who proved in
the constant coefficient case that if $x^0 \in$ sing supp u then sing supp u contains a
bicharacteristic line through x^0. An extension of this result to operators with variable coefficients was announced in [H18, p. 39]. However, the conclusion was only
local which was a great weakness, for if one has concluded that an interval from x^0
to x^1 on a bicharacteristic Γ_0 is contained in sing supp u, it follows that there is
an interval around x^1 on a bicharacteristic Γ_1 contained in sing supp u, but there
is no guarantee that Γ_1 should be a continuation of Γ_0. The proof announced in
[H18] depended on an early local version of the theory of Fourier integral operators. The global theory presented in [H10] was developed precisely to remove this
flaw so that it could be proved directly that a complete bicharacteristic must be
contained in sing supp u. However, just as this machinery was completed the idea
of the wave front set presented in [H6] was conceived. The right statement on the
propagation of singularities is that if $Pu \in C^\infty$ and $(x^0, \xi^0) \in WF(u)$, then the
bicharacteristic strip starting at (x^0, ξ^0) is contained in $WF(u)$, provided that the
principal symbol p is real valued. (If one removes the hypothesis $Pu \in C^\infty$ the
conclusion is that the bicharacteristic strip remains in $WF(u)$ until it encounters
$WF(Pu)$.) A bicharacteristic curve is the base projection of a bicharacteristic strip,
so it is not determined by its starting point whereas a bicharacteristic strip is. This
means that for the improved microlocal version of the theorem on propagation of
singularities the local result implies the global one. A simple proof was outlined in
[H6]. In [H19] the propagation theorem was extended to symbols with Im $p \geq 0$,
and in [DH] an analogue for the case of characteristics where d Re p and d Im p
are linearly independent was established under condition (P). The main point in
[H17] was a fairly complete discussion of propagation of singularities for arbitrary
pseudodifferential operators satisfying condition (P). The results on propagation
of singularities were completed by Dencker [D], replacing for some bicharacteristic
a weaker result in [H17] which gave the same existence theorems though. Some of
the results on propagation of singularities remain valid for operators satisfying only
condition (Ψ), but there are no satisfactory general results in that case.

6. Holmgren's Uniqueness Theorem

The classical uniqueness theorem of Holmgren states that a classical solution of a
linear differential equation $P(D)u = 0$ vanishing on one side of a C^1 surface must
vanish in a neighborhood of every noncharacteristic point. This was extended to
distribution solutions in [H2], and by a simple geometric argument it was concluded
(see [H1, Theorem 5.1]) that if the surface is in C^2 the assertion remains valid at

characteristic points where the curvature of the surface is positive with respect to the corresponding tangential bicharacteristic.

It had been observed already by John [Jo] that the proof of Holmgren's uniqueness theorem could be modified to proving analyticity theorems such as the analyticity of solutions of elliptic differential equations with analytic coefficients. This observation could be reversed when the analytic singularities had been microlocalized to a set similar to the wave front set. The definition of Sato [Sa] (see also [SKK]) mentioned above works for hyperfunctions, whereas a definition of such a set $WF_A(u)$ modelled on the definition of $WF(u)$ introduced in [H20] only works for distributions. Since the equivalence with the definition of Sato in this case has been verified by Bony [Bo] we shall use the notation $WF_A(u)$ for arbitrary hyperfunctions u. (A proof of the equivalence of these definitions and another definition due to Bros and Iagolnitzer [BI] in the case of distributions is also given in [H3, Chap. VIII, IX].) The base projection of $WF_A(u)$ is of course the analytic singular support sing supp$_A u$, the complement of the largest open set where u is a real analytic function.

The connection with Holmgren's uniqueness theorem is given by two facts:

(i) If P is a differential operator with real analytic coefficients and $Pu = 0$, then $WF_A(u)$ is contained in the characteristic set of P.

(ii) If u is a hyperfunction vanishing on one side of a C^2 surface passing through $x^0 \in$ supp u, then $WF_A(u)$ contains (x^0, ν) if ν is conormal to the surface at x^0. Part (i) is a microlocal version of the standard analytic regularity theorem for elliptic differential equations, and part (ii) follows from a part of the arguments used originally to prove Holmgren's uniqueness theorem. If the principal part p of P is real valued then a theorem on propagation of analytic singularities combined with (i) and (ii) proves that if $Pu = 0$ and u vanishes on one side of a C^1 hypersurface through $x^0 \in$ supp u then the surface is characteristic at x^0 and the full bicharacteristic strip through a conormal must remain in $WF_A(u)$ so the base projection stays in supp u. This is a great improvement of [H1, Theorem 5.1] first obtained independently by Kawai in [Ka] and myself in [H20]. Many other improvements of Holmgren's uniqueness theorem have been obtained through a deeper understanding of (i) and (ii); a recent survey is given in [H21].

7. Analytic Hypoellipticity and Propagation of Analytic Singularities

Having made no contributions to this area beyond the first steps taken in [H20] we shall content ourselves here with some references to results which are relevant in connection with the improvements of Holmgren's uniqueness theorem discussed in the preceding section. A differential operator with real analytic coefficients is called analytic hypoelliptic if

$$\text{sing supp}_A u = \text{sing supp}_A Pu, \quad u \in \mathcal{D}'(X), \tag{7.1}$$

and P is said to be microlocally analytically hypoelliptic if

$$WF_A(u) = WF_A(Pu), \quad u \in \mathcal{D}'(X). \tag{7.2}$$

(One sometimes insists on these equalities for all hyperfunctions.) If P is subelliptic in the C^∞ sense then it was proved by Treves [T2] that P is analytically hypoellptic, and by Trépreau [Tre] in even greater generality that P is also microlocally analytically hypoelliptic.

However, operators of the form (4.2) with analytic coefficients satisfying the commutator condition which implies hypoellipticity are not always analytically hypoelliptic even if $X_0 = 0$. (Lower order terms are not expected to affect analytic hypoellipticity.) Simple examples are given in e.g. Christ [Ch]. However, if the characteristic set is a symplectic manifold then microlocal analytic hypoellipticity has been proved, also for some classes of operators with higher order multiplicities for the characteristics (see [Tar], [T3], [Met], [Sj1], and also [DT] for recent results and additional references).

Concerning propagation of analytic singularities we shall content ourselves with referring to the very general results by Kawai and Kashiwara [KK] and Grigis, Schapira and Sjöstrand (see [Sj2]), although these are in no way the last words on the subject.

8. Carleman Estimates

Section 6 in [H1] and Chap. VIII in [H2] were devoted to Carleman estimates of the form

$$\tau \int |D^{m-1}u|^2 e^{2\tau\varphi} dx \le C_1 \int |Pu|^2 e^{2\tau\varphi} \, dx \; .$$

$$+ C_2 \sum_0^{m-2} \tau^{2(m-j)-1} \int |D^j u|^2 e^{2\tau\varphi} \, dx, \quad u \in C_0^\infty(K) \tag{8.1}$$

where P is a differential operator of order m in a neighborhood of the compact set $K \subset \mathbf{R}^n$. Such estimates with $C_2 \ne 0$ were applied to the proof of existence theorems and some very weak results on propagation of singularities, and these results have been made obsolete by those discussed in Section 5. The estimates with $C_2 = 0$,

$$\tau \int |D^{m-1}u|^2 e^{2\tau\varphi} \, dx \le C_1 \int |Pu|^2 e^{2\tau\varphi} \, dx, \quad u \in C_0^\infty(K), \tag{8.1$'$}$$

are still of interest though. It was proved in [H2] that (8.1)$'$ implies that if p is the principal symbol of P then

$$|\xi + i\tau\varphi'(x)|^{2(m-1)} \le C_1 \overline{\{p(x, \xi + i\tau\varphi'(x)), p(x, \xi + i\tau\varphi'(x))\}}/2i\tau, \tag{8.2}$$

$$\text{if } \xi \in \mathbf{R}^n, \, \tau > 0, p(x, \xi + i\tau\varphi'(x)) = 0. \tag{8.3}$$

When (5.2) is fulfilled it is easy to conclude as a limiting case when $\tau \to 0$ that

$$|\xi|^{2(m-1)} \leq C_1 \operatorname{Re}\{\overline{p(x,\xi)}, \{p(x,\xi), \varphi(x)\}\}, \tag{8.2}'$$

$$\text{if } \xi \in \mathbf{R}^n, \ p(x,\xi) = 0. \tag{8.3}'$$

It was proved in [H2] that conversely the conditions (8.2),(8.3) imply (8.1)' (with a larger constant C_1) when(5.2) is valid with $a(x,\xi)$ in C^1 and polynomial in ξ. It was remarked in [H1, p. 343] that (8.2), (8.3) imply (5.1) but that it might be possible to eliminate the extraassumption (5.2) to a large extent. This was done in [H3, Chap. XXVIII] where (5.2) was weakened to

$$|\{\overline{p(x,\xi)}, p(x,\xi)\}| \leq C_3 |p(x,\xi)||\xi|^{m-1}, \tag{8.4}$$

by means of the powerful lower bound for pseudodifferential operators established by Fefferman and Phong [FP].

In the application of the estimate (8.1)' to the proof ofuniqueness theorems only the level sets of φ are important. Replacing φ by an increasing convex function of φsuch as $e^{\lambda\varphi}$ with some large positive λwill add a positive term in the right-hand side of (8.2), (8.2)'unless

$$\langle p'_\xi(x,\xi + i\tau\varphi'(x)), \varphi'(x)\rangle = \{p(x,\xi + i\tau\varphi'(x)), \varphi(x)\} = 0. \tag{8.5}$$

If (8.2), (8.2)' are only assumed to be valid when (8.5) is addedto the hypotheses (8.3), (8.3)' (with $\tau = 0$ in (8.5) in thesecond case), one can modify φ in this way without changing the level sets so that (8.2), (8.2)' become valid with another constant C_1 under the hypotheses (8.3), (8.3)' only. By the standard Carleman argument it follows then that if $Pu = 0$, $u \in H^{loc}_{(m-1)}$ in a neighborhood of x^0 and $u = 0$when $\varphi(x) > \varphi(x^0)$, then $u = 0$ in a neighborhood ofx^0. (The proof uses also that the convexity conditions (8.2),(8.2)' are stable under small perturbations.)

Already in [H2, Section 8.9] some examples based on constructionsby A. Pliš and P. Cohen were given which proved that the convexity assumption in this uniqueness result could not in general be relaxed. A more systematic study of such examples wasgiven in [H22]. However, these constructions relied onperturbations of P by terms of lower but positive order. Much better examples were constructed by Alinhac [Al] who proved that uniqueness fails after addition to P of a suitable C^∞ term of order 0 if (5.1) is not valid, or p has real coefficients and $\varphi(x) \leq \varphi(x^0)$ on a bicharacteristic curvethrough x^0, or the right-hand side of (8.2) vanishes for a suitable family of zeros satisfying (8.3) and (8.5). For first order differential operators it was proved by Strauss and Treves [ST] that there is uniqueness for all non-characteristic surfaces if condition (P) is fulfilled, but recently Colombini and Del Santo [CD] proved that the condition (8.4) cannot be replaced by condition (P) in the uniqueness theorem above; in their example there are no non-trivial solutions of (8.3), (8.3)' satisfying (8.5).

A surprising new discovery concerning the uniqueness of the Cauchy problem was made a few years ago by Robbiano [Ro] who proved that for a differential operator in $\mathbf{R}^{n+1}$ of the form

$$P = D_t^2 - A(x, D_x)$$

which is hyperbolic with respect to t, there is uniqueness for the Cauchy problem on every surface which is cylindrical in the t direction. This would be false in general for lower order terms with coefficients depending on t. A quantitatively more precise form of this result given in [H23] states that there is unique continuation of solutions of the equation $Pu = 0$ across a timelike surface with conormal (τ, ξ) at a point (t, x) provided that

$$27\tau^2/23 - a(x, \xi) < 0;$$

it is of course classical that there is unique continuation acrossspacelike surfaces, that is, surfaces with $\tau^2 - a(x, \xi) > 0$. Very recently Tataru [Tat] has proved that there is uniquecontinuation across all non-characteristic surfaces. This is aspecial case of a general result stating that if the principal symbolis translation invariant along a linear subspace $\mathcal{A}$ and all coefficients are real analytic in the direction of $\mathcal{A}$, then the uniqueness theorem above is valid if the convexity conditions (8.2), (8.2)$'$ are satisfied when to the conditions (8.3), (8.3)$'$ and (8.5) is added that ξ is a conormal of $\mathcal{A}$. In [H24] it is proved that it suffices to assume that the restriction of the principal symbol to the conormal bundle of $\mathcal{A}$ and its parallel spaces is translation invariant, and there are similar improvements of the other results of [Tat] as well.

9. Spectral Asymptotics

When I was a graduate student in Lund the two professors, Lars Gårding and Åke Pleijel, were both working on asymptotic properties of eigenvalues and eigenfunctions of elliptic differential operators P, and so were most of the graduate students. I chose a different direction for my thesis but became of course aware of the state of the field.

Since the work of Carleman in the 1930's the main approach to suchquestions has been to study the kernel of some function of P, suchas the resolvent or the Laplace transform. When pseudodifferentialoperators had appeared and been recognized as a powerful tool for theconstruction of parametrices, it was natural to try their strength inthis field.

Let $P(x, D)$ be an elliptic differential operator of order m with C^∞ coefficients in an open set $\Omega \subset \mathbf{R}^n$ with a self-adjoint extension $\mathcal{P}$ in $L^2(\Omega)$ which is bounded below. Let (E_λ) be the spectral resolution of $\mathcal{P}$ and let $e(x, y, \lambda)$ be the kernel of E_λ. Gårding [Gå1] proved using the simplest asymptotic properties of the resolvent of $\mathcal{P}$ that

$$R(x, \lambda) = \lambda^{-n/m} e(x, x, \lambda) - (2\pi)^{-n} \int_{p(x,\xi)<1} d\xi \tag{9.1}$$

converges to 0 as $\lambda \to +\infty$. Here p is the principalsymbol of P. For operators with constant coefficients he proved in [Gå2] that $R(x, \lambda) = O(\lambda^{-1/m})$. For second order operators with variable coefficients this wasproved by Avakumovič [Av] and in part by Lewitan [Lewi]. By afairly straightforward application of pseudodifferential calculus tothe construction of the resolvent $(P - z)^{-1}$ it was proved in [H25] that $R(x, \lambda) = O(\lambda^{-\theta/m})$ for every $\theta < \frac{1}{2}$ (every $\theta < 1$ if the coefficients of the principal part are constant). The same result was obtained independently with different methods by Agmon and Kannai [AK]; in fact, Agmon had been the first to prove such bounds with a positive θ. It was also proved in [H25] that the value of θ has decisive importance for the summability properties of the eigenfunction expansion with respect to P. The reason why Avakumovič obtained the optimal value $\theta = 1$ was that he used the Hadamard construction of parametrices for second order differential equations which takes advantage of geodesic coordinate systems. In [H9] it was proved that $R(x, \lambda) = O(\lambda^{-1/m})$ in general by applying Fourier integral operators to construct a parametrix for the hyperbolic pseudodifferential operator

$$i\partial/\partial t - P^{1/m}$$

after a reduction to a compact manifold and a positive operator P which makes $P^{1/m}$ a well defined pseudodifferential operator. This construction, which takes into account the geometrical optics description of propagation of singularities, was the starting point for the work on Fourier integral operators described in Section 3. What is required is only an understanding of the operator $e^{-itP^{1/m}}$ for small values of $|t|$. Later work by many mathematicians where this unitary group is studied also for large values of t has led to much deeper understanding of the eigenvalues of elliptic differential operators, in particular the connection between clustering of eigenvalues and closed bicharacteristics. Some of this work is covered by [H3, Chap. XXIX]. (In the first edition there is an error in Theorem 29.1.4 corrected in the second edition.) However, it would carry too far to give a survey of this work here.

References

[AK] S. Agmon and Y. Kannai, On the asymptoticbehavior of spectral functions and resolvent kernels of ellipticoperators. *Israel J. Math.* **5** (1967), 1–30.

[Al] S. Alinhac, Non-unicité du problème de Cauchy. *Ann. of Math.* **117** (1983), 77–108.

[Av] V. G. Avakumovič, Über die Eigenfunktionenauf geschlossenen Riemannschen Mannigfaltigkeiten. *Math. Z.* **65** (1956), 327–344.

[BF1] R. Beals and C. Fefferman, Spatially inhomogeneous pseudo-differential operators I. *Comm. Pure Appl. Math.* **27** (1974), 1–24.

[BF2] ———, On local solvability of linear partial differential equations. *Ann. of Math.* **97** (1973), 482–498.

[BM] D. Bell and S.-E. Mohammed, An extension of Hörmander's theorem for infinitely degenerate second-order operators. *Duke Math. J.* **78** (1995), 453–475.

[Bo] J.-M. Bony, Équivalence des diverses notions de spectre singulier analytique. Sém. Goulaouic-Schwartz 1976–77, Exposé no. III.

[BL] J.-M. Bony and N. Lerner, Quantification asymptotique et microlocalisations d'ordre supérieur I. *Ann. Sci. École Norm. Sup.* **22** (1989), 377–433.

[BI] J. Bros and D. Iagolnitzer, Tuboïdes et structure analytique des distributions. Sém. Goulaouic-Schwartz 1974–75, Exposés XVI et XVIII.

[BG] J. Brüning and V. Guillemin, *Mathematics past and present. Fourier integral operators.* Springer Verlag, 1991.

[Ca] A. P. Calderón, Uniqueness in the Cauchy problem for partial differential equations. *Amer. J. Math.* **80** (1958), 16–36.

[Ch] M. Christ, Certain sums of squares of vector fields fail to be analytic hypoelliptic. *Comm. Partial Differential Equations* **16** (1991), 1695–1707.

[CD] F. Colombini and D. Del Santo, Condition (P) is not sufficient for uniqueness in the Cauchy problem. *Comm. Partial Differential Equations* **20** (1995), 2113–2128.

[D] N. Dencker, On the propagation of singularities for pseudo-differential operators of principal type. *Ark. Mat.* **20** (1982), 23–60.

[DH] J. J. Duistermaat and L. Hörmander, Fourier integral operators II. *Acta Math.* **128** (1972), 183–269.

[DT] M. Derridj and D. S. Tartakoff, Microlocal analyticity for the canonical solution to $\bar\partial_b$ on strictly pseudoconvex CR manifolds of real dimension three. *Comm. Partial Differential Equations* **20** (1995), 1871–1926.

[Eg1] Ju. V. Egorov, The canonical transformations of pseudo-differential operators. *Uspehi Mat. Nauk* **24** (1969), no 5:235–236.

[Eg2] ———, Subelliptic operators. *Uspehi Mat. Nauk* **30** (1975), no 2:59–118, no 3:55–105.

[FP] C. Fefferman and D. H. Phong, On positivity of pseudo-differential operators. *Proc. Nat. Acad. Sci.* **75** (1978), 4673–4674.

[Gå1] L. Gårding, On the asymptotic distribution of the eigenvalues and eigenfunctions of elliptic differential operators. *Math. Scand.* **1** (1953), 237–255.

[Gå2] ———, On the asymptotic properties of the spectral function belonging to a self-adjoint semi-bounded extension of an elliptic differential operator. *Kungl. Fysiogr. Sällsk. i Lund Förh.* **24:21** (1954), 1–18.

[Gr] V. V. Grushin, On the solutions of partial differential equations. *Doklady Akad. Nauk SSSR* **139** (1961), 17–19, (in Russian).

[HN] B. Helffer and J. Nourrigat, *Hypoellipticitémaximale pour des opérateurs polynômes de champs de vecteurs.* Birkhäuser, 1985, Progress in Mathematics **58**.

[H1]   L. Hörmander, Existence, uniqueness and regularity of solutions of linear differential equations. in *Proc. Int. Congr. Math. 1962*, pp. 339–346.

[H2]   ——, *Linear partial differential operators*. Springer Verlag, 1963.

[H3]   ——, *The analysis of linear partial differentialoperators I–IV*. Springer Verlag, 1983–85.

[H4]   ——, Pseudo-differential operators and hypoellipticequations, in *Amer. Math. Soc. Symp. on Singular Integrals*,pp. 138–183, 1966.

[H5]   ——, Hypoelliptic differential operators. *Ann. Inst. Fourier Grenoble* **11** (1961), 477–492.

[H6]   ——, Linear differential operators, in *ActesCongr. Int. Math. Nice* **1** (1970), 121–133.

[H7]   ——, Pseudo-differential operators and non-elliptic boundary problems. *Ann. of Math.* **83** (1966), 129–209.

[H8]   ——, The Weyl calculus of pseudo-differential operators. *Comm. Pure Appl. Math.* **32** (1979), 359–443.

[H9]   ——, The spectral function of an elliptic operator. *Acta Math.* **121** (1968), 193–218.

[H10]   ——, Fourier integral operators I. *Acta Math.* **127** (1971), 79–183.

[H11]   ——, Hypoelliptic second order differential equations. *Acta Math.* **119** (1967), 147–171.

[H12]   ——, On the subelliptic test estimates. *Comm. Pure Appl. Math.* **33** (1980), 339–363.

[H13]   ——, Subelliptic operators. in *Seminaron singularities of solutions of differential equations*, pp. 127–208. Princeton University Press, 1979.

[H14]   ——, Differential operators of principal type. *Math. Ann.* **140** (1960), 124–146.

[H15]   ——, Differential equations without solutions. *Math. Ann.* **140** (1960), 169–173.

[H16]   ——, On the solvability of pseudodifferential equations in *Structure of solutions of differential equations*, ed. M. Morimoto and T. Kawai, World Scientific, 1996, pp. 183–213.

[H17]   ——, Propagation of singularities and semi-global existence theorems for (pseudo-)differential operators of principal type. *Ann. of Math.* **108** (1978), 569–609.

[H18]   ——, On the singularities of solutions ofpartial differential equations, in *Proc. Int. Conf. on Funct. Anal. and Related topics*, Tokyo, April 1969, pp. 31–40.

[H19]   ——, On the existence and the regularity of solutions of linear pseudo-differential equations. *Ens. Math.* **17** (1971), 99–163.

[H20]   ——, Uniqueness theorems and wave front setsfor solutions of linear differential equations with analyticcoefficients. *Comm. Pure Appl. Math.* **24** (1971), 671–704.

[H21] ———, Remarks on Holmgren's uniqueness theorem. *Ann. Inst. Fourier Grenoble* **43** (1993), 1223–1251.

[H22] ———, Non-uniqueness for the Cauchy problem. *Springer Lecture Notes in Math.* **459** (1974), 36–72.

[H23] ———, A uniqueness theorem for second order hyperbolic differential equations. *Comm. Partial Differential Equations* **17** (1992), 699–714.

[H24] ———, On the uniqueness of the Cauchy problem under partial analyticity assumptions in *Geometrical Optics and Related Topics*, F. Columbini and N. Lerner, eds. Birkhäuser, Boston, 1997.

[H25] ———, On the Riesz means of spectral functions and eigenfunction expansions for elliptic differential operators, in *The Belfer Graduate School Science Conference Nov. 1966*, pp. 155–202, 1969.

[HM] L. Hörmander and A. Melin, Free systems ofvector fields. *Ark. Mat.* **16** (1978), 83–88.

[Jo] F. John, On linear differential equations with analytic coefficients. Unique continuation of data. *Comm. Pure Appl. Math.* **2** (1949), 209–253.

[KK] M. Kashiwara and T. Kawai, Microhyperbolic pseudodifferential operators I. *J. Math. Soc. Japan* **27** (1975), 359–404.

[Ka] T. Kawai, On the theory of Fourier hyperfunctions and its application to partial differential equations with constant coefficients. *J. Fac. Sci. Tokyo* **17** (1970), 467–517.

[KN] J. J. Kohn and L. Nirenberg, On the algebra of pseudo-differential operators. *Comm. Pure Appl. Math.* **18** (1965), 269–305.

[Ko] A. N. Kolmogorov, Zufällige Bewegungen. *Ann. of Math.* **35** (1934), 117–117.

[KS] S. Kusuoka and D. Strook, Applications of theMalliavin calculus, Part II. *J. Fac. Sci. Univ. Tokyo.*

[La] P. D. Lax, Asymptotic solutions of oscillatoryinitial value problems. *Duke Math. J.* **24** (1957), 627–646.

[Ler1] N. Lerner, Sufficiency of condition (ψ) forlocal solvability in two dimensions. *Ann. of Math.* **128**(1988), 243–258.

[Ler2] ———, Nonsolvability in L^2 for a first orderoperator satisfying condition (ψ). *Ann. of Math.* **139** (1994), 363–393.

[Ler3] ———, An iff solvability condition for the obliquederivative problem. Séminaire EDP 90–91, École Polytechnique, Exposé 18.

[Lewi] B. M. Lewitan, On the asymptotic behavior ofthe spectral function of a self-adjoint differential equation of thesecond order. *Izv. Akad. Nauk SSSR* **16** (1952), 325–352; II. **16** (1955), 33–58.

[Lew] H. Lewy, An example of a smooth linear partialdifferential equation without solution. *Ann. of Math.* **66**(1957), 155–158.

[Mel] A. Melin, Lower bounds for pseudo-differentialoperators. *Ark. Mat.* **9** (1971), 117–140.

[Met]   G. Métivier, Analytic hypoellipticity foroperators with multiple characteristics. *Comm. Partial Differential Equations* **6** (1981), 1–90.

[Mo]   R. D. Moyer, Local solvability in twodimensions: Necessary conditions for
the principle-type case. Mimeographed manuscript, University of Kansas
(1978).

[NSW]   A. Nagel, E. M. Stein and S. Wainger, Ballsand metrics defined by vector
fields I: Basic properties. *ActaMath.* **155** (1985), 103–147.

[NT1]   L. Nirenberg and F. Treves, Solvability of a firstorder linear partial differential equation. *Comm. Pure Appl. Math.* **16** (1963), 331–351.

[NT2]   ———, On local solvability of linear partial differentialequations. Part I:
Necessary conditions. *Comm. Pure Appl. Math.***23** (1970), 1–38; Part II:
Sufficient conditions. *Comm. Pure Appl. Math.* **23** (1970), 459–509; Cor-
rection. *Comm. Pure Appl. Math.* **24** (1971), 279–288.

[No]   J. Nourrigat, Subelliptic systems. *Comm. in Partial Differential Equations*
15 (1990), 341–405.

[OR]   O. A. Olejnik and E. V. Radkevič, Second order equations with non-negative
characteristic form, in *Matem. Anal.* 1969 (R. V. Gamkrelidze, ed.)
Moscow, 1971, (Russian). English translation Plenum Press, New York,
London (1973).

[Ro]   L. Robbiano, Théorème d'unicité adapté aucontrôle des solutions des
problèmes hyperboliques. *Comm. Partial Differential Equations* **16** (1991),
789–800.

[RS]   L. P. Rothschild and E. M. Stein, Hypoelliptic differential operators and
nilpotent groups. *Acta Math.* **137** (1976), 247–320.

[SC]   A. Sánchez-Calle, Fundamental solutions and geometry of the sum of
squares of vector fields. *Invent. Math.* **78** (1984), 143–160.

[Sa]   M. Sato, Regularity of hyperfunction solutions of partial differential equations, in *Actes Congr. Int. Math. Nice* **2** (1970), 785–794.

[SKK]   M. Sato, T. Kawai and M. Kashiwara, Hyperfunctions and pseudo-
differential equations. *Lecture Notes in Mathematics* **287** (1973), 265–529.

[Sj1]   J. Sjöstrand, Analytic wavefront set and operators with multiple characteristics. *Hokkaido Math. J.* **12** (1983), 392–433.

[Sj2]   ———, Singularités analytiques microlocales. *Astérisque* **95** (1982), 1–166.

[ST]   M. Strauss and F. Treves, First order linear P.D.E.'s and uniqueness of the
Cauchy problem, *J. Differential Equations* **38** (1980), 374–392.

[Tar]   D. Tartakoff, The local real analyticity of solutions to $\Box_b$ and the $\bar{\partial}$-
Neumann problem. *Acta Math.* **145** (1980), 177–204.

[Tat]   D. Tataru, Unique continuation for solutions to PDE's; between
Hörmander's theorem and Holmgren's theorem. *Comm. Partial Differential
Equations* **20** (1995), 855–884.

[Tre]   J.-M. Trépreau, Sur l'hypoellipticité analytique microlocale des opérateurs
de type principal. *Comm. Partial Differential Equations* **9** (1984), 1119–
1146.

[T1]   F. Treves, Opérateurs différentiels hypoelliptiques. *Ann. Inst. Fourier Grenoble* **9** (1959), 1–73.

[T2]   ———, Analytic hypo-elliptic PDE's of principal type. *Comm. Pure Appl. Math.* **24** (1971), 537–570.

[T3]   ———, Analytic hypoellipticity of a class of pseudodifferential operators with double characteristics. *Comm. Partial Differential Equations* **3** (1978), 85–116.

[W]   H. Weyl, Quantenmechanik und Gruppentheorie. *Zeitschrift für Physik* **46** (1927), 1–47, Collected works, Vol. III, pp. 90–135.

Box 118, S-221 00 Lund, Sweden

E-mail address: ivh@maths.lth.se

Complete Mathematical Bibliography
of Lars Hörmander

© Springer International Publishing AG, part of Springer Nature 2018
L. Hörmander, *Unpublished Manuscripts*,
https://doi.org/10.1007/978-3-319-69850-2

Published Papers[1]

[1]   Uniqueness theorems and estimates for normally hyperbolic partial differential equations of the second order, C.R. 12e Congr. Math. Scand. Lund (1953), 105–115, **MR** 16:483b, **Zbl** 56:308.

[2]   A new proof and a generalization of an inequality of Bohr, Math. Scand. **2** (1954), 33–45, **MR** 16:35, **Zbl** 58:255.

[3]   On a theorem of Grace, Math. Scand. **2** (1954), 55–64, **MR** 16:27b, **Zbl** 58:255.

[4]   Sur la fonction d'appui des ensembles convexes dans un espace localement convexe, Ark. för Mat. **3** (1954), 181–186, **MR** 16:831e, **Zbl** 64:105.

[5]   La transformation de Legendre et le théorème de Paley-Wiener, C.R. Acad. Sci. Paris **240** (1955), 392–395, **MR** 16:720a, **Zbl** 64:103.

[6]   Some inequalities for functions of exponential type, Math. Scand. **3** (1955), 21–27, **MR** 17:247d, **Zbl** 65:303.

[7]   On the theory of general partial differential operators, Acta Math. **94** (1955), 161–248, **MR** 17:853d, **Zbl** 67:322.

[8]   (with J.L. Lions) Sur la complétion par rapport à une intégrale de Dirichlet, Math. Scand. **4** (1956), 259–270, **MR** 19:420e, **Zbl** 78:280.

[9]   Local and global properties of fundamental solutions, Math. Scand, **5** (1957), 27–39, **MR** 20:159, **Zbl** 81:96.

[10]   On interior regularity of the solutions of partial differential equations, Comm. Pure Appl. Math. **11** (1958), 197–218, **MR** 21:5064, **Zbl** 81:315.

[11]   On the regularity of the solutions of boundary problems, Acta Math. **99** (1958), 225–264, **MR** 24:A1563, **Zbl** 83:92.

[12]   Definitions of maximal differential operators, Ark. för Mat. **3** (1958), 501–504, **MR** 21:5067, **Zbl** 131:94.

[13]   Differentiability properties of solutions of systems of differential equations, Ark. för Mat. **3** (1958), 527–535, **MR** 21:3673, **Zbl** 131:95.

[14]   On the division of distributions by polynomials, Ark. för Mat. **3** (1958), 555–568, **MR** 23:A2044, **Zbl** 131:119.

[15]   On the uniqueness of the Cauchy problem, Math. Scand. **6** (1958), 213–225, **MR** 21:3674, **Zbl** 88:302.

[16]   On the uniqueness of the Cauchy problem II, Math. Scand. **7** (1959), 177–190, **MR** 22:12306, **Zbl** 90:80.

[17]   Differential operators of principal type, Math. Ann. **140** (1960), 124–146, **MR** 24:A434, **Zbl** 90:81.

[18]   Differential equations without solutions, Math. Ann. **140** (1960), 169–173, **MR** 26:5279, **Zbl** 93:289.

[19]   Null solutions of partial differential equations, Arch. Rat. Mech. Anal. **4** (1960), 255–261, **MR** 22:1760, **Zbl** 98:294.

[20]   Estimates for translation invariant operators in L^p spaces, Acta Math. **104** (1960), 93–140, 22:12389, **Zbl** 93:114, Russian translation in Matematika 1962, **Zbl** 103.326.

[21]   On existence of solutions of partial differential equations., Partial diff. equations and cont. mech., Madison, 1961, pp. 233–240, **MR** 23:A1913, **Zbl** 111.91.

[22]   Hypoelliptic convolution equations, Math. Scand. **9** (1961), 178–184, **MR** 25:3265, **Zbl** 102:110.

[23]   Hypoelliptic differential operators, Ann. Inst. Fourier **11** (1961), 477–492, **MR** 23:A3368, **Zbl** 99:301.

[24]   Weak and strong extensions of differential operators, Comm. Pure Appl. Math. **14** (1961), 371–379, **MR** 24:A2750, **Zbl** 111:292.

[25]   On the range of convolution operators, Ann. of Math. **76** (1962), 148–170, **MR** 25:5379, **Zbl** 109:85.

[26]   Differential operators with nonsingular characteristics, Bull. Amer. Math. Soc. **68** (1962), 354–359, **MR** 27:6026, **Zbl** 113:80, French translation **MR** 28:2465, **Zbl** 231.5011.

[27]   Existence, uniqueness and regularity of solutions of linear differential equations, Proc. Int. Congr. Math., Stockholm, 1962, pp. 339–346, **MR** 31:470, **Zbl** 171:68.

[28]   Supports and singular supports of convolutions, Acta Math. **110** (1963), 279–302, **MR** 27:4070, **Zbl** 188:194.

[29] L^2 estimates and existence theorems for the $\bar{\partial}$ operator, Int. Coll. Diff. Analysis, Bombay, 1964, pp. 65–79, **MR** 32:2713, **Zbl** 197:366.

[30] L^2 estimates and existence theorems for the $\bar{\partial}$ operator, Acta Math. **113** (1965), 89–152, **MR** 31:3691, **Zbl** 158:110.

[31]   The Frobenius-Nirenberg theorem, Ark. för Mat. **5** (1964), 425–432, **MR** 31:2480, **Zbl** 136:91.

---

[1]By **MR** $\mu{:}\nu$ we denote the review on page ν or labelled ν in volume μ of Mathematical Reviews. By **Zbl** $\mu{:}\nu$ we denote a review on page ν or labelled ν in volume μ of Zentralblatt für Mathematik.

[32] (with L. Gårding) *Strongly subharmonic functions*, Math. Scand. **15** (1964), 93–96, Correction in Math. Scand. **18**(1966), 183, **MR** 31:3621, **Zbl** 146:354.

[33] *Pseudo-differential operators*, Comm. Pure Appl. Math. **18** (1965), 501–517, **MR** 31-4970, **Zbl** 125, 334.

[34] *Pseudo-differential operators and non-elliptic boundary problems*, Ann. of Math. **83** (1966), 129–209, **MR** 38:1387, **Zbl** 132:74.

[35] *Pseudo-differential operators and hypoelliptic equations*, Amer. Math. Soc. Symp. Sing. Int. Op. **10** (1966), 138–183, **MR** 52:4033, **Zbl** 167:96..

[36] *On the Riesz means of spectral functions and eigenfunction expansions for elliptic differential operators*, The Belfer Graduate School Science Conference Nov. 1966, 1969, pp. 155–202, **MR** 41:2239. Russian translation 1968, **MR** 40:4618.

[37] L^p *estimates for (pluri-)subharmonic functions*, Math. Scand. **20** (1967), 65–78, **MR** 38:2323, **Zbl** 156:122.

[38] *Generators for some rings of analytic functions*, Bull. Amer. Math. Soc. **73** (1967), 943–949, **MR** 37:1977, **Zbl** 172:417.

[39] *Hypoelliptic second order differential equations*, Acta Math. **119** (1967), 147–171, **MR** 36:5526, **Zbl** 156:107.

[40] *Convolution equations in convex domains*, Inv. Math. **4** (1968), 306–317, **MR** 37:1978, **Zbl** 229:44008.

[41] (with J. Wermer) *Uniform approximation on compact sets in* $\mathbf{C}^n$, Math. Scand. **23** (1968), 5–21, **MR** 40:7484, **Zbl** 181:362.

[42] *On the characteristic Cauchy problem*, Ann. of Math. **88** (1968), 341–370, **MR** 37:6596, **Zbl** 164.407.

[43] *The Cauchy problem for differential equations with constant coefficients*, Springer lecture notes in math. **103** (1969), 60–71, **MR** 41:634, **Zbl** 179:364.

[44] *The spectral function of an elliptic operator*, Acta Math **121** (1968), 193–218, **MR** 58:29418, **Zbl** 164:132.

[45] *On the singularities of solutions of partial differential equations*, Int. Conf. on Funct. Anal. and Rel. Topics, Tokyo, 1969, pp. 31–40, **MR** 42:3396, **Zbl** 191:109.

[46] *On the index of pseudodifferential operators*, Elliptische Differentialgleichungen II, Koll. Aug. 1969, Schriftenreihe der Inst.für Math., Deutsche Akad. d. Wiss. zu Berlin, pp. 127–146, **MR** 58:31292, **Zbl** 188:409. Russian translation in Matematika 14:4(1970), 78–97, **Zbl** 214:100.

[47] *On the singularities of solutions of partial differential equations*, Comm. Pure Appl. Math. **23** (1970), 329–358, **MR** 41:7251, **Zbl** 188:409. Russian translation in Matematika 16:6(1972), 33–59, **Zbl** 239.35009.

[48] *The calculus of Fourier integral operators*, Prospects in Math., Ann. of Math. Studies 70, 1971, pp. 33–57, **MR** 49:5943, **Zbl** 235:47023.

[49] *Fourier integral operators I*, Acta Math. **127** (1971), 79–183, **MR** 52:9299, **Zbl** 212.466. Russian translation in Matematika 16:1(1972), **Zbl** 235.47024.

[50] *Linear differential operators*, Actes Congr. Int. Math., Nice, 1970, pp. 1:121–133, **MR** 58:23766, **Zbl** 223.35083.

[51] *On the* L^2 *continuity of pseudo-differential operators*, Comm. Pure Appl. Math. **24** (1971), 529–535, **MR** 43:6779, **Zbl** 206.393.

[52] *Uniqueness theorems and wave front sets for solutions of linear differential equations with analytic coefficients*, Comm. Pure Appl. Math. **24** (1971), 671–704, **MR** 45:3917, **Zbl** 226.35019, Russian translation in Matematika 17:6(1973), 82–110, **Zbl** 282.35019.

[53] *A remark on Holmgren's uniqueness theorem*, J. Diff. Geom. **6** (1971), 129–134, **MR** 47:9024, 49:9683, **Zbl** 221:35002.

[54] *On the existence and the regularity of solutions of linear pseudo-differential equations*, L'Ens. Math. **17** (1971), 99–163, **MR** 48:9458, **Zbl** 224:35084. Russian translation in Uspehi Mat. Nauk 28:6(1973), 109–164, **MR** 52:9007.

[55] (with J.J. Duistermaat) *Fourier integral operators II*, Acta Math. **128** (1972), 183–269, **MR** 52:9300, **Zbl** 232:47055.

[56] *On the singularities of solutions of partial differential equations with constant coefficients*, Israel J. Math. **13** (1972), 82–105, **MR** 48:11745, 57:16940,**Zbl** 246:35016, 247:35005.

[57] *Oscillatory integrals and multipliers on* FL^p, Ark. för Mat. **11** (1973), 1–11, **MR** 49:5674 **Zbl** 254:42010.

[58] *On the existence of real analytic solutions of partial differential equations with constant coefficients*, Inv. Math. **21** (1973), 151–182, **MR** 49:817, **Zbl** 282:35015.

[59] *Lower bounds at infinity for solutions of differential equations with constant coefficients*, Israel J. Math. **16** (1973), 103–116, **MR** 49:5543, **Zbl** 271:35005.

[60] *The spectral analysis of singularities*, Proc. Fifth National Math. Conf. March 28–31, 1974 (M. Razzaghi, ed.), Shiraz, Iran, 1975, pp. 129–139.

[61] *Non-uniqueness for the Cauchy problem*, Springer Lecture Notes in Math. **459** (1974), 36–72, **MR** 57:7997, **Zbl** 315:35019.

[62] (with S. Agmon) *Asymptotic properties of solutions of differential equations with simple characteristics*, J. d'Analyse Math. **30** (1976), 1–38, **MR** 57:6776, **Zbl** 335.35013.

[63] *The existence of wave operators in scattering theory*, Math. Z. **146** (1976), 69–91, **MR** 52:14691, **Zbl** 319:35059.

[64] *A class of hypoelliptic pseudodifferential operators with double characteristics*, Math. Ann. **217** (1975), 165–188, **MR** 51:13774, **Zbl** 306:35032.

[65] *The boundary problems of physical geodesy*, Arch. Rat. Mech. Anal. **62** (1976), 1–52, Correction **65**(1977), 395. **MR** 58:2902a,b, **Zbl** 331:35020.

[66] *The Cauchy problem for differential equations with double characteristics*, J. d'Analyse Math. **32** (1977), 118–196, **MR** 58:11822, **Zbl** 367:35054.

[67] (with A. Melin) *Free systems of vector fields*, Ark. för Mat. **16** (1978), 83–88, **MR** 58:31284, **Zbl** 383:35013.

[68] *Propagation of singularities and semiglobal existence theorems for (pseudo-)differential operators of principal type*, Ann. of Math. **108** (1978), 569–609, **MR** 81j:35110, **Zbl** 396:35087.

[69] *Spectral analysis of singularities*, Seminar on the sing. of sol. of diff. eq., Ann. of Math. Studies 91, 1979, pp. 3–49, **MR** 80m:35058, **Zbl** 446:47045.

[70] *Subelliptic operators*, Seminar on the sing. of sol. of diff. eq., Ann. of Math. Studies 91, 1979, pp. 127–208, **MR** 82c:35029, **Zbl** 446:35086.

[71] *The Weyl calculus of pseudo-differential operators*, Comm. Pure Appl. Math. **32** (1979), 359–443, **MR** 80j:47060, **Zbl** 388:47032.

[72] (with D. Gilbarg) *Intermediate Schauder estimates*, Arch. Rat. Mech. Anal. **74** (1980), 297–318, **MR** 82a:35038, **Zbl** 454:35022.

[73] *A remark on singular supports of convolutions*, Math. Scand. **45** (1979), 50–54, **MR** 83c:46036, **Zbl** 426:46029.

[74] *On the asymptotic distribution of eigenvalues of pseudodifferential operators in $\mathbf{R}^n$*, Ark. för Mat. **17** (1979), 297–313, **MR** 82i:35140, **Zbl** 436:35064.

[75] *On the subelliptic test estimates*, Comm. Pure Appl. Math. **33** (1980), 339–363, **MR** 81f:35024, **Zbl** 414:35078.

[76] *On the asymptotic distribution of eigenvalues*, A tribute to Åke Pleijel, Uppsala, 1980, pp. 146–154.

[77] *Pseudo-differential operators of principal type*, Nato Adv. Study Inst. on Sing. in Boundary Value Problems, Reidel Publ. Co., Dordrecht, 1981, pp. 69–96, **MR** 83m:35003, **Zbl** 459:35096.

[78] *Symbolic calculus and differential equations*, Proc. 18th Scand. Congr. of Math. Århus 1980, Birkhäuser, 1981, pp. 56–81, **MR** 83e:35134, **Zbl** 473:35079.

[79] *L^2 estimates for Fourier integral operators with complex phase*, Ark. för Mat. **21** (1983), 283–307, **MR** 85h:47058, **Zbl** 533.47045.

[80] *Théorie de la diffusion à courte portée pour des opérateurs à caractéristiques simples*, Sem. Goulaouic, Meyer, Schwartz, 1980–1981 Exp. XIV, pp. 1–18, **Zbl** 482.35009.

[81] *Uniqueness theorems for second order elliptic differential equations*, Comm. Partial Diff. Eq. **8** (1983), 21–64, **MR** 85c:35018, **Zbl** 546:35023.

[82] *Between distributions and hyperfunctions*, Astérisque **131** (1985), 89–106, **MR** 87g:46066, **Zbl** 585:46036.

[83] *Differential operators of principal type and scattering theory*, Proceedings of the 1982 Changchun symposium on differential geometry and differential equations, Science Press, Beijing, China, 1986, pp. 113–184, **Zbl** 689:35099.

[84] *On the Nash-Moser implicit function theorem*, Ann. Ac. Sci. Fenn. **10** (1985), 355–359, **MR** 87a:58025, **Zbl** 591:58003.

[85] *The propagation of singularities for solutions of the Dirichlet problem*, Proc. Amer. Math. Soc. Symp. in Pure Math. **43** (1985), 157–165, **MR** 87g:35137, **Zbl** 618.35015.

[86] *On Sobolev spaces associated with some Lie algebras*, Current topics in partial differential equations, Kinokuniya, Tokyo, 1986, pp. 261–287, **Zbl** 658:46023.

[87] *The lifespan of classical solutions of non-linear hyperbolic equations*, Springer Lecture Notes in Math. **1256** (1987), 214–280, **MR** 88j:35024, **Zbl** 632:35045.

[88] *L^1, L^∞ estimates for the wave operator*, Analyse Mathématique et Applications, Gauthier-Villars, Paris, 1988, pp. 211–234, **MR** 90e:35113, **Zbl** 676:35062.

[89] *Remarks on the Klein-Gordon equation*, Journées Equations aux dérivées partielles Saint-Jean-de-Monts Juin 1987, pp. I-1–I-9, **MR** 89b:35095, **Zbl** 655:35057.

[90] *Pseudo-differential operators of type 1,1*, Comm. Partial Diff. Eq. **13** no 9 (1988), 1085–1111, **MR** 89k:35260, **Zbl** 667:35078.

[91] (with R. Sigurdsson) *Limit sets of plurisubharmonic functions*, Math. Scand. **65** (1989), 308–320, **MR** 91c:32011, **Zbl** 718:32016.

[92] *The fully non-linear Cauchy problem with small data*, Bol. Soc. Brasil. Mat. **20** (1989), 1–27, **MR** 92k:35188, **Zbl** 768:35014.

[93] *The Nash-Moser theorem and paradifferential operators*, Analysis et cetera, Academic Press, 1990, pp. 429–449, **MR** 91k:58009, **Zbl** 711:35001.

[94] *Continuity of pseudo-differential operators of type 1,1*, Comm. Partial Diff. Eq. **14** (1989), 231–243, **MR** 90a:35241, **Zbl** 688:35107.

[95] (with A. Melin) *A remark on perturbations of compact operators*, Math. Scand. **75** (1994), 255–262, **MR** 95m:47015, **Zbl** 824:47008.

[96] *On the fully non-linear Cauchy problem with small data. II*, Microlocal analysis and nonlinear waves, Springer Verlag, 1991, pp. 51–81, IMA Volumes in Mathematics and its Applications vol. 30, **MR** 94c:35127, **Zbl** 783:35036.

[97] *Quadratic hyperbolic operators*, Springer Lecture Notes in Mathematics **1495** (1991), 118–160, CIME conference July 1989, **MR** 93k:35187, **Zbl** 761:35004.

[98] (with G. Grubb) *The transmission property*, Math. Scand. **67** (1990), 273–289, **MR** 92f:58175, **Zbl** 766:35088.

[99] *The wave front set of the fundamental solution of a hyperbolic operator with double characteristics*, J. d'Analyse Math. **59** (1992), 1–36, **MR** 95j:35008, **Zbl** 821:35091.

[100] *Hyperbolic systems with double characteristics*, Comm. Pure Appl. Math. **46** (1993), 261–301, **MR** 94b:35085, **Zbl** 803:35082.

[101] *A remark on the characteristic Cauchy problem*, J. Funct. Anal. **93** (1990), 270–277, **MR** 91m:58154, **Zbl** 724:35060.

[102] *A uniqueness theorem of Beurling for Fourier transform pairs*, Ark. för Mat. **29** (1991), 237–240, **MR** 93b:42016, **Zbl** 755:42009.

[103] *A uniqueness theorem for second order hyperbolic differential equations*, Comm. Partial Diff. Eq. **17** (1992), 699–714, **MR** 93h:35116, **Zbl** 815:35063.

[104] (with Bo Bernhardsson) *An extension of Bohr's inequality*, Boundary value problems for partial differential equations and applications (J.-L. Lions and C. Baiocchi, eds.), Masson, Paris, Milan, Barcelone, 1993, pp. 179–194, **MR** 95e:46052, **Zbl** 803:41030.

[105] *Remarks on Holmgren's uniqueness theorem*, Ann. Inst. Fourier Grenoble **43** no5 (1993), 1223–1251, **MR** 95b:35010, **Zbl** 804:35004.

[106] (with Ragnar Sigurdsson) *Growth properties of plurisubharmonic functions related to Fourier-Laplace transforms*, J. Geometric Analysis **8:2** (1998), 251–311.

[107] *Symplectic classification of quadratic forms, and general Mehler formulas*, Math. Z. **219** (1995), 413–449, **MR** 96c:58172, **Zbl** 829:35150.

[108] *On the solvability of pseudodifferential equations*, Structure of solutions of differential equations (M. Morimoto and T. Kawai, eds.), World Scientific, Singapore, New Jersey, London, HongKong, 1996, pp. 183–213.

[109] *Looking forward from ICM 1962*, Fields Medaillists' Lectures, World Scientific, Singapore, 1997, pp. 86–103.

[110] *On the uniqueness of the Cauchy problem under partial analyticity assumptions*, Geometrical optics and related topics (F. Colombini and N. Lerner, eds.), Birkhäuser, Boston, 1997, pp. 179–219.

[111] *Remarks on the Klein-Gordon and Dirac equations*, Contemporary Math. **205** (1997), 101–125.

[112] *On the Legendre and Laplace transformations*, Ann. Scuola Norm. Sup. Pisa **25** (1997), 517–568.

[113] *On local integrability of fundamental solutions*, Ark. för Mat. **37** (1999), 121–140.

[114] (with J. Boman) *A Paley-Wiener theorem for the analytic wave front set*, Asian J. Math. **3** (1999), 757–769.

[115] *Local P-convexity*, J. d'Analyse Math. **80** (2000), 101–141.

[116] *A counterexample of Gevrey class to the uniqueness of the Cauchy problem*, Math. Res. Letters **7** (2000), 615–624.

[117] *Asgeirsson's mean value theorem and related identities*, J. Funct. Anal. **184** (2001), 377–401.

[118] *A history of existence theorems for the Cauchy-Riemann complex in L^2 spaces*, J. Geometric Analysis **13:2** (2003), 201–229.

[119] *The null space of the $\bar{\partial}$-Neumann operator*, Ann. Inst. Fourier Grenoble **54** (2004), 1305–1369.

[120] *The multinomial distribution and some Bergman kernels*, Contemporary Math. **368** (2005), 249–265.
[121] *Weak linear convexity and a related notion of concavity*, Math. Scand. **102** (2008), 73–100.

PUBLISHED BOOKS

[1] *Linear partial differential operators. Grundlehren d. Math. Wiss. 116*, Springer Verlag, Berlin, Göttingen, Heidelberg Russian translation MIR 1965, **Zbl** 175:392., 1963, **MR** 28:4221, **Zbl** 108:93, second revised printing 1964, **MR** 37:6595, **Zbl** 131:318, third revised printing 1969, **MR** 40:1687, **Zbl** 223:35083 fourth printing 1976, **MR** 53:8622, **Zbl** 321:3500.

[2] *An introduction to complex analysis in several variables.*, D. van Nostrand Publ. Co., Princeton, N. J., 1966, **MR** 34:2933, **Zbl** 176:380; Russian translation MIR 1968, **MR** 39:471, **Zbl** 138:62; Japanese translation 1973. Second extended edition 1973, **MR** 49:9246, **Zbl** 271:32001, and third extended edition 1990, **MR** 91a:32001, **Zbl** 685:32001 published by North Holland Publ. Co., Amsterdam.

[3] *Integrationsteori*, Studentlitteratur, Lund, 1970, Medförfattare Tomas Claesson, **MR** 42:1963, 50:7452, **Zbl** 216:348.

[4] *The analysis of linear partial differential operators. I. Distribution theory. Grundlehren d. Math. Wiss. 256*, Springer Verlag, Berlin, Heidelberg, New York, Tokyo, 1983, **MR** 85g:35002a, **Zbl** 521:35001. Russian translation MIR 1986, **MR** 88a:35003, **Zbl** 619:35001. Second expanded edition and Springer Study edition 1990, **MR** 91m:35001a,b, **Zbl** 712:35001.

[5] *The analysis of linear partial differential operators. II. Differential operators with constant coefficients. Grundlehren d.Math.Wiss. 257*, Springer Verlag, Berlin, Heidelberg, New York, Tokyo, 1983, **MR** 85g:35002b, **Zbl** 52:35002. Russian translation MIR 1986, **MR** 88a:35004, **Zbl** 619:35002. Second revised printing 1990, **Zbl** 687:35002.

[6] *The analysis of linear partial differential operators. III. Pseudo-differential operators. Grundlehren d.Math.Wiss. 274*, Springer Verlag, Berlin, Heidelberg, New York, Tokyo, 1985, **MR** 87d:35002a, **Zbl** 601:35001. Russian translation MIR 1987, **MR** 88m:35001. Corrected second printing 1994, **MR** 95h:35255.

[7] *The analysis of linear partial differential operators. IV. Fourier integral operators. Grundlehren d.Math.Wiss. 275*, Springer Verlag, Berlin, Heidelberg, New York, Tokyo, 1985, **MR** 87d:35002b, **Zbl** 612:35001. Russian translation MIR 1988, **MR** 89k:35002. Corrected second printing 1994.

[8] *Notions of convexity*, Birkhäuser, Boston, Basel, Berlin, 1994, **MR** 95k:00002, **Zbl** 835:32001.

[9] *Lectures on nonlinear hyperbolic differential equations. Mathématiques & Applications vol. 26*, Springer Verlag, Paris, Berlin, Heidelberg, 1997, **MR** 98e:35103, **Zbl** 0881:35001.

LECTURE NOTES

[1] *Föreläsningar över distributionsteori*, Stockholms högskola höstterminen 1958, nytryck 1964.
[2] *Föreläsningar över Fourieranalys*, Stockholms högskola vårterminen 1959.
[3] *Integrationsteori*, Föreläsningar vid Stockholms Universitet höstterminen 1959.
[4] *Funktionalanalys*, Föreläsningar vid Stockholms Universitet vårterminen 1960.
[5] *Fourier integral operators*, Lectures at the Nordic Summer School of Mathematics June 1969.
[6] *Linear functional analysis*, Lectures at Lund University Fall term 1969. Revised and extended edition 1988.
[7] *Tillämpad lineär analys*, Del av andra terminens analyskurs i grundutbildningen i Lund.
[8] *Högre differentialkalkyl*, Föreläsningar vid Lunds Universitet 1973/74 redigerade av Tomas Claesson. Reviderad engelsk översättning Advanced Differential Calculus Lund 1994.
[9] *Implicit function theorems*, Lectures at Stanford University, Summer Quarter 1977. Translation of part of lectures given at Lund University Spring term 1976.
[10] *Analytisk funktionsteori*, Föreläsningar vid Lunds Universitet höstterminen 1979.
[11] *Nonlinear hyperbolic differential equations*, Lectures University of Lund 1986–87.
[12] *Riemannian geometry*, Lectures University of Lund Fall term 1990.
[13] *Integrationsteori*, Revised and extended version of the book [3] with coauthor Tomas Claesson, Lund 1993.
[14] *Lectures on harmonic analysis*, Lectures University of Lund 1994–1995.

Printed in the United States
By Bookmasters